Emerging Topics in Hardware Security

Mark Tehranipoor
Editor

Emerging Topics in Hardware Security

 Springer

Editor
Mark Tehranipoor
University of Florida
Gainesville, FL, USA

ISBN 978-3-030-64450-5   ISBN 978-3-030-64448-2   (eBook)
https://doi.org/10.1007/978-3-030-64448-2

This Springer imprint is published by the registered company Springer Nature Switzerland AG
The registered company address is: Gewerbestrasse 11, 6330 Cham, Switzerland

To the outstanding contributors of this book

Preface

Hardware security, so unknown to many in the domain of cyber security over the past four decades, has now become a household terminology in many organizations, companies, and governments. I vividly recall days in the late 2000 where the questions surrounding the concept of hardware security were simply too many, much has to do with the fact that researchers in the field of information security always assumed that the hardware underlying information systems is secure and trustworthy. With hardware being insecure, all those assumptions and methodologies were challenged. Today, hardware security community comprises thousands of researchers across the globe. Many exciting topic areas are investigated by the community, namely fault injection and mitigation approaches; metrics, policies, assessment, and standards related to hardware security and trust; hardware intellectual property (IP) trust, including watermarking, metering, and trust verification; logic locking; trusted manufacturing, including split manufacturing and 2.5/3D integration; hardware tampering attacks and protection; design and applications of emerging and nano-scale devices for security; sensor-enabled hardware security; machine learning for hardware security; hardware acceleration for security applications; privacy-preserving computing; secure function evaluation; architecture and hardware-enabled system-on-chip (SoC), cyber, and data center security; hardware security primitives, including crypto hardware, physically unclonable functions (PUFs), and true random number generators (TRNGs); hardware design techniques to facilitate software and/or system security; architecture support for security; power and EM side-channel analyses, attacks, and protection; hardware Trojan attacks, detection, and countermeasures; hardware security test and verification; cyber-physical system security; field programmable gate array (FPGA) and SoC security; supply chain risk mitigation (e.g., counterfeit detection and avoidance), and reverse engineering and hardware obfuscation.

The community also constantly looks for and explores new ideas and research directions to improve electronic devices' security and supply chain assurance. Over the past few years, many researchers including the outstanding contributors to this book have been investigating new research ideas with focus on hardware security. This book has put together a comprehensive set of new research directions the community is working on to help educate the researchers and practitioners in the field.

Below is a summary of the content you are about to read:

Chapter 1: Blockchain-Enabled Electronics Supply Chain Assurance. This chapter discusses utilizing a consortium style blockchain to establish an end-to-end supply chain assurance and prevent any unauthorized component from entering the supply chain.

Chapter 2: Digital Twin with a Perspective from Manufacturing Industry. This chapter classifies digital twin in the context of evolution, application, definition, framework, and design process. It also discusses the security issues in digital twin with a perspective from manufacturing industry.

Chapter 3: Trillion Sensors Security. This chapter provides a detailed survey of the applicability and capability of the trillion sensors in diversified application domains.

Chapter 4: Security of AI Hardware Systems. This chapter presents a brief introduction to modern AI systems, reviews reported AI security issues, and discusses possible countermeasures.

Chapter 5: Machine Learning in Hardware Security. This chapter presents the application of machine learning in different hardware security areas, such as IP protection, Trojan detection, side-channel analysis/attacks, hardware security primitives, and architectural vulnerabilities.

Chapter 6: Security Assessment of High-Level Synthesis. This chapter provides a detailed security assessment on the high-level synthesis (HLS) process and shows the potential vulnerabilities during its translation. The chapter also presents IP protection using locking mechanism at the high-level language.

Chapter 7: CAD for Side-Channel Leakage Assessment. This chapter presents the techniques for power side-channel leakage assessment at the pre-silicon and discusses in depth two CAD frameworks called SCRIPT and RTLPSC.

Chapter 8: Post-Quantum Hardware Security. This chapter discusses how the evolution of quantum technology could influence the field of hardware security.

Chapter 9: Post-Quantum Cryptographic Hardware and Embedded Systems. This chapter reviews the progress on acceleration of post-quantum key establishment crypto-systems in hardware and discusses the critical computations in the third round of NIST's PQC standardization competition.

Chapter 10: Neuromorphic Security. This chapter examines the security concerns in emerging neuromorphic systems with emphasis on vulnerabilities arising from devices, circuits, architectures, and supporting subsystems.

Chapter 11: Homomorphic Encryption. This chapter presents a short review of homomorphic encryption, which allows the manipulation of encrypted data without the need to decrypt it first.

Chapter 12: Software Security with Hardware in Mind. This chapter explores various software and hardware vulnerabilities to better understand the potential exploitable scenarios and protect the software from hardware and software perspectives.

Chapter 13: Firmware Protection. This chapter presents the approaches for firmware protection that leverage the intrinsic hardware signatures to bind the firmware with trusted hardware platform.

Chapter 14: Security of Emerging Memory Chips. This chapter discusses the security vulnerabilities of emerging non-volatile memory chips and reviews the existing countermeasures to make the computing systems robust against various attacks.

Chapter 15: Security of Analog, Mixed-Signal, and RF Devices. This chapter provides a comprehensive overview on the field of hardware security, specifically focusing on the side-channel analysis (SCA) and physically unclonable functions (PUFs). It investigates the different mechanisms of analog phenomena-causing vulnerabilities, such as side-channels, counterfeit ICs, and denial-of-service (DoS) attacks.

Chapter 16: Analog IP Protection and Evaluation. This chapter presents the analog IP protection techniques and describes the different evaluation approaches available to determine the resilience offered by these defenses.

Chapter 17: Application of Optical Techniques to Security Threat and Hardware Assurance. This chapter reviews some failure analysis techniques through the perspective of optical attacks, discusses the attack model and the countermeasures, and presents a new approach for hardware Trojan detection.

Chapter 18: Computer Vision for Hardware Security. This chapter provides an overview of basic computer vision concepts and explores hardware security applications.

Chapter 19: Asynchronous Circuits and Their Applications to Hardware Security. This chapter provides an overview on basic concepts of asynchronous design, reviews its applications to hardware security, discusses potential security flaws, and presents a mitigation technique.

Chapter 20: Microfluidic Device Security. This chapter discusses the security of biochips, biosample security, and biochip IP protection.

I hope that this book serves as an invaluable reference to students, faculty, practitioners, and researchers in the field of hardware security and trust.

Gainesville, FL, USA

Mark Tehranipoor

Contents

xxii

Contributors

Rabin Yu Acharya University of Florida, Gainesville, FL, USA

Navid Asadizanjani University of Florida, Gainesville, FL, USA

Reza Azarderakhsh Florida Atlantic University, Boca Raton, FL, USA

A. Sanabria-Borbón Texas A&M University, College Station, TX, USA

Daniel E. Capecci University of Florida, Gainesville, FL, USA

Krishnendu Chakrabarty Duke University, Durham, NC, USA

Koushik Chakraborty Utah State University, Logan, UT, USA

Baibhab Chatterjee Purdue University, West Lafayette, IN, USA

Shigang Chen University of Florida, Gainesville, FL, USA

Sreeja Chowdhury University of Florida, Gainesville, FL, USA

Ana Covic University of Florida, Gainesville, FL, USA

Pinchen Cui Auburn University, Auburn, AL, USA

Debayan Das Purdue University, West Lafayette, IN, USA

Shijin Duan Northeastern University, Boston, MA, USA

Farimah Farahmandi University of Florida, Gainesville, FL, USA

Farah Ferdaus Florida International University, Miami, FL, USA

Domenic Forte University of Florida, Gainesville, FL, USA

Fatemeh Ganji Worcester Polytechnic Institute, Worcester, MA, USA

Pallabi Ghosh University of Florida, Gainesville, FL, USA

Ujjwal Guin Auburn University, Auburn, AL, USA

Miao (Tony) He University of Florida, Gainesville, FL, USA

Muhammad Monir Hossain University of Florida, Gainesville, FL, USA

N. G. Jayasankaran Texas A&M University, College Station, TX, USA

Ramesh Karri New York University, Brooklyn, NY, USA

Waleed Khalil The Ohio State University, Columbus, OH, USA

Brian Koziel Florida Atlantic University, Boca Raton, FL, USA

J. Hu Texas A&M University, College Station, TX, USA

Leonidas Lavdas University of Florida, Gainesville, FL, USA

Zhengang Li Northeastern University, Boston, MA, USA

Tung-Che Liang Duke University, Durham, NC, USA

Xue (Shelley) Lin Northeastern University, Boston, MA, USA

Hangwei Lu University of Florida, Gainesville, FL, USA

Yukui Luo Northeastern University, Boston, MA, USA

Mehran Mozaffari Kermani University of South Florida, Tampa, FL, USA

M. Rafid Muttaki University of Florida, Gainesville, FL, USA

Adib Nahiyan University of Florida, Gainesville, FL, USA

Jungmin Park University of Florida, Gainesville, FL, USA

Nitin Pundir University of Florida, Gainesville, FL, USA

Rajesh J. S. AMD, Santa Clara, CA, USA

J. Rajendran Texas A&M University, College Station, TX, USA

Fahim Rahman University of Florida, Gainesville, FL, USA

M. Tanjidur Rahman University of Florida, Gainesville, FL, USA

Md Tauhidur Rahman Florida International University, Miami, FL, USA

Sanghamitra Roy Utah State University, Logan, UT, USA

Mehdi Sadi Auburn University, Auburn, AL, USA

E. Sánchez-Sinencio Texas A&M University, College Station, TX, USA

Shreyas Sen Purdue University, West Lafayette, IN, USA

Mohammed Shayan New York University, Brooklyn, NY, USA

Haoting Shen University of Nevada, Reno, NV, USA

Mengshu Sun Northeastern University, Boston, MA, USA

Eslam Yahya Tawfik The Ohio State University, Columbus, OH, USA

Mark Tehranipoor University of Florida, Gainesville, FL, USA

Md Sami Ul Islam Sami University of Florida, Gainesville, FL, USA

Haibo Wang University of Florida, Gainesville, FL, USA

Wenhao Wang Northeastern University, Boston, MA, USA

Damon L. Woodard University of Florida, Gainesville, FL, USA

Xiaolin Xu Northeastern University, Boston, MA, USA

Acronyms

2D-FD	2D Fourier shape descriptor
ACM	Access-control memory
AES	Advanced encryption standard
ASLD	All spin logic devices
BL	BitLine
BPA	Birthday paradox attack
CNN	Convolutional neural network
CMOS	Complementary metal oxide semiconductor
DBSCAN	Density-based spatial clustering
DC	Design compiler
DoS	Denial-of-service
DPA	Differential power analysis
DRAM	Dynamic RAM
ECC	Error correction code
EDS	Energy disruptive spectroscopy
E-Field	Electric field
FDTD	Finite difference time domain
FLIR	Forward-looking infrared
FML	Free magnetic layer
FNR	False negative rate
FPR	False positive rate
FRAM	Ferroelectric RAM
FTIR	Fourier transform infrared spectroscopy
GRAA	Generalized RAA
HDD	Hard disc drive
HW	Hamming weight
IC	Integrated circuit
ICC	IC compiler
ICUA	ICs under authentication
KNN	K-nearest neighbor
LBP	Local binary pattern

LDP	Local directional pattern
LGDiP	Local Gabor directional pattern
LLC	Last level cache
LTEM	Law's texture energy measurement
M-Field	Magnetic field
MLC	Multi-level cells
MRAM	Magneto-resistive RAM
MTJ	Magnetic tunnel junction
NIR/IR	Near infrared/infrared
NVM	Non-volatile memory
OS	Operating system
PCA	Principle component analysis
PCB	Printed circuit board
PCM	Phase change memory
PL	Plate line
PZT	Lead zirconate titanate
RAA	Repeat address attack
RAM	Random access memory
RC	Register transfer language compiler
ReRAM	Resistive RAM
RML	Reference magnetic layer
RMS	Root mean square
SAFER	Stuck-at fault error recovery
SEM	Scanning electron microscope
SIFT	Scale-invariant feature transform
SL	SelectLine
SLC	Single-level cell
SMA	Stealth mode attack
SME	Subject matter experts
SNR	Signal-to-noise ratio
SPA	Simple power analysis
SRAM	Static RAM
SSIM	Structural similarity index
STT-MRAM	Spin-transfer torque MRAM
SVM	Support vector machine
SURF	Speeded-up robust features
WER	Write error rate
WL	WordLine

Chapter 1
Blockchain-Enabled Electronics Supply Chain Assurance

Fahim Rahman and Mark Tehranipoor

1.1 Introduction

Driven by the continuous and aggressive scaling of semiconductor fabrication technology, integrated circuits (ICs) have become more complicated than ever. In accordance with Moore's law [2], the total number of transistors on a single chip has roughly doubled every 2 years since the 1960s, while costs have gone down at approximately the same rate. Consequently, consumer electronics such as laptops, smartphones, and even electronic medical instruments are commonly seen and used in everyday life. Moreover, almost all critical infrastructures such as power grid, public transportation systems, and national defense systems are built on numerous electronic devices ranging from high-end digital processors to small controllers, and analog and mixed-signal sensors and systems. Therefore, the security, quality, and assurance of these systems are closely related to the trustworthiness of the underlying integrated circuits.

The security of software, firmware, and communication channels has received a significant amount of attention in the past due to numerous underlying vulnerabilities, threats, and attacks. On the contrary, the security aspect of ICs and electronic systems is generally associated with a limited set of vulnerabilities and attacks, for example, side-channel analysis that exploits the hardware implementation of cryptographic algorithms for leaking secret keys, and invasive/semi-invasive attacks enabling tampering and adversarial reverse engineering. However, the supply chain integrity of ICs and electronic systems is equally important, because hardware

---

Materials of this chapter were previously published in Xu. et al. [1].

F. Rahman (✉) · M. Tehranipoor
Department of ECE, University of Florida, Gainesville, FL, USA
e-mail: fahimrahman@ece.ufl.edu; tehranipoor@ece.ufl.edu

produced from an untrusted supply chain cannot serve as the underlying root of trust. As the globalization of semiconductor industry makes it a joint effort to produce an electronic system, threats arise from various untrusted parties involved in the design, fabrication, development, and distribution of ICs and electronic systems. For example, each component in the system (e.g., various digital ICs, analog devices and sensors, printed circuit boards (PCBs), etc.) may come from a diverse group of suppliers who might often be scattered throughout the globe [3, 4]. Therefore, we need to analyze relevant threats and vulnerabilities at each stage of the life cycle of a component moving through the electronics supply chain.

An untrusted electronics supply chain opens up opportunities for adversaries to introduce counterfeit ICs and systems, such as recycled, remarked, and cloned, as legit ones to the end users. If the counterfeit devices are not detected and prevented while moving through distribution chain, the user may unknowingly use them to build a system with underlying vulnerabilities. It has been reported that electronic companies are losing around $650 billion of global revenue every year because of counterfeiting [5–8]. More severely, although such counterfeit devices (e.g., recycled ICs) may work initially, they may suffer from reduced lifetime, pose reliability risks, and impact computers, telecommunications, automotive, and even military systems in which they are deployed. Around 1% of semiconductor products on the market were believed to be counterfeit in 2013, and this number continues to rise [9]. Further, it is predicted that the tools and technologies used for producing such counterfeit ICs/systems will become increasingly sophisticated as well [4].

It is imperative that one needs to employ an integrated approach to build a trusted electronics supply chain, ensuring the authenticity of the device and system from the device fabrication stage to systems' end-of-life. Although various ad hoc solutions have been proposed till date to detect and avoid counterfeit electronic components. For example, combating die and IC recycling (CDIR) sensors can only detect recycled ICs [10]. Hardware metering [11, 12] and PUFs [13, 14] can only be used to prevent overproduction and cloning, whereas secure split test (SST) can only be used to prevent overproduction and IC piracy by locking the correct function of the design during the test [15–18]. Unfortunately, such individual techniques can only thwart selective threats to some extent and do not offer a holistic solution to create a secure and trusted supply chain especially for both electronic ICs and systems. Therefore, none of them can ensure the trust and integrity of electronics supply chain at the system level [19–21]. In addition, one of the most important features to build a trusted electronics supply chain—tracking and tracing—is not readily established via such techniques. Another critical concern is the management of all necessary information in a trusted and distributed manner, so that only the trusted entities can query and verify authentic devices and systems, as they move through a potentially untrusted channel without creating a single point of data-breaching vulnerability.

Decentralized data management and certification techniques such as blockchain [22] can address the data authenticity and confidentiality concerns and can be used for virtual financial transactions or commodity transportation [23, 24]. A similar approach can be adopted for establishing a trusted supply chain for electronic systems. However, because of the inherently complex nature of the electronics supply chain and associated unique vulnerabilities, a major challenge remains in the suitability for creating a trusted electronics supply chain among the many involved trusted/untrusted/combined entities. In this chapter, we discuss the prospect and necessary steps for establishing a blockchain-enabled trusted electronics supply chain to offer the much-seeking trust and integrity.

1.2 Preliminaries

The complexity of the electronics supply chain renders it hard to track the authenticity of each component (e.g., IC, PCB) that goes into an electronic system since each component travels through different path of the supply chain. Unless all the entities of electronics supply chain, including the distributors, are trusted, the authenticity and integrity of the components and the system remain under question. Different types of counterfeit devices and systems may be present in an untrusted supply chain. Some examples are as follows [4].

- Recycled electronic components [25] are collected from used PCBs that are discarded as electronic-waste (E-waste), repackaged, and sold in the market as new components. Although such devices and systems might still be functional, there exist performance and life expectancy issues due to silicon aging as well as the chip harvesting process.
- Remarked electronic components are those whose marking on the package (or even on the die) is remarked with forged information. New electronic devices could also be remarked with a higher specification, such as from commercial grade part to industrial or defense grade.
- Overproduction is usually done by an untrusted foundry, assembly, or a test site that has access to the original design. These parties could potentially produce more than the contracted amount and sell these chips or systems illicitly.
- Defective and out-of-spec components are devices or systems that do not meet the functional or parametric specifications or grades (i.e., commercial, industrial, or military) but are put into the market as authentic ICs or systems.
- Cloning can be performed by any untrusted entity in the electronics supply chain. A clone is a direct copy of the original design produced without the permission of the original component manufacturer (OCM), as the IP owner. Cloning can be done in two ways: by reverse engineering the IC or system obtained from the market or by directly gaining access to the intellectual property used to develop the electronic system (e.g., masks used during IC fabrication) [26].
- Printed circuit boards (PCBs), as the basic component of electronic systems, are also vulnerable to various attacks, such as reverse engineering, overproduction, and piracy [26].
- System integration is the last step of the electronic supply chain toward building a functional electronic product for the end users. Several vulnerabilities may

emerge in this step, for example, the system integrator may utilize counterfeited PCB boards or ICs in building the electronic systems.

1.2.1 Counterfeit Mitigation Techniques

Most of the proposed techniques to date for combating counterfeit ICs and electronic systems can be broadly classified into two groups: (1) counterfeit detection techniques and (2) counterfeit avoidance techniques.

1.2.1.1 Counterfeit Detection

Common counterfeit detection techniques are traditionally based on inspection and measurement against available golden data to identify counterfeit chips, PCBs, or systems. Due to test data variation and error margin, these techniques generally identify *suspect* components, with some strong level of confidence, that pose high risk of being counterfeit. Physical examinations, such as low-power visual inspection (LPVI) with microscopes, can identify deformed leads or scratches on the package for counterfeit chips. X-ray imaging can also be used to find defects on the die or bond wires of ICs without the need for depackaging. Other detection methods include chemical composition analysis through spectroscopy or imaging using SEM/TEM/FIB [27].

In addition, electrical measurements can be performed to characterize the electrical or functional defects and anomalies of the suspect components. The effectiveness of these methods relies on the changes of electronic parameters since prior usage will either shift the electrical characteristics or degrade the reliability of the devices. Popular methods in this class of detection techniques include parametric tests and functional tests [28–30].

1.2.1.2 Counterfeit Avoidance and Design for Anti-Counterfeit

Most counterfeit detection techniques require known-good or *golden* data to compare against, which is not always readily available. Further, these detection techniques are time consuming and expensive and often cannot be applied to large batches of ICs or systems in an automated fashion. Researchers have proposed different avoidance techniques that can be integrated into the chip and manufacturing process to reduce the risk of the product being counterfeit while entering the market in the first place.

Among popular avoidance techniques, recycling detection sensors, such as the combating die and IC recycling (CDIR) sensor [10], are custom designs for anti-counterfeit (DfAC) that can track the lifetime of ICs by measuring the aging and boot sequences of the system. This can help to decide whether a chip has been previously used, i.e., potential suspect for recycling. Hardware metering [11, 12] and secure split test (SST) [15] enable the design house to lock/unlock the manufactured chips selectively using embedded unique key(s) into the design and test flow. This provides the metering capabilities to the design house and prevents an untrusted foundry from engaging in overproduction and piracy. Hardware watermarking [31] allows designers to embed a signature into their designs to facilitate the proof of IP ownership; however, it does not actively protect against counterfeiting. Researchers have also proposed split manufacturing technique [32] to protect the design from untrusted foundry by partitioning the manufacturing processes between the untrusted (front end of line production) and trusted (back end of line production) facilities. This approach can protect against overproduction and cloning; however, the scalability is extremely limited and requires in-house trusted manufacturing facilities. In addition, unique package IDs, such as DNA marking and nanorods [33, 34], can be used onto the packaging of the devices with no additional design overhead into the die. Such passive techniques can identify recycled and remarked ICs through data-centric enrollment and verification but cannot mitigate other counterfeiting threats. Last but not the least, physical unclonable functions (PUFs) enable interactive authentication by converting the static key on devices into an intrinsic function. In particular, such intrinsic functions leverage the microscopic process variations of electronic devices and thus are unique. The input (challenge) and output (response) behavior of PUFs have been proposed for many applications such as identification, authentication, key generation, and storage [13, 14, 35].

The resistance against known vulnerabilities of existing counterfeit mitigation techniques is summarized in Table 1.1. As we see, none of these methods can

Table 1.1 Comparative threat coverage of existing mitigation techniques [36] and blockchain-enabled electronics supply chain

Mitigation techniques	Overproduction	Recycling	Remarking	Cloning	Out-of-spec /defective
Physical inspection [27]	NA	Low	Low	NA	NA
Electrical measurement [28]	NA	Medium	Medium	NA	Low
Recycling detection sensor [10]	NA	High	High	NA	NA
Secure split test [15]	High	NA	NA	Medium	High
Hardware metering [11]	Low	NA	NA	Low	NA
Hardware watermarking [31]	NA	NA	NA	Medium	NA
Split manufacturing [32]	Low	NA	NA	Low	NA
Package ID-based technique [33]	NA	Medium	Medium	NA	NA
PUF [13, 35]	Low	Low	NA	Low	NA
Blockchain-enabled supply chain [1]	High	High	High	High	High

adequately address all vulnerabilities. For example, *SST* can effectively prevent the overproduction and out-of-spec problems (which are marked as *High*), and it has limited effectiveness in combating the recycling and remarking of ICs (which are marked as *NA*). Keeping these limitations in mind, Xu et al. [1] proposed a blockchain-based framework for the trust and integrity of electronics supply chain with a comprehensive threat evaluation considering different untrusted and trusted entities with a variety of capabilities. The primary objective is to provide a unified solution against listed supply chain threats leveraging existing solutions such as PUFs. In addition, this framework provides secure and distributed tracking and tracing of electronic components, which is not possible with other techniques. Rest of the chapter will contain a detail description of this technique with some emerging solutions.

1.2.2 *Blockchain and Decentralized Ledgers*

Blockchain was first conceptualized by Satoshi Nakamoto in 2008 and then utilized for the digital cryptocurrency named Bitcoin [37]. Blockchain is a distributed database that stores a continuously increasing chain of blocks [38–40]. Since the most well-known and mature blockchain structure has been developed for Bitcoin, we briefly review the background of blockchain with respect to Bitcoin as a case study in this section.

In the Bitcoin scheme, a blockchain is an ordered, back-linked list of blocks of transactions. In most literature, the blockchain is visualized as a vertical stack, in which all blocks are layered vertically, and the first block serves as the stack foundation, as shown in Fig. 1.1. In this visualization, one feature associated with each block is its *height*, that is used to quantify the distance from it to the first block. Within the blockchain, each block can be identified by its header hash and block height number. The header hash of 32-byte length is generated by hashing the block header twice through the SHA256 cryptographic algorithm [41]. Besides the identifier information, each block also refers to a previous block, which is called the parent block. A block keeps the header hash of its parent in its header to link and backtrack. Due to this stacked architecture, each block has just one parent in the blockchain.

Blockchain is believed to have great potential to revolutionize the traditional supply chain of various commodities, e.g., from cryptocurrency to food products [23, 24, 40]. This is because:

- In the blockchain scheme, there is no central administrator (node) as shown in Fig. 1.2a, where the separated nodes are connected via the central node. In a centralized network, the corruption of the administrator will violate the trust and integrity of the whole network. While the nodes of blockchain are connected with each other as shown in Fig. 1.2b, since there is no administrator and thus any single node can broadcast to the whole network.

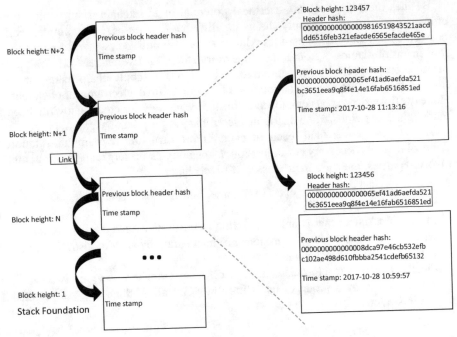

Fig. 1.1 The schematic of vertically layered blockchain structure in Bitcoin scheme, where each block is linked back and refers to a previous block by the header hash value [39]

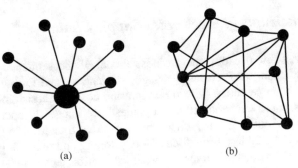

Fig. 1.2 Comparison between centralized and decentralized network. (**a**) In the centralized network, all nodes are connected through the administrator node (denoted with the larger node in the middle). (**b**) In the decentralized network, nodes are directly connected with each other. (**a**) A schematic of centralized network. (**b**) A schematic of decentralized network

- More specifically, in a Bitcoin database, the transaction updates broadcasted by any single node will be verified by all other nodes before it is audited. Therefore, it is ideal to employ such a scheme to ensure the integrity of products in various supply chains.

Besides these applications, a critical potential of blockchain is improving the efficiency of globalized supply chains for different businesses. For example, IBM has deployed blockchain-based tracking service in building systems to record the movement of diamonds from mines to jewelry stores for Everledger [42]. Walmart has also employed a blockchain-based technology for supply chain management for the food industry and distribution [43]. For critical electronic applications, Honeywell Aerospace has employed a similar technology for creating a virtualized marketplace to purchase and distribute electronics and mechanical parts [44].

As one can understand, each blockchain infrastructure and implementation protocol requires necessary customization. Depending on the target applications and involved parties, there are three classes of blockchain:

- Public blockchain that is open to anyone, and any user can participate in the verification of new blocks.
- Private blockchain that is only accessible to those who have the permissions to write and read, and such permissions are maintained by an administrative entity within the private blockchain.
- Finally, consortium blockchain that is different from the previous two types. It is managed by a group of users instead of by all of them or by a selected administrator.

1.3 Blockchain-Enabled Electronics Supply Chain

1.3.1 Participants in Electronics Supply Chain

During the past few decades, the business model of the semiconductor industry has drastically changed. Previously, design, fabrication, and testing were usually completed by a single entity. With the increasing costs of fabrication at advanced process nodes, most semiconductor companies have chosen to operate as fabless design houses and outsource manufacturing to external foundries. In this *horizontal* business model, the fabless firms focus on developing and upgrading their design, while the foundries focus on improving the fabrication technologies. This model dramatically benefits the whole consumer electronics industry, since new products with more features and functionalities can be released with quicker turnaround times.

For such a development model and business practice, it is common for fabricated ICs to go through multiple stages of the electronics supply chain, depending on the functionality and application of the component, before the final product reaches the end user. The participants of the electronics supply chain can be roughly classified into the following categories: IP owner/foundry (fab), distributor, PCB assembler, system integrator, end user, and electronics recycler, as shown in Fig. 1.3.

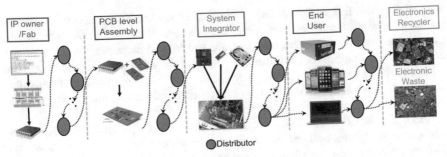

Fig. 1.3 The schematic of electronics supply chain. In each stage, there exist several distributors who connect these major entities

- IP owner refers to the participants that either design the complete IC, PCB, or system by themselves or source various intellectual property (IP) cores from multiple vendors to produce a complete system-on-chip (SoC).
- Foundry (also called fab) is the fabrication facility that gets the design file (e.g., GDSII format for IC, or Gerber format for PCB) from the IP owner and manufactures electronic ICs or PCBs as per its contract with the IP owner. This is the step where the electronic design becomes a physical entity (IC or PCB). Also, manufactured ICs and PCBs are tested and sorted for potential hardware faults, given a physical identity (ECID and marking) at this stage.
- PCB assemblers and system integrators (e.g., original equipment manufacturers in the supply chain) refer to the parties who use ICs and PCBs to build board- or system-level products.
- Distributors include all the possible buyers and sellers of ICs and board-level systems. They act as the transportation channel among the previously described parties. Commonly, there exist one or more distributors between each of the stages (foundry, PCB assemblers, and system integrators) to facilitate the supply of components among various design parties.
- Electronics recyclers are the participants responsible for handling E-waste that is the discarded end-of-life entity of the electronic components and systems.

1.3.2 Integrity Concerns in Electronics Supply Chain

In the existing complex global electronics supply chain, the integrity of the end product can only be assured if all participants are trusted. In such a scenario, all entities, such as IP owners, foundries, PCB assemblers, system integrators, distributors, and end users, would be able to verify the authenticity of an electronic component throughout its lifetime. To attain this goal, a blockchain-based framework can be employed to mitigate the existing vulnerabilities and establish the trusted supply chain, as proposed by Xu et al. [1]. In this work, the authors assume that the five main entities (including IP owner, PCB assembler, system integrator, end user, and electronics recycler) can enroll the associated information of a device/component/system into a secure and trusted database. On the contrary, an entity can enquire the authenticity verification of a component or system without gaining secret information. Any component that is not verified through this framework falls outside of the trusted electronics supply chain, and hence it should be considered as counterfeit or suspect.

From Fig. 1.3, we see that counterfeit electronic chips and systems can be introduced at different stages in the electronics supply chain, either by untrusted distributors or by the main participants such as foundry, PCB assembler, and system integrator. The adversarial role played by each of them is described as follows:

- Distributors widely exist throughout the electronics supply chain and are responsible for mediating the purchasing and selling of components (e.g., between foundries and PCB integrators, PCB integrators, and system integrators). Untrusted distributors can feed counterfeit components, such as recycled or remarked products (collected from the sources located outside of this trusted electronics supply chain) to other entities for higher profit.
- A PCB assembler (or system integrator) can also use recycled components on the PCB (or system); therefore, counterfeit parts are also possibly introduced by them.
- The untrusted manufacturer (fab) can introduce cloned or overproduced components directly into the supply chain by itself or with the help of untrusted distributors.

1.3.3 Additional Challenges in Electronics Supply Chain

Although blockchain (and the family of solutions) has been successfully employed so far to enhance the supply chain integrity of various commodities [42, 43], it is not readily depolyable for the electronics supply chain. Compared to other industries, the semiconductor industry has some unique characteristics. For example, the food supply chain can be monitored by tracking the temperature variations and the time taken for the transit of food commodities [43]. On the contrary, it is impractical to evaluate the integrity of electronic products by the shipping time. Moreover, it is also hard to authenticate electronics from their packaging appearance alone because there have been numerous examples where the success of identifying of a suspect chip solely relies on subject matter experts as well as the margin of error [45].

The merit of blockchain-powered electronics supply chain is that it enables all participants to track, verify, and then choose to deny or accept any single transaction, i.e., an electronic component or system. Correspondingly, the integrity of electronic devices can be guaranteed if they can be tracked throughout the supply chain. To realize such tracking, it is necessary to assign a unique ID for each electronic component. Fortunately, there already exists a unique electronic chip ID (ECID) and/or marking embedded in/on each chip that can be used as identifiers [36]. The ECID is a well-established technique, following the IEEE standard 1149.1, to facilitate the adaptive testing and tracking of ICs. It is commonly utilized in many consumer electronic products, such as iPhone [46]. When carrying an ECID, the chip can be identified and tracked throughout its lifetime. For example, if a chip has been denoted as *E-waste* in the blockchain-based framework, any device found with the same ID should be classified as counterfeit (or at least, *suspect*) since it is very likely of being recycled, remarked, overproduced, or cloned.

To build an authentication infrastructure via blockchain, a database accessible to all the registered participants of the trusted supply chain should be maintained to record the ECIDs of ICs. However, in practice, design houses may prefer to keep a record of their electronic products private. Therefore, it is difficult for a user to check the authenticity of a set of chips if they are not directly bought from these companies. Another limitation is that for an assembler that uses a large number of different chips, it is inconvenient to validate the authenticity of all chips from various companies. These limitations imply that before applying blockchain to track electronic devices, a proper ID database and accessing scheme should be designed first.

1.3.4 *Notation and Terminology*

Here, we list some notations and terminologies often used in this chapter for readers' clarity:

- **Certificate Authority (CA) Network** serves as the consortium blockchain (i.e., the trusted third-party entity) that maintains the electronic chip identification (ECID) information of electronic components in the electronics supply chain. The CA network is responsible for providing the enrollment and verification service to different entities in the electronics supply chain. **CA node** is the primary component of the CA network. Each CA node of the CA network maintains a database that stores the information regarding each chip in the electronic system (e.g., marking, ID, and transaction time, etc.).
- **Marking** provides the device identification and manufacturing traceability information on the package of electronic components. It is usually comprised of several codes denoting wafer fab and assembly plant, date of manufacture, wafer lot, device family and packaging information, etc. [47].
- **ID** denotes the embedded identification of an electronic component. It can be the electronic chip ID (ECID) of an integrated circuit in this work. The ECID of a chip includes the fabrication information, for example, the wafer locations, wafer number, binning information for temperature and speed grade, and any other information deemed appropriate for traceability. **PCB ID (PID)** stands for the unique identification of the PCB board. In this framework, this ID is derived from the IDs of the chips on it, as shown in Fig. 1.8 (described in detail in Sect. 1.4.4.1).

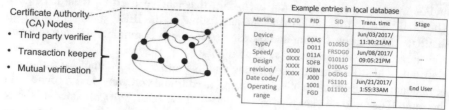

Fig. 1.4 A decentralized "ledger" comprised of several certificate authority (CA) nodes (denoted with the black dot-filled circle). Each CA node keeps a local database for the chip ID enrollment and verification, in which the detailed information such as marking, ECID, PCB ID (PID), system ID (SID), transaction time, and stage of an electronic component are stored. Upon the deployment, this CA network can serve for mutual authentication with each other and provide verification service to different electronics supply chain participants

System ID (SID) is the ID of the electronic system that is comprised of various chips, PCB boards, and operating system (described in detail in Sect. 1.4.4.2).
- **Transaction time** is a record of the time when the CA network receives the enrollment or verification request for a certain ID.
- **Stage** denotes the instant of the electronic life cycle when verification is requested. The CA network can identify the requestor as an entity such as PCB assembler or system integrator, etc. For example, an electronic part with stage "End User" as shown in Fig. 1.4 means that it has been sold and is with the end user. Therefore, any new verification request for the ID (chip-, PCB-, and system-level) related to this product corresponds to counterfeit.

1.3.5 Assumptions

In this chapter, we make the following assumptions:

- The blockchain-enabled framework creates a trusted electronics supply chain only for the entities that are part of the blockchain-enabled electronics supply chain, such as IP owner/Fab, PCB assembler, system integrator, and end user. This allows us to create a peer-to-peer connection among the entities.
- The electronic components, PCBs, and systems can contain and generate necessary identification information. For components that do not have ECID information such as analog ICs, package markings would also suffice.
- The communication between any two Certificate Authority (CA) nodes is secure and is maintained by the CA network. Details of CA network and CA nodes are discussed in Sect. 1.4. This can be ensured by using the appropriate mode of secure communication. Details of such an infrastructure are beyond the scope of this chapter.

- The confidentiality and integrity of communication for all messages in the framework are guaranteed.
- The main entities such as IP owner, PCB assembler, system integrator, and end user have permission to enroll the information of their products to the CA network, and this enrollment is secure.
- All entities have permission to verify the information of electronic components from their upstream entities (by using the CA network), and this verification is secure.
- All distributors (of chip-, PCB-, and system-level) and end users can verify components or systems with the CA network but have no authority to do the enrollment.

1.4 Framework for the Blockchain-Enabled Electronics Supply Chain

1.4.1 Consortium Ledger: The Certificate Authentication Network

It is undesirable to make the ID database of electronics supply chain fully public, as doing so may increase the controlling and accessing complexity of the database. Moreover, making the ID database as publicly accessible may leak the trade secret (e.g., yield information) of semiconductor companies. In practice, the entities who care about the authenticity of electronic chips include:

1. Original component manufacturer (OCM) (e.g., IP owner) who wants to prevent all possible vulnerabilities of electronics supply chain and ensure the economic benefits of their design/products, etc.
2. Original equipment manufacturer (OEM) (e.g., PCB assemblers and system integrators) that do not design but choose to buy chips from the IP owners and distributors and would like to build their products with genuine chips.
3. End users who want to ensure that the electronic products they bought are comprised of authentic electronic components.

Adhering to the "decentralized" feature of the blockchain, we propose to build a consortium blockchain: a networked monitoring system that is comprised of several distributed certificate authority (CA) nodes, as shown in Fig. 1.4. This CA network is decentralized in the sense that (1) every pair of CA nodes are connected and can exchange information with each other, (2) all nodes keep a database for chip ID enrollment and verification, (3) all CA nodes need to reach consensus before adding a block, as denoted by the "mutual verification" operation in the following sections.

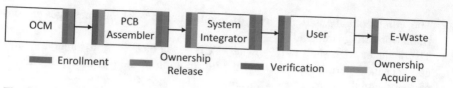

Fig. 1.5 The schematic of a blockchain-enabled electronics supply chain. In which four extra steps are added: enrollment, ownership release, verification, and ownership acquire. These four steps denote the interactive communication between each supply chain entities and the CA network

1.4.2 Blockchain-Enabled Electronics Supply Chain Framework

The blockchain-enabled framework proposed by Xu et al. [1] is as shown in Fig. 1.5; in addition to the normal stages such as PCB assembly and system integration, four more steps: enrollment, ownership release, verification, and ownership acquire are added to enhance the integrity of supply chain. These four steps stand for the interactive communication between various entities and the CA network. The meaning of each step is described as follows:

1.4.2.1 Enrollment

In this framework, enrollment denotes that entities of the electronics supply chain enroll the information of their products into the database of CA network. Specifically, OCM (e.g., IP owner) enrolls the information (e.g., ECID, marking, grade, and the intrinsic ID generated by PUF) of all chips they build, which generates the first block for each hardware device in the CA database. The CA network will store the enrolled chip information among all CA nodes and issue an "enrollment certificate" to the supply chain entity.

1.4.2.2 Ownership Release

When OCM finishes information enrollment, the next step is selling their products. In this process, the OCM will first request the ownership release to CA network with the corresponding chip information and the "enrollment certificate." All CA nodes will *mutually* verify this information and "enrollment certificate"; if authentic, they will issue the "ownership release" certificate (token) to the entity. To finish the transaction while facilitating the verification of PCB assembler (or next-stage distributor), the OCM will sell the chips with the CA-issued "ownership release" token.

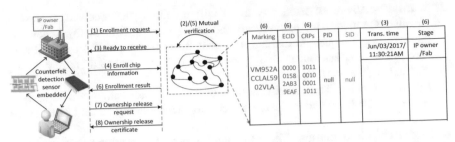

Fig. 1.6 The ID enrollment procedure between IP owner and CA network. If the enrollment is successful, the detailed information of chips will be stored by CA network. Sequential steps are shown in brackets

1.4.2.3 Verification

In this step, the PCB assembler will first conduct the semi-verification of the electronics with CA network, by sending the public information (e.g., marking), and the CA-issued token of chips to the CA nodes. The CA network will make quick search for these information in its database. If found and match, the CA network will then do a "full-verification" with the intrinsic IDs (e.g., challenge and response pairs [CRPs] of PUF) of the chips, which cannot be modified by the PCB assembler.

1.4.2.4 Ownership Acquire

When the CA network confirms the validity of the intrinsic IDs, the "full verification" will pass. The PCB assembler can then send an "ownership acquire" request to the CA network. The CA network will issue an "ownership certificate" to the PCB assembler and change the stage information of the electronic products in its database to "PCB Assembly."

1.4.3 IP Owner and Foundry (OCM)

As the starting point of electronics supply chain where an integrated circuit originates, the IP owner suffers the most economic loss from counterfeited chips. Therefore, in this scheme, IP owner is assumed trusted and in charge of enrolling the information of their chips. The information enrolled by the IP owners include marking, chip ID, grade (military or commercial), CRPs of PUFs, etc. The enrollment flow is as shown in Fig. 1.6.

1. **ID enrollment request**: The IP owner or Fab (OCM) will send ID enrollment request to CA network.

2. **Mutual verification**: Each CA node will broadcast the received request to all other CA nodes for mutual verification. If yes, then go to (3); otherwise, the enrollment request is marked as failed. Note that the OCM can still send enrollment requests, but such requests will only be accepted if they satisfy "mutual verification."

3. **Ready to receive**: The transaction time of the chip information will be updated in the CA database, and a "Ready to receive" decision will be sent to IP owner (or fab).

4. **Enroll chip information**: The IP owner (or fab) will send the information of chips to CA network (all CA nodes), including marking, ECID, grade, and CRPs;

5. **Mutual verification**: Each CA node will broadcast the information it receives to other CA nodes for mutual verification, e.g., whether they also get the verification request for the same IDs.

6. **Enrollment result**: If all CA nodes *mutually* confirm the ID enrollment by OCM, then the enrolled information will be stored in the database, as shown in the table in Fig. 1.6. CA network sends a decision to the IP owner (or fab) about the enrollment. If the enrollment succeeds, the CA network issues an "enrollment complete certificate" to the OCM. The enrollment fails if the enrolled IDs are found pre-existing in the CA database.

7. **Ownership release request**: When the OCM finishes the enrollment, it will consider releasing the ownership of these chips. To complete this, the OCM will send an ownership release request to the CA network, with the chip information and "enrollment complete certificate." The CA network will do a quick search in its database; if the information matches, it will issue an "ownership release" certificate (step (8)) to the OCM.

An example of the enrolled chip information is shown in the table of Fig. 1.6, where the marking, ECID, grade, and intrinsic ID of the chip have been enrolled. Since this chip is newly enrolled into the database, no corresponding PID (*null*) and SID (*null*) will be found. The transaction time ("Trans. time") records the time when this electronic component is enrolled in the CA database. Since this is a newly enrolled chip, the stage record is labeled as "IP owner/Fab." Note that the IC enrollment fails if any of the abovementioned steps do. For example, if the ID enrollment request is not "mutually conducted/sent" by/to all CA nodes, or if the chip IDs already exist in the CA database, the enrollment will fail.

1.4.4 Assembly Stage

In this section, we use "assembly stage" to generally denote the two stages: PCB assembly and system integration as shown in Fig. 1.3.

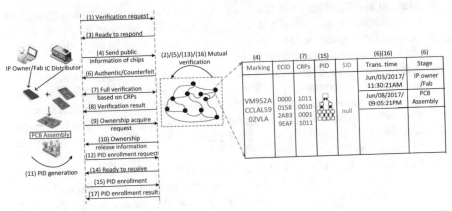

Fig. 1.7 The ID verification and PID enrollment procedure between PCB assembler and CA network. Note that for each verification or enrollment request, a "mutual authentication" will be conducted between all CA nodes, which greatly enhances the security and data integrity

1.4.4.1   PCB Assembly

The first step of building electronic systems is assembling various electronic chips onto a PCB. More specifically, in this step, PCB assemblers buy chips from the OCM (or distributors). These chips are then mounted onto PCBs. Note that after the chips are mounted onto PCBs, the embedded chip ID like ECID can be read out by the PCB assemblers (e.g., through JTAG) and verified with the CA nodes. For example, after getting the ECID information, the PCB assembler can send a verification request to the CA network and get the feedback. The objective of such verification is to detect counterfeit electronic components introduced into electronics supply chain during the distribution stage. We propose a verification procedure as shown in Fig. 1.7. The detailed operation of each step is provided as follows:

1. **Verification request**: The PCB assembler sends ID verification request to CA network.
2. **Mutual verification**: Each CA node will broadcast the ID verification request it received to all other CA nodes and get their feedback (e.g., whether they also get the verification request from the same PCB assembler).
3. **Ready to respond**: All CA nodes check with each other to ensure that all nodes receive the same request. If yes, then go to (4); otherwise, the verification request is marked as failed.
4. **Send public information of chips**: To complete the semi-verification, the PCB assembler sends the public information (e.g., marking, grade, etc.) of chips to CA network for verification. Note that not all these chips will be necessarily used in building electronic products.

5. **Mutual verification**: Each CA node will broadcast the information it received to all other CA nodes and get their feedback (e.g., whether they also get the verification request for the same IDs). If yes, then go to (6); otherwise, the verification request is marked as failed.

6. **Authentic/Counterfeit**: After all CA nodes *mutually* authenticate the information from PCB assembler, the transaction time will be updated, and the stage of these chips will be labeled as "PCB Assembly" if the verification succeeds. The authentication fails if the requested IDs are either not found in the database or found as being used in other PCB boards. The verification results will then be sent to the PCB assembler.

7. **Full verification based on CRPs**: If the semi-verification confirms that the chips are authentic, then the CA network will do a full verification based on the CRPs of PUFs. Note that in this framework, the assumption is that the verification can be done automatically, that is, the PCB assembler has no access or permission to control or change the challenges and responses of PUFs.

8. **Verification result**: The CA network will send the full-verification result to PCB assembler.

9. **Ownership acquire request**: After fully verifying the authenticity of the chips, the PCB assembler can then request the ownership, by sending an ownership acquire request to the CA network.

10. **Ownership release information**: The CA network will issue the ownership release information to PCB assembler.

11. **PID generation**: If the chips are genuine, then the PCB assembler will assemble them in PCB boards, a PCB ID (PID) will be generated based on the rule depicted in Fig. 1.8.

12. **PID enrollment request**: The PCB assembler sends PID enrollment request to CA network.

13. **Mutual verification**: Each CA node will broadcast the PID enrollment request it received to all other CA nodes and get their feedback (e.g., whether they also get the verification request from the same PCB assembler).

14. **Ready to receive**: After all CA nodes *mutually* authenticate this enrollment request, if yes, the CA network sends a "Ready to receive" response to PCB assembler. Otherwise, the verification request is marked as failed.

15. **PID enrollment**: The PCB assembler sends the generated PID and its composition (e.g., the chip IDs that are used to generate this PID) to CA network.

16. **Mutual verification**: Each CA node will broadcast the information it received to all other CA nodes and get their feedback (e.g., whether they also get the verification request for the same IDs). The CA network will also verify the owner of these chips; only if the PCB assembler is the current owner of these chips, the PID enrollment is allowed. After all CA nodes *mutually* authenticate this information, they will update the PID in the database, as shown in Fig. 1.7.

17. **PID enrollment result**: The transaction time and the stage of this chip will be updated, then the CA network sends a decision to the PCB assembler about the success (or failure) for the enrollment.

Fig. 1.8 An example flow of PID generation based on hash tree structure, in which H stands for the hash computation. The root node refers to PID, which is the hashed results of several ECIDs (A, B, C, and D in this example). Based on the algorithm of Merkle tree, SHA-256 protocol is employed as the hash function

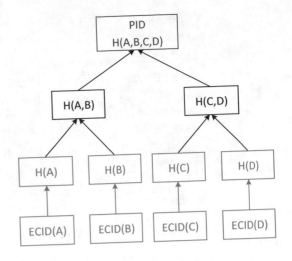

Note that the verification fails if any of the abovementioned steps does, for example, either because the verification is not "mutually conducted/sent" by/to all CA nodes or because the IDs under verification do not exist in the CA database.

Building a PID is advantageous for the tracking and management of electronic components in electronics supply chain for two reasons: (1) When several electronic components are assembled, the labels ("stage = PCB Assembly" in Fig. 1.7) will mark them as in use. (2) When the used parts move forward in the electronics supply chain, a board ID can help managing these parts together, i.e., for verification and deactivation purpose once the system reaches its end-of-life.

As shown in Fig. 1.8, one possible method to build a PID is by organizing the ECID of chips in a "Merkle tree" structure, i.e., each leaf node of the hash tree is filled with a chip ID and the PID is the root of this tree [48]. In this PID generation algorithm, SHA-256 protocol is employed as the hash function. The advantage of using this data structure is that each chip ID (leaf node) can be tracked by computing a number of hash calculations, which is linearly proportional to the logarithm of the number of leaf nodes of the tree. Compared with linear search, this technique greatly decreases the workload for CA network. Once the PCB ID is generated, the "PID enrollment" procedure can be done similarly as that between the IP owner/Fab and CA network. The difference is that for each enrolled PID, the PCB assembler also sends the chip IDs to the CA nodes, and CA nodes will update their database correspondingly to build the relationship between the chip IDs and PCB IDs.

1.4.4.2   System Integration

An example of system integration is as shown in Fig. 1.3, where a computer is comprised of several PCB boards as subcomponents. To facilitate the database management for CA nodes and tracking of all components in the electronics

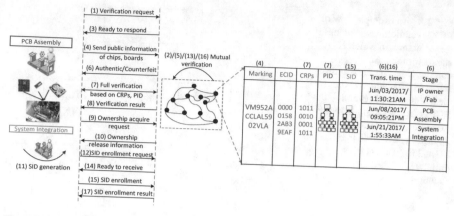

Fig. 1.9 The ID verification and SID enrollment procedure between system integrator and CA network

supply chain, we again propose to build an ID, namely system ID (SID) for each electronic system. Like PID, the SID can be a hashed result of the PCB IDs in this system. The verification and SID enrollment between system integrator and CA network is similar to that of the PCB assembler. Note that the verification and enrollment request from the system integrator changes the stored information in CA network. For example, the SID will be generated and more "transaction time" will be recorded, and the "stage" will be updated as "System Integration," as shown in Fig. 1.9.

1.4.5 End User

When the system integration finishes, the electronic products will be sold to end users (or distributors). Similarly, the end users would like to verify the authenticity of the products with CA network. As shown in Fig. 1.10, the user can first send verification request to the CA nodes and provide some public information of the products. Then, the CA network can make a quick search in the database and do the full verification by checking the authenticity of all electronic components in the product. If the verification result is authentic, the CA network marks the stage of product as user. The user can then send an ownership acquire request to the CA network after confirming the authenticity of the product.

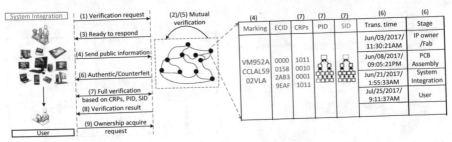

Fig. 1.10 End user verifies the authenticity of the electronic products and then gets the ownership

1.4.6 Distribution Stage

In this work, we use the term "distribution stage" to denote the distribution of components at each stage of the supply chain. As shown in Fig. 1.3, electronic components that have been sold at one stage may be bought or sold again among different chip distributors. The PCB distributors connect PCB assembler and system integrators. The system distributor sells electronic products to end users. Since we assume that the distributors are untrusted, they do not have the authority to enroll any information into the CA network but can send verification requests, if they want to check the authenticity of the products they acquired. One advantage of this regulation is that the "stage" information of electronic components cannot be changed by these distributors. This prevents remarked or recycled chips from re-entering the supply chain.

1.4.7 Electronic Waste

In this work, E-waste stands for the final stage of electronics supply chain, which is the source of many counterfeit components such as recycled chips. In this framework, the electronic recyclers are responsible for collecting and updating electronic components with the "end-of-life" status to CA network, thus preventing them from re-entering the supply chain by marking the stage in the database as "E-waste."

1.5 Evaluation of the Method

As stated earlier in this chapter, there are several known vulnerabilities in the traditional electronics supply chain: overproduction, recycling, remarking, cloning,

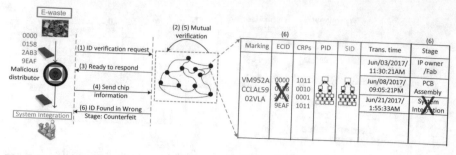

Marking	ECID	CRPs	PID	SID	Trans. time	Stage
					Jun/03/2017/ 11:30:21AM	IP owner /Fab
VM952A CCLAL59 02VLA	0000 0018 9EAF	1011 0010 0001 1011			Jun/08/2017/ 09:05:21PM	PCB Assembly
					Jun/21/2017/ 1:55:33AM	System Integration

Fig. 1.11 The recycled chips (or boards) can be detected by the CA network; even though they are with enrolled IDs stored in CA network, the stage prevents them from being deemed as new devices

etc. In this section, we discuss how each vulnerability can be mitigated with this framework for the integrity of electronics supply chain.

1.5.1 Resistance Against Recycling

Following this framework, the recycled chips, boards, or system would contain IDs that have been enrolled by the IP owner, PCB assembler, and system integrators, respectively. Therefore, they can be prevented from re-entering the electronics supply chain again by verifying with the CA network. An example of recycling detection is as shown in Fig. 1.11, where a recycled chip with an already enrolled ID can be detected by the system integrator since it has an existing ID with the "stage" information as system integration.

1.5.2 Resistance Against Overproduction

In the conventional threat model of electronics supply chain, the foundry is usually untrusted due to threats such as overproduction. In this framework, even if the foundry can manufacture more chips than contracted, such overproduced chips are not allowed to be put into the blockchain-enabled electronics supply chain. As shown in Fig. 1.12, if the overproduced chips enter the electronics supply chain, they will be detected since the ID information is not enrolled and stored in the CA database. In the worst case, the overproduced chips will have the same IDs as that of the genuine chips, and such chips can also be detected by verifying the "stage" information.

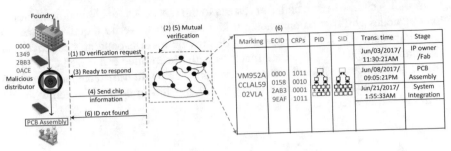

Fig. 1.12 The overproduced chips can enter the electronics supply chain through untrusted entities. However, as the chip buyers can always resort to CA network for verification and tracking, such overproduced chips can be detected

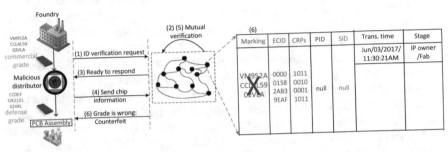

Fig. 1.13 The CA network stores the marking information of the genuine electronic devices, and hence any changes in the marking can be detected

1.5.3   Resistance Against Remarking

In this framework, all important information about an electronic component is recorded. Therefore, the verification information from the CA network would detect the discrepancies for a remarked chip. An example of the remarking detection is as shown in Fig. 1.13, where the marking changes from commercial to defense grade can be detected by the CA network.

1.5.4   Resistance Against Cloning

During the fabrication process, cloned chips can be manufactured in an unauthorized fab through reverse engineering or IP theft. To mitigate this potential vulnerability, we propose to employ PUF in the verification and authentication with CA network. As PUF is built on manufacturing process variations, the input and output (CRPs: challenges and responses) behavior of a cloned chip will not be the same as that of the genuine chip.

In such a case, when an end user resorts to the CA nodes for chip authentication, the CA nodes will first communicate with each other to verify request and search the component ID. The CA network will provide a corresponding challenge and examine the response. If this CRP matches with the enrolled and stored CRP in the ledger, the chip is considered authentic. A cloned chip will be detected and reported if its ID is found in the CA database but the PUF behavior does not match with that of the IP owner's record.

1.6 Conclusion

In this chapter, we discuss the blockchain-based framework proposed by Xu et al.[1] to monitor the integrity of electronics supply chain. We provide analysis of different roles of all entities in the trusted electronics supply chain and discuss the resistance of this framework against some common supply chain threats. This blockchain-enabled framework can effectively mitigate vulnerabilities such as recycling, remarking, overproduction, and cloning. Nevertheless, this framework potentially has some limitations that need to be addressed. For example, overproduced chips can circumvent the monitoring of the framework, when these chips are sold to entities outside the blockchain-enabled supply chain. Another limitation of this framework is the scalability. This scheme achieves the decentralization feature of blockchain but also makes it expensive to manage the database. However, considering the ever-increasing threat in the electronics supply chain, this framework provides the fundamental skeleton to address the issues and allows additional mitigation techniques to be integrated easily.

References

1. X. Xu, F. Rahman, B. Shakya, A. Vassilev, D. Forte, M. Tehranipoor, Electronics supply chain integrity enabled by blockchain. ACM Trans. Des. Autom. Electron. Syst. **24**(3), 1–25 (2019)
2. G.E. Moore et al., Cramming more components onto integrated circuits. Proc. IEEE **86**(1), 82–85 (1998)
3. Defense Science Board, Defense science board task force on high performance microchip supply. Office of the Under Secretary of Defense for Acquisition, Technology, and Logistics (2005)
4. M.M. Tehranipoor, U. Guin, D. Forte, Counterfeit integrated circuits, in *Counterfeit Integrated Circuits* (Springer, Berlin, 2015), pp. 15–36
5. M.M. Tehranipoor, U. Guin, S. Bhunia, Invasion of the hardware snatchers. IEEE Spectr. **54**(5), 36–41 (2017)
6. X. Zhang, M. Tehranipoor, Design of on-chip lightweight sensors for effective detection of recycled ICs. IEEE Trans. Very Large Scale Integr. Syst. **22**(5), 1016–1029 (2013)
7. U. Guin, X. Zhang, D. Forte, M. Tehranipoor, Low-cost on-chip structures for combating die and IC recycling, in *2014 51st ACM/EDAC/IEEE Design Automation Conference (DAC)* (2014)
8. M. Tehranipoor, H. Salmani, X. Zhang, *Integrated Circuit Authentication* (Springer, Cham., 2014)

9. N. Kae-Nune, S. Pesseguier, Qualification and testing process to implement anti-counterfeiting technologies into IC packages, in *Design, Automation & Test in Europe Conference & Exhibition (DATE), 2013* (IEEE, Piscataway, 2013), pp. 1131–1136
10. X. Zhang, M. Tehranipoor, Design of on-chip lightweight sensors for effective detection of recycled ICs. IEEE Trans. Very Large Scale Integr. Syst. **22**(5), 1016–1029 (2014)
11. F. Koushanfar, G. Qu, Hardware metering, in *Proceedings of the 38th Annual Design Automation Conference* (ACM, New York, 2001), pp. 490–493
12. J.W. Lee, D. Lim, B. Gassend, G.E. Suh, M. Van Dijk, S. Devadas, A technique to build a secret key in integrated circuits for identification and authentication applications, in *2004 Symposium on VLSI Circuits, 2004. Digest of Technical Papers* (IEEE, Piscataway, 2004), pp. 176–179
13. G.E. Suh, S. Devadas, Physical unclonable functions for device authentication and secret key generation, in *Proceedings of the 44th Annual Design Automation Conference* (ACM, New York, 2007), pp. 9–14
14. R. Pappu, B. Recht, J. Taylor, N. Gershenfeld, Physical one-way functions. Science **297**(5589), 2026–2030 (2002)
15. G.K. Contreras, M.T. Rahman, M. Tehranipoor, Secure split-test for preventing IC piracy by untrusted foundry and assembly, in *2013 IEEE International Symposium on Defect and Fault Tolerance in VLSI and Nanotechnology Systems (DFT)* (IEEE, Piscataway, 2013), pp. 196–203
16. X. Wang, D. Zhang, M. He, D. Su, M. Tehranipoor, Secure scan and test using obfuscation throughout supply chain. IEEE Trans. Comput. Aided Des. Integr. Circuits Syst. **37**(6), 1867–1880 (2017)
17. M.T. Rahman, D. Forte, Q. Shi, G.K. Contreras, M. Tehranipoor, CSST: preventing distribution of unlicensed and rejected ICs by untrusted foundry and assembly, in *IEEE International Symposium on Defect and Fault Tolerance in VLSI and Nanotechnology Systems (DFT)* (2014), pp. 46–51
18. D. Zhang, M. He, X. Wang, M. Tehranipoor, Dynamically obfuscated scan for protecting IPs against scan-based attacks throughout supply chain, in *IEEE 35th VLSI Test Symposium (VTS)*, 2017
19. C. Lamech, R.M. Rad, M. Tehranipoor, J. Plusquellic, An experimental analysis of power and delay signal-to-noise requirements for detecting Trojans and methods for achieving the required detection sensitivities. IEEE Trans. Inf. Forensics Secur. **6**(3), 1170–1179 (2011)
20. M.T.M. Li, A. Davoodi, A sensor-assisted self-authentication framework for hardware Trojan detection, in *Design, Automation & Test in Europe Conference & Exhibition (DATE)* (2012), pp. 1331–1336
21. K. Xiao, D. Forte, M. Tehranipoor, Efficient and secure split manufacturing via obfuscated built-in self-authentication, in *IEEE International Symposium on Hardware Oriented Security and Trust (HOST)* (2015), pp. 14–19
22. M. Pilkington, Blockchain technology: principles and applications. Browser Download This Paper (2015)
23. N. Subramanian, A. Chaudhuri, Y. Kayıkcı, *Blockchain and Supply Chain Logistics: Evolutionary Case Studies* (Springer, Berlin, 2020)
24. Y. Lu, Blockchain: a survey on functions, applications and open issues. J. Ind. Integr. Manag. **3**(4), 1850015 (2018)
25. N. Tuzzio, K. Xiao, X. Zhang, M. Tehranipoor, A zero-overhead IC identification technique using clock sweeping and path delay analysis, in *Proceedings of the Great Lakes Symposium on VLSI* (2012), pp. 95–98
26. N. Asadizanjani, M. Tehranipoor, D. Forte, PCB reverse engineering using nondestructive x-ray tomography and advanced image processing. IEEE Trans. Compon. Packag. Manuf. Technol. **7**(2), 292–299 (2017)
27. N. Asadizanjani, M. Tehranipoor, D. Forte, Counterfeit electronics detection using image processing and machine learning, in *Journal of Physics: Conference Series*, vol. 787(1) (IOP Publishing, Bristol, 2017), p. 012023

28. K. Huang, J.M. Carulli, Y. Makris, Parametric counterfeit IC detection via support vector machines, in *2012 IEEE International Symposium on Defect and Fault Tolerance in VLSI and Nanotechnology Systems (DFT)* (IEEE, Piscataway, 2012), pp. 7–12
29. M.H. Tehranipour, N. Ahmed, M. Nourani, Testing SOC interconnects for signal integrity using boundary scan, in *Proceedings. 21st VLSI Test Symposium* (IEEE, Piscataway, 2003), pp. 158–163
30. M. Yilmaz, K. Chakrabarty, M. Tehranipoor, Test-pattern selection for screening small-delay defects in very-deep submicrometer integrated circuits. IEEE Trans. Comput. Aided Des. Integr. Circuits Syst. **29**(5), 760–773 (2010)
31. E. Castillo, U. Meyer-Baese, A. García, L. Parrilla, A. Lloris, IPP@ HDL: efficient intellectual property protection scheme for IP cores. IEEE Trans. Very Large Scale Integr. Syst. **15**(5), 578–591 (2007)
32. J.J. Rajendran, O. Sinanoglu, R. Karri, Is split manufacturing secure? in *Proceedings of the Conference on Design, Automation and Test in Europe* (EDA Consortium, San Jose, 2013), pp. 1259–1264
33. M. Miller, J. Meraglia, J. Hayward, Traceability in the age of globalization: a proposal for a marking protocol to assure authenticity of electronic parts, SAE Technical Paper, Tech. Rep., 2012
34. C. Kuemin, L. Nowack, L. Bozano, N.D. Spencer, H. Wolf, Oriented assembly of gold nanorods on the single-particle level. Adv. Funct. Mater. **22**(4), 702–708 (2012)
35. M.T. Rahman, F. Rahman, D. Forte, M. Tehranipoor, An aging-resistant RO-PUF for reliable key generation. IEEE Transactions on Emerging Topics in Computing **4**(3), 335–348 (2015)
36. U. Guin, K. Huang, D. DiMase, J.M. Carulli, M. Tehranipoor, Y. Makris, Counterfeit integrated circuits: a rising threat in the global semiconductor supply chain. Proc. IEEE **102**(8), 1207–1228 (2014)
37. S. Nakamoto, Bitcoin: a peer-to-peer electronic cash system (2008)
38. E. Staff, Blockchains: the great chain of being sure about things. Economist. Retrieved **18** (2016)
39. A.M. Antonopoulos, *Mastering Bitcoin: Unlocking Digital Cryptocurrencies* (O'Reilly Media, Newton, 2014)
40. F. Casino, T.K. Dasaklis, C. Patsakis, A systematic literature review of blockchain-based applications: current status, classification and open issues. Telematics Inform. **36**, 55–81 (2019)
41. M.J. Dworkin, SHA-3 standard: permutation-based hash and extendable-output functions. Tech. Rep., 2015
42. K.S. Nash, IBM pushes blockchain into the supply chain. Wall Street J. (2016)
43. Hyperledger Case Study: Walmart (2020). https://www.hyperledger.org/learn/publications/walmart-case-study
44. Hyperledger Case Study: Honeywell (2020). https://www.hyperledger.org/learn/publications/honeywell-case-study
45. Learn to Know the Difference with AS5553, Learn to know the difference with as5553 (2009). https://escs9120.wordpress.com/
46. Sauriks, ECID—the Iphone Wiki (2009). https://www.theiphonewiki.com/wiki/ECID
47. H. James, T. Cles, Standard linear & logic semiconductor marking guidelines (2002). http://www.ti.com/lit/an/szza020c/szza020c.pdf
48. R.C. Merkle, Method of providing digital signatures. Jan. 5 1982, U.S. Patent 4,309,569

Chapter 2
Digital Twin with a Perspective from Manufacturing Industry

Haibo Wang, Shigang Chen, Md Sami Ul Islam Sami, Fahim Rahman, and Mark Tehranipoor

2.1 Introduction

Digital Twin (DT) has become an important component in programs and initiatives related to Smart Manufacturing, Digital Manufacturing, Advanced Manufacturing, and Industry 4.0 globally. It is a hot topic among researchers, educators, and software vendors, which is evident from that searches of the key word "Digital Twin" have been growing rapidly since 2016.

The advancement of technologies such as smart sensors, Internet of Things (IoT), cloud computing, Artificial Intelligence (AI), Cyber-Physical Systems (CPS), and modeling and simulation makes it possible to realize the "Digital Twin" of a manufacturing product, system, and process. These technologies enable better real-time data collection, computation, communication, integration, modeling, simulation, optimization, and control that are required by Digital Twins.

However, the DT is still in the early stage, and it is not embraced or implemented by the manufacturers as widespread as expected. There is much confusion about what DT is, what it includes, and how to implement it. There is a lack of consensus among researchers and practitioners in different communities, which hinders the acceptance of DT by manufacturers. A commodity DT system for companies, especially, small- and medium-sized companies, has not been available. Driven by the advancement of DT and the benefits of DT based design and manufacturing, the community needs a literature review of current knowledge and exploration associated with DT.

This paper collects and classifies the state-of-the-art literature on DT. In addition, a case study for semiconductor manufacturing is discussed to show the advantages

H. Wang (✉) · S. Chen · M. S. U. I. Sami · F. Rahman · M. Tehranipoor
University of Florida, Gainesville, FL, USA
e-mail: wanghaibo@ufl.edu; sgchen@cise.ufl.edu; md.sami@ufl.edu; fahim034@ufl.edu; tehranipoor@ece.ufl.edu

brought by DT and the research opportunities for DT based semiconductor manufacturing. The contribution of this paper can be summarized as follows.

- This paper introduces the evolution, definition, application, and framework of DT, and DT based production design process.
- It discusses semiconductor manufacturing and its existing problems as a case study to show why DT is important for intelligent manufacturing.
- It presents some challenges and research opportunities for DT based semiconductor design and manufacturing.

2.2 Digital Twin

2.2.1 Evolution of Digital Twin

The Digital Twin (DT) technology could be traced back to 1970. In order to manage the whole systems, NASA created mirrored environment of the physical space to monitor the circumstance that could not be reached in real time (e.g., spacecraft in mission). A successful and famous example is Apollo 13 [83]. When Apollo was launched into the space, one of the oxygen tanks exploded. To fix this problem, the NASA control team built a mirrored system of the spacecraft and its components for testing the possible solutions and finding out the successful one, i.e., an improvised air purifier. When the astronaut followed the instructions to fix the problem, NASA on earth ran simulations to test the procedures at the same time. Eventually, the crew of Apollo 13 got back to earth alive.

The concept of DT was first introduced by Grieves in 2003 [41]. It referred to "a virtual digital representation equivalent to a physical product" at the University of Michigan's Product Lifecycle Management (PLM) course. Its original expectation was that all the data and information of a physical entity could be put together for a higher level of analysis. In the following years, Grieves discussed the Product Lifecycle Management Model (PLM Model) and the Mirrored Spaces Model (MSM) [38], where the later one embodied all functions of a Digital Twin. Grieves termed the concept of Digital Twin as termed information mirroring model in 2006–2010 [39]. It consisted of three critical parts for a Digital Twin, i.e., the physical space, the virtual space, and the linkage or interface between the two spaces. The terminology was changed to Digital Twin in Grieves' book titled, "Virtually perfect: driving innovative and lean products through product lifecycle management" [40], where he quoted his co-author Vickers' way of describing this model Digital Twin. In 2017, Grieves and Vicker [42] proposed the idea of the Digital Twin that links the physical system with its virtual equivalent that can mitigate these problematic issues. They described the Digital Twin concept and its development and showed how it applied across the product lifecycle in defining and understanding system behavior.

In order to reduce costs and resources, NASA has started to investigate the developing DT for its space assets. In 2012, NASA and the US Air Force (USAF) Research Laboratory presented a Digital Twin paradigm for future vehicles [37] that met the requirements of lighter mass, higher loads, and longer service in more severe conditions. The Digital Twin was defined as an integrated multi-physics, multi-scale, probabilistic simulation of an as-built vehicle or system that incorporated the best available physical models, updated sensors data, and historical data to mirror the life and condition of the corresponding flying twin. In the same year, NASA [107] released a road map called "Modeling, simulation, information technology, and processing," and then the term Digital Twin was widely applied for product or shop-floor.

Because the origin of DT is closely related to NASA, DT is mainly applied in aviation (see Table 2.1) and manufacturing (see Table 2.2) fields. The applications in other fields are listed in Table 2.3.

2.2.2 Definitions of DT from Different Perspectives

2.2.2.1 Digital Twin vs. Digital Thread

Digital Twin and Digital Thread are both used in the existing literature. We first discuss these two terms. In 2013, the U.S. Air Force [34] interchangeably used the Digital Thread and the Digital Twin concepts, highlighting that they have a historical memory and the ability of exploiting previous and current knowledge to gain state awareness and system prognosis, thus providing the "agility and tailorability needed for rapid development and deployment." However, in other literature, the Digital Thread concept was distinguished from the DT concept. For instance, authors in [70] claimed that digital thread referred to the "communication framework that allows a connected data flow and integrated view of the asset's data throughout its lifecycle across traditionally siloed functional perspectives." As explained in [60], the digital thread is the communication framework that digitally links all the product data (i.e., model data, product structure data, metadata, effectual data, and process definition data including supporting equipment and tools) to allow each user to access a single, consistent definition of the product and its modifications during its whole lifecycle. For a manufacturer, it provides a single reference point for design, engineering, and manufacturing. Essentially, the digital thread allows linking and integrating all aspects of a system and models from various disciplines through common inputs and data flows, in an always-available, up-to-date single electronic representation that every decision maker involved in the process can access, thus speeding up design time, and enabling trades across traditionally isolated disciplines [110]. The digital thread concept raises the bar for delivering "the right information to the right place at the right time" [70].

However, as they are defined, the digital thread only ensures the connection of heterogeneous elements; it does not have the DT's potential of monitoring,

Table 2.1 Applications of DT in aviation

Prior work	Application	Detail
[136]	Aircraft production	Describing an aircraft DT exploiting an automatic image tracking method to obtain insights about the crack tip deformation and crack growth behavior of aluminum alloy and steel
[15]	Detect fatigue crack	Using a finite element model of an aircraft wing containing shape memory alloy particles embedded in key regions of the aircraft structure
[31]	Detect fatigue crack	High performance fatigue mechanics were used to detect and predict damaged aircraft structures, which is an improved Moving Least Squares (MLS) law for computing fatigue crack growth rates
[61]	Detect fatigue crack	Exploiting a modified dynamic Bayesian network structure to control the state of aircraft wings
[14]	Detect fatigue crack	Combining developments in modeling of fatigue damage, isogeometric analysis of thin-shell structures [53, 139], and structural health monitoring [113], to develop a computational steering framework for fatigue-damage prediction in full-scale laminated composite structures
[129]	Spacecraft lifecycle management	Establishing an ultra-high fidelity simulation model, the Digital Twin, for each space vehicle with an independent tail number to accurately predict the life and damage of a spacecraft structure
[105]	Fault testing	In a NASA-related system, this work studied methods of fault prediction and elimination for such a system based on Digital Twin technology and then applied and tested it
[129]	Aircraft safety and reliability	US Air Force Research Laboratory utilizes an ultra-high fidelity model of individual aircraft by Digital Twin to integrate computation of structural deflections and temperatures in response to flight conditions, with resulting local damage and material state evolution
[95]	Big data-driven smart manufacturing	Presenting the benefits of a data-driven smart manufacturing (BDD-SM) approach exploiting DTs. BDD-SM exploits sensors and the IoT to produce and transport big data. These data can be processed through AI applications and big data analytics executed on the cloud, to monitor the processes, identify failures, and find the optimal solution
[132]	Machine statue prediction	On the basis of the Gaussian–Bernoulli deep Boltzmann machine (GDBM), this work discussed a deep neural network model to optimize the condition prognosis and to predict the future degradation status and remaining service life of a machine
[92]	Jet engine design	General Electric (GE) built Digital Twins of jet engine that can achieve power optimization, monitoring, and diagnostics of jet engine. DXC Technology built a Digital Twin for hybrid car manufacturing process to predict the performance of car before committing the changes in the manufacturing process

Table 2.2 Applications of DT in manufacturing

Prior work	Application	Detail
[28]	3D printing	Using DT to monitor model heat transfer, solidification and residual stresses and distortion for the geometrical conformity, microstructure, and properties of additively manufactured components
[145]	Satellite assembly shop-floor	A DT based smart production management and control for a satellite assembly shop-floor
[102]	Automative	Tesla is working on developing a DT for every car it produces, allowing for synchronous data transmission between cars and its factory
[8, 114]	Product design	Autodesk and Siemens use the DT technology to provide a general guidance for the future design
[134]	Engine maintenance	In 2015, General Electric (GE) company started to realize real-time monitoring, timely inspection, and predictive maintenance of engines based on Digital Twin technology
[45]	Semiconductor	Through a partnership with a center focused on accelerating commercialization of smart sensors, imagers, advanced devices, and 2.5D/3D chip integration, Siemens PLM Software is driving development of Digital Twin technologies for the semiconductor industry
[22]	Automotive	The paper explores the role of Digital Twin in addressing the current challenges of the automotive industry, especially with regards to vehicle product design, manufacturing, sales, and service
[86]	Lifecycle optimization	IBM Engineering Lifecycle Optimization visualizes, analyzes, and gains data insights across design, manufacturing, and operations. Improve the visibility of relationships across engineering data so you can make more effective, timely, and informed decisions and help maintain and demonstrate compliance with regulatory and industry standards. IBM Rational Lifecycle Integration Adapters is a practice application lifecycle management (ALM) in a diverse lifecycle tools environment, leveraging your current lifecycle tools investments
[37]	Vehicle development	NASA and US Air Force applied the DT technology in vehicles development, so they could predict the future performance and status of vehicles by constructing ultra-high fidelity simulation models with the parameters of material properties and manufacturing defects
[73]	Structural simulation	Majumdar et al. developed a DT modeling the way multi-physical environments (such as electrical field) cause microstructural changes in structural composites and hence may affect structural performance
[111]	Sheet metal welding assembly	Higher quality individualized production based on geometry assurance concept

Table 2.3 Other applications of DT

Prior work	Application	Detail
[109]	Smart home	Siano et al. proposed Decision Support and Energy Management Systems (DSEMS) to describe the designing method and testing method for presenting its simulation results and verify its effectiveness
[111]	Real-time geometry assurance	In process design phase, DT is used to optimize tolerances, locator positions, clamping strategies, welding sequence, etc. to obtain good geometrical quality in the final product
[9]	IoT	Microsoft proposed Azure Digital Twins to provide a service for building advanced IoT spatial intelligence solutions. Azure Digital Twins is an IoT service that helps you create comprehensive models of physical environments. Create spatial intelligence graphs to model the relationships and interactions between people, places, and devices
[21]	Geometry modeling	In 2014, Cerrone et al. implemented an accurate prediction of the crack path for each specimen using its as-manufactured geometry, which gave a better definition of Digital Twin through a more intuitive example
[1]	Plant design	With the virtual factory technology, a framework of simulation-based approach was proposed to guide simulation-based plant design and evaluation and optimize plant layout and the production process
[93]	Healthcare	GE Healthcare developed a "Capacity Command Center" that applies simulations and analytics for better decision-making in the Johns Hopkins Hospital in Baltimore. By building a DT of patient pathways, the hospital predicts patient activity and plans capacity according to demand, thus significantly improving patient service, safety, experience, and volume
[101]	Healthcare	MPH and Siemens Healthineers redesigned the radiology department by developing an AI computer model of the radiology department and its operations. The medical DT provided the faithful, realistic 3D animations and produced descriptive and quantitative reports, so as to predict the operational scenarios and instantly evaluate alternative options to find the right solution to transform care delivery
[128]	Healthcare	The Living Heart developed by the French software company Dassault Systemes has been released in 2015 (May) and is currently available for research. It has been the first DT of organs that takes all aspects of the functionality of the organ (including blood flow, mechanics, and electrical impulses) into account
[23]	Healthcare	Siemens Healthineers has exploited the data collected in a vast database containing more than 250 million of annotated images, reports, and operational data. The AI-based DT model was trained to weave together data about the electrical properties, the physical properties, and the structure of a heart into a 3D image
[6]	Automotive	DT used in a driving assistance application
[55]	Education	Driver training service via DT to improve experience and quality of users
[56]	Healthcare	DT to facilitate remote surgery, thus enlarging customer base

maintaining, and optimizing the physical system. For this reason, and after including connection capabilities into the DT model, the digital thread can be considered as a part of DT and performs partial functions of DT.

2.2.2.2   Definition of DT with Different Key Points

DT is in the early stage. The existing literature offers different understandings of DT, emphasizing on different functions and parts according to their research fields [13]. In a narrow sense, the DT is considered as a virtual replica of the physical entity [19, 44, 105]. In a broad sense, some literature [121, 123, 124] believe DT contains physical space, virtual space, and all the components that are necessary for design and manufacturing. There are some literature [20, 41, 77] that consider DT as Digital Thread. They may focus on the information/data processing and delivering function. We list the collected definitions in Table 2.4.

2.3   What a Working DT Would Look Like?

Digital Twin is in the early stage. Many have proposed the general DT frameworks. According to authors' understanding, the frameworks of DT are divided into three or five dimensions, each with different terminologies. In the following, we will explain three-dimensional framework and five-dimensional framework.

2.3.1   Three-Dimensional Framework

As the origin of DT, Grieves [38] suggested three-dimensional framework that has been inherited by most of the existing literature. Although the definitions of these three dimensions are different (see Table 2.5), the function of each dimension is pretty similar. In this survey, we follow Grieves' definition of the three dimensions and call them physical space, virtual space, and links, respectively. Physical space is a complex production environment that consists of machines, materials, rules, and sensors; virtual space models the physical space in the virtual environmental platform to serve applications; links build a bridge between physical space and virtual space.

Figure 2.1 illustrates the basic framework in which the virtual space is mapped to the physical space through the links that exchange data and information [98]. The physical space is a complex, diverse, and dynamic production environment that consists of many factors such as people, machines, material, rules, sensors, and environment. All these factors are necessary for production environment. For instance, the product is composed of a lot of objects and materials that are related to product development and manufacturing, such as production resources

Table 2.4 Definitions of DT from different perspectives

Key point	Reference	Definition
Virtual, Mirror, Replica, and Clone	[129]	A digital information construct about a physical system
	[105]	Virtual representation of a real product
	[86]	Virtual representation of a physical object or system across its lifecycle, using real-time data to enable understanding, learning, and reasoning
	[75]	A virtual representation of the system
	[44]	Digital mirror of the physical world
	[19]	Digital model that dynamically reflects the status of an artifact
	[133]	A digital copy of a physical system
	[117]	Virtual model of a physical asset
	[98]	A replication of real physical production system
	[6]	Cyber copy of a physical system
	[72]	A dynamic digital representation of a physical system
	[71]	A virtual model of physical object
	[11]	Computerized clones of physical assets
	[84]	The virtual and computerized counterpart of a physical system
	[131]	Functional system formed by the cooperation of physical production lines with a digital copy
	[2]	The notion where the data of each stage of a product lifecycle is transformed into information
Integrated systems	[107]	An ultra-realistic integrated multi-physics, multi-scale, probabilistic simulation of a system
	[121, 123, 124]	Integrated multi-physics, multi-scale, and probabilistic simulation composed of physical product, virtual product, data, services, and connections between them
	[100]	A big collection of digital artifacts that has a structure, all elements are connected, and there exists meta-information as well as semantics
	[16, 17]	Comprehensive physical and functional description of a component, product, or system together with all available operational data
	[58]	A systematic approach consisting of sensing, storage, synchronization, synthesis, and service
Ties and links	[41]	Connections of data and information that ties the virtual and the real product together
	[20]	New mechanisms to manage IoT devices and IoT systems-of-systems
	[77]	Technology that links the real and digital worlds

(continued)

Table 2.4 (continued)

Key point	Reference	Definition
Simulation, Test, Prediction	[42]	Reengineering computational model of structural life prediction and management
	[95]	Virtual models for physical objects to simulate their behaviors
	[47]	A safe environment in which you can test the impact of potential change on the performance of a system
	[35]	A simulation based on expert knowledge and real data collected from the existing system

Table 2.5 Definition of three dimensions in DT

[38]	Physical space	Virtual space	Link/interface between physical and virtual space
[95]	Physical world	Virtual world	Data that tie two worlds
[122]	Physical shop-floor	Virtual shop-floor	Shop-floor Digital Twin data
[143]	Physical space	Virtual space	Information processing layer

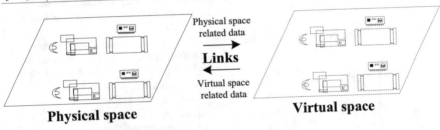

Physical space related data

Links

Virtual space related data

Physical space　　　　**Virtual space**

Fig. 2.1 Three-dimensional framework of DT

(production line equipment, etc.), product data resources, computing resources (high performance computing clusters, etc.), and software resources. All kinds of objects are separated and distributed in different places, and they need to be tracked by sensors. Then, the data of physical world will be collected, integrated, and used for modeling, optimization, and prediction in virtual space.

The virtual space consists of two parts: the virtual environment platform (VMP) and the DT application subsystem (DTAS) for product lifecycle management. The VMP is built to establish a unified 3D virtual model for application and to provide an operating environment for algorithms library. There are interactions between VMP and DTAS. VMP provides various virtual models for DTAS, including polyphysical model, workflow model, simulation model, etc. DTAS accumulates various models, methods, and historical data that are created during the operation into VMP. Meanwhile, the DT, as the real mapping of physical entities, cannot only realize the visualization of products but also realize the simulation of complex systems. When conflicts and disturbances occur in physical space, virtual models can be tested in real time or even predict them, and feed the information back to the physical space.

The links is the third dimension, which is the channel connecting physical space and virtual space, and the bidirectional mapping and inter-operation of physical space and virtual space are realized through the data interaction in this layer. Table 2.6 highlights data exchange protocols in manufacturing environments used by data interaction systems for high level DT communication [62]. The references are classified according to their nearest OSI model layers. Note that from different authors' view, the links may emphasize different functions. Some [44] believe the main function of links dimension is to exchange data between physical space and virtual space, while others believe [143] links dimension is not just a connection of two spaces but can also process, store, and map data.

Table 2.6 Data exchange protocols for DT

OSI layers	References	Rule	Description
Application	[59, 65, 66, 140]	OPC	A series of standards and specifications for industrial telecommunication
	[7, 12, 64, 69, 143]	OPC UA	Machine-to-machine communication protocol for industrial automation. It provides a path forward from the original OPC communications model that would better meet the emerging needs of industrial automation
	[12, 24, 46, 63]	MTConnect	MTConnect is a manufacturing technical standard to retrieve process information from numerically controlled machine tools
	[26, 67]	AMQP	Open standard application layer protocol for message oriented middleware
	[55, 85]	NTP/PTP	Networking protocol for clock/nanosecond/picosecond time synchronization between systems
Transport	[56, 106]	TCP	Main protocol of the Internet protocol suite that establishes a connection between client and server
	[7, 63]	TCP/IP	Communication protocol suite to interconnect network devices
	[56]	UDP	One of the core members of the Internet protocol suite for creating low-latency and loss-tolerating connections
Data link	[81]	Ethernet/IP	Industrial network protocol that adapts the Common Industrial Protocol to standard Ethernet
	[55]	OpenFlow	Protocol to give access to the data plane of network switch

2.3.2   Five-Dimensional Framework

In the previous three-dimensional framework, the connection channel will exchange the data between physical space and virtual space. The five-dimensional framework [120, 124] stresses the importance of data and service functions. The virtual space supports the simulation, decision-making, and control of the physical part. The framework is shown in Fig. 2.2. As we can see, data lies in the center of the DT system because it is a precondition for creating new knowledge. Furthermore, DTs lead to new services that can enhance the convenience, reliability, and productivity of an engineered system. Finally, the connection part bridges the physical part, virtual part, data, and service. The data exchanged among physical space, virtual space, and service platform will be stored in the data part rather than be passed without memorization. In virtual space, the modeling of physical objects is available by obtaining the attributes of the virtual model from the database, and the feedback of models will be stored in the database by using corresponding interfaces. In the DT system, the real-time and historical data of 3D virtual models and physical products are combined to drive the DTAS running synchronously. The service part can enhance the convenience, reliability, and productivity of an engineered system.

Apart from the above frameworks in practical production process, Oracle [87] has proposed Oracle IoT Digital Twin implementation that is developing applications with Oracle Internet of Things cloud service. As shown in Fig. 2.3, the implementation has three pillars, virtual space, predictive twin, and twin projections. In a virtual space, Oracle's device virtualization feature creates a virtual representation of a physical device or an asset in the cloud. A virtual space uses a JSON-based model that contains observed and desired attribute values and also uses a semantic model.

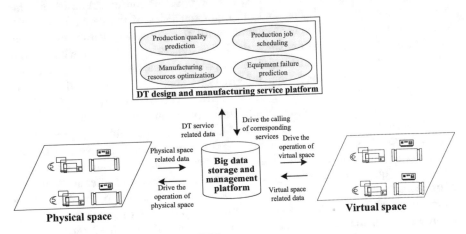

Fig. 2.2 Five-dimensional framework of DT

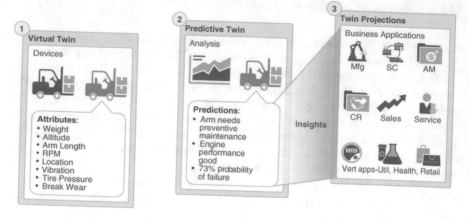

Fig. 2.3 The illustration of Oracle IoT Digital Twin implementation

2.4 DT-driven Product Design

This section introduces the three stages of DT-driven product design.

2.4.1 Conceptual Design

Conceptual design determines the design direction of the entire product, which is the first step of product design process. The purposes of conceptual design include defining the concept, esthetics, and the main functions of the new product. It is worth noting that the designers cannot fulfill the purposes themselves without the feedbacks of other participants, as the designers need to deal with various kinds of data such as customer satisfaction, product sales, product competitiveness, investment plans, and many other information. This data is huge and scattered, which makes it difficult for designers to collect. Through utilizing Digital Twin, all kinds of data in the product's physical space can be classified, integrated, and presented in a vivid way such as 3D modeling. A faithful mapping of the physical product enables the communication between clients and designers to be more transparent and faster by using the real-time transmission data. Customers can have a quick understanding of the product concept and effectively update their requirements. Designers can easily decide where should be improved with its characteristic of having single information source. It can greatly guide the improvement of the new product by making full use of customers' feedback and various problems appeared in customers' usage of the previous generation.

Zheng et al. [142] proposed a generic data-driven cyber-physical approach for personalized smart connected product co-development in a cloud-based environment, which enables individual user innovation in the context of co-development process. A case study of a smart wearable device (i.e., i-BRE respiratory mask) development process is given with general discussions. Driven by smart connected devices, users can take part in the co-development of future products via cloud computing [142]. Tao et al. [125] proposed DT based method to redesign bicycle for customer satisfaction. Liu et al. [66] and Zhao et al. [141] adopted DT methodology to allow rapid designing of individualized requirements.

2.4.2 Detailed Design

Detailed design is the second stage following conceptual design. Its main purpose is to design and construct the product prototype, development tools, and equipment used in the commercial production. Recall that conceptual design is a coarse product design scheme that contains the main functions and characteristics. In the detailed design process, designers need to further refine the product design scheme that includes product functions and appearance, product configuration, design parameters, and test data on the basis of the former stage. The detailed design stage requires repeated simulation tests to ensure that product prototype can achieve the desired performance. However, because of a lack of real-time data and environmental-impacted data, the effect of simulation tests is not obvious. Fortunately, Digital Twin technology can solve this problem well as it can integrate the data collected from the physical production environment in the whole lifecycle of physical objects.

Guo et al. [44] used a modular approach to assist designers in constructing a flexible DT with the purpose of design evaluation in the context of factory design. To assess product effectiveness, process, and servicing decisions, Schleich et al. [82] proposed a comprehensive reference model hinged on the Skin Model Shapes concept, while Tao et al. [123] presented a DT-driven product design method with a bicycle design case study to assist in iterative redesign of existing products. Schluse et al. [103] combined DT with model-based systems engineering and simulation technology in the form of experimental DT, introducing an agile environment process encompassing the entire life cycle. Dias-Ferreira et al. [29] introduced a bio-inspired design framework for dynamic production environments, in which DT can be used to visualize the effectiveness of various interaction patterns. Zhang et al. [140] used DT to provide individualized designing of production lines.

There are communication and computation tools for constructing DT for product design. These technology building blocks reduce the design cost of new products and enable interoperability. Damjanovic-Behrendt and Behrendt [26] adopted the open source approach for the design of a DT demonstrator, while Alam and El Saddik [6] identified basic and hybrid computation–interaction modes with a DT architecture reference model in a telematics-based prototype driving assistance application.

2.4.3 Virtual Verification

The traditional verification process is conducted until the small batch production is carried out, which can be done only after the completion of product design. It will not only extend the production cycle but also greatly increase budget and the cost of time. If designers choose to use Digital Twin model, the quality of any accessories will be predicted before they are actually produced by debugging and predicting directly in the model of Digital Twin. That is, DT enables *virtual verification* that can take full use of the data of equipment, environment, material, customers' physical characteristics, and history data of the last generation. This method can test whether there is a design defect and find the cause of it in a virtual space, and therefore, the redesigning will be fast and convenient. Also, it can greatly improve the design efficiency by avoiding tedious verification and testing. In addition, Digital Twin cannot only describe the behaviors but also propose solutions related to the real system. In other words, it can provide operation and service to optimize the auxiliary system and predict the physical objects based on virtual models. Therefore, by using Digital Twin technology, designers can create vivid simulation scenarios to effectively apply simulation tests on prototypes and accurately predict the actual performance of the physical products as far as possible.

To deal with geometric reconstruction problems, Biancolini and Cella [33] presented a mesh morphing workflow based on radial basis functions for model validation via DT.

2.5 Security Issues in Digital Twin

There are some work that investigates the security issues in DT, which can be categorized into two groups. The first group cares about the data privacy generated or used by DT. Some work [115, 138] stresses that DT technology could help geographically dispersed production, which needs data storage and data analysis among various assets. To this end, the authors in [138] propose an architecture of large-scale DT platform to develop a reliable advanced drive assistance system. The authors in [115] identify the ways and means of collecting, organizing, and storing the data in a secured cloud environment. The work in [25] uses privacy enhancement control to process the privacy data that is exchanged among various assets in automotive industry.

The second group applies DT technology to do testing and simulation in virtual space, such that avoid taking risk doing the same thing on physical space. With the help of the DT technology, security can be enhanced by detecting possible attacks in virtual space. Specifically, the authors of [27] present a DT based framework for the applications in power grid that aids users in assessing the health grid. The work in [135] applies DT technology to Smart Manufacturing for fault diagnosis. Specifically, in their proposed framework, deep transfer learning in the virtual space of DT diagnoses the fault happen in physical space. The authors of [144] also propose an online analysis Digital Twin (OADT) system to realize the new power grid online analysis. The authors in [32] present a framework that allows users to create and execute Digital Twins. The Digital Twins closely match the physical counterparts, such that exploration and test by security professional can be performed without risking negative impacts on physical space. The authors of [36] propose a DT based Industrial Automation and Control System (IACS) security architecture. To prevent the attack from intercepting, modifying, and replaying all the communication between the physical and virtual space, their architecture keeps the consistency for the state replication between the physical and virtual space.

2.6   Semiconductor Manufacturing

2.6.1   Overview of Semiconductor Manufacturing

There are mainly four stages for semiconductor manufacturing. (1) Wafer fabrication. Each job contains a fixed number of wafers. Sometimes up to several thousand identical chips can be made on each wafer by building up the electronic circuits layer by layer in a wafer fabrication facility (called foundry). (2) Wafer probing. The wafers are sent to probe station, where electrical tests identify individual dies that are not likely to be good when packaged. The bad dies are either physically marked or an electronic map is made of them so that they will not be put in a package. (3) Assembly. The probed wafers are sent to an assembly facility where the "good" dies are put into the appropriate package. (4) Test [126, 127, 137]. The packaged dies are sent to a test facility where they are tested in order to ensure that only good products are sent to customers. The whole manufacturing process may require up to 3 months to produce a chip. During manufacturing, the minimum unit is called a lot (denoted as jobs throughout the rest of this paper to conform to the scheduling literature), which consists of a certain number of wafers, at most 25, in one cassette. Note that among these stages, wafer fabrication consists of the most complicated process flow and is the core of semiconductor manufacturing. We start by describing wafer fab operations. From a functional point of view, work areas are the main building blocks of a wafer fab.

We focus on wafer fabrication stage, as it is the most complicated process in the semiconductor manufacturing.

2.6.2   Core Part of Semiconductor Manufacturing: Wafer Fabrication

There are four steps for wafer fabrication. We introduce them in the following.

- Oxidation/Deposition/Diffusion: The purpose is to grow a layer of material on the surface of a cleaned wafer. Note that there are three practical solutions. (1) Oxidation aims at growing a dioxide layer on a wafer. (2) Deposition steps deposit dielectric or metal layers. Deposition can be executed by different processes, such as Physical Vapor Deposition (PVD) or Chemical Vapor Deposition (CVD), Epitaxy, or Metalization. (3) Diffusion is a high temperature process that disperses material on the wafer surface. Note that usually only one of these three operations is used per layer.
- Photolithography: Coating, exposure, developing, and process control are the main steps of the photolithography process. In the first step, the wafer is coated with a thin film of a photosensitive polymer, called photoresist strip. Patterns are produced on the wafer's surface when an IC pattern is transferred from a photo mask ("reticle") onto the photosensitive polymer, which replicates the pattern in the underlying layer. Exposure tools ("steppers") transfer the pattern onto the wafer by projecting light through the reticle to expose the wafer using ultraviolet light. The exposed wafer is then developed by removing polymerized sections of photoresist from the wafer. Every wafer passes through the photolithography area up to 40 times because of the different layers. The photolithography work area is a typical example of a bottleneck because steppers are very expensive.
- Etching: This step is responsible for removing material from the wafer surface. The wafers are partially covered by photoresist strip after the photolithography step. Areas on the wafer that are not covered are then removed from the wafer. We differentiate between wet and dry etching. In the first case, liquids are used, whereas gases are necessary for the latter case.
- Ion implantation: After the wafer is partially etched, the etched parts will be deposited by dopant ions on the surface of the wafer.
- Planarization: This step cleans and levels the wafer surface. A chemical slurry is applied to the wafer and the surface is equalized, which results in the thickness of the wafers being diminished before adding a new layer.

2.7 DT for Semiconductor Manufacturing (Wafer Fabrication): Perspective of Job Scheduling

Apart from the manufacturing technologies existed in the process of semiconductor manufacturing, especially wafer fabrication, we will show that efficient job scheduling is also an important issue to be resolved in DT based semiconductor manufacturing.

2.7.1 Need for Efficient Job Scheduling During Wafer Fabrication

There are six reasons that explain why we need efficient job scheduling.

- Numerical process steps vs. limited number of machines. In typical wafer fabrication, there are dozens of process flows, each containing 300–900 process steps operated by more than 100 machines. The machines are very expensive, with the price ranging from 10^5 to 3×10^7 dollars [79]. Therefore, the number of machines is usually limited. To make full use of the machine, we need efficient scheduling to maximize the throughput of the workload.
- The operation time varies a lot. For different operations on machines, the operation time varies significantly. Some operations need less than 15 min, while others may take 12 h.
- Batching process vs. non-batching process. Some operations involve batch processes, i.e., the joint processing of several jobs on a machine, especially for the long-time operation. Batch processes lead to the formation of long queues in front of non-batching machines and a nonlinear flow of products in the factory.
- Production failure. The probabilistic occurrence of tool failures results in large variability in the time a job spends in process. Therefore, we need to update the scheduling plan in real time to adapt probabilistic failure.
- Auxiliary resource constraint. Auxiliary resources, also called secondary resources, are sometimes necessary to process jobs on machines. Examples are reticles in the photolithography work area of a wafer fab and load boards in the test stage of back-end facilities. Some of these auxiliary resources are quite expensive, i.e., a single reticle can cost more than 100K, so only a very limited number of them are purchased.
- Time constraints between consecutive process steps. For example, there is often a time restriction between operations in the etch work area and oxidation/deposition/diffusion work area (see [104]). Time windows are installed by process engineering department to respect the time constraints. This is important to prevent native oxidation and contamination effects on the wafer surface. More than two consecutive process steps might be involved, and these time constraints might be nested. Jobs with a violation of the recommended time windows have to be scrapped. Rework is often not allowed.

In 300 mm wafer fabs, wafers are transported in FrontOpening Unified Pods (FOUP). Because of the increase in area and weight of the wafers, a manual handling of the wafers has to be eliminated. This fact increases the need for scheduling approaches. Automated material handling is always a critical operation in modern wafer fabs [3, 52, 80].

Table 2.7 Characteristics of scheduling problems found in the different work areas of wafer fabs

Work area	Batch processing	Auxiliary resources
Oxidation	p-Batching with incompatible families	N
Deposition(CVD/PVD)		
Diffusion		
Photolithography	s-Batching with job availability	Reticles
Etch	N	N
Ion implantation	N	N
Planarization	N	N

2.7.2 Taxonomy of Job Scheduling Problem

We list several subproblems in the context of job scheduling problems. Each subproblem may happen in a specific machine area, which is presented in Table 2.7 [79]. Note that there are also other categories of subproblems that need to be resolved, but we only mention some representative problems.

2.7.2.1 Batch Processing Problem

A batch is defined as a group of jobs that have to be processed jointly (see [18]). A batch scheduling problem consists in grouping the jobs on each machine into batches and in scheduling these batches. In the semiconductor manufacturing industry, job scheduling in batches is an important issue as batching is avoidance of setups and/or facilitation of material handling. There are two versions of batching scheduling. One is serial-scheduling (abbreviated as s-batching), in which jobs may be batched and processed one by one. The processing time of a batch is the sum of the processing times of all jobs that form the batch. The other is parallel-batching (abbreviated as p-batching). In this case, jobs in a batch can be processed simultaneously. Hence, the processing time of the batch is given by the maximum processing time of jobs contained in the batch. In fact, p-batching scheduling is more important in semiconductor manufacturing field. An example is diffusion furnaces in wafer fabs. Here, the jobs are assigned to incompatible job families. While several jobs can be processed at the same time, jobs of different families cannot be processed together due to the chemical nature of the process. The processing time of all jobs within a family is the same. Therefore, batching jobs with incompatible families is a special case of p-batching. More literature about batching in general and p-batching for semiconductor manufacturing can be found in [94] and [74], respectively. Batching problems can be found mainly in the oxidation/deposition/diffusion area of wafer fabs.

2.7.2.2   Scheduling Under Constraint of Auxiliary Resources

Auxiliary resources are typically related to parallel machine environments, such as the photolithography process. Stepper processing requires that both the job and the correct reticle are available at the same time. For each layer of the wafer, it needs different reticles for photolithography, as reticle requirements are typically both product-dependent and layer-dependent. Since the number of reticles in a wafer fab is limited, we need to consider the constraint of auxiliary resources when dealing with the parallel machine scheduling problem associated with steppers.

2.7.2.3   Multiple Orders per Job Scheduling Problem

In today's wafer fabrication, different orders from different customers may be grouped into one or more FOUPs to form production jobs. Consider a 300 mm manufacturer. Due to the decreased line widths and more area per wafer in 300 mm wafer fabs, fewer wafers are needed to fill an IC order of a customer, which can reduce the consumption of the FOUPs. It is desirable as FOUPs are expensive and a large number of FOUPs have the potential to cause the Automated Material Handling Systems (AMHS) to become overloaded. Grouping orders into one or more FOUPs can also save cost of time as some tools have the same processing times regardless of the number of wafers in the batch. Therefore, 300 mm manufacturers often have the need and the incentive to group orders from different customers into one or more FOUPs to form production jobs. These jobs have to be scheduled on the various types of tool groups in the wafer fab and processed together. This class of integrated job formation and scheduling problems is called multiple orders per job (MOJ) scheduling problems (see [97]).

2.7.3   Solution Techniques for Job Scheduling Problems

2.7.3.1   Batching Problem

As p-batching is widely used in semiconductor manufacturing, there are many literature studying this problem. Choi et al. [54] suggested different heuristics with the objective of minimizing the total tardiness (TT) of all jobs. Note that tardiness of job j, denoted as T_j, is calculated by $T_j = \max\{C_j - d_j, 0\}$, where C_j is the completion time of job j and d_j is the due data of job j. Balasubramanian et al. [10] extended solution techniques for min-TT presented by Mehta and Uzsoy [76] to that for minimizing total weighted tardiness (TWT) problem. Dispatching rules are used to form batches, and genetic algorithms assign jobs or batches to machines and sequence them on each single machine. This approach was extended to min-TWT problem by Monch et al. [78], assuming that the release time of each job j, i.e., r_j, is given. This problem is harder because the given release dates r_j of the jobs require a decision on whether it makes sense to wait for future jobs arrivals in the case of incomplete batches or to start a non-full batch. Robinson et al. [99] pointed out the importance to incorporate the release dates in batch scheduling problems. The proposed procedure was based on the time window decomposition approach suggested by Ovacik and Uzsoy [88]. Heuristics are used to form batches from the jobs that are available within the time window. Tangudu and Kurz [119] proposed a branch-and-bound algorithm with effective dominance rules for the same scheduling problem that was able to solve problems involving up to 32 jobs in reasonable time.

2.7.3.2 Scheduling Under Constraint of Auxiliary Resources

Park et al. [90] studied reticle management issues by investigating different storage and inspection policies and the relationship between cycle time and product mix via simulation analysis. They found that in the case of a homogeneous product mix, jobs wait due to reticle unavailability and/or machine availability constraints. This highlights the importance of incorporating an appropriate treatment of auxiliary resources in making scheduling decisions. Akcali and Uzsoy [4] investigated the photolithography work center scheduling problem by investigating the setting of job start and finish times over the duration of a production shift. The authors assumed that an operation that requires a specific reticle type can be assigned to at most r machines, where r is the existing number of reticles of the corresponding type that are available in their simulation-based analysis. Akcali et al. [5] examined the effects of three test-run policies and two dedication policies in photolithography scheduling when test runs are necessitated prior to a reticle change.

2.7.3.3 Multiple Orders per Job Scheduling Problem

Qu et al. [97] presented the first MOJ analysis of the scheduling problem with the objective of min-TT. This initial optimization-based study was later extended to examine larger test instances problems of min-TWC (total weighted completion time) and min-TWT problems in Qu and Mason [96]. While these first two single machine MOJ studies focused on both the job and item MOJ processing environments, MOJ scheduling on a single batch processing machine (e.g., a diffusion oven) is studied for the min-TWT problem. But it assumed that each job is compatible in a batch. A more practical solution without this assumption was studied by Erramilli and Mason [96].

Jia and Mason [51] examined parallel machine MOJ min-TWC problem via optimization and constructive heuristic-based solution approaches. Alternately, Jampani and Mason [48] presented column generation based heuristics for the same problem. Laub et al. [57] investigated two-machine flow shop MOJ scheduling problems with combinations of item-type and job-type machines. Finally, Jampani and Mason [49] examined a mini-fab-based complex job shop MOJ scheduling problems with the objective of min-TWC via a network flow based model and column generation heuristics. Integrated heuristic approaches, i.e., constraint programming and Ant Colony Optimization (ACO), were used to solve the same problem by Jampani et al. [50].

2.7.4 Why Do We Employ DT for Efficient Job Scheduling?

So far, most of the researched scheduling problems have only a single criterion. However, in many real-world situations, multiple objectives are important. As we have mentioned in Sect. 2.7.1, there are many factors that affect the production efficiency. Improving one objective may conflict with others. In addition, we need to coordinate each step to make them a unified entity.

Within the vision of Industry 4.0 and Cyber-Physical Production Systems, complex problems are due to planning, scheduling, and control of production, and logistic processes are derived by data-driven decisions in the nearer future. Thus, new processes and interoperable systems must be designed, and existing ones have to be improved, since Industry 4.0 has placed extremely high expectations on production systems to have substantial increase in productivity, resource efficiency, and level of automation. The deliverance of these expectations lies in the ability of manufacturing companies to accurately and timely schedule the job processing.

Current simulation systems such as Discrete Event Simulation (DES) [89] are suitable to model the reality in a manufacturing system with high fidelity. Such models are easy to parameterize, and they are able to consider several influences including stochastic behavior. However, simulation models are challenged when it comes to operational decision support in manufacturing as well as logistics. The simulation models are very complex and need huge amount of production data and up to hours for the execution of simulation experiments. In addition, the current scheduling problems may not be practical as they have strong assumptions. The input for the problem may change due to unexpected production failure. A better approach is to employ Digital Twin, so as to integrate the methods and algorithms from (big) data analytics and AI during the implementation of the "digital production twin" for different purposes, e.g., predictive maintenance or workforce scheduling. The Digital Twin represents the behavior of the corresponding real object or process and is compared with it at (mostly regular) defined points in time. DT can provide a virtual twin of the physical production environment. It is equipped with advanced data analyzing ability, which can make better decisions in real time to help improve the production efficiency.

2.8 Building a Mature DT System in Semiconductor Fields: Challenges and Research Opportunities

DT based semiconductor design and manufacturing systems are expected to be invented. However, semiconductor design and manufacturing is more complicated compared to common products like cars and bicycles. Although many literature have formally proposed a DT system for a certain product, currently there is no literature claiming that it has already invented such a system in semiconductor field. We expect the progress to be made for even a stage of semiconductor design in the near future. In the following, we propose some challenges and research opportunities in the way of completing DT systems in semiconductor fields.

2.8.1 Semiconductor Production Digitalization for Design

Production digitalization is the first function of DT-driven semiconductor production system. To build a DT-driven semiconductor production system, the first function needs to be fulfilled is production digitalization. It enables designers to react better to shift consumer trends, where DT is used to digitalize process models. Below we introduce the current works on production digitalization.

2.8.1.1 What Are the Current Methods for Manufacturing Digitalization?

Lu and Xu [68] introduced a cloud-based manufacturing system architecture to achieve on-demand production, thus achieving better business flexibility. Bao et al. [12] proposed a DT modeling and operating construction approach in an aircraft structural parts machining cell case study, while Tan et al. [118] proposed a DT construction framework that models IoT data into a simulation. For DT application in shop-floors, Ding et al. [30] used DT technologies to enhance interconnection and interoperability between cyber and physical shop-floors, whereas Zhang et al. [123] presented a novel production system architecture that also supported job scheduling in an aircraft engine manufacturing case study.

2.8.2 Semiconductor Manufacturing Modeling Strategies

To enhance the quality of output, DT modeling technologies need to be built to suit diverse manufacturing conditions. An important question to be addressed for manufacturing modeling through a Digital Twin could be, for example, whether certain scheduling parameters can be enhanced and better parameter values can be identified consistently. However, in a cleaning area of a large 300 mm fab comprising more than 100 wet benches, furnaces, and metrology tools, for example, commercially available scheduling tools typically run at a frequency of once every 10 min, whereby the scheduling procedure runs most of this time and the remaining time is needed for data input and output. This basically means that the scheduler runs almost continuously and hence also the Digital Twin, i.e., the simulation model of the cleaning area (in which the scheduler would have to run equally frequently) will inherently not be able to run faster than real time. Optimization of scheduling parameters, in the sense of what the best parameter values are under which circumstances, will therefore be possible only retrospectively by comparing the (simulated) performance associated with different scheduler settings for different historical patterns.

2.8.2.1 A Possible Way?

Shao et al. [108] suggested parallel execution of different scenarios to adjust the parameters in real time. Specifically, multiple instances of the scheduling solution will be required, basically equivalent to the number of instances that would be required to compare different scenarios on a cloud infrastructure, posing challenges to the licensing models currently practiced by commercial vendors of scheduling solutions.

2.8.2.2 What Are the Current Methods for Manufacturing Modeling Strategies?

Luo et al. [69] proposed a multi-domain unified modeling method as a cyber-physical mapping strategy also used for fault prediction and diagnosis. Zheng et al. [143] introduced parametric virtual modeling and construction flow of DT application subsystems to fulfill the case of a welding production line. In an aircraft assembly context, Guo et al. [43] improved competitiveness with digital coordination model, utilizing DT to accomplish better flexible assembly accuracy and efficiency. Sharif Ullah [130] created a semantic modeling methodology to compute virtual abstractions for material removal processes.

2.8.3 Production Optimization: Improving the Manufacturing Efficiency

The factors in the semiconductor manufacturing process are complicated (e.g., temperature, pressure, wafer flatness, process accuracy, input parameter settings, and measurement errors). Various random interference values affect the process, making the process difficult to stabilize. Process control methods can ensure the process is stabilized at a target value.

2.8.3.1 Traditional Statistical Process Control (SPC) and Advanced Process Control

SPC uses control charts to monitor whether a serious variation has occurred in the process. The system is not adjusted if it stays within predetermined boundaries. If the process exceeds the boundaries, then machines must be stopped and checked until the outliers have been found. However, this approach cannot be applied to semiconductor manufacturing; semiconductor production must be continuous and no downtime can be tolerated. If the upper and lower control limits are set too narrowly, false alarms will increase and the time of wafer output will be prolonged, delaying delivery and causing substantial losses for the semiconductor foundry. Thus, basic SPC cannot solve generated drift in semiconductor processes.

Advanced process control can use differences in premeasured batches of wafers. The system can consider uncontrollable interference values to determine process parameters. The process can achieve its required target stability value. In a conventional control system, control parameters must be set, and the settings of those parameters deeply affect the outcomes. Such control methods typically make assumptions regarding the model inputs and outputs (IO model) and the controller, use a certain type of interference distribution, set the same combination of control parameter values, and investigate long-term stability to find a steady-state region.

2.8.3.2 Problems for Semiconductor Manufacturing

Semiconductor industries undergo rapid product changes; a semiconductor manufacturing system has on average 300–500 manufacturing system processes. The process factors are uncountable; stability before film processing affects the yield after each layer of the product is fabricated. Interference disturbance is difficult to estimate and predict for each batch, even with mathematical models. A control system can assume a fixed interference value for a long control parameter in a steady-state situation, but that assumption does not match the actual process conditions. Process interference values often vary with time. The question of how to determine the parameters of the controller is crucial. Likewise, the process of dynamically adjusting the control parameters from batch to batch is crucial.

In semiconductor manufacturing, photolithography processes are among the most expensive and most critical processes. The stability of the process depends on the stability of the line width. However, as technology improves and product features shrink, the line width diminishes and the tolerable degree of variation is also reduced; hence, the process stability control grows ever more difficult. In addition to controlling the output value of the line width of the outside layer and the rear layer processed in the previous step, the system must control layer overlay errors (i.e., errors regarding the relative alignment of layers). These errors may affect the functionality of the semiconductor elements, such as electrical conductivity and insulation. To minimize errors, the semiconductor elements can be produced to achieve the desired functionality. Because lithography machines are expensive, the lithography step is a bottleneck in the production flow. Using a batch method can control the value of an output control process to some target value. By reducing the process variation, one can increase the stability of the process. Heavy machines substantially reduce the rework ratio of defective products, and this can reduce the bottleneck effect of lithography, thus reducing the cycle time of the wafers.

2.8.3.3 What Are the Current Methods for DT Based Production Optimization?

Optimization of production aspects such as manufacturing speed and machine control was studied. In the dyeing and finishing industry, Park et al. [91] proposed a service-oriented platform to enhance performance measures and achieve cost reduction through optimization algorithms. Moreno et al. [81] showcased a DT for a sheet metal punching machine to optimize NC machining programs, while Zhao et al. [141] demonstrated a joint optimization DT model for coordinating micro-punching processes to boost punching speed. Using geometric assurance DT, Tabar et al. [116] reduced computation speed for weld points. Liu et al. [65] described a DT based machining process evaluation method for a marine diesel engine manufacturing process. Liu et al. [66] researched on a DT hot rolling production scheduling model and provided decision-aiding support. Coronado et al. [24] presented manufacturing execution system as a core DT technology for production control and optimization, allowing easy implementation and lowering costs. Soderberg et al. [112] applied real-time geometrical quality control for welded components through DT to enhance production quality for a range of welding processes.

2.8.4 Identifying Useful Data

DT cannot run normally without data, more precisely, the useful data. It needs to be quality data that is noise free with a constant, uninterrupted data stream. If the data is poor and inconsistent, it runs the risk of the Digital Twin under performing as it is acting on poor and missing data. The quality and number of IoT signals is an essential factor for Digital Twin data. Planning and analysis of device use is needed to identify whether the right data is collected and used for efficient use of a Digital Twin.

2.8.5 Summarizing Disparate Source of Data

Begin by connecting disparate sources of data with a summary of results and statistics to demonstrate value through connecting information. Consolidating to a single source of truth for this data is not needed right away. Meaningful insights can be gleaned by analyzing summary-level data across domains. There are many tools that can enable views and analytics from disparate databases.

2.8.6 Bring More External Data

Establish an intelligent data lake that can connect more and more sources of data and bring in additional context and structure to that data. This will be an ever-evolving source of data that can be sliced and viewed in different ways to continually improve predictive and prescriptive analytics. DT with Big Data capabilities allows management to make informed decisions via its decision-aiding functionalities.

2.8.7 Privacy and Security

Within an industry setting, it is clear that the privacy and security associated with Digital Twins is a challenge because of the vast amount of data they use and the risk this poses to sensitive system data. To overcome this, data analytics and IoT must follow the current practices and updates in security and privacy regulations. Security and privacy considerations for Digital Twins data contribute to tackling trust issues with Digital Twins. With the advancements in technology like blockchain, several opportunities emerge to ensure security with the main focus on looking for solutions that help secure Digital Twins

2.9 Conclusion

This paper reviews the existing literature about the Digital Twin, including its evolution, application, definition, framework, and design process. We also discuss some security issues existed in Digital Twin. We also investigate the semiconductor manufacturing process and provide a specific view from semiconductor manufacturing to see why DT is necessary for improving the production efficiency and quality. Several research opportunities and challenges are presented as inspirations for future work.

References

1. A simulation-based approach for plant layout design and production planning
2. M. Abramovici, J.C. Göbel, P. Savarino, Reconfiguration of smart products during their use phase based on virtual product twins. CIRP Ann. **66**(1), 165–168 (2017)
3. G.K. Agrawal, S.S. Heragu, A survey of automated material handling systems in 300-mm semiconductorfabs. IEEE Trans. Semicond. Manuf. **19**(1), 112–120 (2006)
4. E. Akcali, R. Uzsoy, A sequential solution methodology for capacity allocation and lot scheduling problems for photolithography, in *Twenty Sixth IEEE/CPMT International Electronics Manufacturing Technology Symposium (Cat. No. 00CH37146)* (IEEE, Piscataway, 2000), pp. 374–381
5. E. Akcalt, K. Nemoto, R. Uzsoy, Cycle-time improvements for photolithography process in semiconductor manufacturing. IEEE Trans. Semicond. Manuf. **14**(1), 48–56 (2001)
6. K.M. Alam, A. El Saddik, C2PS: a digital twin architecture reference model for the cloud-based cyber-physical systems. IEEE Access **5**, 2050–2062 (2017)
7. A. Ardanza, A. Moreno, Á. Segura, M. de la Cruz, D. Aguinaga, Sustainable and flexible industrial human machine interfaces to support adaptable applications in the industry 4.0 paradigm. Int. J. Prod. Res. **57**(12), 4045–4059 (2019)
8. Autodesk, Dna for Digital Twin. https://www.autodesk.com/campaigns/digital-twin
9. M. Azure, Azure Digital Twins Preview. https://azure.microsoft.com/en-us/services/digital-twins/
10. H. Balasubramanian, L. Mönch, J. Fowler, M. Pfund, Genetic algorithm based scheduling of parallel batch machines with incompatible job families to minimize total weighted tardiness. Int. J. Prod. Res. **42**(8), 1621–1638 (2004)
11. A. Banerjee, R. Dalal, S. Mittal, K.P. Joshi, Generating digital twin models using knowledge graphs for industrial production lines, in *UMBC Information Systems Department* (2017)
12. J. Bao, D. Guo, J. Li, J. Zhang, The modelling and operations for the digital twin in the context of manufacturing. Enterp. Inf. Syst. **13**(4), 534–556 (2019)
13. B. Barricelli, E. Casiraghi, D. Fogli, A survey on digital twin: definitions, characteristics, applications, and design implications. IEEE Access **7**, 167653–167671 (2019)
14. Y. Bazilevs, X. Deng, A. Korobenko, F. Lanza di Scalea, M. Todd, S. Taylor, Isogeometric fatigue damage prediction in large-scale composite structures driven by dynamic sensor data. J. Appl. Mech. **82**(9), 091008 (2015)
15. B. Bielefeldt, J. Hochhalter, D. Hartl, Computationally efficient analysis of SMA sensory particles embedded in complex aerostructures using a substructure approach, in *ASME 2015 Conference on Smart Materials, Adaptive Structures and Intelligent Systems*. American Society of Mechanical Engineers Digital Collection (2015)
16. S. Boschert, C. Heinrich, R. Rosen, Next generation digital twin, in *Proceedings of TMCE*. Las Palmas de Gran Canaria, Spain (2018), pp. 209–218
17. S. Boschert, R. Rosen, Digital twin the simulation aspect, in *Mechatronic Futures* (Springer, Berlin, 2016), pp. 59–74
18. P. Brucker, A. Gladky, H. Hoogeveen, M.Y. Kovalyov, C.N. Potts, T. Tautenhahn, S.L. Van De Velde, Scheduling a batching machine. J. Sched. **1**(1), 31–54 (1998)
19. K. Bruynseels, F. Santoni de Sio, J. van den Hoven. Digital twins in health care: ethical implications of an emerging engineering paradigm. Front. Genet. **9**, 31 (2018)
20. A. Canedo, Industrial IoT lifecycle via digital twins, in *Proceedings of the Eleventh IEEE/ACM/IFIP International Conference on Hardware/Software Codesign and System Synthesis* (2016), pp. 1–1
21. A. Cerrone, J. Hochhalter, G. Heber, A. Ingraffea, On the effects of modeling as-manufactured geometry: toward digital twin. Int. J. Aerosp. Eng. **2014**, 439278 (2014)
22. T. Consultancy Services, *Digital Twin in the Automotive Industry: Driving Physical-Digital Convergence*. https://www.tcs.com/content/dam/tcs/pdf/Industries/manufacturing/abstract/industry-4-0-and-digital-twin.pdf

23. C. Copley, *Medical Technology Firms Develop 'Digital Twins' for Personalized Health Care* (2018). https://www.theglobeandmail.com/business/article-medical-technology-firms-develop-digital-twins-for-personalized/

24. P.D.U. Coronado, R. Lynn, W. Louhichi, M. Parto, E. Wescoat, T. Kurfess, Part data integration in the shop floor digital twin: mobile and cloud technologies to enable a manufacturing execution system. J. Manuf. Syst. **48**, 25–33 (2018)

25. V. Damjanovic-Behrendt, A digital twin-based privacy enhancement mechanism for the automotive industry, in *2018 International Conference on Intelligent Systems (IS)* (IEEE, Piscataway, 2018), pp. 272–279

26. V. Damjanovic-Behrendt, W. Behrendt, An open source approach to the design and implementation of digital twins for smart manufacturing. Int. J. Comput. Integ. Manuf. **32**(4–5), 366–384 (2019)

27. W. Danilczyk, Y. Sun, H. He, ANGEL: An intelligent digital twin framework for microgrid security, in *2019 North American Power Symposium (NAPS)* (IEEE, Piscataway, 2019), pp. 1–6

28. T. Debroy, W. Zhang, J. Turner, S.S. Babu, Building digital twins of 3D printing machines. Scripta Mat. **135**, 119–124 (2017)

29. J. Dias-Ferreira, L. Ribeiro, H. Akillioglu, P. Neves, M. Onori, BIOSOARM: a bio-inspired self-organising architecture for manufacturing cyber-physical shopfloors. J. Intel. Manuf. **29**(7), 1659–1682 (2018)

30. K. Ding, F.T. Chan, X. Zhang, G. Zhou, F. Zhang, Defining a digital twin-based cyber-physical production system for autonomous manufacturing in smart shop floors. Int. J. Prod. Res. **57**(20), 6315–6334 (2019)

31. L. Dong, R. Haynes, S.N. Atluri, On improving the celebrated Paris' power law for fatigue, by using moving least squares. CMC: Comput. Mat. Continua **45**(1), 1–15 (2015)

32. M. Eckhart, A. Ekelhart, Towards security-aware virtual environments for digital twins, in *Proceedings of the 4th ACM Workshop on Cyber-Physical System Security* (2018), pp. 61–72

33. M. Evangelos Biancolini, U. Cella, Radial basis functions update of digital models on actual manufactured shapes. J. Comput. Nonlinear Dyn. **14**(2), 021013 (2019)

34. U.A. Force, Global horizons final report: United States air force global science and technology vision (2013). https://www.hsdl.org/?view&did=741377

35. T. Gabor, L. Belzner, M. Kiermeier, M.T. Beck, A. Neitz, A simulation-based architecture for smart cyber-physical systems, in *2016 IEEE International Conference on Autonomic Computing (ICAC)* (IEEE, Piscataway, 2016), pp. 374–379

36. C. Gehrmann, M. Gunnarsson, A digital twin based industrial automation and control system security architecture. IEEE Trans. Ind. Inf. **16**(1), 669–680 (2019)

37. E. Glaessgen, D. Stargel, The digital twin paradigm for future NASA and US air force vehicles, in *53rd AIAA/ASME/ASCE/AHS/ASC Structures, Structural Dynamics and Materials Conference 20th AIAA/ASME/AHS Adaptive Structures Conference 14th AIAA* (2012), pp. 1818

38. M.W. Grieves, Product lifecycle management: the new paradigm for enterprises. Int. J. Prod. Develop. **2**(1–2), 71–84 (2005)

39. M. Grieves, *Product Lifecycle Management: Driving the Next Generation of Lean Thinking* (McGraw-Hill, New York, 2006), pp. 95–120

40. M. Grieves, *Virtually Perfect: Driving Innovative and Lean Products Through Product Lifecycle Management* (Space Coast Press, Merritt Island, 2011)

41. M. Grieves, Digital twin: manufacturing excellence through virtual factory replication. White Paper **1**, 1–7 (2014)

42. M. Grieves, J. Vickers, Digital twin: Mitigating unpredictable, undesirable emergent behavior in complex systems, in *Transdisciplinary Perspectives on Complex Systems* (Springer, Berlin, 2017), pp. 85–113

43. F. Guo, F. Zou, J. Liu, Z. Wang, Working mode in aircraft manufacturing based on digital coordination model. Int. J. Adv. Manuf. Technol. **98**(5–8), 1547–1571 (2018)

44. J. Guo, N. Zhao, L. Sun, S. Zhang, Modular based Flexible digital twin for factory design. J. Ambient Intell. Humaniz. Comput. **10**(3), 1189–1200 (2019)
45. A. Han, *Partnership to Create Digital Twin for Semiconductor Manufacturing* (2019). https://www.automationworld.com/products/software/news/13318328/partnership-to-create-digital-twin-for-semiconductor-manufacturing
46. M. Helu, A. Joseph, T. Hedberg Jr, A standards-based approach for linking as-planned to as-fabricated product data. CIRP Ann. **67**(1), 487–490 (2018)
47. V. Hempel, *Healthcare Solution Testing for Future, Digital Twins in Healthcare* (2019). https://www.dr-hempel-network.com/digital-health-technolgy/digital-twins-in-healthcare/
48. J. Jampani, S.J. Mason, Column generation heuristics for multiple machine, multiple orders per job scheduling problems. Ann. Oper. Res. **159**(1), 261–273 (2008)
49. J. Jampani, S.J. Mason, A column generation heuristic for complex job shop multiple orders per job scheduling. Comput. Ind. Eng. **58**(1), 108–118 (2010)
50. J. Jampani, E.A. Pohl, S.J. Mason, L. Monch, Integrated heuristics for scheduling multiple order jobs in a complex job shop. Int. J. Metaheuristics **1**(2), 156–180 (2010)
51. J. Jia, S.J. Mason, Semiconductor manufacturing scheduling of jobs containing multiple orders on identical parallel machines. Int. J. Prod. Res. **47**(10), 2565–2585 (2009)
52. J.A. Jimenez, G.T. Mackulak, J.W. Fowler, Levels of capacity and material handling system modeling for factory integration decision making in semiconductor wafer fabs. IEEE Trans. Semicond. Manuf. **21**(4), 600–613 (2008)
53. J. Kiendl, K. Bletzinger, J. Linhard, R. Wüchner, Isogeometric shell analysis with Kirchhoff–Love elements. Comput. Methods Appl. Mech. Eng. **198**(49–52), 3902–3914 (2009)
54. Y.-D. Kim, B.-J. Joo, S.-Y. Choi, Scheduling wafer lots on diffusion machines in a semiconductor wafer fabrication facility. IEEE Trans. Semicond. Manuf. **23**(2), 246–254 (2010)
55. H. Kim, H. Shin, H.-S. Kim, W.-T. Kim, VR-CPES: a novel cyber-physical education systems for interactive VR services based on a mobile platform. Mob. Inf. Syst. **2018**, 8941241 (2018)
56. H. Laaki, Y. Miche, K. Tammi, Prototyping a digital twin for real time remote control over mobile networks: Application of remote surgery. IEEE Access **7**, 20325–20336 (2019)
57. J.D. Laub, J.W. Fowler, A.B. Keha, Minimizing makespan with multiple-orders-per-job in a two-machine flowshop. Eur. J. Oper. Res. **182**(1), 63–79 (2007)
58. J. Lee, E. Lapira, B. Bagheri, H. Kao, Recent advances and trends in predictive manufacturing systems in big data environment. Manuf. Lett. **1**(1), 38–41 (2013)
59. J. Lee, S.D. Noh, H.-J. Kim, Y.-S. Kang, Implementation of cyber-physical production systems for quality prediction and operation control in metal casting. Sensors **18**(5), 1428 (2018)
60. C. Leiva, ItBase, Enabling the Digital Thread: Unify Design, Manufacturing and ERP in a Closed Loop Digital Thread that Streamlines Operations, Improves Quality and Boosts Productivity (2018). https://info.ibaset.com/enable-the-digital-thread?utm_campaign=eBook:%20Enabling%20the%20Digital%20Thread&utm_source=In-Text%20Link%20for
61. C. Li, S. Mahadevan, Y. Ling, S. Choze, L. Wang, Dynamic Bayesian network for aircraft wing health monitoring digital twin. Aiaa J. **55**(3), 930–941 (2017)
62. K. Lim, P. Zheng, C. Chen, A state-of-the-art survey of digital twin: techniques, engineering product lifecycle management and business innovation perspectives. J. Intell. Manuf. **31**, 1–25 (2019)
63. C. Liu, H. Vengayil, R.Y. Zhong, X. Xu, A systematic development method for cyber-physical machine tools. J. Manuf. Syst. **48**, 13–24 (2018)
64. C. Liu, H. Vengayil, Y. Lu, X. Xu, A cyber-physical machine tools platform using OPC UA and MTConnect. J. Manuf. Syst. **51**, 61–74 (2019)
65. J. Liu, H. Zhou, X. Liu, G. Tian, M. Wu, L. Cao, W. Wang, Dynamic evaluation method of machining process planning based on digital twin. IEEE Access **7**, 19312–19323 (2019)
66. J. Liu, H. Zhou, G. Tian, X. Liu, X. Jing, Digital twin-based process reuse and evaluation approach for smart process planning. Int. J. Adv. Manuf. Technol. **100**(5–8), 1619–1634 (2019)

67. R. Lovas, A. Farkas, A.C. Marosi, S. Ács, J. Kovács, Á. Szalóki, B. Kádár, Orchestrated platform for cyber-physical systems. Complexity **2018**, 1–16 (2018)
68. Y. Lu, X. Xu, Cloud-based manufacturing equipment and big data analytics to enable on-demand manufacturing services. Rob. Comput. Integ. Manuf. **57**, 92–102 (2019)
69. W. Luo, T. Hu, C. Zhang, Y. Wei, Digital twin for CNC machine tool: modeling and using strategy. J. Amb. Intell. Hum. Comput. **10**(3), 1129–1140 (2019)
70. S. Luściński, Digital twinning for smart industry, in *3rd EAI International Conference on Management of Manufacturing Systems*. European Alliance for Innovation (EAI) (2018)
71. M.M. Mabkhot, A.M. Al-Ahmari, B. Salah, H. Alkhalefah, Requirements of the smart factory system: a survey and perspective. Machines **6**(2), 23 (2018)
72. A. Madni, S.M. Lucero, Leveraging digital twin technology in model-based systems engineering. Systems **7**(1), 7 (2019)
73. P.K. Majumdar, M. FaisalHaider, K. Reifsnider, Multi-physics response of structural composites and framework for modeling using material geometry, in *54th AIAA/ASME/ASCE/AHS/ASC Structures, Structural Dynamics, and Materials Conference* (2013), pp. 1577
74. M. Mathirajan, A.I. Sivakumar, A literature review, classification and simple meta-analysis on scheduling of batch processors in semiconductor. Int. J. Adv. Manuf. Technol. **29**(9–10), 990–1001 (2006)
75. M. Maybury, Global horizons final report: United States air force global science and technology vision. *US Air Force, Washington, DC, Report No. AF/ST TR* (2013), pp. 13–01
76. S.V. Mehta, R. Uzsoy, Minimizing total tardiness on a batch processing machine with incompatible job families. IIE Trans. **30**(2), 165–178 (1998)
77. F. Michelfeit, *Exploring the Possibilities Offered by Digital Twins in Medical Technology*. Communications at RSNA (2018)
78. L. Mönch, R. Drießel, A distributed shifting bottleneck heuristic for complex job shops. Comput. Ind. Eng. **49**(3), 363–380 (2005)
79. L. Mönch, J. Fowler, S. Dauzere-Peres, S. Mason, O. Rose, A survey of problems, solution techniques, and future challenges in scheduling semiconductor manufacturing operations. J. Sched. **14**(6), 583–599 (2011)
80. J. Montoya-Torres, A literature survey on the design approaches and operational issues of automated wafer-transport systems for wafer fabs. Prod. Plan. Control **17**(7), 648–663 (2006)
81. A. Moreno, G. Velez, A. Ardanza, I. Barandiaran, Á.R. de Infante, R. Chopitea, Virtualisation process of a sheet metal punching machine within the industry 4.0 vision. Int. J. Interact. Design Manuf. **11**(2), 365–373 (2017)
82. J. Moyne, J. Iskandar, Big data analytics for smart manufacturing: case studies in semiconductor manufacturing. Processes **5**(3), 39 (2017)
83. NASA, The Ill-fated Space Odyssey of Apollo 13. https://er.jsc.nasa.gov/seh/pg13.htm. Accessed 16 Oct 2019
84. E. Negri, L. Fumagalli, M. Macchi, A review of the roles of digital twin in cps-based production systems. Procedia Manuf. **11**, 939–948 (2017)
85. N. Nikolakis, K. Alexopoulos, E. Xanthakis, G. Chryssolouris, The digital twin implementation for linking the virtual representation of human-based production tasks to their physical counterpart in the factory-floor. Int. J. Comput. Integ. Manuf. **32**(1), 1–12 (2019)
86. I.W.I. of Things, *Digital Twin: Helping Machines Tell Their Story*. https://www.ibm.com/internet-of-things/trending/digital-twin
87. Oracle, *About the Oracle IoT Digital Twin Implementation*. https://docs.oracle.com/en/cloud/paas/iot-cloud/iotgs/oracle-iot-digital-twin-implementation.html
88. I.M. Ovacik, R. Uzsoy, Rolling horizon procedures for dynamic parallel machine scheduling with sequence-dependent setup times. Int. J. Prod. Res. **33**(11), 3173–3192 (1995)
89. C.M. Overstreet, Model specification and analysis for discrete event simulation. PhD Thesis, Virginia Polytechnic Institute and State University, 1982

90. S. Park, J. Fowler, M. Carlyle, M. Hickie, Assessment of potential gains in productivity due to proactive reticle management using discrete event simulation, in *Proceedings of the 31st Conference on Winter Simulation: Simulation—A Bridge to the Future-Volume 1* (1999), pp. 856–864

91. K.T. Park, S.J. Im, Y.-S. Kang, S.D. Noh, Y.T. Kang, S.G. Yang, Service-oriented platform for smart operation of dyeing and finishing industry. Int. J. Comput. Integ. Manuf. 32(3), 307–326 (2019)

92. A. Mussomeli, M. Cotteleer, A. Parrott, Industry 4.0 and the digital twin-manufacturing meets its match, Deloitte University Press, 2017

93. K. Polyniak, J. Matthews, *The Johns Hopkins Hospital Launches Capacity Command Center to Enhance Hospital Operations* (2016). https://www.hopkinsmedicine.org/news/media/releases/the_johns_hopkins_hospital_launches_capacity_command_center_to_enhance_hospital_opera

94. C.N. Potts, M.Y. Kovalyov, Scheduling with batching: a review. Eur. J. Oper. Res. 120(2), 228–249 (2000)

95. Q. Qi, F. Tao, Digital twin and big data towards smart manufacturing and industry 4.0: 360 degree comparison. IEEE Access 6, 3585–3593 (2018)

96. P. Qu, S.J. Mason, Metaheuristic scheduling of 300-mm lots containing multiple orders. IEEE Trans. Semicond. Manuf. 18(4), 633–643 (2005)

97. P. Qu, S. Mason, E. Kutanoglu, Scheduling jobs containing multiple orders, in *Proceedings International Conference on Modeling and Analysis of Semiconductor Manufacturing (MASM)* (2002), pp. 264–269

98. K. Rajratna, V. Bavane, S. Jadhao, R. Marode, *Digital Twin: Manufacturing Excellence Through Virtual Factory Replication* Global Journal of Engineering Science and Researches (2018), pp. 6–15

99. J. Robinson, J.W. Fowler, J.F. Bard, The use of upstream and downstream information in scheduling semiconductor batch operations. Int. J. Produ. Res. 33(7), 1849–1869 (1995)

100. R. Rosen, G. Von Wichert, G. Lo, K.D. Bettenhausen, About the importance of autonomy and digital twins for the future of manufacturing. IFAC-PapersOnLine 48(3), 567–572 (2015)

101. S. Scharff, *From Digital Twin to Improved Patient Experience, Siemens Healthineers* (2010). https://www.siemens-healthineers.com/news/mso-digital-twin-mater.html

102. B. Schleich, N. Anwer, L. Mathieu, S. Wartzack, Shaping the digital twin for design and production engineering. CIRP Ann. 66(1), 141–144 (2017)

103. M. Schluse, M. Priggemeyer, L. Atorf, J. Rossmann, Experimentable digital twins-streamlining simulation-based systems engineering for industry 4.0. IEEE Trans. Ind. Inf. 14(4), 1722–1731 (2018)

104. W. Scholl, J. Domaschke, Implementation of modeling and simulation in semiconductor wafer fabrication with time constraints between wet etch and furnace operations. IEEE Trans. Semicond. Manuf. 13(3), 273–277 (2000)

105. G. Schroeder, C. Steinmetz, C.E. Pereira, I. Muller, N. Garcia, D. Espindola R. Rodrigues, Visualising the digital twin using web services and augmented reality, in *2016 IEEE 14th International Conference on Industrial Informatics (INDIN)* (IEEE, Piscataway, 2016), pp. 522–527

106. K. Senthilnathan, I. Annapoorani, Multi-port current source inverter for smart microgrid applications: a cyber physical paradigm. Electronics 8(1), 1 (2019)

107. M. Shafto, M. Conroy, R. Doyle, E. Glaessgen, C. Kemp, J. LeMoigne, L. Wang. *Modeling, Simulation, Information Technology & Processing Roadmap*. National Aeronautics and Space Administration, Washington (2012)

108. G. Shao, S. Jain, C. Laroque, L.H. Lee, P. Lendermann, O. Rose. Digital twin for smart manufacturing: The simulation aspect, in *2019 Winter Simulation Conference (WSC)* (IEEE, Piscataway, 2019), pp. 2085–2098

109. P. Siano, G. Graditi, M. Atrigna, A. Piccolo, Designing and testing decision support and energy management systems for smart homes. J. Amb. Intell. Human. Comput. 4(6), 651–661 (2013)

110. D.J. Siedlak, O.J. Pinon, P.R. Schlais, T.M. Schmidt, D.N. Mavris, A digital thread approach to support manufacturing-influenced conceptual aircraft design. Res. Eng. Design **29**(2), 285–308 (2018)

111. R. Söderberg, K. Wärmefjord, J.S. Carlson, L. Lindkvist, Toward a digital twin for real-time geometry assurance in individualized production. CIRP Ann. **66**(1), 137–140 (2017)

112. R. Söderberg, K. Wärmefjord, J. Madrid, S. Lorin, A. Forslund, L. Lindkvist, An information and simulation framework for increased quality in welded components. CIRP Ann. **67**(1), 165–168 (2018)

113. H. Sohn, C.R. Farrar, F.M. Hemez, D.D. Shunk, D.W. Stinemates, B.R. Nadler, J.J. Czarnecki, A Review of Structural Health Monitoring Literature: 1996–2001 *Los Alamos National Laboratory* (2003), pp. 1–7

114. B. Stackpole, *Digital Twins Land a Role In Product Design* (2015). https://www.digitalengineering247.com/article/digital-twins-land-a-role-in-product-design/

115. N. Susila, A. Sruthi, S. Usha, Impact of cloud security in digital twin, in *Advances in Computers*, vol. 117 (Elsevier, Amsterdam, 2020), pp. 247–263

116. R.S. Tabar, K. Wärmefjord, R. Söderberg, A method for identification and sequence optimisation of geometry spot welds in a digital twin context. Proc. Instit. Mech. Eng. Part C J. Mech. Eng. Sci. **233**(16), 5610–5621 (2019)

117. B.A. Talkhestani, N. Jazdi, W. Schloegl, M. Weyrich, Consistency check to synchronize the digital twin of manufacturing automation based on anchor points. Procedia CIRP **72**(1), 159–164 (2018)

118. Y. Tan, W. Yang, K. Yoshida, S. Takakuwa, Application of IoT-aided simulation to manufacturing systems in cyber-physical system. Machines **7**(1), 2 (2019)

119. S. Tangudu, M. Kurz, A branch and bound algorithm to minimise total weighted tardiness on a single batch processing machine with ready times and incompatible job families. Prod. Plann. Control **17**(7), 728–741 (2006)

120. F. Tao, M. Zhang, Digital twin shop-floor: a new shop-floor paradigm towards smart manufacturing. IEEE Access **5**, 20418–20427 (2017)

121. F. Tao, Y. Cheng, J. Cheng, M. Zhang, W. Xu, Q. Qi, Theories and technologies for cyber-physical fusion in digital twin shop-floor. Comput. Integr. Manuf. Syst. **23**, 1603-161 (2017)

122. F. Tao, M. Zhang, J. Cheng, Q. Qi, Digital twin workshop: a new paradigm for future workshop. Comput. Integr. Manuf. Syst. **23**(1), 1–9 (2017)

123. F. Tao, J. Cheng, Q. Qi, M. Zhang, H. Zhang, F. Sui, Digital twin-driven product design, manufacturing and service with big data. Int. J. Adv. Manuf. Technol. **94**(9–12), 3563–3576 (2018)

124. F. Tao, M. Zhang, Y. Liu, A. Nee, Digital twin driven prognostics and health management for complex equipment. CIRP Ann. **67**(1), 169–172 (2018)

125. F. Tao, F. Sui, A. Liu, Q. Qi, M. Zhang, B. Song, Z. Guo, S.C.-Y. Lu, A. Nee, Digital twin-driven product design framework. Int. J. Prod. Res. **57**(12), 3935–3953 (2019)

126. M.N.M. Tehranipour, N. Ahmed, Testing SoC interconnects for signal integrity using boundary scan, in *Proceedings. 21st VLSI Test Symposium* (IEEE, Piscataway, 2003), pp. 158–163

127. X.Z.M. Tehranipoor, H. Salmani, *Integrated Circuit Authentication* (Springer, Berlin, 2014)

128. The Living Heart Project (2019). https://www.3ds.com/products-services/simulia/solutions/life-sciences/the-living-heart-project/

129. E.J. Tuegel, A.R. Ingraffea, T.G. Eason, S.M. Spottswood, Reengineering aircraft structural life prediction using a digital twin. Int. J. Aerosp. Eng. **2011**, 32011 (2011)

130. A.S. Ullah, Modeling and simulation of complex manufacturing phenomena using sensor signals from the perspective of industry 4.0. Adv. Eng. Inf. **39**, 1–13 (2019)

131. J. Vachálek, L. Bartalský, O. Rovný, D. Šišmišová, M. Morháč, M. Lokšík, The digital twin of an industrial production line within the industry 4.0 Concept, in *2017 21st International Conference on Process Control (PC)* (IEEE, Piscataway, 2017), pp. 258–262

132. J. Wang, K. Wang, Y. Wang, Z. Huang, R. Xue, Deep Boltzmann machine based condition prediction for smart manufacturing. J. Amb. Intell. Human. Comput. **10**(3), 851–861 (2019)

133. K. Wärmefjord, R. Söderberg, L. Lindkvist, B. Lindau, J.S. Carlson, Inspection data to support a digital twin for geometry assurance, in *ASME 2017 International Mechanical Engineering Congress and Exposition*. American Society of Mechanical Engineers Digital Collection (2017)

134. G. Warwick, *GE Advances Analytical Maintenance with Digital Twins*. Aviation Week & Space Technology (2015), pp. 10–19

135. Y. Xu, Y. Sun, X. Liu, Y. Zheng, A digital-twin-assisted fault diagnosis using deep transfer learning. IEEE Access **7**, 19990–19999 (2019)

136. J. Yang, W. Zhang, Y. Liu, Subcycle fatigue crack growth mechanism investigation for aluminum alloys and steel, in *54th AIAA/ASME/ASCE/AHS/ASC Structures, Structural Dynamics, and Materials Conference*, (2013), pp. 1499

137. M.T.M. Yilmaz, K. Chakrabarty, Test-pattern selection for screening small-delay defects in very-deep submicrometer integrated circuits. IEEE Trans. Comput. Aided Design Integr. Circ. Syst. **29**(5), 760–773 (2010)

138. S. Yun, J. Park, W. Kim, Data-centric middleware based digital twin platform for dependable cyber-physical systems, in *2017 Ninth International Conference on Ubiquitous and Future Networks (ICUFN)* (IEEE, Piscataway, 2017), pp. 922–926

139. A.J. Zakrajsek, S. Mall, The development and use of a digital twin model for tire touchdown health monitoring, in *58th AIAA/ASCE/AHS/ASC Structures, Structural Dynamics, and Materials Conference* (2017), p. 0863

140. H. Zhang, Q. Liu, X. Chen, D. Zhang, J. Leng., A digital twin-based approach for designing and multi-objective optimization of hollow glass production line. IEEE Access **5**, 26901–26911 (2017)

141. R. Zhao, D. Yan, Q. Liu, J. Leng, J. Wan, X. Chen, X. Zhang, Digital twin-driven cyber-physical system for autonomously controlling of micro punching system. IEEE Access **7**, 9459–9469 (2019)

142. P. Zheng, X. Xu, C.-H. Chen, A data-driven cyber-physical approach for personalised smart, connected product co-development in a cloud-based environment. J. Intell. Manuf. **31**, 1–16 (2018)

143. Y. Zheng, S. Yang, H. Cheng, An application framework of digital twin and its case study. J. Amb. Intell. Human. Comput. **10**(3), 1141–1153 (2019)

144. M. Zhou, J. Yan, D. Feng, Digital twin framework and its application to power grid online analysis. CSEE J. Power Energy Syst. **5**(3), 391–398 (2019)

145. C. Zhuang, J. Liu, H. Xiong, Digital twin based smart production management and control framework for the complex product assembly shop-floor. Int. J. Adv. Manuf. Technol. **96**(1–4), 1149–1163 (2018)

Chapter 3
Trillion Sensors Security

Pinchen Cui, Ujjwal Guin, and Mark Tehranipoor

The advancement of ubiquitous computing in the Internet of Things (IoT) and Cyber-Physical Systems (CPS) applications boosted the number of connected devices increased in the last decade. The prevalent use of IoT applications with pervasive sensing enables billions of devices connected to the Internet, and each device may have multiple sensors attached, which leads us to the age of trillion sensors. However, the widespread use of sensors in critical infrastructures and applications, such as smart grid, smart city, industry 4.0 (smart manufacture), and healthcare, also raises additional security concerns. It is necessary to understand the infrastructure consisting of trillions of sensors to identify and thwart the potential threats before they occur. In this chapter, we aim to provide a detailed survey of the applicability and capabilities of the trillion sensors in diversified application domains. We present a comprehensive taxonomy of sensors introduced in [3] that have been widely adopted in IoT/CPS applications. We illustrate the IoT/CPS architecture and the underlying vulnerabilities in the trillion sensors era. We analyze and discuss future directions and solutions for solving security issues.

3.1 Introduction

The rapid growth of IoT/CPS applications leads us to a trillion sensors era, and the number of devices used by the IoT/CPS applications is expected to reach one trillion

---

P. Cui · U. Guin (✉)
Auburn University, Auburn, AL, USA
e-mail: ujjwal.guin@auburn.edu

M. Tehranipoor
University of Florida, Gainesville, FL, USA
e-mail: tehranipoor@ece.ufl.edu

© The Author(s), under exclusive license to Springer Nature Switzerland AG 2021
M. Tehranipoor (ed.), *Emerging Topics in Hardware Security*,
https://doi.org/10.1007/978-3-030-64448-2_3

by 2035 [123]. The trillion geographically distributed sensors enable real-time monitoring and real-time data analytics. Therefore, fault detection and diagnosis can be achieved, control decisions can be performed, resource sharing and resource management can be achieved, and system security can be optimized. The needs and opportunities for sensors are identified in many different application domains, such as environmental pollution monitoring, personal health monitoring, manufacturing supervision, energy harvesting, food delivery, global disasters, and aging infrastructure monitoring. Peter et al. projected a need for 45 trillion networked sensors in 20 years [37].

With the advent of large scale use of smartphones, the sensor industry experiences an outstanding growth. From 2007 through 2014, the mobile sensor market has grown nearly 200%, which was not envisioned by any market research organization [23]. Modern smartphone devices adopt acceleration, pressure, IR, humidity, temperature, light, and proximity sensors to support different applications. It is reported that the rise of demand for microphones, acceleration sensors, magnetic sensors, and gyroscopes becomes 10 billion in 2014 [24]. With the further progress of virtual reality and the Internet of Things techniques in the coming decade, similar growth is expected to be sustained in the following decade.

A typical smartphone uses around 15 sensors, a modern vehicle uses approximately 200 sensors, a smart home system adopts around 100 sensors, and a personnel may use 10 wearable medical sensors ($WMSs$) [24]. With the advancement in *mHealth* or *eHealth* technologies, the number of $WMSs$ is on the rise. The cost of sensors becomes the key driving factor for their rapid development in different market segments. Besides, the companies like GE provide support to the Industrial IoT applications to increase production efficiency, improve execution, and optimize business process through advanced real-time data analytics further pushing the use of application-specific sensors for industrial process monitoring [6]. The further development of expensive/critical industrial process/machines also requires large scale real-time data collection and analysis, and hence, the increase in the demand for various types of sensors [117] was predicted.

Figure 3.1 summarizes sensors forecast of different companies. GE has shown a plan for 10 trillion pollution monitoring printed sensors for 2025, while Texas Instruments projected 13 trillion Internet-connected devices by 2025 expecting sensors/MEMS to enable technology. Cisco delivered a forecast of 1 trillion networked sensors by 2020, whereas Bosch presented a vision of 7 trillion sensors by 2017. Intel introduced sensors for context-aware computing, and such systems are expected to absorb a trillion sensors by 2020–2022. Harbor Research considers smart systems to be the biggest business opportunity in the history of business, which will fusion computing, communication, and sensing into one platform [121].

The application of sensors in different connected applications poses several challenges and limitations. The widespread use of sensors demands a lower price for sensors, which results in low power budget, low die area allocation, and low computational power. Further, these sensors have to send the data to a server,

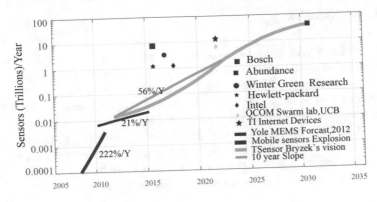

Fig. 3.1 Projected growth of sensors [23]

which raises the need for the development of low resource communication technologies/protocols where security is often compromised. Designing such sensors under severe resource constraints is a challenging task that might raise piracy for such designs in the coming decade. Huge market demand can give rise to massive scale sensor device counterfeiting through recycling, remarking, shipping rejected devices, and many other ways affecting the global IC supply chain [35, 130].

The application domains of sensors are widely varied. Sensors used in different applications are selected based on different assessment parameters like functionality, price, power, and performance. Identifying the sensors and sensor technologies used in different applications is necessary. A comprehensive taxonomy of sensors is needed, which can present an overview of all the sensors used in different applications. Also, based on the understanding of these sensors, sensor-based systems' security can be better ensured.

3.2   Taxonomy of Sensors

Sensors can be classified into different application domains, and each application domain requires domain-specific sensors with the specified area, power, and cost constraints. These sensors can be further classified into different categories according to the technologies used to develop them or based on their operating principles. Generally, sensors are developed by manipulating a particular physical property or a set of properties of some objects/materials. A sensor operates by converting energy from one form to another to detect any physical phenomenon, and the final form of energy is usually electrical. In the following subsections, we present a detailed taxonomy of sensors based on applications, technologies, and operating principles.

3.2.1 Applications

The International Technology Roadmap for Semiconductors (ITRS) has identified sensor-based IoT applications as the next driving factor for the semiconductor industry [12]. Companies like Libelium, GE, Intel, Cisco, and AT&T have successfully implemented sensor-based industrial and domestic applications [6, 11, 41]. The TSensors movement has identified some market segments with large volume sensor use potentials [23]. Based on the initiatives taken by the major companies and the market segments identified by the TSensors movement, we have recognized seven major application domains where trillions of sensors will be required (see Fig. 3.2). The application domains are discussed in the following section.

(a) Computing: The sensor market for the mobile communications is one of the key driving forces for the rapid growth of sensors. High-end mobile devices need more than 15 sensors on average [24]. Accelerometers, GPS, gyroscopes, touch, microphones, and proximity sensors are commonplace in modern cell phones, tablets, PCs, and toys. Different types of motion sensors are used in virtual reality devices to capture and recognize all the motions conveyed from the user's limbs. Also, temperature, pressure, light, sound, and moisture sensors enable the computing devices to have an awareness of the surrounding environment.

(b) Healthcare: In the past few years, we have seen an explosive growth of wearable medical devices in the consumer market to promote healthcare. Freefall detection technologies are built around accelerometers. Medical fridges are based on light, temperature, humidity, impact, and vibration sensors. Electrocardiography (ECG), pulse, and respiration sensors have contributed to athlete care and patient surveillance equipment. Ultraviolet (UV) sensors enable the devices to emit radiation alarms. Occlusion sensors are adopted in ambulatory

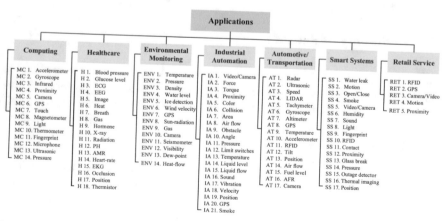

Fig. 3.2 Trillion sensors application domains [3]

infusion pumps, insulin pumps, and enteral feeding pumps to measure occlusions (blockages) in silicone or polyvinyl chloride (PVC) tubing.

(c) **Environmental Monitoring:**   Agriculture systems are significantly benefited from environmental monitoring through geographically distributed sensors. IoT-based irrigation systems have been deployed in Malaysia, which allowed the farmers to control the water supplies to their fields [68]. A large number of water-level and temperature sensors are distributed over the farming land and enable real-time data analytics. This system has almost doubled rice production in Malaysia since its deployment. Cisco and Sprint have worked together to build a truly connected city in Kansas [72]. A smart lighting system with light and video sensors save a significant amount of energy in the city. There is also a considerable potential in developing air quality monitoring and forest fire detection systems based on gas and temperature sensors.

(d) **Industrial Automation:**   Industrial automation relies on the sensors, and the ubiquitous sensing and big data analytics are opening new applications for industrial automation. Intel and Kontron have built a platform to monitor factory equipment using sensors and real-time data analytics, which provides accurate fault detection and automated maintenance notification; thus, the unplanned factory downtime can be minimized [14]. Both GE and Libelium have identified location intelligence as a critical factor in factory automation, where GE adopted a GPS-based approach, and Libelium has adopted passive tags (RFID+NFC) and active tags (ZigBee, WiFi, Bluetooth) [11, 108]. Machine health monitoring systems built with temperature, pressure, current, vibration, and gas sensors can improve manufacturing systems' efficiency and reliability.

(e) **Automotive/Transportation:**   Smart sensing and low-cost communication technologies have paved the way for intelligent transportation. Video camera and proximity sensors with computer vision have enabled autonomous driving. A modern vehicle incorporates around 30 sensors to facilitate safety features. Headway RADAR sensors are equipped with cars and trucks to detect the distance between vehicles and large objects. Adaptive cruise control and collision avoidance systems rely on the sensors as well. Airflow meters are used to measure the airflow consumption of automobile engines. LIDAR sensors are used in autonomous cars to provide a 360° vision of the surrounding environment.

(f) **Smart Systems:**   Smart systems integrate functions of sensing, actuation, and control to describe and analyze a situation. These systems make decisions based on the available data in a predictive or adaptive manner, thereby performing smart actions. A smart home is a prime example of smart systems, where temperature and humidity sensors enable the automated control of the household air conditioner. Water leak, smoke detector, glass break, light, sound, and fingerprint sensors lead the way for smart home development. On the other hand, by extracting energy data from smart sensors and meters, efficient energy measurement and management can be performed in smart grid [119].

(g) **Retail Service:**   Smart retail service, like Amazon Go, uses motion sensors and image sensors to automate their service [112]. Many other retail services, such as Walmart and postal deliveries, use RFID for tracking deliveries [81].

RFID tags based anti-theft are common in retail services as well. The tags are deactivated upon the purchase. An alarm will be generated if anyone moves the product outside the store without deactivating the tag.

3.2.2 Operating Principles

Sensors are an integral part of all electronic control applications. In electronic systems, sensors provide measurable electrical output (current/voltage) by converting energies according to physical phenomena. Energy can be converted from chemical, mechanical, optical, and thermal to electrical energy. The characteristics of materials used to develop sensors may vary widely. Based on the operating principle of sensors (energy conversion topology), we have identified seven main sensor categories, which we believe are critical to the development of trillions of sensor applications. The classification is shown in Fig. 3.3.

(a) **Electrical/Electronic:** Electrical sensors examine the changes in electrical or magnetic signals based on environmental input. Generally, any environmental input is converted into a voltage for measurement in any electrical/electronic sensor. Metal detectors, radar systems, voltmeters, and ohmmeters are a few simple examples where electrical sensors are used. Applications such as smart grids or smart meters require sensors to sense different electrical parameters like current, voltage, capacitance, and resistance [44].

(b) **Electromechanical:** The interaction of electronics, machinery, light, or fluids together constitutes an electromechanical system. They redirect light, pump and mix fluids, and detect molecules, heat, pressure, or motion. Consumer electronics and hand-held devices use many electromechanical sensors, such as accelerometer, gyroscope, motion, position, pressure, touch, and force sensors. Artificial retina and a hearing-aid transducer are revolutionizing the healthcare applications. Pressure sensors, inertial sensors, and chemical sensors have been deployed in industrial, automotive, and aerospace applications.

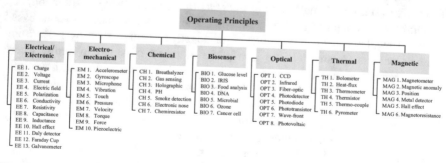

Fig. 3.3 Trillion sensors classification based on operating principles [3]

(c) **Chemical:** Any particular target chemical substance present in the desired medium will make the chemical sensors respond to produce the desired signal output for any required analyte concentration [33]. However, the chemical sensors' performance is limited by specific characteristics, such as selectivity, sensitivity, response time, and package size. Breathalyzer and pH sensors are used for biological monitoring. Smoke detectors are conventional in any household, industry, or office. Toxic chemical sensors play an essential role in environmental monitoring and industrial manufacturing.

(d) **Biosensors:** Biosensors are devices comprising biological elements and physio-chemical detectors for detecting analytes, a substance whose chemical composition is being identified and measured. Biosensors such as glucose level sensors, hormone or enzyme detectors, cancer cell detectors, and blood pressure sensors are widely deployed in wearable medical devices and other healthcare systems. Also, the instruments have a wide range of applications ranging from clinical to environmental and agricultural. For instance, biosensors are also adopted in the food industry.

(e) **Optical:** In the optical sensors, the light rays are converted into electronic signals. Common optical sensors such as photo-electric, photo-diode, photo-voltaic, and photoresistive sensors are used in lighting control applications, while cameras use charge-coupled devices (CCD) and CMOS sensors. Infrared sensors are used to measure the temperature in industrial applications. Fiber-optic sensors are used in electrical switching equipment to transmit light from an electrical arc flash to a digital protective relay, enabling the fast tripping of a breaker to reduce the energy in the arc blast [149].

(f) **Thermal:** Thermal sensors convert thermal energy into electronic signals. Bolometers are used to detect light in the far-infrared and mm-waves. Pyrometers can determine the temperature of any distant surface. Temperature sensors, thermocouples, heat-flux sensors, and thermistors are used in industrial applications and consumer electronics.

(g) **Magnetic:** Magnetometers are used to detect the magnetic field's direction in the environment, such as the earth's magnetic field. Magnetic sensors such as eddy current sensors, Hall effect sensors, magnetic field anomaly detectors, magneto-resistance sensors, and magnetometers are used in mapping, positioning, and non-contact switching applications [84]. Magnetic angle sensors and linear Hall sensors can measure the steering angle and steering torque in the Electric Power Steering (EPS). The magnetic anomaly detectors are used in the military to detect submarines.

3.2.3 Technology

Modern IoT/CPS systems require smart sensors where embedded microprocessors and communication modules are indispensable parts of the sensing devices [124]. Although numerous sensing technologies are available, few technologies have a

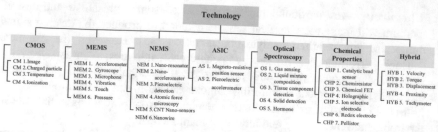

Fig. 3.4 Trillion sensors classification based on technologies [3]

greater impact on the overall sensor-based applications due to the complexity of integration. Figure 3.4 summarizes some of these technologies.

(a) Discrete CMOS: CMOS sensors are perhaps most suitable for consumer electronics and computing devices. The progress of CMOS image sensors leads to the global video/camera industry. CMOS temperature sensors and ionization detectors are also finding their ways into standard chips/ASICs [21]. On-chip CMOS temperature sensors are used for power management. Low power consumption and the ease of integrating such sensors in system-on-chips (SoC) make them suitable for any SoC design.

(b) MEMS: Microelectromechanical Systems (MEMS) have become the driving force of consumer electronics and portable devices. Accelerometer, gyroscope, microphones, pressure, and touch sensors have revolutionized the mobile computing devices by adding more functionalities and providing opportunities in numerous application domains [11]. Accelerometers are also used in inertial navigation systems for aircraft and missiles. In addition, they can also be used to detect vibrations of rotating machines. Gyroscopes can measure and maintain orientation. MEMS pressure sensors are used in the development of capacitive touch sensing applications for touch screens.

(c) NEMS: The ultra-high-frequency nano-resonators are used in ultra-high sensitive sensing, molecular transportation, molecular separation, high-frequency signal processing, and biological imaging [8]. Nanoelectromechanical systems (NEMS) based sensors are popular in medical diagnostic applications [93]. Nanowire sensors have been adopted in life sciences and medicine applications [104]. Nanowire, nano-resonators, and nano-rods can help to build in complex medical diagnostic applications. Carbon nanotubes (CNT) can potentially develop electromechanical sensors for implantable devices, which can be used in minimally invasive diagnosis, health monitoring, drug delivery, and many other intra-corporal tasks [64].

(d) ASIC: The development of application-specific integrated circuits (ASIC) for specific applications can optimize the sensor performance, but the associated cost may be high, making it suitable for applications where size and performance are decisive factors rather than the price. Companies like Bosch are developing MEMS-based ASIC solutions for the automotive industry [46].

(e) **Optical Spectroscopic:**   When interacting with the light of certain wavelengths, most liquids and gases provide unique signatures, a function of the molecular structure of the particular substance. This is the underlying principle of any optical spectroscopic sensors. These sensors are being used in numerous applications ranging from industrial process control to component detection techniques used in tumor delineation [31]. Solid detection, gas or liquid composition measurement, and tissue composition detection are some other examples where optical spectroscopy can be used.

(f) **Electrochemical:**   Electrochemical sensors are primarily used to detect toxic gas and oxygen. Each sensor is designed to be sensitive/selective to the specific gas. The chemiresistor is the backbone component for electronic nose [148]. Pellistors are widely used to detect gases. Chemical FETs are field-effect transistors that can be used to detect atoms, molecules, and ions in liquids and gases. The distance device identification in industrial applications and anti-counterfeiting applications are enabled by holographic sensors [140].

(g) **Hybrid:**   Hybrid sensors like tachymeters, torque, velocity, and pressure sensors are widely used in automotive and control applications where multiple sensing mechanisms are integrated into an embedded control system for detection and measurement purposes.

3.3   Infrastructure for Sensor-Based Applications

Sensors have been integrated into almost all electronic devices like mobile devices, virtual reality devices, toys, stand-alone medical diagnostic devices, electrical kitchen utensils, and others. While consumer electronics is undoubtedly the most extensive application of sensors right now, with the rapid growth of IoT/CPS applications, it is relatively safe to say that sensor-based IoT/CPS applications will be the largest destination of sensors in the future. To assess the vulnerabilities in sensor-based applications, we need to analyze the system architectures in which a large number of sensors will be deployed. This section presents the standard architectures for IoT/CPS systems using a large number of sensors. Although the difference between IoT and CPS is not well defined, and many see them as two different explanations of the same thing, IEEE has endeavored to differentiate them in [95]. The IoT system starts from the level where a single "thing" is identified using a unique global identifier and can be accessed from anywhere and anytime. The information provided by "things" or devices can be sensor data or static data stored in its memory. If the "things" in this IoT system are networked together to control a particular scenario in a coordinated way, then the IoT system can be considered to grow to a CPS level.

Since Kevin Ashton first coined the term Internet of Things in 1999 [50], numerous researchers have come forward with their views about the ideal architecture for IoT. Several reference models have been proposed to come up with a generalized reference model that can be adopted by different players in the market, thus developing products that can operate in systems built by different companies. Although the reference models vary widely, the base layer in any IoT reference model remains the same across all the models that deal with sensing.

Perhaps, the three-layer model proposed by Gubbi et al. was among the first IoT reference models proposed by the researchers [50]. This simplistic model is an extension of wireless sensor networks (WSN). It models IoT as a combination of WSN and cloud computing that can offer different end-user applications. The bottom layer consists of ubiquitous sensing devices that feed data to the cloud. The applications are built on the data stored in the cloud storage and the directly fetched data from the sensing devices. Rafiullah et al. proposed a five-layer IoT framework [75]. They envisioned the IoT as an information network where numerous data sensed by IoT devices will be collected through the network layer in a database, and applications will be built on that data. Several applications and services will be combined to develop business models in this framework's upper layer. In this model, the base layer is called the perception layer composed of physical objects and sensors. Atzori et al. proposed another five-level IoT framework where complex IoT systems have been decomposed into simplified applications consisting of an ecosystem of more specific and well-defined components [13]. The idea is to create many abstraction levels to hide issues that are not pertinent to a developer or a programmer. The base layer again is composed of the sensing objects enabled by identification, sensing, and communication technologies.

Similar architectures have been proposed for CPS. Lee et al. presented a five-layer architecture for industry 4.0 based CPS, where the bottom layer incorporates sensor network and tether-free communication [83]. Rad et al. proposed a four-layer CPS architecture for precision agriculture where sensing is at the base layer of the framework [109]. Cisco, IBM, and Intel presented a seven-layer IoT framework in IoT World Forum 2014 for the industry's reference framework for IoT/CPS. In this model, the data flow is usually bidirectional, but the dominant data flow direction is determined by the nature of the applications [132]. In this model, the base layer is called physical devices and controllers or edge, which includes sensors, actuators, machines, and smart devices. So, whichever framework is chosen to build up an IoT system, sensing devices will be at the bottom layer of the whole architecture.

In the Cisco IoT reference model shown in Fig. 3.5, communications and connectivity of the IoT systems are provided on the connectivity layer. IoT has already kicked off with many traditional devices not fully IP-enabled or devices that require additional communication gateway devices for external connectivity. Note that modern IoT devices will have an integrated sensing mechanism and communication modules. Data generated by these devices will be preprocessed at the gateway, and a considerable chunk of the data will be discarded because of limited network resource and storage facility. Threats can come at any stage of this system—during sensing, passing the data to the gateway, preprocessing the data at the gateway, or inside the information network. The information network's security issues are already well defined, and a lot of research work has been carried out to solve these issues in the last few decades. In this chapter, we primarily focus on the threats that can appear at the edge devices. As the sensing mechanism and

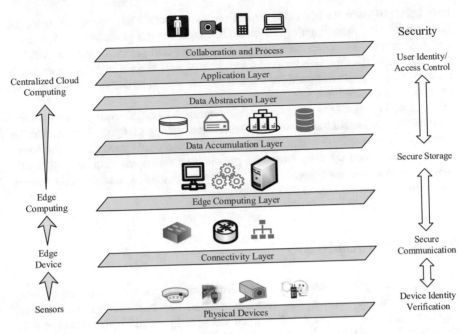

Fig. 3.5 Seven-layer IoT framework by Cisco [132]

communication modules are to be merged in a single IoT device, perhaps a closer look at the communication between the IoT device and the gateway will better evaluate the threat models.

The connectivity layer provides three basic data transmission types: between the devices and the network, across networks, and between the network and low-level information processing at the edge computing layer. The edge nodes must be equipped with sensing mechanism for generating data, analog to digital conversion circuitry, and the ability to be queried/controlled over the network. The communication scenario between the IoT devices and the Internet/data server may vary based on applications. For instance, WMS-based applications might use a hand-held device like a smart phone as the gateway device [150]. A more extensive application like a weather monitoring system might use dedicated gateway devices for specific regions. For some applications, the gateway might transmit the raw data received, and in some other cases, the gateway might filter some data before transmission to reduce the load on the network and the data storage [132].

For communication between the IoT devices and the gateway, several technologies are available, and the technologies that have caught the most attention are Bluetooth Low Energy (BLE), ZigBee, Z-$Wave$, Near Field Communication (NFC), WiFi, Thread, and Cellular ($2G/3G/4G$) [146]. Analyst firm ABI Research claimed that Bluetooth smart home devices would show a 75% growth rate between 2016 and 2021 [69]. $ZigBee$ and $Thread$ will lead with 34% volume share of the home automation and 29% of the smart lighting markets by this time [69]. Some protocols have been proposed for IoT communication, most notably the *6LowPAN* protocol [96]. An IoT device designer's goal would be to minimize the area overhead, energy, and cost of the associated communication module and select a suitable protocol that minimizes the data overhead. As the devices will use various communication technologies, any hub designed for a smart system should be capable of handling all types of communications.

Considering the Cisco seven-layer IoT reference model's requirements, a standard IoT device should have a sensing mechanism, analog to digital data conversion circuitry, a communication module, and probably a security module. The security module will ensure operator trust and prevent any unwanted data leakage under suspicious queries. Incorporating all these modules in a resource-constrained environment presents several design challenges.

3.4 Security Analysis

Ensuring security becomes one of the primary challenges for connected systems with thousands of sensors and receives tremendous attention in the research community. In this section, we illustrated the security of sensor-based systems in the following aspects: device identity verification, secure communication, secure storage, and access control.

3.4.1 Device Identity Verification

Due to the globalization in product manufacturing and its resulting horizontal integration, it is becoming extremely challenging to ensure the authenticity of electronic systems. At present, electronic products are manufactured, distributed, and sold globally. The electronic systems are assembled using the components sourced from all across the globe. It is now virtually impossible to find out the origin of electronic products, and track their route in the supply chain. Ensuring the security and integrity of the electronic parts has become increasingly difficult due to the widespread infiltration of untrusted hardware, specifically, counterfeit and cloned parts. Information Handling Services reported that the potential annual risk of the global supply chain from counterfeiting is at $169 billion and increasing [67]. The far-reaching penetration of non-authentic systems and counterfeit parts into the electronics supply chain has been reported in [52, 53, 55, 130, 131].

An electronic supply chain generally consists of four different stages, such as design, manufacturing, distribution, and resign/end-of-life. In each stage, multiple independent entities with different motives interact, which complicates the supply chain and exposes potential vulnerabilities. The supply chain of IoT/CPS sensor devices will not be different from the traditional one. From the design perspective, the original system designer needs to develop the hardware and firmware for the sensor nodes and invest in further research and development. Third-party intellectual properties (IPs) can be added to the design to reduce the time to market a product. The finalized design will go to the foundries or assemblies for manufacturing, and the manufactured products will come to the market through the distributors.

Any of the involved entities in the supply chain might try to manipulate the vulnerabilities to gain some unfair financial benefits. For example, a rogue designer in the design house might intentionally put a bug into the design. A third-party IP vendor might put Trojans in their IPs for malicious intentions. The design house might overuse third-party IPs to avoid licensing fees. An untrusted manufacturing unit can pirate a sensor's design during manufacturing, thus avoiding R&D cost. Design can also be reconstructed through reverse engineering. The stolen design can be sold to other competitors or modified to insert malicious circuits. In today's IP based SoC design, the system integrator can steal, overuse, or modify third-party IPs and thus gain various benefits [32, 51, 57].

3.4.1.1 Attack Vectors

The competitive market for IoT devices will put much pressure on the design companies, and time to market for IoT/CPS products will shrink considerably over time. Sensor-based applications will require a large number of sensors and associated chips to be supplied in a very short time. The increasing demand for newer low-cost and resource-constrained embedded systems forces the system design companies to look for low-cost chips through the semiconductor supply chain. Recycled sensors and associated networking and processing chips collected from the electronic waste, commonly known as e-waste, might find their ways to the supply chain because of lower cost. This further complicates the supply chain and leads to additional security concerns. It might give rise to design cloning, design piracy, and counterfeit sensors. Counterfeit sensors in an IoT/CPS environment might compromise the security of the system, be used to leak valuable information, or disrupt operation through unexpected device failure.

The adversary can corrupt the supply chain and sensor-based systems with different motives and using various methods:

Piracy: An adversary can penetrate the IoT/CPS market with sensors that have been rejected after manufacturing tests [129, 147] or have been remarked with a higher grade. We have observed rejected ICs in the defense supply chain [130]. This is very plausible that we will see rejected sensors in our critical applications, such as the smart grid. Due to lower cost, these sensors might attract the system design companies, especially those who produce low-end devices for sensor-based applications. In addition, an adversary can mark the sensor devices with a higher grade than they are in reality for financial benefits. Such a remarking process can cause severe risk if the sensors are operated in harsh environments where a real higher grade sensor is necessary. We have seen a similar trend in the IC supply chain, where new ICs are often remarked with higher grades to make a profit [130]. Remarked sensors in sensor-based IoT/CPS applications that are operated in severe environmental conditions might cause a system failure in critical conditions.

The diversity of IoT-based applications is already giving rise to small scale companies that specialize in specific products. Especially smart systems and healthcare solutions are so varied in nature that these are encouraging the development of different start-ups [1]. Moreover, modern self-driving cars incorporate multiple IoT solutions to enable self-driving capabilities and continuously improve system performance. As the original component manufacturers (OCM) continuously improve the specification, performance, and cost of their products, we believe similar situations, like the semiconductor industry, will occur soon. This can open new financial opportunities for the counterfeiters to pirate the design and manufacture chips and make a stock of such chips that they can sell at a very high rate when the original chips are discontinued.

Reverse Engineering: Considering the design complexities of a low-end resource-constrained sensor node and probable large scale demand for such devices in IoT/CPS applications, reverse engineering will attract counterfeiters to gain undue profits. Avoiding the R&D cost through reverse engineering, counterfeiters might be able to supply these sensors and associated networking and processing chips at a lower cost than the original manufacturers, which will result in a considerable loss of revenue for the industry. We have seen a similar trend for electronic systems [55, 131]. IoT/CPS sensor nodes are embedded systems with sensing and communicating capabilities (discussed in Sect. 3.3). As stated earlier, the exponential growth of new IoT solutions in the healthcare and smart systems domain, and demand for chips fabricated in older technologies in the field of automobiles, industrial control systems, and airplanes will justify the cost associated with reverse engineering.

Several destructive and non-destructive methods are available today to produce a 3D layout of an IC or printed circuit boards (PCB) with superfine resolution. Scanning Electron Microscope (SEM) or Transmission Electron Microscopes (TEM) can be used to capture the inner layer of an integrated circuit. X-ray tomography can be used to extract the 3D layout of a fully functional PCB to make clones [10]. Such cloning has been performed successfully, even in an academic research laboratory for feasibility studies [9]. Full chip layout can be constructed by extracting the layout of each layer of the design through etching and scanning with electron microscopes. The recent advancement of imaging technology makes it feasible to construct the inner details of an electronic product.

Counterfeiting: A standard IoT/CPS sensor node incorporates sensor devices, embedded processors, and memory. Any of these devices can come through recycling. Recycling ICs from electronic waste has become a colossal problem [52–54, 56, 130]. These recycled ICs have fewer remaining useful lifetimes than any fresh ICs, and the process through which they are collected also reduces their life further. These ICs come to the electronic supply chain through the grey market, and incorporating them in IoT/CPS infrastructure can cause frequent chip failures, thus increasing the cost of system maintenance. Low-cost recycled sensors and associated chips might find their ways into large scale sensor-based applications. As these applications are expected to perform uninterrupted for a long time to reduce maintenance costs, a shorter lifetime of the recycled chips might disrupt such applications. Using recycled chips in low-cost medical devices can cause false diagnoses, leading to severe health risks.

Physical Attacks: As the sensor nodes are spread geographically in most IoT/CPS applications, they can also be prone to physical attacks. The sensor data can be corrupted by manipulating the sensor environment by any adversary. Attackers can physically damage IoT devices to disrupt the availability of service. As in a connected environment, the sensor data can be used to control some other applications, and these dependent applications can be attacked through physical attacks on the sensor nodes. For example, if a sprinkler system in a smart home environment operates based on the temperature feedback from a temperature sensor, an adversary can manipulate the sensor environment to feed a false temperature data into the sprinkler system, and thus he can turn it on. In an industrial IoT application, manipulating a few sensor data can disrupt the whole control system if necessary security measures are not taken.

3.4.1.2 Countermeasures

The detection of non-authentic electronic devices in the IoT/CPS infrastructure manifests a variety of challenges. It is not easy to detect whether devices or appliances have been cloned or manufactured with pirated designs. For example, it is hard to detect an eavesdropping microphone embedded in a smart bulb or smart television. For an IoT infrastructure, rogue employees can fulfill their malicious intensions by replacing an authentic device with its cloned counterpart. Moreover, if the protection cannot fully cover all regions of travel, devices with mobility like drones would have a higher potential risk of getting intercepted. Besides, it is hard to keep verifying many low-cost and low power devices at a regular interval, even if many traditional verification mechanisms are provided.

To thwart the potential risks on the sensors, the importance of device authenticity verification needs to be performed. In the following subsections, we first introduced the traditional approaches for ID verification. Then, we discussed the state-of-art blockchain-based approaches.

The authenticity of a low-cost sensor device can be guaranteed by verifying an unclonable device ID. These IDs can be constructed from an onboard SRAM memory (SRAM PUF [49, 141]) to avoid costly tamperproof non-volatile memory. Physically unclonable functions (PUFs) have the potential to generate unique, unclonable bits and received significant attention in the hardware security community for the traceability of ICs. PUFs use inherently uncontrollable and unpredictable manufacturing process variations to generate random and unclonable bits. Over the years, several PUF architectures have been proposed, which include the arbiter PUF [48], ring oscillator PUF [126], and SRAM PUF [49, 141], among others.

Since IoT/CPS sensor nodes have SRAM-based memory and embedded processors, SRAM PUFs can offer a better solution to produce device IDs with no additional cost or complexity as compared to other options. In addition, DNA markings are commercially available to provide traceability for electronic components [94]. Fake devices can be identified using the fast or detailed authentication process. The fluorescent marks under UV light are verified in the fast authentication. A decision of a valid mark is taken based on the color produced. Detailed authentication is performed in specific laboratories, which is inefficient and costly [118]. Further, if counterfeiters add a different mechanism to the chip, which can produce similar fluorescence, DNA marking will be ineffective for fast authentication, which only verifies a specific color.

The integration of blockchain in the supply chain receives increasing attention recently, since the inherent properties and features of blockchain could significantly enhance the traceability, transparency, and reliability of the supply chain [34, 106]. Different researchers discussed, proposed, and analyzed various blockchain-based frameworks to refine the traceability for supply chain [2, 19, 20, 29, 30, 76, 90, 133–135]. By leveraging the blockchain, the traceability of food [19, 133, 134], healthcare [20, 30], and post-delivery supply chain [135] could be enhanced. Contrary to the traditional blockchain-based tracking (e.g., food and healthcare products), electronic devices possess an advantage of integrating unclonable ID, which can be generated from a PUF embedded into the device, and thus can enable efficient and low-cost tracking (e.g., registration, verification, and status update).

Note that the blockchain was first introduced in Bitcoin [97] and is now widely used in different cryptocurrencies. Blockchain is known as a distributed and shared digital ledger, where all the transactions and records are hashed and stored in the chain to provide both integrity and transparency. Specific blockchains also support the smart contracts [5, 7], which allows the user to run Turing-complete scripts on the chain. Using a smart contract (also known as chaincode in HyperLedger) enables the user to store and manage data inside of the blockchain. Various applications, such as Filecoins [45] and Storj [125], have been proposed.

Blockchain can eliminate the need for a central or local database to store device IDs, which helps solve the single point of failure problem. The inherent immutability feature of blockchain ensures data integrity and prevents adversaries from tampering the data records. However, if a system is built upon the traditional centralized database, all the data records could be altered and affected once it is compromised. In addition, the distributors and customers can directly verify the device using the blockchain without querying the specific manufacturer's database.

Guin et al. [58] introduces a blockchain-based framework to prevent and detect counterfeit IoT/CPS edge devices. The proposed framework has four main components: the sensor nodes contain a physically unclonable function (PUF) to produce a unique unclonable ID, global blockchain to store device IDs securely, local blockchain to the IDs of devices used in an IoT/CPS applications, and a secure communication protocol [59, 91] to access IDs. This system needs manufacturers to

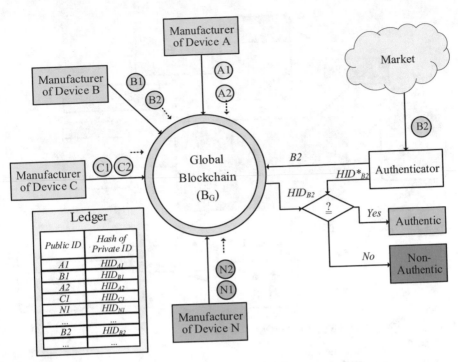

Fig. 3.6 Blockchain for verifying the identity of IoT/CPS sensor nodes [58]

upload a cryptographic hash of the ID from PUF into the global blockchain so that an entity in the supply chain can verify the true identity of a device. The registration of various devices and ID verification is shown in Fig. 3.6. The global blockchain infrastructure (B_G) stores the information regarding the manufacturers along with the edge device IDs. Once a device is registered, anyone can access its identity from anywhere. Manufacturers need to upload a cryptographic hash of the ID to prevent it from being cloned. During the registration of a device in the IoT infrastructure, it is necessary to verify whether the hash of device ID is present in B_G. The local blockchain provides the necessary support to authenticate every device in a regular interval when an IoT device (sensor nodes) is deployed in the field. This prevents an adversary (e.g., a rogue or a compromised employee) from putting a counterfeit device into the IoT/CPS infrastructure. The communication between the gateway and an edge node has to be very low cost. To provide that support, a low-cost communication protocol [59, 91] is used, which uses a secret key for transferring the PUF response to the gateway to verify the edge node's identity in the field. Note that this secret key can also be generated from the PUF. A local blockchain contains the hash of device ID for identity verification [59].

Xu et al. provided a comprehensive solution and summary of using blockchain to improve and secure the integrity of electronic supply chain [145]. Islam et al. proposed a method that uses PUF and blockchain to enhance the authenticity and

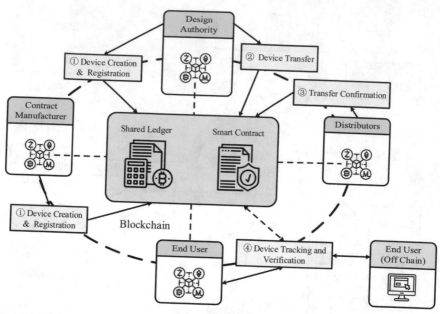

Fig. 3.7 Blockchain for supply chain provenance [35]

traceability of parts in the supply chain [70]. However, the device ownership transfer in this framework is simply triggered and controlled by device owners. This design may lead to potential security issues. Human errors, delivery and management failures, in-transit thefts, and dishonest participants are still threatening the supply chain even with blockchain implementation for tracking [137]. Wrong electronic parts could be accidentally sent, which come with inappropriate ownership transferred. Logistics and transportation could be delayed or failed due to external causes (e.g., natural disasters), the receiver may cancel the original order even the ownerships have already been transferred in the blockchain. The adversaries could steal electronic parts during transportation. These stolen parts are still valid since ownerships are already transferred to a trusted participant in the blockchain. Moreover, a receiver of parts can deny the transfer or acceptance of part after the ownership transfer is completed. Therefore, directly transferring ownerships within one transaction creates irreversible results in blockchain that can cause additional security and management risks, especially in the scenarios as mentioned earlier.

To provide reliable traceability and secure ownership transfer, Cui et al. proposed a blockchain-based provenance framework using confirmation-based ownership transfer [35]. The proposed architecture is based on a permissioned blockchain (e.g., HyperLedger) with the use of a non-resource intensive consensus algorithm, where most of the previous works were implemented via Proof of Work (PoW) based permissionless blockchain (e.g., Ethereum). The features of consortium blockchain and HyperLedger eliminate the cost of a transaction fee and improve the efficiency by using a non-resource intensive consensus algorithm. Figure 3.7 shows the architecture of the blockchain-based provenance framework [35]. The framework is built upon a consortium-based blockchain consisting of four types of nodes: design authority, contract manufacturer, distributor, and end-user/customer. Four operations are designed to enable traceability using smart contracts. The design authority and contract manufacturer could register the devices into the blockchain using a device registration function. The devices' transfer needs to be performed using a device transfer transaction. To provide additional security, the new owner of the devices needs to send a transfer confirmation to complete the ownership transfer. After the sender sends the ownership transfer transaction, an additional confirmation transaction from the receiver is required. The ownership transfer will be completed once the mutual agreement between the sender and the receiver is reached. This will automatically tag the items that are missing during the transportation. Finally, the end user could track the devices' trace and verify the authenticity using the device tracking and verification function.

3.4.2 Secure Communication

Recent studies have shown that modern IoT/CPS sensor nodes are severely resource-constrained, limiting the ability of these devices to use standard cryptographic protocols to communicate securely. Trappe et al. showed that the power constraint in IoT edge devices limits the encryption/encoding functionality of the sensor nodes, leading to poorly encrypted communication or no encryption all [136]. In a study, HP revealed that almost 70% of tested IoT devices did not encrypt communications to the Internet and the local network, while half of the device's mobile applications performed unencrypted communications with the cloud, the Internet, and local network [111]. Almost 60% of the devices under study did not use encryption while downloading software updates. Symantec also found 19% of the devices under test use no encryption during communication with the cloud server or the back-end applications and even fewer use encryption while communicating in the local network [16]. These lead to opportunities for adversaries to attack such low-cost sensor nodes.

No use of costly cryptoprimitives during the communication results in inadequate security for the edge devices. Consequently, ensuring data integrity, maintaining the confidentiality of the communication, authenticating the edge devices, and controlling access to the devices become very challenging. The devices might be vulnerable to different kinds of denial-of-service (DoS) attacks [143]. Sleep deprivation attacks can severely affect the IoT/CPS application's operation by draining the energy source of power-constrained edge devices [18]. Attacks such as hello flooding, Sinkholes, Wormholes, and Sybil attack, which are prevalent in any wireless sensor network, will also be present in any IoT/CPS application [73, 92, 105, 120]. In the gateway-centric IoT/CPS model, security measures would most likely have to be implemented in the gateway devices to provide security against these attacks due to the resource constraints in the IoT/CPS edge devices. These security threats are well documented in the domain of wireless sensor networks. One can look at these articles to have a greater understanding of those attacks, which is clearly out of the scope of this chapter.

In this section, we will focus on the limitations of standard IoT devices and the threats that might arise due to the large volume of sensors in both—the software and hardware levels.

3.4.2.1 Traditional Approaches

Sensors in an IoT/CPS network generate real-time data and send it to a designated server through gateways. The gateways have ample resources to deploy required security protocols to send the data to the server over the public Internet, whereas the edge nodes in an IoT/CPS system are resource-constrained, which prohibits them from using standard security protocols. Note that any secure data transfer protocol should have confidentiality, message integrity, and end-point authentication [80].

- *Confidentiality:* Only the sender and the receiver should extract the meaning of the transferred data. Symmetric key cryptography and public key cryptography are two widely used methods for ensuring confidentiality in communication. This can be achieved through symmetric ciphers (e.g., advanced encryption standard (AES) [36]) and asymmetric ciphers (Rivest–Shamir–Adleman (RSA) [113] and elliptic curve cryptography (ECC) [62]).
- *Message Integrity:* The receiver needs to verify the integrity of the received data. In other words, there should be an easy process to verify whether the data has been altered during the transmission. Message authentication codes (MAC) are generally used for the verification of message integrity. Keyed-hash message authentication code (HMAC) [98] is widely popular that uses a secure hash function (e.g., SHA-2 and SHA-3 [99]).
- *End-point Authentication:* The receiver must verify that the transfer request was initiated by the sender (not by an adversary). This can be achieved through digital signatures [89] generally constructed using RSA or ECC.

All these cryptographic primitives are computationally heavy, which require considerable processing power and energy. Hardware implementation of the above cryptographic primitives with a separate chip can be prohibitively costly for an IoT/CPS sensor node. So, a software implementation of the security primitives is more practical. However, a sensor node's computing capabilities are minimal and consist of an 8 MHz micro-controller with less than 128 KB of instruction memory and approximately 10 KB of RAM [114].

3.4.2.2 Low-Cost Approaches

Resource constraints in IoT/CPS sensor nodes have fueled the development of low resource communication and computation techniques for these applications. These approaches can be described using whether they have used PUFs or RFIDs.

PUF-Based Authentication Protocols: Researchers have proposed and developed several PUF-based authentication protocols for applications that use resource constraint devices. Majority of these protocols consist of a *prover* (a resource-constrained device) and a *verifier* (a resource-rich gateway server). The *prover* is a node (e.g., sensors) that responds to the *verifier*'s (e.g., routers) request for its authorization in the system. During the verification, the *verifier* sends a random challenge to the *prover*. The *prover* acknowledges it by sending a valid response in return. The authors in [60] combined a delay PUF-based authentication protocol with an HB-based protocol (named after the authors) [65] to remove security vulnerabilities of these individual protocols. Later, they proposed a protocol to reduce power and area overhead using a 2-level noisy PUF instead of using area and power-intensive cryptographic modules [61]. Katzenbeisser et al. proposed a logically re-configurable PUF proposed in [74] to reduce excessive area requirement [115]. The above protocols use a challenge-response pair (*CRP*) only once to prevent the replay attack.

Management and storage of *CRPs* at the *verifier's* end can be costly when thousands of devices connected to an IoT/CPS applications. A large number of CRPs need to store for a single device so that one CRP is used only once to address replay attacks. Unfortunately, this does not completely address the security issues. An adversary can model a PUF mathematically and predict the responses once he/she acquired large CRPs by listening the communication channel. Rührmair et al. presented a modeling attack for several PUF designs, including arbiter PUFs and ring oscillator PUFs [116]. The *converse PUF-based* authentication protocol has been proposed to eliminate modeling attacks [77]. A hardware and software co-verification based authentication scheme has also been proposed, where the firmware hash along with PUF response is used to detecting software and hardware impersonation attacks [66]. Chatterjee et al. proposed an authentication scheme that combines PUF, identity-based encryption (IBE), and keyed-hash function to eliminate the need for explicitly storing *CRPs* [28]. However, Braeken et al. [22] proved that it is vulnerable to the man-in-the-middle (MITM) attack, impersonation, and replay attacks [40].

A preshared key-based host identity protocol is proposed to authenticate an IoT edge node [47]. However, an adversary can gain access to the network if the shared key gets compromised. Kothmayr et al. proposed a two-way authentication protocol that uses datagram transport layer security (DTLS) based on the X.509 certificate [78]. Porambage et al. proposed an implicit certificate-based two-phase authentication protocol [107]. This protocol can support distributed IoT applications since the certificates are more light weight than the protocol proposed in [78]. This implicit certificate-based protocol can be suitable for highly resource-constrained sensor nodes. Unfortunately, the protocol can be vulnerable to replay, DoS, and MITM attacks [27]. To mitigate these vulnerabilities, Turkanović et al. presented a four-step authentication model, where a remote user negotiates a session key with a sensor node without requesting it to the gateway [138]. Challa et al. proposed a signature-based user authenticated key agreement scheme using ElGamal and ECC-based signature for device authentication [27]. Although this protocol is advantageous compared to [107] and [138], it requires high computational power. Hash function-based and chaos-based privacy-preserving schemes have been proposed for smart home systems [122]. Wazid et al. proposed three-factor remote user authentication for a hierarchical IoT network [142].

The majority of these protocols use a strong-PUF that requires hardware modification, which may not be feasible for a resource-constrained sensor node. Adding extra hardware may lead to an increased cost for a sensor, which is highly discouraging for the broad adoption of low-cost sensors. Therefore, a better approach would be to utilize existing hardware resources to implement security features for light-weight devices.

RFID-Based Approaches: In this section, we will explore different radio-frequency identification (RFID)-based authentication protocols, which will be suitable for the IoT/CPS sensor nodes. RFID tags are a prime example of resource-constrained devices containing a limited number of logic gates. RFID-based protocols used light-weight cryptographic primitives [15, 71]. Several light-weight RFID protocols have been proposed [38, 63, 79, 87, 127, 128, 139]. Most of these protocols utilize cryptographic hash functions, random number generators, and XOR functions compared to other computationally expensive symmetric key and public key cryptographic primitives.

Henrici et al. have presented a protocol that uses cryptographic hash functions and XOR operators to encrypt the communication between the RFID tag and the reader [63]. Lim et al. proposed a protocol where the reader and the RFID tag use random number generators, hash functions, and XOR primitives to authenticate each other [87]. Tan et al. proposed a serverless authentication protocol that also utilizes a known hash function between the reader and the tags and XOR primitives. All these protocols are light-weight but have some security flaws that the adversaries can manipulate to attack the system. For instance, the protocol proposed by Henrici et al. maintains a session number to synchronize the tag and the reader, which can be easily manipulated by an adversary by interrogating the tag in an intermediate step of the authentication process and thus desynchronizing the tag and the reader [63]. In Lim et al.'s protocol, the total number of authentication session requests is limited, making it vulnerable to denial-of-service (DoS) attacks [87]. In the protocol proposed by Tan et al., the tag returns a static form of data based on its ID and a secret, which can be utilized to track the tag by any adversaries. These naive security flaws have kept the development of such protocols ongoing [128].

3.4.3 Secure Storage

Whether audio, video, text, discrete, and stream IoT data are generated by various IoT/CPS sensor nodes or devices and are likely to have inconsistent formats and semantics, managing the large volume of heterogeneous data introduces a series of technical challenges. Pruning, compression, labeling data generated by digital sensors, automobiles, and electrical meters are often required processing before it can be used and stored, and may require significant processing power cost. After the data is collected and processed, another hard requirement is to provide data storage for the massive amount of data. Few organizations and enterprises can house all the IoT data collected from their networks [82]. Centralized data storage requires and relies on bandwidth and availability; migrating to distributed and decentralized storage may be a better option.

It is challenging to ensure the data's integrity and freshness when the data is accumulated and stored in the central database. The use of blockchain can be an appropriate approach to address data provenance and data integrity. The cryptographically secure hash function-based blockchain systems inherently address data integrity. The timestamp information in the data can be useful for the data freshness perspective.

Different researchers are already working on addressing data security issues using blockchain. Liang et al. [86] introduced a blockchain-based data assurance system for drones, including drones, control system, blockchain network, and cloud server. The drone collects images or video data and receives commands from the control system. The cloud server logs the control system's command records and stores the raw data in a database. The blockchain stores the hashed data records to provide data integrity and generate receipts for data validations. The implementation of a blockchain network in the proposed system is based on Tierion and Chainpoint [26]. Blockchain-based IoT data storage systems can always provide additional and reliable data protection. A similar example is [88], which also used blockchain to perform data verification and assure data integrity. They used blockchain to store the hashes of uploaded data and established and evaluated a proof-of-concept system based on Ethereum and *InterPlanetary File System (IPFS)* protocol [17]. To ensure the time synchronization of IoT services, and guarantee the freshness and reliability of data, a blockchain-based time synchronization method is proposed in [43], where an IoT node can verify and synchronize time by reading the time consensus result in the blockchain.

3.4.4 Access Control

Access management in an IoT system is a challenging problem. It is important to maintain proper access to devices and data as they directly determine the system's security; however, the traditional centralized access control systems cannot fulfill all the IoT system requirements. Some of the weaknesses of the centralized access control systems are summarized in [103] as follows:

1. It is hard to achieve end-to-end security.
2. Centralization creates a single point of failure on whose availability and reliability determine the entire system's security.
3. As a remote/local central entity manages the access control, the user cannot be involved in controlling his or her data.
4. Centralized service providers can make illegitimate use of data (e.g., Prism Program [101]).
5. Running a centralized access control system for a large number of devices is expensive.
6. For some IoT scenarios that require device mobility or collaborative management, a centralized access control system is not well fitted.

These limitations lead to the proposal of using blockchain to provide an available and reliable access control mechanism.

A blockchain-based access control system has better support for mobility, accessibility, scalability, and transparency. Ouaddah et al. [102] briefly demonstrate an access control framework named FairAccess, in which the blockchain is used to store, manage, and enforce the access policies. A simple case study is implemented on the Raspberry Pi, and the framework can securely manage a camera's access. However, the limitation of FairAccess is that it requires the tokens as the cost and fee for creating and verifying the policies. A similar framework was introduced in [42], where a framework with no tokens and fewer transactions are involved. Ding et al. [39] proposes to use blockchain as an attribute authority and key generation center for IoT infrastructure. The proposed access control management framework is decentralized and scalable and further increases the system's robustness.

Novo [100] implements a prototype system based on CoAP to provide large scale access management in the IoT system. The proposed framework consists of three parts: IoT network, management hub, and blockchain network. All data access inside the IoT network is recorded, verified, and managed by the blockchain. Since the edge devices or sensors may not afford the overhead of blockchain operations, the management hub works as the agent (miner) of the blockchain network. Also, the design and evaluation of a blockchain-based access control system for IoT are illustrated in [144]. It shows that the system performs with acceptable processing delay and latency.

Preserving privacy for the IoT system is another focus of access control. Ensuring the availability and transparency of data access is important, but it is more important to prevent the reading or leaking of others' data and configurations. A general platform based on blockchain, IPFS [4], is designed to provide privacy for IoT data. By utilizing a private sidechain, all the data operations are logged and validated. In the work of Cha et al. [25], IoT service providers can obtain user consent on privacy policies without modifying legacy IoT devices using blockchain and smart contracts. The user can query the Blockchain Connected Gateway to check the device information or access the device's data without connecting to the device directly.

3.5 Conclusion

This chapter presents an overview of sensor applications, current infrastructure, and security challenges. Large scale deployment of sensors demands simplification of sensor hardware for reducing cost and optimize security requirements. Due to IoT and CPS's widespread application, the number of connected sensor nodes in these systems grows exponentially. The sensors are embedded everywhere—consumer electronics, automobiles, wearables, machines, supermarkets, fields, warehouses, drones, industrial robots, and shipping containers—we are virtually heading for a trillion sensor supply chain. Maintaining security across all these various sensors becomes our paramount objective.

Identification and authentication of each sensor node connected in any IoT/CPS application with unclonable IDs is one solution that has been considered to ensure IC supply chain security. PUFs have emerged as a low-cost solution for creating unclonable IDs. However, the unreliability in standard PUF outputs has limited their applications. Developing authentication protocols that can deal with unreliable IDs is necessary. Also, blockchain can help ensure the traceability of electronic parts in the supply chain. Developing new PUF architectures, exploiting the current architectures to increase the reliability, and using PUFs in the blockchain are ongoing research directions.

Most of the light-weight encryption schemes and communication protocols that have been developed for IoT/CPS sensor nodes have loopholes due to resource limitations. These loopholes can be manipulated by an adversary to attack such systems. Authentication of the sensor nodes and ensuring the integrity of the generated data are still a challenge. Developing encryption schemes that require low power, memory, and computational resources is still an ongoing research area. The sensor nodes incorporate small batteries, and they are expected to operate uninterrupted for an extended period. So, power consumption must be minimized in every possible step—during data generation and data transmission. The associated protocols have to be very light weight. However, avoiding security features altogether in the sensor nodes, which is the current trend in IoT system development considering the generated data is of limited value to the attackers, can lead to disastrous effects, shown by recent studies. Developing light-weight and secure communication protocols is a challenge that has to be solved to provide a secure and trusted IoT/CPS environment.

The resource-constrained IoT/CPS sensor nodes are vulnerable to both software- and hardware-based attacks. The limited computational power, low energy resource, and memory do not allow these devices to use standard cryptographic protocols during data communication and edge device authentication. So, these sensor nodes are vulnerable to many software-based attacks like data sniffing, false packet injection, sleep deprivation, and DoS attacks. Piracy, reverse engineering, recycling, and tampering are the forms of hardware-based attacks that pose severe threats to such systems' security. Among trillions of sensors, even a tiny percentage of compromised devices can target different systems and cause significant damage to the applications. So, any solution to secure these sensors applications must be comprehensive that ensures security at both the software and hardware levels.

References

1. 64 Healthcare IoT Startups in Patient Monitoring, Clinical Efficiency, Biometrics, and More. https://www.cbinsights.com/research/iot-healthcare-market-map-company-list/
2. S.A. Abeyratne, R.P. Monfared, Blockchain ready manufacturing supply chain using distributed ledger. International Journal of Research in Engineering and Technology. 5(9), 1–10 (2016)
3. M. Alam, M.M. Tehranipoor, U. Guin, TSensors vision, infrastructure and security challenges in trillion sensor era. J. Hardware Syst. Secur. 1(4), 311–327 (2017)
4. M.S. Ali, K. Dolui, F. Antonelli, IoT data privacy via blockchains and IPFS, in *Proceedings of the Seventh International Conference on the Internet of Things, IoT'17*, (2017), pp. 14:1–14:7
5. E. Androulaki, A. Barger, V. Bortnikov, C. Cachin, K. Christidis, A.D. Caro, D. Enyeart, C. Ferris, G. Laventman, Y. Manevich, S. Muralidharan, C. Murthy, B. Nguyen, M. Sethi, G. Singh, K. Smith, A. Sorniotti, C. Stathakopoulou, M. Vukolic, S.W. Cocco, J. Yellick, Hyperledger fabric: a distributed operating system for permissioned blockchains, Proceedings of the thirteenth EuroSys conference, 1–15 (2018)
6. M. Annunziata, P.C. Evans, Industrial internet: Pushing the boundaries. General Electric Reports 488–508 (2012)
7. Anonymous, White paper: Next-generation smart contract and decentralized application platform. https://github.com/ethereum/wiki/wiki/White-Paper
8. B. Arash, J.W. Jiang, T. Rabczuk, A review on nanomechanical resonators and their applications in sensors and molecular transportation. Appl. Phys. Rev. 2(2), 021301 (2015)
9. N. Asadizanjani, S. Shahbazmohamadi, M. Tehranipoor, D. Forte, Analyzing the impact of x-ray tomography for non-destructive counterfeit detection, in *Proceedings of International Symposium on Testing Failure Analysis* (2015), pp. 1–10
10. N. Asadizanjani, S. Shahbazmohamadi, M. Tehranipoor, D. Forte, Non-destructive PCB reverse engineering using x-ray micro computed tomography, in *41st International Symposium for Testing and Failure Analysis, ASM* (2015), pp. 1–5
11. A. Asin, D. Gascon, 50 sensor applications for a smarter world. Libelium Comunicaciones Distribuidas, Tech. Rep. (2012)
12. S.I. Association, et al., International technology roadmap for semiconductors 2.0 (2015). http://public.itrs.net/
13. L. Atzori, A. Iera, G. Morabito, The internet of things: a survey. Comput. Netw. 54(15), 2787–2805 (2010)
14. Automating Field Service with the Internet of Things (IoT). Cisco, Case Study
15. G. Avoine, P. Oechslin, RFID traceability: a multilayer problem, in *Financial Cryptography*, vol. 3570 (Springer, Berlin, 2005), pp. 125–140
16. M.B. Barcena, C. Wueest, Insecurity in the internet of things. Security response, symantec, 20 (2015)
17. J. Benet, Ipfs-content addressed, versioned, p2p file system (2014). arXiv:1407.3561
18. T. Bhattasali, R. Chaki, S. Sanyal, Sleep deprivation attack detection in wireless sensor network (2012). arXiv:1203.0231

19. K. Biswas, V. Muthukkumarasamy, W.L. Tan, Blockchain based wine supply chain traceability system, in *Future Technologies Conference* (2017)
20. T. Bocek, B.B. Rodrigues, T. Strasser, B. Stiller, Blockchains everywhere-a use-case of blockchains in the pharma supply-chain, in *2017 IFIP/IEEE Symposium on Integrated Network and Service Management (IM)* (IEEE, Piscataway, 2017), pp. 772–777
21. S. Boppel, A. Lisauskas, M. Mundt, D. Seliuta, L. Minkevicius, I. Kasalynas, G. Valusis, M. Mittendorff, S. Winnerl, V. Krozer, et al.: CMOS integrated antenna-coupled field-effect transistors for the detection of radiation from 0.2 to 4.3 THZ. IEEE Trans. Microwave Theory Tech. **60**(12), 3834–3843 (2012)
22. A. Braeken, PUF based authentication protocol for IoT. Symmetry **10**(8), 352 (2018)
23. J. Bryzek, Roadmap for the trillion sensor universe. Berkeley, vol. 2 (2013)
24. J. Bryzek, Trillion sensors: foundation for abundance, exponential organizations, internet of everything and mHealth. SENSOR MAGAZINE, Trade Journal Rep. (2014)
25. S.C. Cha, J.F. Chen, C. Su, K.H. Yeh, A blockchain connected gateway for BLE-based devices in the internet of things. IEEE Access, IEEE **6**, 24639–24649 (2018)
26. Chainpoint, https://tierion.com/chainpoint/
27. S. Challa, M. Wazid, A.K. Das, N. Kumar, A.G. Reddy, E.J. Yoon, K.Y. Yoo, Secure signature-based authenticated key establishment scheme for future IoT applications. IEEE Access, IEEE **5**, 3028–3043 (2017)
28. U. Chatterjee, V. Govindan, R. Sadhukhan, D. Mukhopadhyay, R.S. Chakraborty, D. Mahata, M.M. Prabhu, Building PUF based authentication and key exchange protocol for IoT without explicit CRPS in verifier database. IEEE Trans. Dependable Secure Comput. IEEE **16**(3) 424–437 (2018)
29. S. Chen, R. Shi, Z. Ren, J. Yan, Y. Shi, J. Zhang, A blockchain-based supply chain quality management framework, in *2017 IEEE 14th International Conference on e-Business Engineering (ICEBE)* (IEEE, Piscataway, 2017), pp. 172–176
30. K.A. Clauson, E.A. Breeden, C. Davidson, T.K. Mackey, Leveraging blockchain technology to enhance supply chain management in healthcare: an exploration of challenges and opportunities in the health supply chain. Blockchain in Healthcare Today. Partners in Digital Health **1**(3) 1–12 (2018)
31. O.M. Conde, A. Eguizabal, E. Real, J.M. López-Higuera, P.B. Garcia-Allende, A.M. Cubillas, Optical spectroscopic sensors: from the control of industrial processes to tumor delineation, in *2013 6th International Conference on Advanced Infocomm Technology (ICAIT)* (IEEE, Piscataway, 2013), pp. 91–92
32. G.K. Contreras, M.T. Rahman, M. Tehranipoor, Secure split-test for preventing IC piracy by untrusted foundry and assembly, in *2013 IEEE International Symposium on Defect and Fault Tolerance in VLSI and Nanotechnology Systems (DFT)* (IEEE, Piscataway, 2013), pp. 196–203
33. N.R. Council, et al., *Expanding the Vision of Sensor Materials* (National Academies Press, Washington, 1995)
34. M. Crosby, P. Pattanayak, S. Verma, V. Kalyanaraman, et al., Blockchain technology: beyond bitcoin. Appl. Innov. **2**(6–10), 71 (2016)
35. P. Cui, J. Dixon, U. Guin, D. DiMase, A blockchain-based framework for supply chain provenance. IEEE Access **7**, 157113–157125 (2019)
36. J. Daemen, V. Rijmen, *The Design of Rijndael: AES-the Advanced Encryption Standard* (Springer, Berlin, 2013)
37. P.H. Diamandis, S. Kotler, Abundance: The Future is Better Than You Think (Simon and Schuster, New York, 2012)
38. T. Dimitriou, A lightweight RFID protocol to protect against traceability and cloning attacks, in *First International Conference on Security and Privacy for Emerging Areas in Communications Networks, 2005. SecureComm 2005* (IEEE, Piscataway, 2005), pp. 59–66
39. S. Ding, J. Cao, C. Li, K. Fan, H. Li, A novel attribute-based access control scheme using blockchain for IoT. IEEE Access **7**, 38431–38441 (2019)

40. D. Dolev, A. Yao, On the security of public key protocols. IEEE Trans. Inf. Theory **29**(2), 198–208 (1983)
41. J. Duffy, At&t allies with CISCO, IBM, Intel for city IoT. Network World. https://www.networkworld.com/article/3019433/atandt-allies-with-cisco-ibm-intel-for-cityiot.html
42. C. Dukkipati, Y. Zhang, L.C. Cheng, Decentralized, blockchain based access control framework for the heterogeneous internet of things, in *Proceedings of the Third ACM Workshop on Attribute-Based Access Control* (2018)
43. K. Fan, S. Wang, Y. Ren, K. Yang, Z. Yan, H. Li, Y. Yang, Blockchain-based secure time protection scheme in IoT. IEEE Internet Things J. **6**, 4671–4679 (2018)
44. H. Farhangi, The path of the smart grid. IEEE Power Energy Mag. **8**(1), 18–28 (2010)
45. Filecoin, https://filecoin.io/
46. A. Fischer, Bosch designs application-specific integrated circuits for MEMS sensors in Dresden. http://www.bosch-presse.de/pressportal/de/en/bosch-designs-application-specific-integrated-circuits-for-mems-sensors-in-dresden-42032.html
47. O. Garcia-Morchon, S.L. Keoh, S. Kumar, P. Moreno-Sanchez, F. Vidal-Meca, J.H. Ziegeldorf, Securing the IP-based internet of things with hip and DTLS, in *Proceedings of the Sixth ACM Conference on Security and Privacy in Wireless and Mobile Networks* (ACM, New York, 2013)
48. B. Gassend, D. Clarke, M. Van Dijk, S. Devadas, Silicon physical random functions, in *Proceedings of the ACM Conference on Computer and Communications Security (CCS)* (ACM, New York, 2002)
49. J. Guajardo, S.S. Kumar, G.J. Schrijen, P. Tuyls, FPGA intrinsic PUFs and their use for IP protection, in *International Workshop on Cryptographic Hardware and Embedded Systems* (Springer, Berlin, 2007)
50. J. Gubbi, R. Buyya, S. Marusic, M. Palaniswami, Internet of things (IoT): a vision, architectural elements, and future directions. Future Gener. Comput. Syst. **29**(7), 1645–1660 (2013)
51. U. Guin, M.M. Tehranipoor, Obfuscation and encryption for securing semiconductor supply chain, in *Hardware Protection through Obfuscation* (Springer, Berlin, 2017), pp. 317–346
52. U. Guin, D. DiMase, M. Tehranipoor, Counterfeit integrated circuits: detection, avoidance, and the challenges ahead. J. Electron. Test. **30**(1), 9–23 (2014)
53. U. Guin, K. Huang, D. DiMase, J. Carulli, M. Tehranipoor, Y. Makris, Counterfeit integrated circuits: a rising threat in the global semiconductor supply chain. Proc. IEEE **102**(8), 1207–1228 (2014)
54. U. Guin, X. Zhang, D. Forte, M. Tehranipoor, Low-cost on-chip structures for combating die and IC recycling, in *Proceedings of the 51st Annual Design Automation Conference* (ACM, New York, 2014), pp. 1–6
55. U. Guin, S. Bhunia, D. Forte, M. Tehranipoor, SMA: a system-level mutual authentication for protecting electronic hardware and firmware. IEEE Trans. Dependable Secure Comput. **14**, 265–278 (2016)
56. U. Guin, D. Forte, M. Tehranipoor, Design of accurate low-cost on-chip structures for protecting integrated circuits against recycling. IEEE Trans. Very Large Scale Integ. Syst. **24**(4), 1233–1246 (2016)
57. U. Guin, Q. Shi, D. Forte, M. Tehranipoor, FORTIS: a comprehensive solution for establishing forward trust for protecting IPs and ICs. ACM Trans. Des. Autom. Electron. Syst. **21**, 63 (2016)
58. U. Guin, P. Cui, A. Skjellum, Ensuring proof-of-authenticity of IoT edge devices using blockchain technology, in *IEEE International Conference on Blockchain* (2018)
59. U. Guin, A. Singh, M. Alam, J. Canedo, A. Skjellum, A secure low-cost edge device authentication scheme for the Internet of things, in *International Conference on VLSI Design* (2018)
60. G. Hammouri, B. Sunar, PUF-HB: a tamper-resilient HB based authentication protocol, in *International Conference on Applied Cryptography and Network Security* (2008), pp. 346–365

61. G. Hammouri, E. Öztürk, B. Sunar, A tamper-proof and lightweight authentication scheme. Pervasive Mobile Comput. **4**(6), 807–818 (2008)
62. D. Hankerson, A.J. Menezes, S. Vanstone, *Guide to Elliptic Curve Cryptography* (Springer, Berlin, 2006)
63. D. Henrici, P. Muller, Hash-based enhancement of location privacy for radio-frequency identification devices using varying identifiers, in *Proceedings of the Second IEEE Annual Conference on Pervasive Computing and Communications Workshops* (IEEE, Piscataway, 2004), pp. 149–153
64. C. Hierold, A. Jungen, C. Stampfer, T. Helbling, Nano electromechanical sensors based on carbon nanotubes. Sens. Actuators A Phys. **136**(1), 51–61 (2007)
65. N.J. Hopper, M. Blum, Secure human identification protocols, in *International Conference on the Theory and Application of Cryptology and Information Security* (Springer, Berlin, 2001), pp. 52–66
66. M. Hossain, S. Noor, R. Hasan, HSC-IoT: a hardware and software co-verification based authentication scheme for internet of things, in *International Conference on Mobile Cloud Computing, Services, and Engineering (MobileCloud)* (2017), pp. 109–116
67. IHS iSuppli, Top 5 most counterfeited parts represent a $169 billion potential challenge for global semiconductor market (2011)
68. Intel IoT. Increasing Food Production with the Internet of Things. Intel, Case Study (2016), https://www.mouser.com/pdfdocs/increasing-food-production-iot-paper.pdf
69. IoT to Account for 28% of Wireless Connectivity IC Market by 2021; Driven by Fast-Growing Smart Home, Wearables, and Beacons. https://www.abiresearch.com/press/iot-account-28-wireless-connectivity-ic-market-202/
70. M.N. Islam, S. Kundu, Enabling IC traceability via blockchain pegged to embedded PUF. ACM Trans. Des. Autom. Electron. Syst. **24**(3), 36 (2019)
71. A. Juels, RFID security and privacy: a research survey. IEEE Journal Sel. Areas Commun. **24**(2), 381–394 (2006)
72. Bob Bennett, Kansas City & Cisco: Engaging the 21st Century Citizen. Cisco (2016). https://giotportaldevstorage.blob.core.windows.net/portalproductioncontainer/documents/P9_Kansas_City_Case_study.pdf
73. C. Karlof, D. Wagner, Secure routing in wireless sensor networks: attacks and countermeasures. Ad Hoc Netw. **1**(2), 293–315 (2003)
74. S. Katzenbeisser, Ü. Kocabaş, V. Van Der Leest, A.R. Sadeghi, G.J. Schrijen, C. Wachsmann, Recyclable PUFs: logically reconfigurable PUFs. J. Cryptograph. Eng. **1**(3), 177 (2011)
75. R. Khan, S.U. Khan, R. Zaheer, S. Khan, Future internet: the internet of things architecture, possible applications and key challenges, in *2012 10th International Conference on Frontiers of Information Technology (FIT)* (IEEE, Piscataway, 2012), pp. 257–260
76. H.M. Kim, M. Laskowski, Toward an ontology-driven blockchain design for supply-chain provenance. Intell. Syst. Account. Financ. Manag. **25**(1), 18–27 (2018)
77. Ü. Kocabaş, A. Peter, S. Katzenbeisser, A.R. Sadeghi, Converse PUF-based authentication, in *International Conference on Trust and Trustworthy Computing* (Springer, Berlin, 2012), pp. 142–158
78. T. Kothmayr, C. Schmitt, W. Hu, M. Brünig, G. Carle, DTLS based security and two-way authentication for the internet of things. Ad Hoc Netw. **11**(8), 2710–2723 (2013)
79. L. Kulseng, Z. Yu, Y. Wei, Y. Guan, Lightweight mutual authentication and ownership transfer for RFID systems, in *2010 Proceedings IEEE INFOCOM* (IEEE, Piscataway, 2010), pp. 1–5
80. J.F. Kurose, K.W. Ross, *Computer Networking: A Top-Down Approach*, vol. 4 (Addison Wesley Boston, 2009)
81. K. Salmon, Kurt Salmon RFID in Retail Study. Corporate site (2016). https://easyscan.dk/wp-content/uploads/2018/03/rfid-retail_study_-kurt-salmon.pdf
82. I. Lee, K. Lee, The internet of things (IoT): applications, investments, and challenges for enterprises. Bus. Horizons **58**, 431–440 (2015)
83. J. Lee, B. Bagheri, H.A. Kao, A cyber-physical systems architecture for industry 4.0-based manufacturing systems. Manuf. Lett. **3**, 18–23 (2015)

84. J. Lenz, S. Edelstein, Magnetic sensors and their applications. IEEE Sens. J. **6**(3), 631–649 (2006)
85. M. Li, A. Davoodi, M. Tehranipoor, A sensor-assisted self-authentication framework for hardware Trojan detection, in *Design, Automation & Test in Europe Conference & Exhibition (DATE)* (2012), pp. 1331–1336
86. X. Liang, J. Zhao, S. Shetty, D. Li, Towards data assurance and resilience in IoT using blockchain, in *IEEE Military Communications Conference (MILCOM)* (2017)
87. T.L. Lim, T. Li, T. Gu, Secure RFID identification and authentication with triggered hash chain variants, in *14th IEEE International Conference on Parallel and Distributed Systems, 2008. ICPADS'08* (IEEE, Piscataway, 2008), pp. 583–590
88. B. Liu, X.L. Yu, S. Chen, X. Xu, L. Zhu, Blockchain based data integrity service framework for IoT data, in *IEEE International Conference on Web Services (ICWS)* (2017)
89. G. Locke, P. Gallagher, FIPS PUB 186-3: digital signature standard (DSS). Federal Inf. Process. Standards Publ. **3**, 186–3 (2009)
90. Q. Lu, X. Xu, Adaptable blockchain-based systems: a case study for product traceability. IEEE Softw. **34**(6), 21–27 (2017)
91. Mahmod, M.J.A., Guin, U.: A robust, low-cost and secure authentication scheme for IoT applications. Cryptography **4**(1), 8 (2020)
92. M. Meghdadi, S. Ozdemir, I. Güler, A survey of wormhole-based attacks and their counter-measures in wireless sensor networks. IETE Techn. Rev. **28**(2), 89–102 (2011)
93. M.T. Michalewicz, A. Sasse, Z. Rymuza, Quantum tunneling NEMS devices for bio-medical applications. Quantum Precision Instruments (2016). https://www.quantum-pi.com/PAPERS/Quantum-Pi-TCF_2007.pdf
94. M. Miller, J. Meraglia, J. Hayward, Traceability in the age of globalization: a proposal for a marking protocol to assure authenticity of electronic parts, in *SAE Aerospace Electronics and Avionics Systems Conference* (2012). https://doi.org/10.4271/2012-01-2104
95. R. Minerva, A. Biru, D. Rotondi, Towards a definition of the internet of things (IoT). IEEE Internet Initiative, IEEE **1**(1), 1–86 (2015)
96. G. Mulligan, The 6LoWPAN architecture, in *Proceedings of the 4th Workshop on Embedded Networked Sensors* (ACM, New York, 2007), pp. 78–82
97. S. Nakamoto, Bitcoin: a peer-to-peer electronic cash system (2008). http://bitcoin.org/bitcoin.pdf
98. J.M. Turner, The Keyed-Hash message authentication code (HMAC). Federal Information Processing Standards Publication **198**(1), (2008)
99. PUB, FIPS, Secure Hash Standard (SHS). Federal Information Processing Standards Publication **180**(4), (2012)
100. O. Novo, Blockchain meets IoT: an architecture for scalable access management in IoT. IEEE Internet Things J. **5**(99), 1184–1195 (2018)
101. NSA prism program taps in to user data of Apple, Google and others. https://goo.gl/2RCCQB
102. A. Ouaddah, A.A.E. Kalam, A.A. Ouahman, Fairaccess: a new blockchain-based access control framework for the internet of things. Secur. Commun. Netw. (2016)
103. A. Ouaddah, H. Mousannif, A.A. Elkalam, A.A. Ouahman, Access control in the internet of things: big challenges and new opportunities. Comput. Netw. **112**, 237–262 (2017)
104. F. Patolsky, G. Zheng, C.M. Lieber, Nanowire sensors for medicine and the life sciences. Nanomedicine (Lond). **1**(1), 51–65 (2006). https://doi.org/10.2217/17435889.1.1.51. PMID: 17716209.
105. A. Perrig, J. Stankovic, D. Wagner, Security in wireless sensor networks. Commun. ACM **47**(6), 53–57 (2004)

106. M. Pilkington, Chapter 11: Blockchain technology: principles and applications, in *Research Handbook on Digital Transformations* (2016), pp. 225–253. https://doi.org/10.4337/9781784717766.00019

107. Porambage, P., Schmitt, C., Kumar, P., Gurtov, A., Ylianttila, M.: Two-phase authentication protocol for wireless sensor networks in distributed IoT applications, in *Wireless Communications and Networking Conference (WCNC), 2014* (IEEE, Piscataway, 2014), pp. 2728–2733

108. GE, Predix–The platform for the Industrial Internet. GE, Platform Brief (2016). https://ecosystems4innovating.files.wordpress.com/2016/11/predix-the-platform-for-theindustrial-internet-whitepaper.pdf

109. C.R. Rad, O. Hancu, I.A. Takacs, G. Olteanu, Smart monitoring of potato crop: a cyber-physical system architecture model in the field of precision agriculture. Agric. Agric. Sci. Procedia **6**, 73–79 (2015)

110. M.T. Rahman, D. Forte, Q. Shi, G.K. Contreras, M. Tehranipoor, CSST: preventing distribution of unlicensed and rejected ICS by untrusted foundry and assembly, in *IEEE International Symposium on Defect and Fault Tolerance in VLSI and Nanotechnology Systems (DFT)* (2014), pp. 46–51

111. K. Rawlinson, HP study reveals 70 percent of internet of things devices vulnerable to attack. http://www8.hp.com/us/en/hp-news/press-release.html?id=1744676#.WUrrwWgrKM8

112. D. Reisinger, Amazon's cashier-free store might be easy to break. http://fortune.com/2017/03/28/amazon-go-cashier-free-store/

113. R.L. Rivest, A. Shamir, L. Adleman, A method for obtaining digital signatures and public-key cryptosystems. Commun. ACM **21**(2), 120–126 (1978)

114. R. Roman, C. Alcaraz, J. Lopez, A survey of cryptographic primitives and implementations for hardware-constrained sensor network nodes. Mobile Netw. Appl. **12**(4), 231–244 (2007)

115. U. Rührmair, M. van Dijk, PUFs in security protocols: attack models and security evaluations, in *2013 IEEE Symposium on Security and Privacy (SP)* (IEEE, Piscataway, 2013), pp. 286–300

116. U. Rührmair, F. Sehnke, J. Sölter, G. Dror, S. Devadas, J. Schmidhuber, Modeling attacks on physical unclonable functions, in *Proceedings of the 17th ACM Conference on Computer and Communications Security* (ACM, New York, 2010), pp. 237–249

117. A. Saxena, Digital twin enabling PHM at industrial scales. 19th Nordic Seminar on Railway Technology (2016). https://www.ltu.se/cms_fs/1.147575!/file/Saxena%20DigitalTwin_NSRT2016.pdf

118. Semiconductor Industry Association (SIA), Public Comments - DNA Authentication Marking on Items in FSC5962 (2012). https://www.semiconductors.org/wp-content/uploads/2018/07/Nov-15-2012-Defense-Logistics-Agency-Response-from-SIA-FINAL.pdf

119. F. Shrouf, J. Ordieres, G. Miragliotta, Smart factories in industry 4.0: a review of the concept and of energy management approached in production based on the internet of things paradigm, in *2014 IEEE International Conference on Industrial Engineering and Engineering Management* (2014)

120. V.P. Singh, S. Jain, J. Singhai, Hello flood attack and its countermeasures in wireless sensor networks. Int. J. Comput. Sci. Issues **7**(11), 23–27 (2010)

121. Smart systems and services growth opportunities. http://harborresearch.com/wp-content/uploads/sites/8/2016/02/HRI_ThingWorx\discretionary-Reprort_Smart\discretionary-Services\discretionary-Business\discretionary-Model\discretionary-Innovation.pdf

122. T. Song, R. Li, B. Mei, J. Yu, X. Xing, X. Cheng, A privacy preserving communication protocol for IoT applications in smart homes. IEEE Internet Things J. **4**(6), 1844–1852 (2017)

123. P. Sparks, The route to a trillion devices. White Paper, ARM (2017)

124. B.F. Spencer, M.E. Ruiz-Sandoval, N. Kurata, Smart sensing technology: opportunities and challenges. Struct. Control. Health Monit. **11**(4), 349–368 (2004)

125. Storj, https://storj.io/

126. G. Suh, S. Devadas, Physical unclonable functions for device authentication and secret key generation, in *Proceedings of ACM/IEEE on Design Automation Conference* (2007)

127. H.M. Sun, W.C. Ting, A gen2-based RFID authentication protocol for security and privacy. IEEE Trans. Mobile Comput. **8**(8), 1052–1062 (2009)
128. C.C. Tan, B. Sheng, Q. Li, Secure and serverless RFID authentication and search protocols. IEEE Trans. Wireless Commun. **7**(4), 1400–1407 (2008)
129. M. Tehranipoor, H. Salmani, X. Zhang, *Integrated Circuit Authentication* (Springer, Berlin, 2014)
130. M.M. Tehranipoor, U. Guin, D. Forte, *Counterfeit Integrated Circuits: Detection and Avoidance* (Springer, Berlin, 2015)
131. M.M. Tehranipoor, U. Guin, S. Bhunia, Invasion of the hardware snatchers. IEEE Spectrum **54**(5), 36–41 (2017)
132. The Internet of Things Reference Model. Cisco Systems (2014)
133. F. Tian, An agri-food supply chain traceability system for china based on RFID & blockchain technology, in *2016 13th International Conference on Service Systems and Service Management (ICSSSM)* (IEEE, Piscataway, 2016), pp. 1–6
134. F. Tian, A supply chain traceability system for food safety based on HACCP, blockchain & internet of things, in *2017 International Conference on Service Systems and Service Management* (IEEE, Piscataway, 2017), pp. 1–6
135. K. Toyoda, P.T. Mathiopoulos, I. Sasase, T. Ohtsuki, A novel blockchain-based product ownership management system (POMS) for anti-counterfeits in the post supply chain. IEEE Access **5**, 17465–17477 (2017)
136. W. Trappe, R. Howard, R.S. Moore, Low-energy security: limits and opportunities in the internet of things. IEEE Secur. Privacy **13**(1), 14–21 (2015)
137. B. Tukamuhabwa, M. Stevenson, J. Busby, Supply chain resilience in a developing country context: a case study on the interconnectedness of threats, strategies and outcomes. Supply Chain Manag. Int. J. **22**(6), 486–505 (2017)
138. M. Turkanović, B. Brumen, M. Hölbl, A novel user authentication and key agreement scheme for heterogeneous ad hoc wireless sensor networks, based on the internet of things notion. Ad Hoc Netw. **20**, 96–112 (2014)
139. I. Vajda, L. Buttyán, et al.: Lightweight authentication protocols for low-cost RFID tags, in *Second Workshop on Security in Ubiquitous Computing–Ubicomp*, vol. 2003 (2003)
140. F.D.C. Vasconcellos, A.K. Yetisen, Y. Montelongo, H. Butt, A. Grigore, C.A. Davidson, J. Blyth, M.J. Monteiro, T.D. Lowe Wilkinson, C.R. Lowe, Printable surface holograms via laser ablation. ACS Photon. **1**(6), 489–495 (2014)
141. W. Wang, A. Singh, U. Guin, A. Chatterjee, Exploiting power supply ramp rate for calibrating cell strength in SRAM PUFs, in *IEEE Latin-American Test Symposium* (2018)
142. M. Wazid, A.K. Das, V. Odelu, N. Kumar, M. Conti, M. Jo, Design of Secure User Authenticated Key Management Protocol for Generic IoT Networks. IEEE Internet Things J. **5**, 269–282 (2018)
143. A.D. Wood, J.A. Stankovic, A taxonomy for denial-of-service attacks in wireless sensor networks, in *Handbook of Sensor Networks: Compact Wireless and Wired Sensing Systems* (2004), pp. 739–763
144. R. Xu, Y. Chen, E. Blasch, G. Chen, Blendcac: a blockchain-enabled decentralized capability-based access control for IoTs, in *2018 IEEE International Conference on Blockchain (Blockchain-2018)* (2018)
145. X. Xu, F. Rahman, B. Shakya, A. Vassilev, D. Forte, M. Tehranipoor, Electronics supply chain integrity enabled by blockchain. ACM Trans. Des. Autom. Electron. Syst. **24**(3), 31:1–31:25 (2019)
146. Z. Yang, Y. Yue, Y. Yang, Y. Peng, X. Wang, W. Liu, Study and application on the architecture and key technologies for IoT, in *International Conference on Multimedia Technology (ICMT'11)* (IEEE, Piscataway, 2011), pp. 747–751
147. M. Yilmaz, K. Chakrabarty, M. Tehranipoor, Test-pattern selection for screening small-delay defects in very-deep submicrometer integrated circuits. IEEE Trans. Comput.-Aided Des. Integr. Circuits Syst. **29**(5), 760–773 (2010)

148. J. Yinon, Peer reviewed: detection of explosives by electronic noses. ACS Publications (2003). https://pubs.acs.org/doi/pdf/10.1021/ac0312460
149. M. Zeller, G. Scheer, Add trip security to arc-flash detection for safety and reliability, in *Power Systems Conference, PSC'09* (IEEE, Piscataway, 2009), pp. 1–8
150. Z.K. Zhang, M.C.Y. Cho, S. Shieh, Emerging security threats and countermeasures in IoT, in *Proceedings of the 10th ACM Symposium on Information, Computer and Communications Security* (ACM, New York, 2015), pp. 1–6
151. D. Zhang, M. He, X. Wang, M. Tehranipoor, Dynamically obfuscated scan for protecting IPs against scan-based attacks throughout supply chain, in *IEEE 35th VLSI Test Symposium (VTS)* (2017)

Chapter 4
Security of AI Hardware Systems

Haoting Shen

4.1 Artificial Intelligence System

Since the breakthrough on speech recognition and pattern recognition obtained by applying deep neural network, machine learning-based artificial intelligence (AI) has become the hottest technology trend in both academia and industry. Due to the dominance of machine learning in AI research, the AI we discuss in this chapter is specifically machine learning-based. Originally, AI technique was driven by algorithms and data to solve various problems such as pattern recognition, natural language processing, decision making, and so on. But later, remarkable progresses in parallel hardware [11, 18, 25] and scalable software systems [10, 15, 34] make it possible for AI to process "big data" and solve real-world problems. Many disciplines including data storage, mobile applications, smart sensors, internet of things, and novel computer architectures are also involved during the building of powerful AI systems with broad impacts in our society. Successful implementations of advanced AI systems are promised to greatly promote current industry production and business activities and, in addition, will help to bring about novel industries and society services, such as unmanned vehicle, smart city, and automatic decision systems. Because recent developments allow developers to build advanced machine learning systems to process massive amount of data, semiconductor devices are required as matches [4], including short-term and long-term memories, logic circuits, and networking equipment. Highly customized chip designs are also desired to accelerate the data processing for specific AI application. According to the detailed tasks of AI systems at different stages, the required devices will be distinct. For these applications, the reliability and security of AI systems are

H. Shen (✉)
University of Nevada, Reno, NV, USA
e-mail: hshen@unr.edu

© The Author(s), under exclusive license to Springer Nature Switzerland AG 2021
M. Tehranipoor (ed.), *Emerging Topics in Hardware Security*,
https://doi.org/10.1007/978-3-030-64448-2_4

essential as critical system failures can cause tremendous property damage and even serious casualty losses. A natural question thus arises: are these AI systems secure? In this chapter, we will discuss the hardware-related security concerns of modern AI systems.

4.1.1 An Example of Neural Network System

Here we use a simple neural network as an example to very briefly explain how a machine learning system is used. As shown in Fig. 4.1, the neural network is consisted of three layers: input layer, hidden layer, and output layer. In each layer, there are a number of nodes (circles in Fig. 4.1). The values of input nodes are obtained from inputs as x_1, x_2, and x_3. The values of hidden nodes can be labeled as z_1 to z_4, and the outputs are y_1, y_2, and y_3. The connections between nodes in neighbor layers are illustrated as arrows in Fig. 4.1, which are measured as *weights* (w). For example, the connection between x_3 and z_2 is written as w_{32}. The values of the nodes can be calculated based on the previous node values and the weights of connections. For example,

$$z_2 = x_1 \cdot w_{x1,z2} + x_2 \cdot w_{x2,z2} + x_3 \cdot w_{x3,z2} + b_{z2} , \tag{4.1}$$

and

Fig. 4.1 A simple example of neural network with one hidden layer

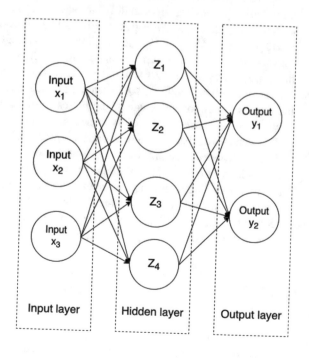

$$y_1 = z_1 \cdot w_{z1,y1} + z_2 \cdot w_{z2,y1} + z_3 \cdot w_{z3,y1} + z_4 \cdot w_{z4,y1} + b_{y1}, \qquad (4.2)$$

where b_{z2} is the bias. At the output layer, when the node values are obtained in the same way, activation function is typically applied in addition to introduce non-linearity for final outputs. Common activation functions can be identity ($f(x) = x$), binary step ($f(x) = 0$ for $x < 0$ and 1 for $x \geq 0$), logistic ($f(x) = 1/(1 + e^{-x})$), and more. Take logistic function as an example:

$$Output_1 = f(y_1) = \frac{1}{1 + e^{-y_1}}. \qquad (4.3)$$

To obtain the desired outputs upon the inputs, the parameters in this neural network system, including the weights and the bias, have to be optimized. The optimization process is known as *training*. In most cases, we first build the neural network structure and decide the number of layers and nodes. Then we start with arbitrary weights and feed the neural network with data as inputs. When the outputs are obtained, they are compared with pre-established criteria (e.g., labels on data). According to the comparison results, the weights and bias will be tuned for the next round, until optimized. For deep learning system, the amount of required data for training is tremendous and the parameter iterative optimization takes the most time during the implementation. Once the AI system is trained and passes necessary testing, it is ready to be used in specific application scenarios, also called as inference phase. In some applications, the AI systems keep learning while performing inference (e.g., recurrent neural networks).

4.1.2   Hardware in AI Systems

In current algorithms used by AI systems, matrix multiplications are the core computation. Due to the large scale of sophisticated AI systems and the demands on massive data processing, sufficient understanding in hardware components becomes important to better utilize the computational resources. Obviously, the hardware requirements on AI systems are different for training and inference. During the training, the system needs to process big data so powerful servers or cloud-based computing systems that provide access to vast stores of data are preferred. While for inference, the systems deal with less data but have to provide responses in rapid and robust ways. In addition, the available computational resources during the inference can be notably diverted. For example, high performance central processing unit (CPU) or graphic processing unit (GPU) may be available on autonomous passenger vehicles, while on lightweight edge devices, only limited circuit areas and power consumptions are acceptable. Due to the characters and requirements of training and inference, the former one is typically performed in data centers, while the latter one is usually carried out on edge computing systems. A forecast report made in 2018 [4] estimated that the hardware used for AI systems in data centers and edge

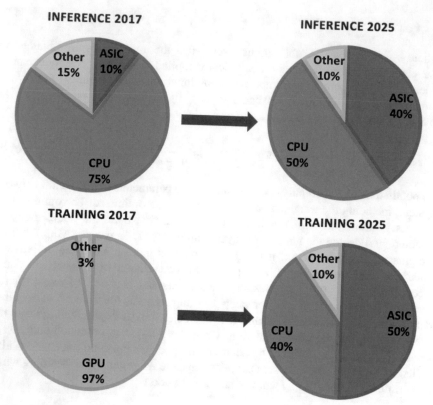

Fig. 4.2 The shift of preferred architectures for computing in data centers. Data source: artificial intelligence hardware: new opportunities for semiconductor companies (reproduced from [4])

nodes would shift from generalized chips such as CPU and GPU to customizable ones such as application specific integrated circuit (ASIC) and field programmable gate array (FPGA), as shown in Figs. 4.2 and 4.3.

Besides the hardware performing matrix multiplications for neural network data processing, other components for data collection, storage, transferring, and action taking are also necessary to build comprehensive AI systems that cover perception, training, inference, and actions. To build reliable and secure AI systems, research in security of all these components are imperative.

4.2 Security Threats and Countermeasures

As AI is playing more and more important roles in our lives, many aspects in AI systems are valuable targets for the adversaries. According to the targets, attacks on AI systems can be roughly categorized into two classes: one is aiming

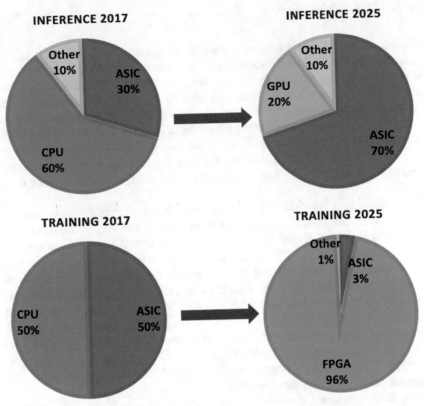

Fig. 4.3 The shift of preferred architectures for computing at edge nodes. Data Source: artificial intelligence hardware: new opportunities for semiconductor companies (reproduced from [4])

to compromise the integrity of the decision process, and the other is aiming to extract sensitive/valuable data from the AI system without changing the behavior of the system. For example, data poisoning [5, 19, 22, 32], evasion attack [20], and impersonate attack [6] try to tamper the training result by manipulating the training data, while inversion attack seeks to learn private input data based on the AI system outputs. Due to the potential damage and possible serious consequences made by these attacks, academic and industry have spent considerable effort on the countermeasure study. However, the study from the hardware security aspect is severely insufficient. In this section, we will discuss the security threats of AI systems and the possible countermeasures case by case, with emphases on hardware.

4.2.1 Data Poisoning

Data poisoning attack aims to corrupt the learned models in AI systems by injecting maliciously altered training data and/or removing valid training data. As one of the most threatening attacks against AI systems, effective data poisoning attacks have been demonstrated on different machine learning systems, including logistic regression [22], support vector machine (SVM) [5], Principal Component Analysis (PCA) [7], clustering [8], deep neural network (DNN) [33], etc. One simple example is shown in Fig. 4.4, where the attack is against a classifier by adding some small part of new training points. As shown, the decision boundary is significantly changed after the malicious data injection, resulting in a degraded performance of the trained classifier.

To prevent data poisoning attacks, the research community proposed two basic solutions: data sanitization and learning algorithm improvement. The main idea of data sanitization is to detect abnormal data and remove it from the input training data, such that the purity of the training data is ensured and the expected learning results will not be altered (Fig. 4.4 Right). However, the detection of poisoned data is not always efficient [19]. If the data is manipulated in more concealed forms, or a comprehensive understanding of the true data is in a lack, data sanitization may not work. The second solution is to enhance the system robustness by improving learning algorithm. For example, Liu et al. proposed techniques integrating low-rank matrix approximation and principle component regression to prune adversarial instances [21]. Another group from IBM proposed a data provenance based approach to mitigate the poisoning attacks [3].

Solutions mentioned above are based on data and algorithms. While from the hardware aspect, promising techniques that can be used to prevent data poisoning are data source tracking and data integrity checking. Given the recent trend, it can be safely predicted that there will be more and more data for the AI training collected

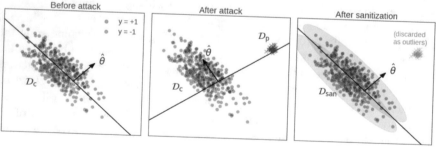

Fig. 4.4 Left: With pure data, the machine learning model can learn parameters $\hat{\theta}$ fitting the data D_c properly. The decision boundary learned through a linear SVM is shown on synthetic data. **Middle:** a small amount of poisoned data D_p is enough to significantly change the learned $\hat{\theta}$ and the decision boundary. **Right:** in this example, the poisoned data D_P can be easily differentiated from the true data. After ignoring D_p, the model can be properly trained, resulting in desired $\hat{\theta}$ and decision boundary (with permission from the authors of [19])

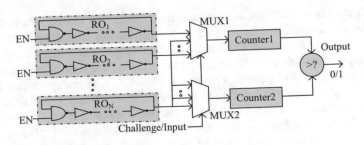

Fig. 4.5 Ring oscillator PUF

by ubiquitous devices in automatic ways. If we can trace down the data source and check the data integrity, the data generated by unauthorized devices will be excluded during the training. To enable the tracking, a reliable and unique device identity is necessary. One candidate technique is physical unclonable function (PUF) that provides unique and unpredictable features. As one type of popular hardware security primitive, PUF is widely implemented in various security applications, such as authentication and verification [35]. The schematic of a typical RO-PUF is shown in Fig. 4.5. It has N identical ring oscillators (ROs), two multiplexers, two counters, and one comparator. Each of the ROs run with slightly different frequencies due to manufacturing process variation. A challenge is applied to both MUXes so that one pair of ROs is selected in order to extract the frequency difference within the pair and generate a 1-bit response. The number of oscillations of each of the RO pairs is counted for a fixed time interval (known as comparison time) using the counters. The outputs of the counters are then compared using the comparator, and the comparator result is set to "0" or "1" based on which oscillator from the selected RO pair is faster. The random and unpredictable PUF features are from the uncontrollable variations of device parameters, such as the variations of transistor channel length and threshold voltages. PUFs typically work in challenges–response mode. Given different challenges, a PUF will provide responses accordingly. The challenge–response pairs (CRPs) for each PUF are unique (different from other PUFs) and unpredictable. To serve as device identity for data tracking, weak PUF designs that provide a relatively limited number of CRPs can be used. Although the weak PUFs offer less CRPs compared to strong PUFs, they usually consume less power. The number of possible CRPs can be considered as the key space. As long as the adversary is not allowed to running exhausting test, the key space provided by a weak PUF should be sufficient to serve for identity verification purpose. It is worth noting that machine learning assisted attacking on PUFs has been demonstrated years ago [17]. According to the reports, a small amount of CPRs learnt by the attacker will allow an efficient PUF modeling and result in a successful prediction of PUF behavior (i.e., leakage of most CPRs). Therefore, PUF designs that are resilient against modeling attack [24] should be implemented.

In practice, if the device used to collect data can communicate with the network, then the verifier can send challenges to the device, wait for the responses, and verify. If the verification passes, data sent from the device will be accepted. While if the device function is limited, without communication capability but only with data collection and sending, challenges need to be generated in an internal way. For example, the challenge can be generated based on the data, a time stamp, and the device public ID (e.g., serials number). The data, time stamp, device ID, and the PUF response are sent together to the server. If an adversary alters the data or other information, s/he won't be able to provide a consistent PUF response that matches the altered information as the PUF CPRs are kept secret (only shared with the authorized verifiers). Considering that internal challenge PUF design does not require communication circuit, it might be more friendly for lightweight affordable devices. At the same time, without communication circuit, there will not be a portal that adversaries can use to actively collect CPRs by sending challenges and observing the responses. Although it is still possible to obtain CRPs by reading the information used to generate challenges, conducting the challenges, and reading the responses, the CRP collection rate is limited in this way, making the modeling attacks take longer if possible.

4.2.2 Evasion Attack

Different from data poisoning, evasion attacks occur during the inference stage. In evasion attacks, the adversary's goal is to manipulate data samples such that they will be misclassified [20]. Such attacks have been reported or demonstrated on pattern recognition [16], spam detection [31], PDF malware detection [29], etc. One famous example is that a stop sign can be slightly changed when humans still see it as a stop sign for sure, while the trained pattern recognition model on an autonomous vehicle classifies it as a yield sign (shown in Fig. 4.6) [13]. To perform such attacks, the adversary needs to know at least one of the followings: the training data set or part of it; the feature used to represent sample raw data; the learning algorithms and the decision function; the parameters of the trained classifier model such as weights; and the inputs and feedbacks given by the classifier. Based on the knowledge about the machine learning system, the adversary further needs to modify the input data or the features related to the input data [6].

Currently, the popular research direction to prevent the evasion attack is developing approaches that can effectively verify the integrity of data. While the researchers are still looking for a satisfying algorithm for this purpose, PUF-based hardware solution is promising for application scenarios where data is generated or collected by devices equipped with PUF. When the data is obtained, hash function is performed first, to generate a nonce to be used as input challenges to PUF module. The response provided by the PUF then can be sent with the data together for data integrity check. If an adversary wants to alter the data and provide correct PUF response, s/he will need to break the PUF secrete CRPs first, which makes the attack much more difficult.

Fig. 4.6 The left image shows real graffiti on a stop sign, which is not appearing suspicious to most humans. The right image shows a physical perturbation applied to a stop sign for graffiti mimic to "hide in the human psyche." (with permission from the authors of [13])

4.2.3 Inversion Attack

Instead of trying to impact the machine learning systems, inversion attacks aim to steal valuable information from the system, as there are a large amount of subsistent data flowing in. The adversary manages to access the machine learning operation results (e.g., through API) and perform an inversion operation based on the knowledge about the machine learning system [14]. Different attacking models have been reported. For example, one group demonstrated membership inference attacks given data records and black-box access to one machine learning model [27]. In short words, the attacker trained an attack model by observing the inputs and outputs from the target model, to make predictions whether given data records were in the training data set for the target model. In another attacking research [30], the attack was performed on FPGA-based convolutional neural network accelerators, through power side-channel analysis. When the programmed FPGA was processing input data, power trace was collected by oscilloscope, went through low-pass filter, and aligned for further fitting and analysis, as shown in Fig. 4.7 Although the attack requires physical access to the accelerator hardware for power analysis and knowledge about the structure of the neural network, it could reconstruct the input images that might include sensitive information.

Such attacks cause wide concerns in privacy, as more and more machine learning systems are being deployed in fields like health care and biometrics-based authentications. The user information as inputs for such systems can be patient medical data or directly related to sensitive personal data. Although the inputs to the machine learning systems are usually protected with strict access control policies or strong encryption strategies because of privacy concerns, the inversion attacks extract secretes through the information obtained during the data processing carried out the machine learning systems while circumvent the access control or data encryption.

According to the techniques used during inversion attacks, data encryption, or more specifically, homomorphic data encryption [9], is one potential candidate, which enables calculations on encrypted data. Meanwhile, operating encrypted cipher texts and then decrypting them will give the same results obtained from

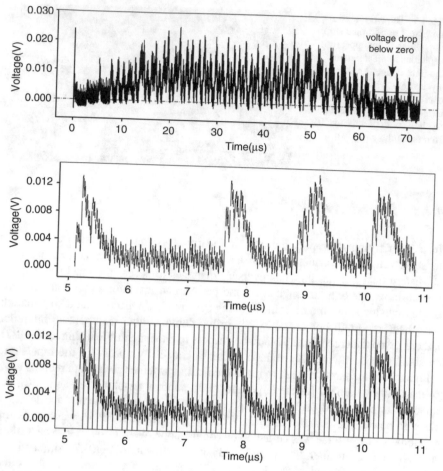

Fig. 4.7 Power extraction on traces collected from a FPGA-based machine learning system (with permission from the authors of [27])

directing operation of plain texts. For lightweight devices connected to machine learning systems, designs to accelerate homomorphic encryption in hardware [12, 23] have been proposed as well. For side-channel analysis assisted inversion attack, countermeasures developed by circuit security community can be employed. For example, masking design [28], which aims to minimize correlation between the power consumption and the processed information, can be used to impede the attack. It should be advised that hardware security designs often come with overheads in power consumption and chip area [36] and thus should be carefully optimized and balanced according to the specific requirements on performance and security [37].

4.3 System Level Hardware Solutions for AI Systems

4.3.1 Trusted Execution Environments

In the last two decades, cloud computation and data service have thrived across the market. Benefitting from this, many companies or institutions do not need to build their own servers but can easily obtain the support from cloud service providers at an affordable cost. Nowadays, a wide range of AI applications are running on such public clouds. Along with the convenience and efficiency, clouds and the complex software stack behind also rise opportunities for adversaries, as rigorous supervision on users becomes difficult for a system running on a distributed set of servers and the service is open to the public. One possible solution to handle this risk is building Trusted Execution Environment (TEE), which only allows trusted codes and data within to perform sensitive computation.

For example, ARM provides a TEE solution called *"TrustZone"* (shown in Fig. 4.8). The TrustZone technology provides the infrastructure foundations while allows system on chip (SoC) designers to choose different components that are most suitable to ensure the security for specific application scenarios. While the basic idea is consistent [2]: all of the hardware and the software on the system of chip (SoC) are partitioned into two worlds—normal world and secure world. Hardware logic present in TrustZone enabled bus fabric ensures that the normal world components are not allowed to access the resources in the secure world. To optimize the security, performance, and power consumption, system architecture is carefully designed in both hardware and software architectures. In terms of hardware architecture, the design covers buses, processors, and debug provision.

For buses, there is an extra control signal for each read and write channel on the main system bus that is known as the non-secure bit. Secure peripheral buses, such as interrupt controllers, timers, and user I/O devices, allow the extension of security environment. For processors, each physical processor core provides two

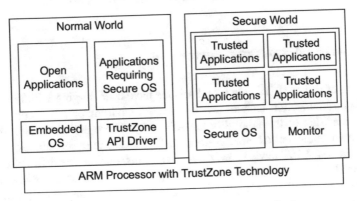

Fig. 4.8 Simplified schematic of ARM processor with TrustZone technology

virtual cores, such as non-secure and secure. The two virtual ones execute in a time-sliced mode, switching through monitor mode. To comply with the TrustZone design, the hardware also provides two virtual memory management units for non-secure and secure virtual processors, respectively, so each virtual processor has a local set of translation tables to independently control the address and address mapping. Debug is also important for developers, usually after the device is applied in the field. However, debug access can be used for some simple hardware attacks. To handle the risk, the TrustZone debug design separates the access control into independently configurable views of secure privilege invasive (JTAG) debug, secure privileged non-invasive (trace) debug, secure user invasive debug, and secure user non-invasive debug. This makes it possible to make normal world debug visible while screening the secure world debug.

In software, the TrustZone implements concurrent general operating system (e.g., Android, Linux, etc.) and security subsystem (e.g., customized Linux) in normal and secure worlds, respectively. To ensure the secure booting of both systems, every step during the booting can be cryptographically verified. The booting starts from read-only memory to initialize essential peripherals, followed by the booting of secure world. A public key from vendor can be used to initiate the booting. Unique keys for device parts can also be realized through one-time programmable (OTP) or PUF techniques.

Intel also introduced Software Guard Extension (SGX) technology that employs hardware security designs. The secured domains are called *secure enclaves*. "When an enclave is instantiated, the hardware provides protections (confidentiality and integrity) to its data, when it is maintained within the boundary of the enclave [1]." In this way, the sensitive data being processed inside SGX is protected even the operating system is compromised. To perform the attestation, the SGX architecture produces an assertion conveying the identities of the software environment to be attested, data that the software environment associated with itself, and a cryptographic binding to the platform trusted computing base (TCB, in charge of protecting the secrets, includes the processor's firmware, hardware, and the software inside the enclave). It also provides solutions for remote attestations.

4.3.2 IoT-AI Accelerator-Server

As internet of things (IoTs) and cloud service have been surging with the development of communication technology, layered AI environments become more and more popular due to the flexibility and diversity. There are three essential parts in such environments: AI accelerators, IoT devices, and servers, as shown in Fig. 4.9. Here, the IoT devices are on the edge, responsible for data collection by sensors and/or taking actions through actuators. Thanks to Moore's law, the cost of a system on chip has been greatly reduced and can be deployed on many IoT devices. Lightweight machine learning accelerator module, non-volatile memory (NVM), user interface, and input/output (I/O) portals are also optionally provided on the IoT devices. IoT devices are typically distributed. On a machine learning

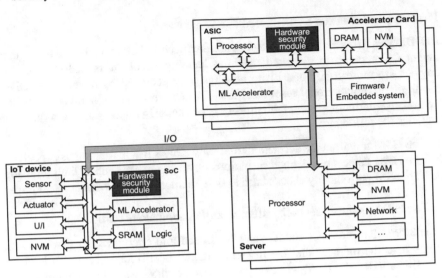

Fig. 4.9 Layered AI environments including three essential parts: IoTs, AI accelerators, and severs

accelerator card, there are an application specific integrated circuit (ASIC) chip including processor and accelerator and memories including DRAM and NVM, and embedded operation system or firmware is installed. Multiple accelerator cards can be clustered to perform intensive training tasks. While the server's architecture is following traditional computing system, consisting of a general purpose processor, memories, network components, and other units if needed, servers are used for general computing tasks and controls. According to the application scenarios, servers can be clustered as well.

In one white paper released by Rambus [26], hardware security modules (HSM, also named as Hardware Root of Trust) are suggested to be implemented in accelerators and IoT devices, as shown in Fig. 4.9. According to the attack models and security analysis, implementing hardware security modules in IoTs and accelerators is sufficient to protect the system. In addition, TEE techniques can always be deployed on the servers. To understand the security mechanism, let us describe the work flows in different use cases.

Accelerator Reset On the booting of accelerator, the HSM will be activated first to start a robust secure boot function. Once it is done, the HSM will ensure secure boots on other components. Operations are as below:

1. The HSM holds processor in reset.
2. The HSM boots itself securely.
3. The HSM verifies its signed hash of the boot code in NVM.
4. The HSM checks the hash in the boot image.
5. If the checking passes, the HSM releases the reset and processor continues to boot.

Firmware updating can be monitored in a similar way.

IoT Devices Monitoring. IoT devices are deployed in field, mostly in a distributed manner. Compared to clustered machine learning accelerators and servers, IoT devices are more easily and likely be physically accessed by adversaries. Given the limited computational resources, IoT devices are more vulnerable to attacks. Therefore, continuous system monitoring is particularly important. The monitoring includes the following:

1. The HSM monitors the accelerator and only allows if it is running as expected.
2. The HSM detects pre-defined suspicious physical attacks such as fault injection.
3. The HSM periodically checks hash known memory status to detect data tampering/altering.
4. The HSM monitors network traffic, reporting anomalous traffic.

In addition, the inference model shall be signed and encrypted, so that even the attack read the memory and the inference model is still kept secret. To load the inference model, the HSM will decrypt it, generate the hash value, and compare it with safely stored records.

Training Data Protection Training data is value asset and can be sensitive. Typical threats to training data include tampering and theft. To protect the training data, signing and authenticating are required before the data access can be granted. The protocols can be listed as follows:

1. When not in use, training data is stored on servers and encrypted. So even the servers are compromised, data breaching should not occur.
2. The encrypted data on servers is also signed by hash.
3. When the machine learning accelerators receive encrypted data from server, the data is hashed to verify the signature. If passes, the data is decrypted for training.

4.3.3 Summary

As AI is playing a more and more significant role in our lives, the security of AI systems is attracting attention from both academia and industry. Although tremendous efforts have been devoted, the game of offence and defence never ends. Compared to the research performed on algorithm and programming, study on hardware security of AI systems is in a serious lack. In hardware aspect, the AI systems are facing both general hardware threats and AI-specific risks. In this chapter, we started from several reported attacks aiming against AI systems and discussed countermeasures at different levels. In future, we are looking forward to seeing more systematic threat modeling and comprehensive solutions to be released.

References

1. I. Anati, S. Gueron, S. Johnson, V. Scarlata, Innovative technology for CPU based attestation and sealing, in *Proceedings of the 2nd International Workshop on Hardware and Architectural Support for Security and Privacy*, vol. 13 (ACM, New York, 2013), p. 7

2. ARM Limited, ARM security technology, building a secure system using Trust-Zone technology (2009). https://static.docs.arm.com/genc009492/c/PRD29\discretionary-GENC\discretionary-009492C_trustzone_security_whitepaper.pdf. Accessed July 30, 2020

3. N. Baracaldo, B. Chen, H. Ludwig, J.A. Safavi, Mitigating poisoning attacks on machine learning models: a data provenance based approach, in *Proceedings of the 10th ACM Workshop on Artificial Intelligence and Security* (2017), pp. 103–110

4. G. Batra, Z. Jacobson, S. Madhav, A. Queirolo, N. Santhanam. Artificial-intelligence hardware: new opportunities for semiconductor companies. McKinsey & Company Report (2019). https://www.mckinsey.com/industries/semiconductors/our-insights/artificial-intelligence-hardware-new-opportunities-for-semiconductor-companies#. Accessed 09 Jan. 2020

5. B. Biggio, B. Nelson, P. Laskov, Poisoning attacks against support vector machines (preprint, 2012). arXiv:1206.6389

6. B. Biggio, I. Corona, D. Maiorca, B. Nelson, N. Šrndić, P. Laskov, G. Giacinto, F. Roli, Evasion attacks against machine learning at test time, in *Joint European Conference on Machine Learning and Knowledge Discovery in Databases* (Springer, Berlin, 2013), pp. 387–402

7. B. Biggio, L. Didaci, G. Fumera, F. Roli, Poisoning attacks to compromise face templates, in *2013 International Conference on Biometrics (ICB)* (IEEE, Piscataway, 2013), pp. 1–7

8. B. Biggio, K. Rieck, D. Ariu, C. Wressnegger, I. Corona, G. Giacinto, F. Roli, Poisoning behavioral malware clustering, in *Proceedings of the 2014 Workshop on Artificial Intelligent and Security Workshop* (2014), pp. 27–36

9. I. Damgård, V. Pastro, N. Smart, S. Zakarias, Multiparty computation from somewhat homomorphic encryption, in *Annual Cryptology Conference* (Springer, Berlin, 2012), pp. 643–662

10. J. Dean, S. Ghemawat, MapReduce: simplified data processing on large clusters. Commun. ACM **51**(1), 107–113 (2008)

11. J. Dean, G. Corrado, R. Monga, K. Chen, M. Devin, M. Mao, M. Ranzato, A. Senior, P. Tucker, K. Yang, Q. Le, A.Y. Ng, Large scale distributed deep networks, in *NIPS'12* (2012). http://papers.nips.cc/paper/4687-large-scale-distributed-deep-networks.pdf J. Dean, G. Corrado, R. Monga, K. Chen, M. Devin, M. Mao, M. Ranzato, A. Senior, P. Tucker, K. Yang, Q. Le, A.Y. Ng, Large Scale Distributed Deep Networks. In NIPS '12. http://papers.nips.cc/paper/4687-large-scale-distributed-deep-networks.pdf

12. Y. Doröz, E. Öztürk, B. Sunar, Accelerating fully homomorphic encryption in hardware. IEEE Trans. Comput. **64**(6), 1509–1521 (2014)

13. K. Eykholt, I. Evtimov, E. Fernandes, B. Li, A. Rahmati, C. Xiao, A. Prakash, T. Kohno, D. Song, Robust physical-world attacks on deep learning visual classification, in *Proceedings of the IEEE Conference on Computer Vision and Pattern Recognition* (2018), pp. 1625–1634

14. M. Fredrikson, S. Jha, T. Ristenpart, Model inversion attacks that exploit confidence information and basic countermeasures, in *Proceedings of the 22nd ACM SIGSAC Conference on Computer and Communications Security* (2015), pp. 1322–1333

15. J.E. Gonzalez, Y. Low, H. Gu, D. Bickson, C. Guestrin, Powergraph: distributed graph-parallel computation on natural graphs, in *Presented as part of the 10th USENIX Symposium on Operating Systems Design and Implementation (OSDI'12)* (2012), pp. 17–30

16. I.J. Goodfellow, J. Shlens, C. Szegedy, Explaining and harnessing adversarial examples (preprint, 2014). arXiv:1412.6572

17. G. Hospodar, R. Maes, I. Verbauwhede, Machine learning attacks on 65nm Arbiter PUFs: accurate modeling poses strict bounds on usability, in *2012 IEEE International Workshop on Information Forensics and Security (WIFS)* (IEEE, Piscataway, 2012), pp. 37–42

18. N.P. Jouppi, C. Young, D.H. Yoon, et al., In-datacenter performance analysis of a tensor processing unit, in *Proceedings of the 44th Annual International Symposium on Computer Architecture (ISCA'17)* (ACM, New York, 2017), pp. 1–12. https://doi.org/10.1145/3079856. 3080246

19. P.W. Koh, J. Steinhardt, P. Liang, Stronger data poisoning attacks break data sanitization defenses (preprint, 2018). arXiv:1811.00741

20. P. Laskov, M. Kloft, A framework for quantitative security analysis of machine learning, in *Proceedings of the 2nd ACM Workshop on Security and Artificial Intelligence* (2009), pp. 1–4

21. C. Liu, B. Li, Y. Vorobeychik, A. Oprea, Robust linear regression against training data poisoning, in *Proceedings of the 10th ACM Workshop on Artificial Intelligence and Security* (2017), pp. 91–102

22. Z. Mengchen, B. An, W. Gao, T. Zhang, Efficient label contamination attacks against black-box learning models, in *(IJCAI)* (2017), pp. 3945–3951

23. A.C. Mert, E. Öztürk, E. Savaş, Design and implementation of encryption/decryption architectures for BFV homomorphic encryption scheme. IEEE Trans. Very Large Scale Integr. Syst. **28**(2), 353–362 (2019)

24. P.H. Nguyen, D.P. Sahoo, C. Jin, K. Mahmood, U. Rührmair, M. van Dijk, The interpose PUF: secure PUF design against state-of-the-art machine learning attacks. IACR Trans. Cryptograph. Hardware Embed. Syst. **4**, 243–290 (2019)

25. R. Raina, A. Madhavan, A.Y. Ng, Large-scale deep unsupervised learning using graphics processors, in *Proceedings of the 26th Annual International Conference on Machine Learning (ICML'09)* (ACM, New York, 2009), pp. 873–880. https://doi.org/10.1145/1553374.1553486

26. Rambus Inc., Hardware security for AI accelerators. https://go.rambus.com/hardware-security-for-ai-accelerators

27. R. Shokri, M. Stronati, C. Song, V. Shmatikov, Membership inference attacks against machine learning models, in *2017 IEEE Symposium on Security and Privacy (SP)* (IEEE, Piscataway, 2017), pp. 3–18

28. V. Sundaresan, S. Rammohan, R. Vemuri, Defense against side-channel power analysis attacks on microelectronic systems, in *2008 IEEE National Aerospace and Electronics Conference* (IEEE, Piscataway, 2008), pp. 144–150

29. N. Šrndic, P. Laskov, Detection of malicious PDF files based on hierarchical document structure, in *Proceedings of the 20th Annual Network & Distributed System Security Symposium* (Citeseer, 2013), pp. 1–16

30. L. Wei, B. Luo, Y. Li, Y. Liu, Q. Xu, I know what you see: power side-channel attack on convolutional neural network accelerators, in *Proceedings of the 34th Annual Computer Security Applications Conference* (2018), pp. 393–406

31. G.L. Wittel, S.F. Wu, On attacking statistical spam filters, in *First Conference on Email and Anti-Spam CEAS* (2004)

32. G. Xu, H. Li, H. Ren, K. Yang, R.H. Deng, Data security issues in deep learning: attacks, countermeasures, and opportunities. IEEE Commun. Mag. **57**(11), 116–122 (2019)

33. C. Yang, Q. Wu, H. Li, Y. Chen, Generative poisoning attack method against neural networks (preprint, 2017). arXiv:1703.01340

34. M. Zaharia, M. Chowdhury, T. Das, A. Dave, J. Ma, M. McCauly, M.J. Franklin, S. Shenker, I. Stoica, Resilient distributed datasets: a fault-tolerant abstraction for in-memory cluster computing, in *Presented as Part of the 9th USENIX Symposium on Networked Systems Design and Implementation (NSDI'12)* (2012), pp. 15–28

35. J.-L. Zhang, G. Qu, Y.-Q. Lv, Q. Zhou, A survey on silicon PUFs and recent advances in ring oscillator PUFs. J. Comput. Sci. Technol. **29**(4), 664–678 (2014)

36. Reference[60]

37. Reference[61]

Chapter 5
Machine Learning in Hardware Security

Shijin Duan, Zhengang Li, Yukui Luo, Mengshu Sun, Wenhao Wang, Xue (Shelley) Lin, and Xiaolin Xu

5.1 IP Protection

Hardware *intellectual property* (IP) is commonly used in modern semiconductor industry, which can be of different formats like logic, cell, block, or *integrated circuit* (IC) layout designed and owned by an IP vendor [1]. Due to the faster time-to-market demands and the increasing complexity of the modern *system on chip* (SoC), it has been challenging for a single manufacturing company or a design house to handle the entire *application-specific integrated circuit* (ASIC) design flow. Instead, the modern globalized semiconductor supply chain is composed of different stages like design, verification, validation, and fabrication. To facilitate the globalization, most design houses purchase IP blocks from third-party IP vendors for their design and then send out their design packages to a fabrication company to produce the chip.

5.1.1 Security Issues in Hardware IP and Countermeasures

Since the globalization trend and regional laws diminish the designer's controllability on the third-party IPs, the IP vendor and fabrication company may be located in different countries or regions. As a result, an untrusted participator can potentially exist in every stage of the supply chain. In this context, four major security issues that threat the hardware IPs have been identified: (1) **unauthorized reuse**, (2) **piracy**, (3) **reverse engineering**, and (4) **malicious modifications** or **hardware**

S. Duan · Z. Li · Y. Luo · M. Sun · W. Wang · X. (Shelley) Lin · X. Xu (✉)
Department of Electrical and Computer Engineering, Northeastern University, Boston, MA, USA
e-mail: xue.lin@northeastern.edu; x.xu@northeastern.edu

© The Author(s), under exclusive license to Springer Nature Switzerland AG 2021   111
M. Tehranipoor (ed.), *Emerging Topics in Hardware Security*,
https://doi.org/10.1007/978-3-030-64448-2_5

Trojan, as described in Sect. 5.2. To address the IP issues, *hardware obfuscation* techniques have been studied in the past decade [1].

5.1.2 Hardware Obfuscation

Hardware obfuscation is a method of modifying a circuit design, which systematically transforms the circuit functionality to prevent reverse engineering. Figure 5.1 illustrates the primary goals of hardware obfuscation: hiding the design intent to prevent the black-box usage. The taxonomy of the obfuscation technique can be described as below:

- **Register transfer-level (RTL) obfuscation**: RTL IP, known as **soft IP**, uses hardware description language (HDL) to describe an IP, such as Verilog or VHDL. Some popular obfuscation techniques for the soft IPs are IP encryption, RTL locking, and white-box obfuscation [2]. The basic idea of those methods is transferring the RTL design to an unintelligible and unreadable form.
- **Gate-level obfuscation**: A gate-level IP, known as **firm IP**, represents a design in a logic circuit diagram. The essential obfuscation technique in this IP level is logic locking. With systematic design approaches and practical tools, logic locking is becoming a promising method. Detailed introduction and discussion are given in Sect. 5.1.3.
- **Layout-level obfuscation**: A layout-level IP, referred as **hard IP**, comes in a form of a physical layout of a design. 2.5D/3D IC obfuscation technology can help the designer perform wire-lifting on a layout design to apply split manufacturing. Another similar obfuscation method in this level design is camouflaging, which allocates some configurable cells in the design module to disable the regular operation.

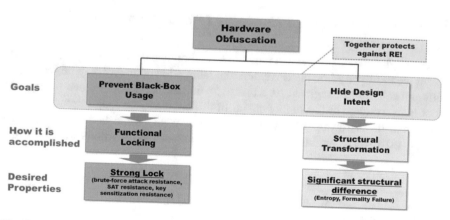

Fig. 5.1 Primary goals of hardware obfuscation and properties [1]

5.1.3 Logic Locking

There exist two types of logic locking: **combinational logic locking** and **sequential logic locking**. The combinational logic locking inserts some gates (XOR or NXOR, etc.) into the original circuit to hide the correct functionality and structure of a hardware IP.

In a sequential circuit, the finite-state machine (FSM) is commonly used to control the logic. The sequential logic locking aims at the state-space obfuscation by modifying the associated combinational logic. This modification can directly impact the state-transition function of sequential logic. Only if the user can provide the *primary inputs* (key), the circuit functionality will be unlocked and operate correctly; otherwise, it will stay in an obfuscation mode with an incorrect behavior.

Figure 5.2 illustrates the combinational logic locking method. The red logic gates in Fig. 5.2b are called key-gates. To lock a design with N-bit key, at least N key-gates are needed. In this example, the original inputs are A, B, C, and D, while K1 and K2 denote the key-inputs. The designated circuit functionality can only be unlocked while setting the correct key, which is $K1K2 = 01$ in this example. Here, we can summarize that the XOR key-gate represents a key-bit "0," and the NXOR key-gate represents a key-bit "1."

For the obfuscation FSM (OFSM) illustrated in Fig. 5.3, it will be forced to reset to the initial state (obfuscation mode) after the start-up. An N-bit "key" will be sent

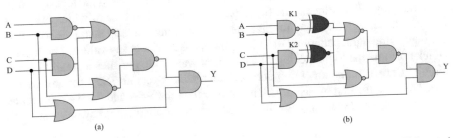

Fig. 5.2 The combinational logic locking study example. (**a**) The original circuit. (**b**) Obfuscated circuit with two key-gates

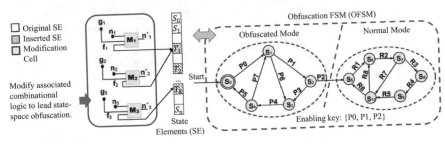

Fig. 5.3 An example circuit with sequential logic locking. In this example, we modify its state-transition function through associated combinational logic

in sequence, which controls the OFSM to go through the complete state-transition sequence. When the correct input sequence is provided, the normal OSFM mode will be activated. In Fig. 5.3, the normal mode enabling "key" is P0 $\Rightarrow$ P1 $\Rightarrow$ P2. Designers can determine the obfuscation state between P0 and P2 by modifying the associated combinational logic of FSM to a different state-transition function. Furthermore, adding flip-flop (FF) can insert state in the state elements, and only one primary input pattern should be designed to reach the normal mode of the OFSM.

5.1.4 Possible Attacks on Logic Locking and Countermeasures

There exist various specific attacks that target logic locking methods. These methods can be generally classified as (1) **functional attack** and (2) **structural attack**.

5.1.4.1 Conventional Functional Attacks

The conventional functional attacks include *key sensitizing attack* (KSA) [3], *Boolean satisfiability attack* (SAT-attack) [3, 4], and *bypass attack* [5]. All these attacks assume that the adversary can obtain an unlocked version of the circuits to be attacked.

The KSA and the SAT-attack analyze the functional behavior of the locked design to extract keys. In particular, the SAT-attack, the algorithm flow of which is shown in Fig. 5.4, can significantly minimize the key searching space by using the so-called distinguishing input patterns (DIPs) to rule out the incorrect key values. The bypass attack corrupts the logic locking in a different way, which creates a small "bypass circuit" to reverse incorrect outputs once an attacker gets all DIPs for any wrong key. The cost of using bypass attack is smaller than the KSA and the SAT-attack, as the latter will keep searching for the correct (or the approximate) key.

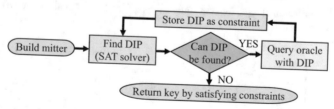

Fig. 5.4 The SAT-attack algorithm flow

5.1.4.2 Countermeasures Against the SAT-Attack

As the representative attack against logic locking, SAT-attack has threatened various existing logic locking techniques [3]. A straightforward method to mitigate SAT-attacks is using a large key search space, which thus makes it challenging for the SAT solver to find the right key. Furthermore, SAT-resistant techniques have been incorporated into the logic locking design, called anti-SAT blocks. The anti-SAT block takes inputs from the internal nodes of the original circuit along with a partial key value to activate the correct functionality of the circuit, which exponentially increases the key search space [1].

5.1.4.3 Threats from Machine Learning (ML)

With the exploration in the logic locking field, several SAT-resistant logic locking strategies have exponentially increased the difficulty of attacks. Researchers now start to consider the threat from various ML methods as those methods have excellent properties in statistical analysis, classification, and optimization. Those properties may enhance the SAT-solver ability or create ML-based SAT-solver to destroy the protection of SAT-resistant logic locking. ML-based logic locking attack is an active research field. In the next section, we will introduce the detail of ML-based logic locking and discuss possible countermeasures.

5.1.5 ML-Based Functional Attack on SAT-Resistant Logic Locking

It has been explored that the SAT-attack can be converted to a specific active learning model [6]. For the active learning, the main goal is to find a querying strategy that minimizes the number of queries required to learn the target function [7]. Among various strategies, the one called uncertainty sampling or query by disagreement (QBD) can describe the SAT-attack [8].

Following the definition in Sect. 5.1.4.1, the active learning (with QBD strategy)-based SAT-attack means that no more disagreements can be found in the strategy. In other words, the SAT-attack is precise, and it will not terminate unless the correct key is found. However, if we stop the active learning at a suitable step before the result converges to a "perfect" target function F_0, it can return an approximately (App) correct key. As a result, this type of attack is named AppSAT-attack.

Assuming that the learner queries a black box, the aim is to find F_0 within a hypothesis space F. In every iteration, the learner stores DIPs as additional constraints for the next iteration. If we obverse the query set L at the step i and the step j with $i < j$, when the corresponding version space V, a subspace of F, meets the condition $|V_j| < |V_i|$, it means that the V is converged to F_0 [7].

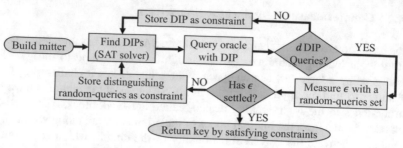

Fig. 5.5 The AppSAT-attack algorithm flow, which can return an approximate key in any early step

The question now becomes to find a suitable step to stop the querying that can achieve the best benefit from the performance and the functional accuracy. One way to formulate the approximate strategy is the probably approximately correct (PAC) learning [9]. It pre-defines an error-approximation (ϵ-approximation) to determine whether to stop or not. If ϵ is less than a certain threshold, we define it as "ϵ settled," then the algorithm will return the approximate key.

As illustrated in Fig. 5.5, the AppSAT generates another random query set after every d number of DIPs iterations to measure the ϵ. If the "ϵ settled" condition is not satisfied, this random query set will be stored as a constraint. This technique can be referred as the query reinforcement [7], which significantly improves the attack performance.

The AppSAT is a powerful attack that can significantly reduce the running time comparing with the conventional SAT-attack, even against the circuits that employ the SAT-resilient scheme, such as anti-SAT and SARLock. AppSAT has a systematic way to generate an approximated "wrong" key and a minimum size of DIPs, and they can associate with the bypass technique to create a minimum overhead bypass circuit to achieve the correct functionality.

5.1.6 ML-Based Structural Attack on SAT-Resistant Logic Locking

All functional attacks on logic locking can be regarded as a key retrieval attack, which focuses on activating the exact functionality of the designated circuit. Differently, the structural attack concentrates on (1) removing the anti-SAT structure [10] and (2) reversing the logic locking camouflage. One representative work of the second scheme is the so-called Structural Analysis uses Machine Learning (SAIL), which is a pure structural attack [11]. In this attack, the key-gates will be detected and removed after reversing the obfuscation structure, and thus an unlocked chip is not required to obtain the golden outputs.

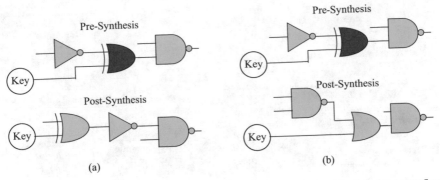

Fig. 5.6 An example of XOR-based logic locking resynthesis result. (**a**) The Level-2 camouflage example. (**b**) The Level-3 camouflage example

The circuit camouflage can be classified into three levels. Level-1: No change applies to the logic locking design, as Fig. 5.2b shows. Level-2: The insert key-gate does not change, and however, the neighboring structure is modified due to the logic simplification, as the example in Fig. 5.6a shows. Level-3: The basic key-gates and logic gates are transformed into a new structure to hide the key inputs. With the statistical analysis of the benchmark results [11], most key inserted localities' pre-synthesis and post-synthesis results do not have any structural change (Level-1). Only a few localities have Level-3 changes, which means that the insert key-gates have been obfuscated by the resynthesis function. We can conclude that changes induced by logic locking are local and limited.

Designs before and after logic locking resynthesis create the training set to train a ML model in order to reverse the logic locking circuit. However, fetching the original pre-obfuscation netlist is challengeable. A critical scheme called pseudo-self-referring in paper [11] can solve this problem. The solution is training the particular circuit, setting the obtained logic locking netlist as a pseudo-golden circuit, and then running the logic locking tool for multiple rounds to create a set of logic locking circuits. In this way, the machine learning method can capture the circuit's resynthesis rules and characteristics.

Two key models are trained with different variables, as shown in Fig. 5.7. The "Change Prediction Model" is trained with [Locality, Change Indicator Boolean], while the "Reconstruction Model" is trained with [Post-Synthesis Locality, Pre-Synthesis Locality]. The first model determines whether localities change or not, and the second model is used to predict the Pre-Synthesis Locality by giving a Post-Synthesis Locality. Multiple locality sizes (from 3 gates locality up to 10 gates locality) may change the efficiency of training. As reported in [11], the Random Forest method gives the best results for the "Change Prediction Model." The "Reconstruction Model" obtains the results from a multilayer neural network and applies standard cumulative confidence voting ensemble scheme to combine different locality size models into the final result. In the conventional resynthesis rule, the SAIL attack can recover around 84% (up to 95%) transformations

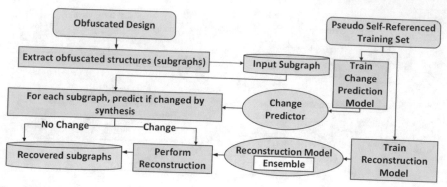

Fig. 5.7 Overview of the SAIL attack flow [11]

introduced by logic locking, in average. Consequently, the employment of ML method drives the logic locking research into a new era, and future studies may focus on the specific synthesis rule to increase the resistance against ML-oriented attacks.

5.1.7 Future Work Against ML-Based Attacks

The current countermeasures against possible logic locking attacks focus on the key retrieval (functional) attack. Many existing works have discussed where to insert key-gates and how to increase the key-guessing cost exponentially [12–16]. On the other side, ML-based attack schemes and algorithms are being proposed to either speed up the conventional SAT attacks [7] or reverse engineer the obfuscated circuit [11]. Some emerging obfuscation techniques may extend the logic locking study in the future, such as SAIL-resistant obfuscation in the resynthesis process.

5.2 ML-Assisted HW Trojan Detection

In modern semiconductor business, the design and manufacturing of chips are usually separated. Many large chip designers are only responsible for designing. They will hand over the designed chips to multilayer manufacturers for production. During the production, the circuits of the chips may be purposely implanted with hardware Trojans by the untrusted manufacturing plant or vendors. A hardware Trojan can possibly steal secret or even remotely control the circuit, which brings great challenges to the security of modern electronic systems. In this chapter, we will first introduce the basic concepts of hardware Trojans, and then we will review several machine learning (ML)-assisted methods to detect the potential hardware Trojans in the chip circuits.

5.2.1 Preliminaries of Hardware Trojan

A hardware Trojan is inserted to the original hardware design to bring malicious function changes [17]. Some Trojans, which are triggered under certain conditions, even hijack the original system to implement certain attacks [17]. Generally speaking, a hardware Trojan is mainly composed of two parts: the trigger unit and the payload unit [18]. Note that not all hardware Trojans contain the trigger units, based on this rule, the hardware Trojans can be roughly divided into two categories, unconditionally triggered and conditionally triggered [17].

Unconditionally triggered hardware Trojans are always working since it implanted into the target system (or circuit) [19]. Conditionally triggered hardware Trojans can be further divided into two types according to the sources trigger them: externally triggered and internally triggered [19]. Externally triggered hardware Trojans are activated by receiving external control signals from attackers, while internally triggered hardware Trojans are triggered by specific signal patterns or physical factors (like temperature, time, etc.) in the systems (or circuit) [18].

Once a hardware Trojan is triggered, its payload unit will operate in its predetermined functionality, i.e., to steal secret information from the systems. According to whether the behaviors of the payload units will change the system outputs or not, payload units can be roughly divided into two categories, explicit and implicit [20]. Explicit payload units often change the logic of the original circuits to introduce errors to the system outputs. Implicit payload units do not directly change the original circuit logic but occupy some resources of the original system (such as storage, calculation units, or battery resource) to cause errors. Besides, implicit payload units could also be used to leakage of side-channel information of the circuits, which could be used for malicious purposes [17].

5.2.2 Overview of the Hardware Trojan Detection Based on Machine Learning (ML) Algorithms

There exist many different types of hardware Trojan detection methods. Most of them are not ML assisted in the old days. Adding the testing circuits in advance, i.e., during the design stage, is one of the conventional methods for Trojan detection in the past time. For example, Li et al. [21] used comparators and registers on each critical data path to detect very small time gaps caused by the hardware Trojan. However, the hardware overhead for this method is extremely high, which makes it less efficient.

With the rapid development of ML algorithms and neural networks, combining the hardware Trojan detection schemes with ML algorithms and neural networks is becoming a hot topic [22]. As a result, ML-assisted hardware Trojan detection methods have much lower costs on time and resource consumption, which makes them a promising method for Trojan detection [22]. Today, plenty of ML-assisted hardware Trojan detection methods have been developed. According to whether the ML-assisted hardware Trojan detection methods will cause damage to the chip, they can be divided into two categories, non-destructive and destructive [23, 24].

There are two main types of non-destructive methods: logic test and side-channel analysis [19]. The logic test is to use the input and output (I/O) behavior of an IC to detect whether there are hardware Trojans. This is one of the most straightforward testing among all hardware Trojan detection schemes. However, its accuracy is limited in detecting the Trojans with hidden implicit payload units. The usage of ML algorithms greatly overcomes this problem. For example, multilayer neural networks could help the detection schemes in analyzing the inner logic of smaller units of the circuits, even at gate level [25]. Side-channel analysis, as the other non-destructive method for Trojan detection, checks the side-channel parameters such as power, delay, transient and quiescent current, and maximum frequency [26]. These parameters will be changed due to the existence or triggering of hardware Trojans. However, some newly developed hardware Trojans with more hidden working physical parameters are more challenging to detect. Thus, using the binary classification ML algorithms could help to find out the very small changes in these physical parameters to make the detection more accurate [26].

One of the most common destructive methods is using image recognition to compare the detailed structure on the scanning electron microscope (SEM) images of the chips. Comparing with non-destructive methods, the destructive methods will have a higher test accuracy. However, the destructive methods are complicated, time-consuming, and will cause irreversible damage to circuits and chips. Thus, using the multi-class classification ML algorithms would greatly help to save time in the gate-level image recognition, thus reducing the overall costs [27].

5.2.3 Case Study of ML-Assisted Hardware Trojan Detection

In this part, we will discuss some ML-assisted hardware Trojan detection schemes to show how they work in detail.

5.2.3.1 ML-Assisted Logic Test

Analyzing the logic of the circuits is a very efficient way to check whether there is hardware Trojan or not. Although the implantation of the hardware Trojan may not always change the overall logic of the entire circuits, it inevitably changes some of the logic of the non-critical data paths. Using one or more layer neural network to analyze the whole data paths of the circuits could help to detect hardware Trojan.

Hasegawa et al. [25] use multilayer neural networks to analyze the gate-level logic for the whole system to detect the potential hardware Trojans. By selecting a specific node in the circuit, they divided the testing circuit into multiple functional areas according to the distance from the selected node. The number of area

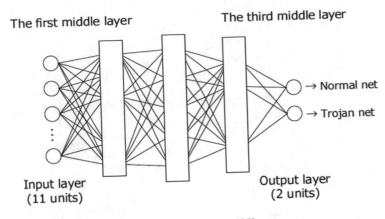

Fig. 5.8 The overall structure of 3-layer neural network [25]

divisions could be adjusted according to the complexity of the circuits. As shown in Figure 5.8, multiple inputs are used to test each functional area and generate area scores. The final score of the whole testing circuit will be a weighted calculation of these area scores, and the specific weights come from the benchmarks used for training a 3-layer neural network. There will be threshold value to the score of the testing circuits to help classify them into two categories, normal net or Trojan net. As a result of this experiment, a detection correct rate of 100% at most and 89% on average has been achieved.

This kind of tests can often achieve acceptable test accuracy with relatively lower time and resources consumption. Increasing the number of layers of the neural networks may help to increase the average testing correct rate but will also lead to more time and resources usage. Thus, the number of layers used to do the test will be a trade-off between accuracy and efficiency.

5.2.3.2 ML-Assisted Side-Channel Analysis

When hardware Trojans are working, they will either occupy some resources of the original circuits, such as RAM and register, or bring greater time delay and larger power consumption, which implies that the physical characteristics of the original circuits will be changed. Therefore, recording the real-time physical characteristics of the testing circuits could help to find out the abnormal fluctuations at runtime. Then, using ML algorithm to analyze these abnormal fluctuations can help to increase the accuracy of hardware Trojan detection.

Iwase et al. [26] use support vector machine (SVM) to analyze the runtime power to detect the potential hardware Trojans. As shown in Fig. 5.9, they first collect power consumption data from different working systems when the Trojans

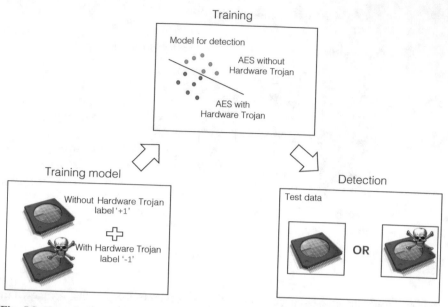

Fig. 5.9 The overall structure of SVM-assisted power consumption test

are triggered or not triggered. Then, the collected power consumption data will be converted from time domain into frequency domain using discrete Fourier transform (DFT). These converted data will be used to train an SVM model for binary classification (with Trojan or Trojan-free). The trained model can be used for other test samples. The experimental results demonstrate that all 12 kinds of hardware Trojans implanted in the Advanced Encryption Standard (AES) are detected.

This kind of tests can often achieve a higher accuracy than logic test, while at the cost of more complicated implementation. Usually, the costs of this kind of tests are in the acceptable range of the industry, which make ML-assisted side-channel analysis a promising method for Trojan detection [28].

5.2.3.3 ML-Assisted Cross-Sectional Photography

The most intuitive manifestation of a circuit being implanted with hardware Trojans is that there exist unusual circuits in the original design. Therefore, using the scanning electron microscope (SEM) to photograph the testing circuits and their corresponding standard circuits could help to find the abnormal layout to detect hardware Trojans. However, the abnormal layout in the SEM photographs may be caused by not only the hardware Trojans but also the process variation from the manufacturing [29]. This may lead to the misjudging of the hardware Trojans. Therefore, to avoid this kind of misjudging, using the ML-assisted image recognition algorithm is necessary.

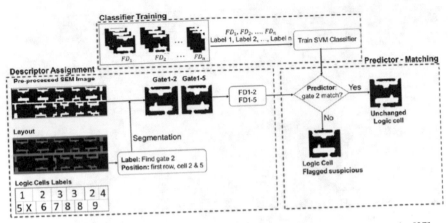

Fig. 5.10 The overall structure of multi-class SVM-assisted cross-sectional photography [27]

Vashistha et al. [27] use multi-class SVM to analyze the cross-sectional photog-raphy of the testing circuits to detect the potential hardware Trojans. As shown in Fig. 5.10, the package die of the standard chips and the testing chips will be removed to do cross-sectional photography to get the SEM images of the whole chips.

Then, these SEM images will be preprocessed into binary data, which could be used in the multi-class SVM algorithm. The standard chips will be used to train the SVM model with different classes of logic gates. Then, the testing chips will be tested by the SVM model to verify whether they are the logic gates they should be in the original design. The experimental results demonstrate that a correct detection rate of over 95% among millions of logic gate of 80 different ICs has been achieved [27].

The ML-assisted cross-sectional photography is currently outperforming when compared with other hardware Trojan detection methods, i.e., with the highest detection accuracy. However, the implementation of this method is relatively more expensive, which requires to completely remove the package of the chips and to use the SEM. Besides, it usually takes several hours to analyze a single chip [27]. All these factors make this method an inefficient way to be used in practice.

5.2.4 Summary

At present, most conventional ML-assisted detection methods are not effective enough to detect multiple hardware Trojans with diverse functions and implanting flexibility [19]. Using more complex ML algorithms or neural networks with more layers to improve the analysis ability of detection has become a promising direction [18]. Besides, using multiple single tests like logic test and side-channel analysis for joint detection is also adopted by many researchers. As shown in

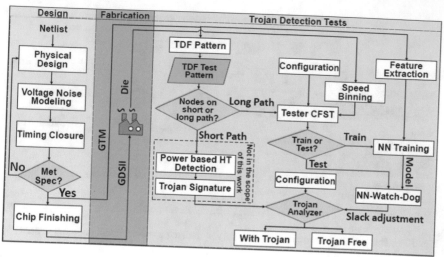

Fig. 5.11 The overall structure of multi-factor ML-assisted detection scheme [30]

Fig. 5.11, Vakil et al. [30] come up with a comprehensive detection scheme that tests the logic, time delay, and power consumption at the same time. However, they still could not reduce the costs of time and resources into a reasonable range for the industry [30]. Therefore, how to balance the costs and efficiency of currently existed or future hardware Trojan detection methods should also be further considered. Generally speaking, the ML-assisted hardware Trojan detection approaches have shown significant potential and thus are worth further investigation.

5.3 ML-Assisted Side-Channel Analysis

Side-channel attack (SCA) aims at extracting the secret of a computing system through indirectly retrieving information leaking from the physical implementation rather than analyzing the cryptographic algorithm itself, as depicted in Fig. 5.12. The leaked information such as power [31–34], memory access [35], and execution time [35] constructs a *trace* as the input of the SCA approach.

The most powerful type of SCA is profiled SCA that considers the worst-case security analysis. Profiled SCA includes a profiling phase before the attacking phase to extract available information, including the input and secret key, of an additional device similar to the target one. Template attack [36] is the most common profiled SCA and is considered as the most powerful SCA in theory. However, this method is restricted by assumptions about probability density distributions, and the accuracy drops drastically when the number of features exceeds that of observed traces due to the ill-conditioning of the covariance matrix [32].

Fig. 5.12 Side-channel
attack

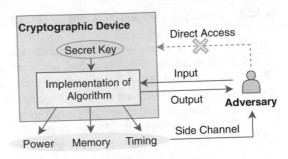

Machine learning techniques are able to deal with higher dimensional features and extend the applicable range of SCA. Even in unrestricted scenarios where there are sufficient traces to build precise leakage models for template attack, machine learning can still perform better [37].

5.3.1 Profiled SCA

Besides, in traditional methods like template attack, profiling is also utilized in machine learning approaches. In SCA, the mapping y maps the plaintext or the ciphertext t and the key k^* to a value related to the deterministic part of the measured leakage x. A profiled attack is conducted in two phases as follows [37]:

(1) *Profiling phase*: Provided N traces $\mathbf{x}_{p_1}, \ldots, \mathbf{x}_{p_N}$, plaintexts/ciphertext $t_{p_1}, \ldots, t_{p_N}$, and the secret key k_p^*, the adversary can calculate $y(t_{p_1}, k_p^*), \ldots, y(t_{p_N}, k_p^*)$.
(2) *Attacking phase*: Given Q traces $\mathbf{x}_{a_1}, \ldots, \mathbf{x}_{a_Q}$, (independent from the profiling traces) and plaintexts/ciphertext $t_{a_1}, \ldots, t_{a_Q}$, the goal is to make predictions about $y(t_{a_1}, k_a^*), \ldots, y(t_{a_N}, k_a^*)$, where k_a^* is the secret key on the target device.

5.3.2 Feature Selection in SCA

The raw input data is often numerous and thus preprocessed before SCA to extract the most relevant information and thus mitigating the computation burden and confusion for the classifiers. This section introduces several data preprocessing techniques based on feature selection and dimensionality reduction.

5.3.2.1 Principal Component Analysis (PCA)

Principal component analysis (PCA) projects the input features into a set of new uncorrelated variables called principal components that are linear combinations of the original variables with the maximum variance. In SCA, the traces can be projected with the first several principal components. PCA demonstrates that traces are less separable when they are connected to keys with different values for lower bits, indicating that the lower bits are more difficult to predict [32]. Following an enhanced brute-force process, the attack strategy updates the prediction by inverting the value of the most difficult bit, i.e., the first bit of the key, then flipping the second bit, and so on, until the whole key is correctly predicted.

5.3.2.2 Kullback–Leibler (KL) Divergence

Kullback–Leibler (KL) divergence [38] measures the difference between two probability distributions and is employed in SCA to extract distinct and non-varying features. When applied at instruction level [33], two types of KL divergence are evaluated with transform coefficients that map power traces into time–frequency region. One is between-class KL divergence D_{KL}^B and the other within-class KL divergence D_{KL}^W between different programs in the same instruction class. A local maxima D_{KL}^B value indicates distinct feature points, and D_{KL}^W value less than a threshold means that the features are not varying in the non-stationary environment.

5.3.2.3 Other Approaches

In addition to the abovementioned approaches, Pearson's correlation coefficient can also be adopted to select from a trace the points that have the highest correlation with the key. In classification, components in each trace with the largest Pearson's correlation coefficients are first selected [31].

5.3.3 Classification for SCA

SCA has been widely studied based on Hamming weight (HW) classes with 8-bit intermediate variable resulting in 9 binomially distributed classes. This section provides introduction of classifiers mostly based on HW for SCA on the basis of preprocessed data.

5.3.3.1 Support Vector Machine (SVM)

Support Vector Machine (SVM) represents a family of classifiers based on kernel functions that transform linearly inseparable data to linearly separable data in a higher dimensional space [39]. Common-used kernels include the linear kernel and the radial basis function (RBF) kernel. SVM is capable for a binary classification, and problems with multiple classes can be split into sub-problems with two classes that are handled with one-vs-one strategy. In situations with low noise and 9 HW classes, SVM enjoys higher accuracy than other methods like Random Forest and Rotation Forest [37]. SVM with RBF kernel also outperforms linear discriminant analysis (LDA), quadratic discriminant analysis (QDA), and Naive Bayes in instruction disassembly [33].

A specific form of SVM called Least Squares Support Vector Machine (LS-SVM) reformulates the classifier with a least squares loss function and equality constraints to reduce the computational complexity. LS-SVM with RBF kernel is demonstrated to be more suitable than that with linear kernel for nonlinear problems like deciding whether the Hamming weights are even or odd [31].

5.3.3.2 Random Forest and Rotation Forest

Random Forest is an ensemble of decision trees using random selection of features at each node and voting for the most popular class [40]. It can be incorporated with PCA to enhance the accuracy [32].

Rotation Forest splits the decision trees into subsets on each of which PCA is performed and the coefficients compose a rotation matrix [41]. It performs slightly worse than SVM, but more stable under various noise levels and numbers of classes [37].

5.3.3.3 Other Approaches

Naive Bayes as a simple classifier might outperform SVM and Template Attack in the 256 class case. It relies on an assumption that the features are mutually independent given the target class and is extremely fast to be a first indicator of the expected results from machine learning methods for SCA [42].

Deep learning models like multilayer perceptron (MLP) and Convolutional Neural Networks (CNNs), as a recent widely employed type of machine learning methods, can also be improved with SMOTE technique but suffer from small number of features [34]. MLP consists of fully connected layers, while CNN is mostly composed of convolutional layers, and both models need a large number of features for good performance.

5.3.4 SCA Against Machine Learning Models

In addition to acting as the method of SCA, a machine learning model itself can be attacked. For CNNs, the feature maps and parameters are numerous and thus stored in off-chip memory on the hardware accelerator. Treating the memory trace as the side channel, SCA against a CNN model aims at composing a CNN duplicate with similar accuracy, which is a form of reverse engineering [35]. Specifically, the overall structure is reflected from the read-after-write (RAW) dependency, and the size of parameters and input/output feature maps in each layer can be inferred from the access modes of read/write operations. When zero pruning techniques are applied to feature maps, the number of zeros in feature maps can further assist to infer the values of weight parameters.

5.3.5 Hierarchical Attack and Solution to Imbalanced Classes

Machine learning techniques were introduced into SCA firstly for binary classification and then extended to 9 Hamming weight (HW) classes. Nevertheless, HW may not apply when the observed leakage of traces is nonlinear. A better approach is considering 2^n uniformly distributed classes instead of $n+1$ binomial classes where n is the number of bits of the intermediate state, making each class corresponding to one key guess and therefore increasing the information gain. Actually, the accuracy of 16 uniform classes is higher than that of 9 binomial classes, even though the latter scenario has a higher signal-to-noise ratio (SNR) and correlation [37]. The study of 16 classes for a 4-bit intermediate variable is manageable, but an 8-bit word will generate 256 classes and rises the computational complexity significantly. This could be solved by a divide-and-conquer approach [42], where the data is first split into 9 HW classes and then dividing into all the leaf classes. As shown in Fig. 5.13, HW follows a binomial distribution, and the number of leaves in each HW class is different. For HW $= 0$ and HW $= 8$, there is only one leaf, while the number of leaves can reach 70 for HW $= 4$. Combining the standard attack with hierarchical attack makes a structured attack that will further result in higher accuracy.

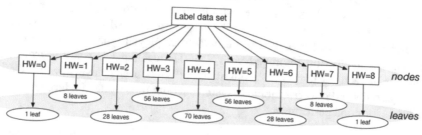

Fig. 5.13 Hierarchical classification with HW classes as intermediate nodes (Adapted from [42])

The data imbalance issue in HW classes could be addressed by incorporating data resampling techniques into machine learning classifiers. One oversampling method called Synthetic Minority Oversampling Technique (SMOTE) generates synthetic minority class instances to help with SCA metrics like success rate and guessing entropy [34].

5.4 ML-Based Attacks Against HW Security Primitives

As hardware devices permeate our daily life with an emerging trend, the security issue is inevitably getting cautions, so that users can process private information without worries about eavesdropping by adversaries or malicious applications. The hardware security proposed various strategies to ensure the secret isolated from malicious access. Besides ingenious algorithms and architectures are developed to provide security, primitives are also another significant branch to make users' information secretive. For example, physical unclonable functions (PUFs) are a fundamental primitive commonly used in producing unique ID (i.e., digital fingerprint) for a certain semiconductor device. We will take as the main example the PUFs, in this section, to show how hardware security primitives can interact with machine learning techniques.

5.4.1 PUF Overview

Physical unclonable functions (PUFs) are proposed for high-security applications, in which people must create or store confidential information keeping away from malicious eavesdropping, like creating or storing sensitive information in a smartcard. They are triggered by an input sequence (a.k.a. challenge) and will give a corresponding output (a.k.a. response) after through complicated microstructure interactions. Formally, this pair of challenge and response is called Challenge–Response Pair (CRP). PUFs are hard to be predicted because those microscopic variations mainly depend on various random physical factors when manufacturing, which makes every device with a unique PUF. However, since the physical characteristics of a circuit (e.g., propagation delay) could be affected by the fluctuation of environmental factors like temperature, PUFs might not be 100% stable or repeatable when used under different circumstances (i.e., the same challenge input might lead to different responses from the pre-measured ones), although the error rate is tolerated at a significantly low level. Many types of PUFs have been raised and classified since its principle was established [43]. This section focuses on two most famous PUFs, namely Arbiter PUF and Ring Oscillator PUF (RO PUF), as examples to briefly show how PUFs work. Moreover, these two PUF schemes will be used to introduce how machine learning algorithms can be used to emulate the PUFs' behavior and thus to predict CRPs with a high probability. We refer the interested readers to [44] for other PUF schemes.

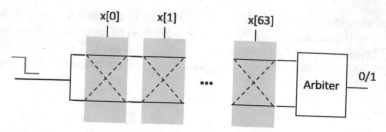

Fig. 5.14 The structure of a 64-bit Arbiter-based PUF

5.4.1.1 Arbiter-Based PUF

Arbiter PUF is a silicon PUF based on multiplexers and arbiters. The schematic of an Arbiter PUF is shown in Fig. 5.14. The Arbiter PUF circuit takes a certain-bit challenge (64 bits in this case) as the input and will output 1-bit response by comparing the delay difference between the two signal paths with the same layout. For the i-th MUX block, $MUX[i]$, the input $x[i]$ will decide whether the signals coming from $MUX[i-1]$ go along the solid lines or the dashed lines, so that the 64-bit challenge will allocate two symmetric signal paths for the input pulse signal. Finally, an arbiter (e.g., latch) will output the response according to the order of falling edge comings; for example, logic 1 will be the output if falling edge comes to the upper port first, and logic 0 will be the output otherwise.

Each component of these two paths will have slightly different propagation delays, since every MUX delay difference is randomly upon the device manufacturing and is thus unpredictable. As a result, the response of one input challenge is unpredictable and unknown in advance, but deterministic (i.e., a pair of the response and challenge (CRP) is expected to be fixed for a certain PUF instance). This kind of PUFs is formally classified as Strong PUFs [43], which have an exponential number of CRPs. The Strong PUFs can be attacked by Machine Learning (ML), because of two facts: (1) the responses of PUFs are only 0 and 1, which makes it easy to turn into a classification problem and (2) strong PUF circuits can be modeled easily, since the components (propagation delays) are additive.

5.4.1.2 Ring Oscillator PUF

Different from the Arbiter PUF, Ring Oscillator PUF (RO PUF) does not use symmetric path delay difference to decide responses; instead, a RO PUF leverages the slightly frequency differences between a group of ROs. The structure of a RO PUF is shown in Fig. 5.15. In this circuit, n ROs with same structures are placed in parallel. Considering the process variations from the manufacturing procedure, these ROs are with slightly different inherent frequencies. The input (challenge) bits will select two ROs and compare their frequencies using two counters within

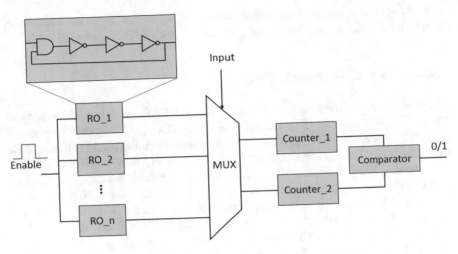

Fig. 5.15 The structure of a Ring Oscillator PUF

a certain period (controlled by an Enable signal). Finally, a Comparator (similar to the Arbiter) is utilized to decide the output according to the frequency difference.

An obvious advantage of RO PUF, compared to Arbiter PUF, is no need for symmetric delay paths, which will make the circuit construction more flexible, i.e., be implemented with reconfigurable devices like FPGA [45]. However, since the bits of MUX input are small (only $\lceil log_2 n \rceil$), the available number of CRPs is not as many as that from an Arbiter PUF. Hence, RO PUFs are formally classified as a Weak PUF, which is usually utilized as primitives in the cases where not that many CRPs are needed.

5.4.2   Machine Learning Attacks on PUFs

The most intuitive attacking method against a PUF is to record all its challenge–response pairs (CRPs) and then mimics its behavior for authentication or verification purpose. Consider that even a small 32-bit Arbiter PUF indeed has 4.3×10^9 CRPs, which might be feasible to measure and store in a memory. However, a 64-bit Arbiter PUF will have more than 1.84×10^{19} CRPs, which is impractical to enumerate and record within an acceptable time duration and storage, needless to say a larger PUF (like 128-bit). Considering this issue, modeling the behavior of a PUF instead of storing all its CRPs becomes a promising solution. In more detail, a successful PUF model can predict the responses for any given challenges with a high prediction rate. For the modeling attacks with machine learning, the common assumptions are that one attacker (called "adversary") can make a way to collect some CRPs of a certain PUF instance, with which he/she can derive (or "train") a numerical model with these CRPs to predict other CPRs of the PUF [46]. Still, like previous sections, we will take several examples to show how this attacking works. There are numbers of literature introducing machine learning attacks for various PUFs, like [47–49].

5.4.2.1 Machine Learning Methods

Many common ML methods can be used in training CRPs models, such as Support Vector Machine (SVM) [50], Neural Networks (NN) [51], Logistic Regression (LR) [52], Evolution Strategies (ES) [53], Sequence Learning (SL) [54], etc. These methods are all widely used, and the readers can easily find bunches of detailed publishments analyzing them. For brevity, we simply exhibit the ES method, which is relatively new in such ML modeling attacks, to explain how a PUF is attacked with ML algorithms. Then, we will introduce how those ML methods perform in attacking PUFs, using two basic Arbiter-based PUFs, namely standard PUFs and Feed-Forward (FF) PUFs as examples, which both belong to Strong PUFs. As for Weak PUFs (e.g., RO PUFs), they usually have very few and fixed CRPs or have surrounding control logic; in general, attackers cannot easily access many available CRPs from external ports, so they are not suitable for ML attacks. Note that, for readers interested, in addition to ML-based attacks, other non-ML attacking methods against PUF are also mentioned, such as Quick Sort for RO PUFs in paper [47].

Evolution Strategies (ESs) [55] are inspired by the natural evolution of individual population regarding to the environment. In the PUF attacking scenario, an individual of the population is the specific component propagation delays of an individual PUF instance. The objective of the evolution modeling is to optimize the individuals (i.e., the propagation delays) to get close to the real PUF behavior as good as possible. For a single evolutionary cycle (called a generation), ES will select the best individual(s), which can predict CRPs best, among several individual candidates, and make the evolution. After some generations, the predicted challenge–response behavior will gradually approach the target PUF's performance. So, finally, when the modeling procedure stops at an appropriate threshold, compared to its predecessors, the newest generation will more approximate the target PUF. ES is a randomized method, without needs for "linear separable" (like SVM) or "differentiable" (like LR), so it is well suited for PUF circuits, which are easy to be parameterized, since they are composed by simple operations, such as addition and multiplication.

5.4.2.2 Attacks on Standard PUFs

Standard Arbiter PUF is exactly the structure we discussed in Fig. 5.14. Its CRP generating procedure can be mimicked with an additive linear delay model $t = sgn(\mathbf{w}^T \boldsymbol{\phi})$. t represents the output (response) of a Standard Arbiter PUF, and $sgn(\cdot)$ stands for the $sign$ function, i.e., acting as the Arbiter. The parameter vector $\mathbf{w}$ encodes the delays for the sub-components in the delay chain, and the feature vector $\boldsymbol{\phi}$ represents the challenge. Since the standard Arbiter PUFs are both linear and

Table 5.1 LR on Arbiter PUFs with 64 and 128 stages (i.e., with bit length of 64 and 128), for noise-free, simulated CRPs [47]

ML method	Bit length	Prediction rate	CRPs	Training time
LR	64	95%	640	0.01 s
		99%	2555	0.13 s
		99.9%	18,050	0.60 s
LR	128	95%	1350	0.06 s
		99%	5570	0.51 s
		99.9%	39,200	2.10 s

differentiable, the paper [47] utilized different ML methods including SVM, LR, and ES to evaluate the performance of ML modeling attacks.

The result showed that LR behaved the best[47]. Table 5.1 shows the LR algorithm performance, and the prediction rates are estimated with 10,000 CRP test set. The first prediction rate target, 95%, is roughly the environment stability of a 64-bit Arbiter PUF when exposed to a temperature variation of 45°C and voltage variation of ±2% (as aforementioned, PUFs cannot run with a 100% stability). The other two prediction rates, 99% and 99.9%, are usually the criteria of optimized ML results. We can briefly analyze this result to see how LR method performs. For a certain prediction rate, when the Arbiter PUF bit length increases (i.e., from 64 to 128), the needed CRPs for training are also roughly doubled. For a certain bit length PUF, the needed CRPs and training time will significantly increase as a higher prediction rate requires many more to achieve. For example, the tiny increase from 99% to 99.9% will consume more than 7 times CRPs to achieve, which is a typical characteristic of ML algorithms (usually, the needed training set will greatly increase when the success rate gets closer to 100%).

5.4.2.3 Attacks on Feed-Forward PUFs

Feed Forward (FF) Arbiter PUFs are a variant of the Arbiter PUFs with some more complicated structures. For example, the signal selection path (e.g., 1 or 0) of some MUXs in the FF Arbiter PUFs is not directly decided by the input challenge bits, but as a function output of delay differences in the earlier part of the circuit. In more detail, some arbiters are used to complete the delay difference functions, and their outputs are fed into MUXs in a feed-forward loop. The paper [47] utilized SVM, LR, and ES for attacking FF Arbiter PUFs. However, SVM and LR were unable to achieve satisfactory results, while ES finally tackled complex FF PUF architectures with up to 8 FF loops. This is because FF PUFs contain nonlinearities and the gradient is also not easy to calculate, which makes SVM and LR behave worse.

We omit the experiment details and show the final results in Table 5.2. We can simply see some differences between this result and the one from the Standard PUFs. Unlike Standard PUFs, the needed CRPs for attacking a FF PUF do not

Table 5.2 ES on Feed-Forward Arbiter PUFs for noise-free, simulated CRPs. Prediction rates are for the best of a total of 40 trials on a single, randomly chosen PUF instance. Training times are for a single trial [47]

Bit length	FF loops	Prediction rate	CRPs	Training time
64	6	97.72%	50,000	07:51 min
	7	99.38%	50,000	47:07 min
	8	99.50%	50,000	47:07 min
	9	98.86%	50,000	47:07 min
	10	97.86%	50,000	47:07 min
128	6	99.11%	50,000	3:15 h
	7	97.43%	50,000	3:15 h
	8	98.97%	50,000	3:15 h
	9	98.78%	50,000	3:15 h
	10	97.31%	50,000	3:15 h

increase evidently to achieve a certain prediction rate while the bit length increases. However, the training time indeed increases significantly by almost 4 times. When the FF loops are too many, like 10 loops, the prediction rate will slightly decrease, but the training time will not get affected by the use of more FF loops. All these differences are due to the different PUF structures, as well as the applicable ML algorithms (e.g., LR can perfectly deal with Standard PUFs while behave poorly in FF PUFs).

5.4.2.4 Walk into Reality

Until now, our discussions mainly concentrate on the theory and experiments. It will be better if we could put some realistic examples (e.g., ML attacks on commercial hardware platforms) in the analysis. In 2015, Becker et al. successfully applied the first ML attacks on a commercial strong PUF-based RFID tag, which demonstrates the gap between the promised and achieved security with strong PUFs in practice [56]. The main feature of the PUF-based RFID tags is that each tag can be authenticated based on the built-in 4-way XOR Arbiter PUF. As shown in their results, the employed 4-way PUF can be tackled using ML algorithms within the measurement time of 200 ms; furthermore, an RFID smartcard emulator with hardware costs of less than $25 can then be used to "clone" the PUF. The parameterized XOR PUFs were believed to be stable and secure to ML attacks, in theory and experiments. However, they developed a novel divide-and-conquer ML attack that can successfully emulate the XOR PUFs. What is more, the scalability is also exciting, of which there is only a linear increase as the CRPs of XOR PUFs increase. This research shows that though some architectures perform well in experiments, they might not be really secure in practice. The hardware security of strong PUFs (e.g., XOR-based PUFs) is still in great risk, and industries still have a long way to go for their very reliable hardware security targets. This paper leveraged the reliability of PUFs for modeling attacks, meaning that, as this kind of modeling attacks is raised in practice, it is more challenging to put into Strong PUFs as a secure solution.

5.4.2.5   Cooperation with Side Channels

As can be seen from the above, machine learning (ML) is an effective method to attack PUFs, especially Strong PUFs. However, we also have to admit that ML is still not a one-size-fit-all solution for PUF attacks, since the computing complexity of modeling a PUF will significantly increase as the size of PUFs increases [57]. For example, experiments in [58] have shown that when there are more than 256 bits in Lightweight PUFs or 6 XORs in XOR Arbiter PUFs (another two variants of Strong PUFs), the ML modeling will encounter the bottleneck because of the high computational complexity. To mitigate this issue, researchers then figured out a new solution that combines the ML modeling method with conventional Side-Channel Attacks (SCAs) on circuits. In more detail, in such hybrid attacks [59], SCA is used to extract additional circuit characteristics like power consumption, or delay difference, which is then fed to the ML algorithms to reduce the modeling complexity.

There are several works focusing on this kind of hybrid attacking (ML + SCA), e.g., [58–60]. Although SCA has various types of attacks, we briefly discuss the main ideas of SCA as aforementioned. Also, it is worth to mention that SCA is only used to decrease the workload of ML modeling, and thus the main task is still using ML for building PUF models. The main idea raised in [58] is that some sub-responses like "all-1s" and "all-0s" can be filtered out as side-channel information. After this, the composed XOR PUF or Lightweight PUF can be divided into several individual basic Arbiter PUFs. Then, as presented in previous sections, Arbiter PUFs can be easily attacked by ML modeling. This is like the *divide-and-conquer* thoughts utilized in conventional encryption attacks. Furthermore, [60] improved this hybrid method with asymptotic performance analysis, showing that the number of used/required CRPs is merely linear in the same parameter.

5.4.3   Another Hardware Primitive

In addition to PUFs, there also exist many other hardware security primitives, such as random number generator (RNG). The RNGs can be classified into two categories: true random number generator (TRNG) and pseudo random number generator (PRNG). TRNG is a circuit that can generate random numbers from some inherent randomness resources, such as thermal noises [61], process variations of transistors [62], special chaotic topologies [63], etc. Since TRNG leverages the true randomness in physical structures, its behavior (i.e., the next random number output) is hard to measure and predict. PRNG is another vital random number generator, which mainly leverages complicated mathematical chaotic behaviors. There are plenty of literature raising various PRNGs used in different environments, such as [64–66]. Although the outputs of PRNG are random and even can pass statistical tests, they are not truly random since the number generating only relies on the designed algorithm and the initial value (called *seed*) [67]. Landing on this characteristic, many attacking methods have been raised, such as Direct Crypt-analytic Attack, Input-Based Attacks, and State Compromise Extension Attacks summarized in [67]. Also, some machine learning algorithms (e.g., Artificial Neural Network) were proposed in recent years to emulate and predict the PRNG behaviors [68, 69]. The results from [68] showed that even a simplistic neural network (a 6-30-20-1 network used in this paper) can effectively learn whatsoever probabilistic associations in a training dataset. When the testing data have the same statistical properties as the training data, the neural network will be able to do prediction or association as effectively as the information content of the training data allows.

5.4.4 Summary

Hardware devices, as the main carrier of secret information and the basic element of all electronic systems, whose security is playing an important role in the cybersecurity. Meanwhile, with the rapid development of ML algorithms and application, using ML techniques to evaluate hardware security primitives is also becoming an emerging topic. In this chapter, we take PUF and RNG as examples to briefly introduce the idea that how machine learning is used to evaluate these hardware security primitives. In conclusion, we are not determining that all PUFs and RNGs can be attacked by ML algorithms, or there exists one ML algorithm that can suit all PUF or RNG attacks. Instead, we would like to emphasize that machine learning, as a common method used for learning tasks, is definitely a good way to evaluate the security of emerging hardware primitives. Sometimes ML attacks still need to cooperate with other attack methods like side-channel analysis, to mitigate the computational workload. However, in practice, some researches already showed that those PUFs that seem secure in experiments might not be difficult to crack. Another point we need to stress is that, instead of destructive applications like attacking, machine learning algorithms can also play a constructive role for PUFs, i.e., improving the reliability of most delay-based PUFs [70]. At last, we introduce another important hardware primitive, namely Pseudo Random Number Generator (PRNG), which also faces unique vulnerabilities under ML-based attacks. Reasonably, using ML algorithms to explore hardware security primitives will surely bring stable and prosperous development to hardware security.

5.5 Hardware-Related System/Architecture Security

5.5.1 System Security: Malware Detection

In recent years, new advances in digital electronics and wireless communications have aroused a widespread development of embedded systems ranging from micro-sensors, cell phones to military applications. The constant interaction between physical and cyber worlds becomes stronger than ever. In this process, security becomes an important issue in many mission and safety-critical systems [71]. Malware, a severe threat to system security, is a piece of code designed by attackers to perform various malicious activities in computer systems [72]. Many efforts have been taken to defend against malware. The most commonly existing malware detection approaches are software-based solutions such as signature-based detection and semantics-based, which are ineffective for resource-constrained embedded systems due to the limited computing resources [73]. In response to this, Hardware-assisted Malware Detection (HMD) methods combined with Machine Learning (ML) models are proposed as an effective way to protect the security of computing systems and detect pattern of malicious applications [74].

5.5.1.1 Malware and Its Classification

Malicious software, referred to as malware, is any software intentionally designed to cause damage to a computer, server, client, or computer network, which is an ever-growing security threat these days, and the detection of malware remains an important area of research [75]. Malware can be generally classified into four cases as below:

Virus A computer virus is basically a software program which can spread itself and inject itself into executable file to run. Virus normally requires some kind of user intervention and performs a harmful action, such as destroying data.

Trojan This term is derived from Greek mythology of the Trojan horse used to invade the city of Troy by stealth. In system security, a Trojan horse is a malware program that gets installed into a system by pretending to be a legitimate software program. Generally, a Trojan horse does not try to replicate itself like virus but instead carries a hidden destructive function that can be activated when the application is started, which allows hackers to control the computer remotely. While Trojan horses are not easily detected, computers affected by Trojan may appear to run slower than usual due to heavy processor workload or network usage.

Rootkit A rootkit is malware which is hard to detect and actively tries to hide itself from the user, the operating system, and anti-malware programs. Rootkits keep the concealment by modifying the host's operating system and prevent the harmful process from being read in the list of processes. Rootkits can also help to hide other harmful programs like viruses.

Some types of rootkits contain routines to evade identification and even the removal attempts. An early example of this behavior is building a pair of ghost-jobs. Each ghost-job can detect if the other had been killed and build a new copy of the paired ghost-job within a few milliseconds, which makes the harmful software hard to stop.

Backdoor A backdoor is a malware type that bypass normal authentication procedures to access a system. When a system has been compromised, one or more backdoors may be installed, giving perpetrators the ability to remotely perform system commands and update malware in the future, which are generally invisible to the users. Once the system is infested by backdoors, detection is difficult as files tend to be highly obfuscated. Backdoors may be installed by different methods, especially as a component of hybrid malware. For example, backdoor Trojan horse, a highly dangerous and widespread kind of hybrid malware, which can set backdoor when the Trojan horse starts and open infected machines to external control through the Internet.

5.5.1.2 Overview of Hardware-Assisted Malware Detection with Machine Learning

Traditional malware detection techniques, both static and dynamic, are implemented in software. But the pure software implementation for malware detection has non-negligible vulnerability. First, software-based malware detection techniques often incur significant computational overheads to the system [76], which make it ineffective for some resource-constrained system. Second, software for malware detection is susceptible since malware can often disable the software-based detection mechanism [77].

To overcome these problems, Hardware-assisted Malware Detection (HMD) has been studied in many works [73–75, 78–81], which is not vulnerable to disabling caused by malware and can reduce the latency of detection process with small hardware costs. The early works of HMD, like [78–80], are focused on the specific classes of malware. For instance, Copilot proposed by Petroni et al. [80] can detect kernel-level rootkits by checking if the snapshots are different from their expected values. Later, more works demonstrated that malware can be distinguished by classifying the detected abnormal conditions using Machine Learning (ML) techniques in low-level feature spaces captured by Hardware Performance Counters (HPCs) [73, 82, 83]. The HPCs are a set of special-purpose registers built into modern microprocessors to store the counts of hardware-related activities. The trace of hardware-related events captured by HPCs can be used to conduct low-level performance analysis including malware detection. These ML-based malware detection methods can be implemented in microprocessor with a significantly low overhead. Because the detection processes inside the hardware are really fast with only several clock cycles. However, there are still mainly two problems with malware detection: First, each ML method performs differently with different types of malware, and there is no ML classifier that gets the best results in all malware classes. Second, the number of HPCs influences the effect of ML classifier a lot. Due to the physical limitations in microprocessors, small system with limited HPCs may not get a good performance with original ML-based HMD methods. In response to this, Sayadi et al. [81] present a two-stage machine learning-based approach to integrate different kinds of ML classifier, which adapts to different malware types. At the last part of Sect. 5.5.1, we will discuss this method as an example of ML-based HMD.

5.5.1.3 Commonly Used Machine-Learning Methods in HMD

HMD has been studied widely in these years. Different types of ML classifiers have been used in this field. In this section, we provide introduction of several important ML classifiers that are mostly used in ML-based HMD.

Multi-class Support Vector Machine (MSVM)
In Sect. 5.3.3.1, we have discussed SVM, a powerful ML classifier, which is a supervised learning model used for classification and regression analysis. Multi-class support Vector Machine (MSVM) is an extension of SVM, aiming to assign labels to instances with multiple classes. The main approach for achieving this goal is to separate the multi-class problem into multiple binary classification problems [84]. Generally, there are two methods to construct a MSVM: the first one is to build binary classifiers that differentiate between one label and the others, and the second one is to build binary classifiers between every pair of classes. For the first method, the inference of classification is done by comparing the output score of each class, and the class with the highest output score is the result. For the second method, the inference is done by "voting." Every binary classifier assigns one winner between two instances, and the final result is the class with maximum "winner tickets."

As for HMD, MSVM can be used in the classification of malware to give a preliminary result. In the work of Sayadi et al. [74], the output of multi-class SVM is corresponding to a set of feasible applications based on different ML classifiers in order to choose the best method to perform the malware detection.

Multinomial Logistic Regression (MLR)
Multinomial logistic regression (MLR), also called multinomial regression, is used to predict a nominal dependent variable given one or more independent variables. Generally, it can be considered as an extension of binomial logistic regression (BLR). The most straightforward implementation of MLR with K classes is to run $K - 1$ independent BLR for every class. Considering one of the classes as the main class during the operation, the other $K - 1$ classes can be regressed with the main class separately.

In the work of Sayadi et al. [81], MLR is used in the first stage of the framework to pre-estimate the type of malware and select the most suitable ML classifier for the specific situation.

Principal Component Analysis (PCA)

PCA was invented in 1901 by Karl Pearson, which is defined as an orthogonal linear transformation that transforms the data to a new coordinate system. It can be used to reduce the number of dimensions while capturing most of the data variation in the process of transformation. Generally, people use PCA to project original features into a new dimensional space to determine the most important features along different principal component dimensions. Combined with HMD, PCA can be employed to extract important features from HPCs to simplify the input of the malware detector [74, 81].

Ensemble Learning

Ensemble learning is a branch of ML, which uses multiple learning algorithms to obtain better predictive performance. It can improve the accuracy of ML model by generating a set of base learners and combining their outputs for the final decision [85]. There are three types of ensemble learning: Bagging (i.e., Random Forest), Boosting (i.e., Adaptive Boosting (AdaBoost) and Gradient Boosting Decision Tree (GBDT)), and Stacking.

In HMD, ensemble learning can combine different types of ML classifiers to improve the malware detection accuracy. Sayadi et al. [81] use AdaBoost to intensify the HMD performance with limited HPCs.

5.5.1.4 Case Study of ML-Based Hardware-Assisted Malware Detection

In this part, we will discuss one ML-based Hardware-assisted Malware Detection, 2SMaRT (Two-Stage Machine Learning-Based Approach for Run-Time Specialized Hardware-Assisted Malware Detection), to show how it works in detail.

As what has been mentioned in Sect. 5.5.1.2, there are two challenges from the original HMD works: the first is the limited HPCs in some small systems, and the second is that the performance of single malware detector is highly correlated to the class of malware infecting the system. To deal with these problems, 2SMaRT [81] presents a framework with two stages *Application Type Prediction* and *Effective Machine Learning Techniques* as shown in Fig. 5.16, using the first stage to initially classify the malware and then using the corresponding specialized malware detector to double check the result in the second stage. Besides that, PCA is used to extract the main data from HPCs and, thus, decreases the demanded number of HPCs from 44 to 16 (4 common HPCs for the first stage and 12 custom HPCs for the second stage).

Initially, at the beginning of the first stage, the system is unaware of existence of malware in the application. To predict the most possible malware, MLR technique is used here to convert the basic binary malware classification to a multi-class problem. Here, the input of the MLR only includes features from 4 HPCs which are selected by PCA as the common features. The output of MLR is corresponding to 5 individual classes: "benign," "Virus," "Trojan," "Rootkit," and "Backdoor." During runtime, the probability of each class is calculated and the MLR classifier

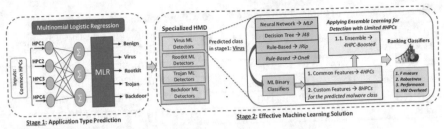

Fig. 5.16 Overview of 2SMaRT. A two-stage malware detection approach (Adapted from [81])

selects the class that achieves the highest probability to be the preliminary result. The combination of PCA and MLR guarantees high resource efficiency of the first stage.

To improve the detection performance, second stage of detection is cascaded as specialized HMD which uses various types of ML techniques for per-class analysis. These ML classifiers are chosen based on the predicted class of malware by the MLR. As discussed before, the performance of a single malware detector is highly correlated to the class of malware infecting the system. Here, four specialized ML classifiers (J48, JRip, MLP, and OneR) are picked to deal with the different malware types. An extra set of custom HPCs with the total number of 12 is provided to the different specialized ML classifiers (each classifier is associated with maximum 8 selected HPCs according to PCA from the total 12 custom HPCs) to get a more accurate result.

Besides the specialized ML classifiers, ensemble learning is also performed as an optional choice in the second stage. Adaptive Boosting (AdaBoost, one method of the ensemble learning) is deployed to construct the final classifier and analyze its impact on the performance improvement of malware detection. In AdaBoost, each base classifier is trained on a weighted form of the training set in which the weights depend on the performance of the previous base classifier. Once all the base classifiers are trained, they are combined to produce the final classifier [81].

The part of the results is shown in Table 5.3. In this table, $\#HPC$ means the total number of HPCs that has been used in the corresponding experiments (common HPCs + custom HPCs) and the $Boosted$ means the final classifier of AdaBoost. The result is presented with F_1 score form, which is a commonly used metrics for HMD evaluation. The F_1 score can be calculated as $\frac{2TP}{2TP+FP+FN}$, where TP means true positive, FP means false positive, and FN means false negative.

We can see that, although the boosted detector only has 4 common HPCs as the input, it achieves higher or mostly similar F_1 score to 16/8 HPC-based detectors. By applying AdaBoost at the end of the 2SMaRT, 98.9% F_1 score (MLP in Trojan) and almost 92% F_1 score on average across all ML classifiers are achieved. 2MSaRT compensates the possible negative impact of HPCs reduction and eliminates the need for multiple runs of application to capture the required HPCs while keeping a high malware detection performance.

Table 5.3 F measure of 2SMaRT detectors with and without boosting (Adapted from [81])

#HPC	16	8	4	4-Boosted	16	8	4	4-Boosted
Class	Backdoor				Rootkit			
J48	86.7	79.6	80.4	85.5	94.6	87.7	85.75	91.2
JRip	90.5	90	87.8	87.6	84.1	82.5	80.8	91.5
MLP	94.4	92.4	89.5	90	82.9	82.35	93.8	79.8
OneR	94	94	94	93.8	73.2	73.2	73.18	85.99
	Virus				Trojan			
J48	94.7	94.5	93.2	96.5	98.8	98	93.2	97.3
JRip	93.6	93.1	93	96.5	98.9	98.2	93.3	94
MLP	68.1	67.6	94.7	95.4	98.6	96.7	98.9	98.9
OneR	97.1	90.2	89	94.8	92.7	92.7	92.7	92.7

5.6 Future of ML-Involved Hardware Security

With the developments of hardware design and manufacturing methodologies, as well as the emerging of different more practical and efficient Machine Learning algorithms, it is predictable that the future integration of hardware security and Machine learning will become more tight and common. From another perspective, with the demand for high-performance computing capabilities, the complexity of hardware platforms will continue growing up, and more security concerns will inevitably appear. Machine learning, as one of the most powerful analysis tools, will be playing a more important role in hardware security area with bringing more *intelligence*. We conclude that studying the double-edged sword effect of machine learning on hardware security will be a fruitful area.

References

1. S. Bhunia, M. Tehranipoor, *Hardware Security: A Hands-On Learning Approach* (Morgan Kaufmann, Burlington, 2018)
2. M. Brzozowski, V.N. Yarmolik, Obfuscation as intellectual rights protection in VHDL language, in *6th International Conference on Computer Information Systems and Industrial Management Applications (CISIM'07)* (IEEE, Piscataway, 2007), pp. 337–340
3. P. Subramanyan, S. Ray, S. Malik, Evaluating the security of logic encryption algorithms, in *2015 IEEE International Symposium on Hardware Oriented Security and Trust (HOST)* (IEEE, Piscataway, 2015), pp. 137–143
4. M. El Massad, S. Garg, M.V. Tripunitara, Integrated circuit (IC) decamouflaging: reverse engineering camouflaged ICs within minutes, in *NDSS* (2015), pp. 1–14
5. X. Xu, B. Shakya, M.M. Tehranipoor, D. Forte, Novel bypass attack and BDD-based tradeoff analysis against all known logic locking attacks, in *International Conference on Cryptographic Hardware and Embedded Systems* (Springer, Berlin, 2017), pp. 189–210
6. M. Li, K. Shamsi, T. Meade, Z. Zhao, B. Yu, Y. Jin, D.Z. Pan, Provably secure camouflaging strategy for IC protection. IEEE Trans. Comput. Aided Desig. Integ. Circuits Syst. **38**, 1399–1412 (2017)

7. K. Shamsi, M. Li, T. Meade, Z. Zhao, D.Z. Pan, Y. Jin, Appsat: approximately deobfuscating integrated circuits, in *2017 IEEE International Symposium on Hardware Oriented Security and Trust (HOST)* (IEEE, Piscataway, 2017), pp. 95–100

8. D.D. Lewis, J. Catlett, Heterogeneous uncertainty sampling for supervised learning, in *Machine Learning Proceedings 1994* (Elsevier, Amsterdam, 1994), pp. 148–156

9. A. Ehrenfeucht, D. Haussler, M. Kearns, L. Valiant, A general lower bound on the number of examples needed for learning. Inf. Comput. **82**(3), 247–261 (1989)

10. M. Yasin, B. Mazumdar, O. Sinanoglu, J. Rajendran, Removal attacks on logic locking and camouflaging techniques. IEEE Trans. Emerg. Topics Comput. **8**, 517–532 (2017)

11. P. Chakraborty, J. Cruz, S. Bhunia, Sail: machine learning guided structural analysis attack on hardware obfuscation, in *2018 Asian Hardware Oriented Security and Trust Symposium (AsianHOST)* (IEEE, Piscataway, 2018), pp. 56–61

12. Y. Xie, A. Srivastava, Anti-sat: mitigating sat attack on logic locking. IEEE Trans. Comput. Aided Des. Integr. Circuits Syst. **38**(2), 199–207 (2018)

13. M. Yasin, A. Sengupta, M.T. Nabeel, M. Ashraf, J. Rajendran, O. Sinanoglu, Provably-secure logic locking: from theory to practice, in *Proceedings of the 2017 ACM SIGSAC Conference on Computer and Communications Security* (2017), pp. 1601–1618

14. M. Yasin, B. Mazumdar, J.J.V. Rajendran, O. Sinanoglu, Sarlock: sat attack resistant logic locking, in *2016 IEEE International Symposium on Hardware Oriented Security and Trust (HOST)* (IEEE, Piscataway, 2016), pp. 236–241

15. J. Rajendran, Y. Pino, O. Sinanoglu, R. Karri, Security analysis of logic obfuscation, in *Proceedings of the 49th Annual Design Automation Conference* (2012), pp. 83–89

16. B. Shakya, X. Xu, M. Tehranipoor, D. Forte, Cas-lock: a security-corruptibility trade-off resilient logic locking scheme. IACR Trans. Cryptogr. Hardware Embed. Syst. **2020**, 175–202 (2020)

17. X. Wang, M. Tehranipoor, J. Plusquellic, Detecting malicious inclusions in secure hardware: challenges and solutions, in *2008 IEEE International Workshop on Hardware-Oriented Security and Trust* (2008), pp. 15–19

18. F. Wolff, C. Papachristou, S. Bhunia, R.S. Chakraborty, Towards Trojan-free trusted ICs: problem analysis and detection scheme, in *2008 Design, Automation and Test in Europe* (2008), pp. 1362–1365

19. J. Rajendran, E. Gavas, J. Jimenez, V. Padman, R. Karri, Towards a comprehensive and systematic classification of hardware Trojans, in *Proceedings of 2010 IEEE International Symposium on Circuits and Systems* (2010), pp. 1871–1874

20. Y. Jin, Y. Makris, Hardware Trojan detection using path delay fingerprint, in *2008 IEEE International Workshop on Hardware-Oriented Security and Trust* (2008), pp. 51–57

21. J. Li, J. Lach, At-speed delay characterization for ic authentication and Trojan horse detection, in *2008 IEEE International Workshop on Hardware-Oriented Security and Trust* (2008), pp. 8–14

22. S. Narasimhan, S. Bhunia, Hardware Trojan detection, in *Introduction to Hardware Security and Trust* (Springer, Berlin, 2012), pp. 339–364

23. N. Vashistha, M.T. Rahman, H.-T. Shen, D.L. Woodard, N. Asadizanjani, M.M. Tehranipoor, Detecting hardware Trojans inserted by untrusted foundry using physical inspection and advanced image processing. J. Hardw. Syst. Secur. **2**, 333–344 (2018)

24. Zetec, Nondestructive testing overview: a comprehensive guide to NDT. Website (2019). https://www.zetec.com/resources/nondestructive-testing-overview/

25. K. Hasegawa, M. Yanagisawa, N. Togawa, Hardware Trojans classification for gate-level netlists using multi-layer neural networks, in *2017 IEEE 23rd International Symposium on On-Line Testing and Robust System Design (IOLTS)* (2017), pp. 227–232

26. T. Iwase, Y. Nozaki, M. Yoshikawa, T. Kumaki, Detection technique for hardware Trojans using machine learning in frequency domain, in *2015 IEEE 4th Global Conference on Consumer Electronics (GCCE)* (2015), pp. 185–186

27. N. Vashistha, Trojan scanner: detecting hardware Trojans with rapid imaging combined with image processing and machine learning (2018)

28. S. Narasimhan, D. Du, R.S. Chakraborty, S. Paul, F.G. Wolff, C.A. Papachristou, K. Roy, S. Bhunia, Hardware Trojan detection by multiple-parameter side-channel analysis. IEEE Trans. Comput. **62**(11), 2183–2195 (2012)

29. D.G. Drmanac, F. Liu, L.-C. Wang, Predicting variability in nanoscale lithography processes, in *2009 46th ACM/IEEE Design Automation Conference* (IEEE, Piscataway, 2009), pp. 545–550

30. A. Vakil, F. Behnia, A. Mirzaeian, H. Homayoun, N. Karimi, A. Sasan, LASCA: learning assisted side channel delay analysis for hardware Trojan detection. e-prints, arXiv:2001.06476 (2020)

31. G. Hospodar, B. Gierlichs, E. De Mulder, I. Verbauwhede, J. Vandewalle, Machine learning in side-channel analysis: a first study. J. Cryptogr. Eng. **1**(4), 293 (2011)

32. L. Lerman, G. Bontempi, O. Markowitch, Side channel attack: an approach based on machine learning, in *Center for Advanced Security Research Darmstadt* (2011), pp. 29–41

33. J. Park, X. Xu, Y. Jin, D. Forte, M. Tehranipoor, Power-based side-channel instruction-level disassembler, in *2018 55th ACM/ESDA/IEEE Design Automation Conference (DAC)* (IEEE, Piscataway, 2018), pp. 1–6

34. S. Picek, A. Heuser, A. Jovic, S. Bhasin, F. Regazzoni, The curse of class imbalance and conflicting metrics with machine learning for side-channel evaluations. IACR Trans. Crypto. Hardware Embedded Syst. **2019**(1), 1–29 (2019)

35. W. Hua, Z. Zhang, G.E. Suh, Reverse engineering convolutional neural networks through side-channel information leaks, in *2018 55th ACM/ESDA/IEEE Design Automation Conference (DAC)* (IEEE, Piscataway, 2018), pp. 1–6

36. S. Chari, J.R. Rao, P. Rohatgi, Template attacks, in *International Workshop on Cryptographic Hardware and Embedded Systems* (Springer, Berlin, 2002), pp. 13–28

37. S. Picek, A. Heuser, A. Jovic, S.A. Ludwig, S. Guilley, D. Jakobovic, N. Mentens, Side-channel analysis and machine learning: a practical perspective, in *2017 International Joint Conference on Neural Networks (IJCNN)* (IEEE, Piscataway, 2017), pp. 4095–4102

38. S. Kullback, R.A. Leibler, On information and sufficiency. Annal. Math. Stat. **22**(1), 79–86 (1951)

39. V. Vapnik, *The Nature of Statistical Learning Theory* (Springer Science & Business Media, Berlin, 2013)

40. L. Breiman, Random forests. Mach. Learn. **45**(1), 5–32 (2001)

41. J.J. Rodriguez, L.I. Kuncheva, C.J. Alonso, Rotation forest: a new classifier ensemble method. IEEE Trans. Pattern Anal. Mach. Intell. **28**(10), 1619–1630 (2006)

42. S. Picek, A. Heuser, A. Jovic, A. Legay, Climbing down the hierarchy: hierarchical classification for machine learning side-channel attacks, in *International Conference on Cryptology in Africa* (Springer, Berlin, 2017), pp. 61–78

43. T. McGrath, I.E. Bagci, Z.M. Wang, U. Roedig, R.J. Young, A puf taxonomy. Appl. Phys. Rev. **6**(1), 011303 (2019)

44. G.E. Suh, S. Devadas, Physical unclonable functions for device authentication and secret key generation, in *2007 44th ACM/IEEE Design Automation Conference* (IEEE, Piscataway, 2007), pp. 9–14

45. S. Morozov, A. Maiti, P. Schaumont, A comparative analysis of delay based PUF implementations on FPGA. IACR Cryptol. ePrint Arch. **2009**, 629 (2009)

46. X. Xu, S. Li, R. Kumar, W. Burleson, When the physical disorder of cmos meets machine learning, in *Low Power Semiconductor Devices and Processes for Emerging Applications in Communications, Computing, and Sensing* (2018)

47. U. Rührmair, J. Sölter, F. Sehnke, X. Xu, A. Mahmoud, V. Stoyanova, G. Dror, J. Schmidhuber, W. Burleson, S, Devadas, PUF modeling attacks on simulated and silicon data. IEEE Trans. Inf. Forensics Secur. **8**(11), 1876–1891 (2013)

48. Q. Ma, C. Gu, N. Hanley, C. Wang, W. Liu, M. O'Neill, A machine learning attack resistant multi-PUF design on FPGA, in *2018 23rd Asia and South Pacific Design Automation Conference (ASP-DAC)* (IEEE, Piscataway, 2018), pp. 97–104

49. J. Delvaux, Machine-learning attacks on polypufs, OB-PUFS, RPUFS, LHS-PUFS, and PUF–FSMS. IEEE Trans. Inf. Forensics Secur. **14**(8), 2043–2058 (2019)

50. J.A.K. Suykens, J. Vandewalle, Least squares support vector machine classifiers. Neural Proces. Lett. **9**(3), 293–300 (1999)
51. L.K. Hansen, P. Salamon, Neural network ensembles. IEEE Trans. Patt. Anal. Mach. Intell. **12**(10), 993–1001 (1990)
52. R.E. Wright, Logistic regression, in *Reading and Understanding Multivariate Statistics* (American Psychological Association, Washington, 1995), pp. 217–244
53. H.-G. Beyer, H.-P. Schwefel, Evolution strategies–a comprehensive introduction. Nat. Comput. **1**(1), 3–52 (2002)
54. M. McCloskey, N.J. Cohen, Catastrophic interference in connectionist networks: the sequential learning problem, in *Psychology of Learning and Motivation*, vol. 24 (Elsevier, Amsterdam, 1989), pp. 109–165
55. T. Back, *Evolutionary Algorithms in Theory and Practice: Evolution Strategies, Evolutionary Programming, Genetic Algorithms* (Oxford University Press, Oxford, 1996)
56. G.T. Becker, The gap between promise and reality: On the insecurity of XOR arbiter PUFs, in *International Workshop on Cryptographic Hardware and Embedded Systems* (Springer, Berlin, 2015), pp. 535–555
57. X. Xu, U. Rührmair, D.E. Holcomb, W. Burleson, Security evaluation and enhancement of bistable ring PUFs, in *International Workshop on Radio Frequency Identification: Security and Privacy Issues* (Springer, Berlin, 2015), pp. 3–16
58. A. Mahmoud, U. Rührmair, M. Majzoobi, F. Koushanfar, Combined modeling and side channel attacks on strong PUFs. IACR Cryptol. ePrint Arch. **2013**, 632 (2013)
59. X. Xu, W. Burleson, Hybrid side-channel/machine-learning attacks on pufs: a new threat? in *2014 Design, Automation & Test in Europe Conference & Exhibition (DATE)* (IEEE, Piscataway, 2014), pp. 1–6
60. U. Rührmair, X. Xu, J. Sölter, A. Mahmoud, M. Majzoobi, F. Koushanfar, W. Burleson, Efficient power and timing side channels for physical unclonable functions, in *International Workshop on Cryptographic Hardware and Embedded Systems* (Springer, Berlin, 2014), pp. 476–492
61. C.S. Petrie, J.A. Connelly, A noise-based IC random number generator for applications in cryptography. IEEE Trans. Circ. Syst. I Fund. Theory Appl. **47**(5), 615–621 (2000)
62. X. Xu, V. Suresh, R. Kumar, W. Burleson, Post-silicon validation and calibration of hardware security primitives, in *2014 IEEE Computer Society Annual Symposium on VLSI* (IEEE, Piscataway, 2014), pp. 29–34
63. S. Best, X. Xu, An all-digital true random number generator based on chaotic cellular automata topology, in *2019 IEEE/ACM International Conference on Computer-Aided Design (ICCAD)* (IEEE, Piscataway, 2019), pp. 1–8
64. M. Matsumoto, T. Nishimura, Mersenne twister: a 623-dimensionally equidistributed uniform pseudo-random number generator. ACM Trans. Model. Comput. Simul. (TOMACS) **8**(1), 3–30 (1998)
65. L. Blum, M. Blum, M. Shub, A simple unpredictable pseudo-random number generator. SIAM J. Comput. **15**(2), 364–383 (1986)
66. M.W. Thomlinson, D.R. Simon, B. Yee, Non-biased pseudo random number generator (1998). US Patent 5778069
67. J. Kelsey, B. Schneier, D. Wagner, C. Hall, Cryptanalytic attacks on pseudorandom number generators, in *International Workshop on Fast Software Encryption* (Springer, Berlin, 1998), pp. 168–188
68. F. Fan, G. Wang, Learning from pseudo-randomness with an artificial neural network–does god play pseudo-dice? IEEE Access **6**, 22987–22992 (2018)
69. T. Fischer, Testing cryptographically secure pseudo random number generators with artificial neural networks, in *2018 17th IEEE International Conference on Trust, Security and Privacy in Computing and Communications / 12th IEEE International Conference on Big Data Science and Engineering (TrustCom/BigDataSE)* (IEEE, Piscataway, 2018), pp. 1214–1223

70. X. Xu, W. Burleson, D.E. Holcomb, Using statistical models to improve the reliability of delay-based PUFS, in *2016 IEEE Computer Society Annual Symposium on VLSI (ISVLSI)* (IEEE, Piscataway, 2016), pp. 547–552

71. H. Sayadi, H. Farbeh, A.M.H. Monazzah, S.G. Miremadi, A data recomputation approach for reliability improvement of scratchpad memory in embedded systems, in *2014 IEEE International Symposium on Defect and Fault Tolerance in VLSI and Nanotechnology Systems (DFT)* (IEEE, Piscataway, 2014), pp. 228–233

72. N. Patel, A. Sasan, H. Homayoun, Analyzing hardware based malware detectors, in *2017 54th ACM/EDAC/IEEE Design Automation Conference (DAC)* (IEEE, Piscataway, 2017), pp. 1–6

73. H. Sayadi, N. Patel, S. Manoj P.D., A. Sasan, S. Rafatirad, H. Homayoun, Ensemble learning for effective run-time hardware-based malware detection: a comprehensive analysis and classification, in *2018 55th ACM/ESDA/IEEE Design Automation Conference (DAC)* (IEEE, Piscataway, 2018), pp. 1–6

74. H. Sayadi, H.M. Makrani, O. Randive, S. Manoj P.D., S. Rafatirad, H. Homayoun, Customized machine learning-based hardware-assisted malware detection in embedded devices, in *2018 17th IEEE International Conference on Trust, Security and Privacy in Computing and Communications/12th IEEE International Conference on Big Data Science and Engineering (TrustCom/BigDataSE)* (IEEE, Piscataway, 2018), pp. 1685–1688

75. Z. Xu, S. Ray, P. Subramanyan, S. Malik, Malware detection using machine learning based analysis of virtual memory access patterns, in *Design, Automation & Test in Europe Conference & Exhibition (DATE), 2017* (IEEE, Piscataway, 2017), pp. 169–174

76. G. Jacob, H. Debar, E. Filiol, Behavioral detection of malware: from a survey towards an established taxonomy. J. Comput. Virol. **4**(3), 251–266 (2008)

77. F. Xue, Attacking antivirus, in *Black Hat Europe Conference* (2008)

78. J. Demme, M. Maycock, J. Schmitz, A. Tang, A. Waksman, S. Sethumadhavan, S. Stolfo, On the feasibility of online malware detection with performance counters. ACM SIGARCH Comput. Archit. News **41**(3), 559–570 (2013)

79. M. Ozsoy, C. Donovick, I. Gorelik, N. Abu-Ghazaleh, D. Ponomarev, Malware-aware processors: a framework for efficient online malware detection, in *2015 IEEE 21st International Symposium on High Performance Computer Architecture (HPCA)* (IEEE, Piscataway, 2015), pp. 651–661

80. N.L. Petroni Jr, T. Fraser, J. Molina, W.A. Arbaugh, Copilot – a coprocessor-based kernel runtime integrity monitor, in *USENIX Security Symposium*, San Diego (2004), pp. 179–194

81. H. Sayadi, H.M. Makrani, S.M.P. Dinakarrao, T. Mohsenin, A. Sasan, S. Rafatirad, H. Homayoun, 2smart: a two-stage machine learning-based approach for run-time specialized hardware-assisted malware detection, in *2019 Design, Automation & Test in Europe Conference & Exhibition (DATE)* (IEEE, Piscataway, 2019), pp. 728–733

82. A. Tang, S. Sethumadhavan, S.J. Stolfo, Unsupervised anomaly-based malware detection using hardware features, in *International Workshop on Recent Advances in Intrusion Detection* (Springer, Berlin, 2014), pp. 109–129

83. K.N. Khasawneh, M. Ozsoy, C. Donovick, N. Abu-Ghazaleh, D. Ponomarev, Ensemble learning for low-level hardware-supported malware detection, in *International Symposium on Recent Advances in Intrusion Detection* (Springer, Berlin, 2015), pp. 3–25

84. K.-B. Duan, S.S. Keerthi, Which is the best multiclass svm method? An empirical study, in *International Workshop on Multiple Classifier Systems* (Springer, Berlin, 2005), pp. 278–285

85. E. Aghaei, G. Serpen, Ensemble classifier for misuse detection using n-gram feature vectors through operating system call traces. Int. J. Hybrid Intell. Syst. **14**(3), 141–154 (2017)

Chapter 6
Security Assessment of High-Level Synthesis

M. Rafid Muttaki, Nitin Pundir, Mark Tehranipoor, and Farimah Farahmandi

6.1   Introduction

With rapid technological development to remain competitive in the semiconductor industry, less time has been allocated to design, testing, and verification [1, 2]. However, as it can be estimated from the ongoing trend of designs becoming more complex than ever and the technology becoming smaller and smaller. It will demand more time into the planning, design, and verification phase. Investing more time on every step of the process contradicts the industry's progression towards aggressive time-to-market. To address this issue, high-level synthesis (HLS) has become a popular choice for different government and industry entities.

HLS is an automated design process that interprets a high-level description of the desired behaviors and creates corresponding Register Transfer Level (RTL) hardware to implement the behavior. HLS enables designers to raise the design abstraction which in turn produces design specifications at C/C++ level with reduced design complexity for both application-specific integrated circuits (ASIC) and field-programmable gate arrays (FPGAs). As a result, HLS has become an efficient way to lower development costs, verification efforts, and time-to-market for industrial and government applications.

HLS can benefit different entities in various ways. For example, government entities have been relying on third-party IPs for a long time due to the increasing complexity of modern system-on-chips (SoC) and smaller development teams. As a result, the security of the SoC has been susceptible to vulnerabilities such as hardware Trojans and malicious functionality that can be inserted into third-party IPs. The use of HLS can address the security issues of third-party IPs by reducing

M. R. Muttaki · N. Pundir (✉) · M. Tehranipoor · F. Farahmandi
University of Florida, Gainesville, FL, USA
e-mail: m.muttaki@ufl.edu; nitin.pundir@ufl.edu; tehranipoor@ece.ufl.edu;
frarahmandi@ufl.edu

M. Tehranipoor (ed.), *Emerging Topics in Hardware Security*,
https://doi.org/10.1007/978-3-030-64448-2_6

time spent in the development phase and generating those IPs in-house with the help of abstract modules as well as HLS.

HLS benefits the industry by reducing design and verification efforts. Nvidia, which is a leading semiconductor company in GPU manufacturing, has recently reported that its design complexity (transistor count per chip) is increasing 1.4 times per generation [3]. On the other hand, the number of developers on a project remains constant which causes increased effort for each member of the team. This situation will account for increased time for design and verification shortly with the increase of complex SoC designs if not addressed promptly. To address this, Nvidia's video team turned to HLS and by incorporating HLS, they could simplify their design by five times, which in turn reduced design and verification time by 40%.

The modern IC supply chain is filled with untrusted agents that can cause various forms of security threats at different design stages. Such forms of threats include overproduction, IP piracy [4–7], hardware Trojans [8–12], counterfeit [2, 13–16], and reverse engineering [17]. To counter these threats locking the design with key gates has been proposed. Design locking can be done at various stages of the design. One way to lock the design and obfuscate its functionality is to insert key gates in the gate-level netlist of the design. The design is not functional unless the correct keys are applied. In most of the existing techniques, key gates are inserted at the gate-level when the design is fully optimized. As a result, there is not much flexibility in choosing the location of key gates to protect desired functionality/critical information in the design. It has been shown that this type of design locking and obfuscation is susceptible to various attacks such as SAT attacks [18], removal attacks [19], machine learning attacks [20], and reverse engineering attacks [17]. To address these vulnerabilities, we propose locking at higher levels of abstractions (e.g., C/C++/SystemC).

Locking/obfuscation can be done at the C/C++ level of the design by analyzing design assets and critical operations using formal methods and hiding its security-critical operations by inserting locking keys. This obfuscation type provides the designer access to high-level specifications, design assets, and dependency between different functionalities relating to the design. For this reason, comprehensive and optimal locking for different designs can be done in terms of attack resiliency, area and power overhead analysis, and the required size of the locking key. As different entities are relying on high-level synthesis for designing various hardware components due to its benefits, a comprehensive high-level locking technique may prove to be a useful measure to counter against security threats. This technique will be more and more beneficial as more design houses will use HLS to reduce their design and verification costs. However, for incorporating this type of technique, a secure translation from the higher abstraction levels to HDL is a must. To address this, a high-level security assessment is provided in this chapter to ensure that no security vulnerability can arise while using HLS.

In the rest of this chapter, we first briefly discuss the conventional ASIC design steps in Sect. 6.2. In Sects. 6.3 and 6.4, we describe the HLS based design flow and the definition of obfuscation and why it is necessary, respectively. Finally, we present a framework of hardware obfuscation using HLS and perform an overall security assessment on HLS in Sects. 6.5 and 6.6.

6.2 Conventional ASIC Design Flow

We provide a brief description of the conventional ASIC design flow to identify the elementary differences between the HLS and the traditional design flow. A conventional ASIC design flow goes through a series of steps, i.e., from specification to tape-out to produce the intended IC [21]. In this section, a brief overview will be given on the major steps.

6.2.1 Defining Specification

It is of utmost importance to foresee the trends that would be relevant 1–2 years down the line to define the specification of the product that will be worked upon. To learn the trends, market surveys with potential customers, as well as comments from technology experts are necessary. The next step is to identify high-level product specifications such as computation algorithm, clock frequencies, communication protocol, power supply, package type, the targeted area and power overheads, etc. Developing a list of thorough specification sets a solid foundation for the ASIC design.

6.2.2 Architecture Selection

After setting the specifications, the entire ASIC or SOC's functionality is divided into multiple functional blocks. Different architectures are reviewed based on their pros and cons while considering performance implications, technical feasibility, and resource allocation in terms of both cost and time. Good architecture is selected based on extracting the best performance of the chip minimizing the hardware resources to keep the cost within the allocated budget.

6.2.3 RTL Design and Verification

The RTL design step involves writing codes of the data flow of each functional block in a hardware description language, i.e., Verilog, VHDL, or System Verilog. The functional blocks comprise combinational logic (i.e., combinational gates such as OR, AND, and NAND), sequential elements, finite state machines (FSMs), arithmetic logic blocks, etc. On the other hand, the verification team needs to create test benches to be able to test the design for all possible corner cases to ensure the correct functionality, which is defined in the specification phase. However, with the increasing complexity of modern ASIC designs, the solution space for functional verification has become much smaller than before. As a result, designers are leaning towards use case-based verification, which works with each property/specification of the design. The correctness of the design is validated by proving the compatibility of each design specification.

6.2.4 Physical Design

The physical design comprises synthesis, floor planning, placement, clock tree synthesis, and detail routing. The synthesis step takes in the RTL code along with physical libraries of the standard cells including delay information (.lib files), physical dimensions, and other constraint files and converts the behavioral code into a gate-level netlist. In the floor planning step, the entire die area is divided into physical partitions, and their shapes are molded considering area requirements. Pins and ports are also assigned a rough location, to be further refined depending on the Place and Route results. During the placement step, minimizing the wire lengths, the optimal placement of the cells is ensured for faster timing specification convergence. In the clock tree synthesis step, an optimal clock tree is used to achieve expected clock latency with minimized clock skew. In the detailed routing, signal nets are routed in such a way that design rule check (DRC) violations are minimized. Finally, physical and timing verification performs design rule check (DRC), layout versus schematic (LVS), electromigration, electrostatic discharge (ESD) violations, and timing check to find if each specification is met.

6.2.5 Tape-Out

After meeting specifications at the post-layout stage, the photomask of the approved design file along with database files is sent to the foundry for fabrication. The conventional ASIC design flow is shown in Fig. 6.1.

6.3 HLS Based Design Flow

As we mentioned before, HLS is an automatic process that enables the transformation of a high-level (C/C++) description of hardware into RTL modules. In other words, the main difference with conventional ASIC flow is that RTL modules are not implemented manually and HLS generates them. The input of HLS is an

Fig. 6.1 Conventional ASIC
design steps

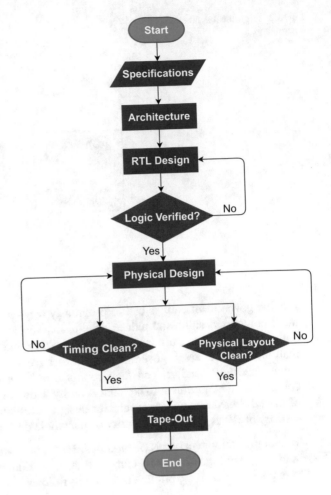

untimed description of functionality, and the tool used for HLS can assist with the
selection of an architecture that will optimize the performance, area, and power of
the implementation. Before moving into how HLS works and the inherent steps
of HLS, we need to understand why different entities prefer HLS compared to
conventional ASIC flow.

- HLS is best suited for applications with complex algorithmic content, hardware
  whose specifications evolve quickly, and systems with much shortened time-to-
  market. As a result, applications such as cryptography, machine learning, and
  hardware accelerators in the cloud environment use HLS.
- Function as a service (FaaS) such as Amazon, Nimbix, CAD developers, and
  software companies use HLS to keep pace with the rapidly growing product
  field. Also, product specification with higher abstraction makes it easier for the
  developers to realize, and with HLS more complex designs can be handled by
  smaller hardware teams.

Fig. 6.2 High-level synthesis
design steps

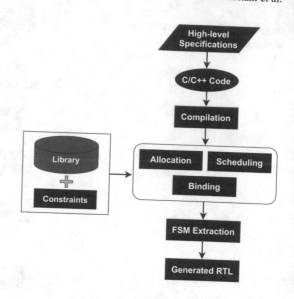

- Finally, government entities had to previously rely on third-party IPs to implement complex designs and reduce time-to-market. However, relying on third-party IPs introduce trust issues since these IPs may come with deliberate malfunction. Moreover, government entities have smaller development teams which meant IP vendors, and they cannot develop everything in-house. This could cause a major threat to national security. However, with the increased use of the HLS government can accelerate design development without risking the security of IPs as well as the security of the whole design.

To automatically generate RTL modules, HLS comprises a series of individual steps as shown in Fig. 6.2, namely compilation, allocation, scheduling, binding, and FSM extraction, which are briefly discussed as follows.

6.3.1 HLS Steps

- *Compilation:* In the first step, HLS converts the functional specification of the encoded design into a formal/intermediate representation. Control and data dependencies are identified and an exhaustive control data flow graph (CDFG) is generated. HLS can also take advantage of the C/C++ compiler to optimize the high-level specification when generating the formal/intermediate representation of the design. Some of the optimizations include dead code elimination, constant propagation, variable optimization, etc. Depending on the algorithmic flow and user constraints, HLS may also use various optimizations to improve the RTL architecture in terms of throughput, area, latency, or power [22]. Some of these optimizations include loop unrolling, pipelining, resource sharing, etc.

- *Allocation:* The next step of HLS is to define and assign various hardware resources to operations. HLS analyzes various operations and maps them to different resources such as multipliers, adders, digital signal processing (DSPs), etc. The mapping of hardware resources is done from the available resource pool, i.e., technology library files or hardware specification.
- *Scheduling:* The next step is scheduling, which determines which design operations should occur when. It identifies the data dependence between the various operations and schedules them accordingly. If there is no dependence between operations, they can be scheduled in parallel to improve performance. Based on the clock frequency of the targeted design, it could also decide whether the operation should be a single cycle, multicycle, or multiple operations could be scheduled within a single cycle.
- *Binding:* The final step is binding upon scheduling. The hardware resources are allocated to the operations in this phase. Depending on the availability of hardware resources and operations scheduling information, HLS determines which hardware resource should perform which operation. Occasionally, due to hardware resource constraints, HLS can adjust the scheduling of operations. Once the binding has been successfully completed, the final RTL will be produced.
- *FSM Extraction:* In the FSM extraction step, the control signals are identified which controls the algorithmic data flow of the design. The control and data flow logic are merged together to generate the final RTL model.

After the RTL generation using High-level synthesis, the subsequent steps remain the same as previously shown in Fig. 6.1. An example shown in Fig. 6.3 contains the intermediate steps of high-level synthesis.

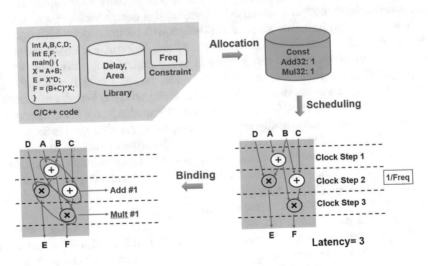

Fig. 6.3 High-level synthesis example containing the intermediate steps

6.4   What Is Obfuscation and Why Is It Necessary?

The foundry is one of the main entities in the integrated circuit (IC) design flow. An Untrusted foundry can provide unauthorized access to various adversaries (within or outside the foundry) to perform IC reverse engineering and overproduction. To address these security and trust vulnerabilities, different techniques have been proposed by the researchers using some form of locking. The threat model in this scenario is the untrusted foundry that has access to the GDS II file containing layout information and working IC for applying inputs and observing corresponding outputs. The locking proposition revolves around the idea of without the knowledge of the secret key, the unauthorized users cannot access the IC. Obfuscation can be defined as a form of locking to ensure the security of the design. Obfuscation at the high-level synthesis involves algorithmic locking, i.e., arithmetic, control flow, etc., at the C/C++ form of the design.

To secure the IPs against the mentioned attacks, locking has been used at different stages of design during the last decade. A widely used method is to insert secret keys at the gate-level netlist of the design to lock the design [23, 24]. However, with the emergence of SAT attack and machine learning-based attacks, the secret keys can be easily extracted to unlock the design [18, 25, 26]. Another approach is to use RTL hardening techniques by adding extra connections among the functional units [27]. Although this approach is more potent than gate-level locking techniques, inserting locking keys to obfuscate constant values and branches is challenging as the design is already optimized.

To increase the security of the design, obfuscation at the higher levels of abstraction can be considered as a better choice as the designer has complete access to unoptimized functionality, security-critical operations, and information, as well as the dependency between assets. For example, if a design includes any kind of cryptographic algorithm, the asset of the design will be the encryption keys. The security-critical operations will be any form of data/information flow through the key encryption module. As a result, there is every chance to perform obfuscation keeping design specifications, i.e., overhead (area, power, timing), resiliency, and locking key size into consideration. Therefore, along with the benefits of reducing design and verification efforts that HLS provides, it can also be used to obfuscate the design behavior and functionality effectively at C/C++. As discussed in Sect. 6.3.1, HLS involves several steps for creating CDFG, allocating operations to functional blocks, mapping operations to clock cycles, optimizing the number of functional units necessary to finally generating the RTL. These automated steps are inherent to the HLS process, and so it is very difficult to read or reverse engineer the generated RTL code due to the complex nature of each of these steps.

In the HLS obfuscation, the locking keys are inserted at the C/C++ level of the design. Then the design is transformed into RTL using HLS. The generated RTL design then undergoes another synthesis using the intended technology library to generate the gate-level netlist. As a result, firstly, the designer has the opportunity to perform locking key insertions effectively considering optimum output corruptibility. Secondly, two levels of transformations and optimizations enable better blending of the locking keys with the FSM states of the design. It creates logic and FSM obfuscations facilitating better SAT resiliency. Additionally, multiple transformations and optimizations hide the locking keys in a better way providing the design more resiliency towards machine learning and removal attacks.

6.5 Secure High-Level Synthesis for Hardware Obfuscation

In this section, a framework for obfuscation at the C/C++ level will be discussed. To protect the design, the followings steps need to be taken:

- **Step 1:** The framework takes the design specifications in C/C++, the size of the obfuscation key (Key_{Spec}), information on the desired overhead of the obfuscation (e.g., SAT attack resiliency, machine learning attack resiliency, area, power overhead, etc.) as inputs. In the first step, designers need to identify the design assets and critical operations of the design, such as encryption/decryption key, random numbers, plaintext form of a critical data, original equipment manufacturer (OEM) keys, original component manufacturer (OCM) keys, configuration bits that need to be protected against unauthorized access and reverse engineering.
- **Step 2:** In the second step, intermediate representation (IR) of the C/C++ code will be analyzed along with the CDFG to track the usage of the already identified assets at the previous step and to achieve optimum corruptibility of the outputs as maximum corruptibility makes it easier for SAT solver to discard incorrect keys. To achieve this, critical paths of the design and the blocks that are mostly used for correct functionality are identified by analyzing the CDFG and added to the potential candidate for locking.
- **Step 3:** In this step, the obfuscation keys are used to lock the obfuscation candidates. The details on each of these candidates are as follows:
  - *Function Calls:* The functions at the C/C++ level by either using fake functions, fake arguments, or by locking inside operations. The application of the locking case will depend on whether the function involves assets/critical operations or not. For critical operations, the function can be called multiple times and so the inside operation of the function can be locked. For other cases, fake function/fake arguments can be added to protect single instances of the function for incorrect keys.
  - *Control Flow Obfuscations:* Control flow of the design can be hidden by adding extra branches, changing the number of iteration loop, changing the array indexes, creating fake states similar to FSM obfuscation [28, 29], and skipping states. From the above, which candidate can be chosen will depend on the control flow and the analysis from steps 1 and 2 of the intended design.
  - *Arithmetic Operations:* Arithmetic operations can be locked by changing the type of operation.

- *Memory Unit:* The read and write indices can be modified in case of incorrect keys. Global variables also can be obfuscated by a part of the key.
- *Constant Values:* Constant values, i.e., encryption keys, configuration bits, and design assets, can be changed by bitwise operations (e.g., XOR, OR, AND, etc.) with the obfuscation key [30].

- **Step 4:** In this step, we will determine the number of key bits needed for obfuscation based on potential obfuscation points and the type of obfuscation. As mentioned earlier, the framework will take a certain key size (Key_{Spec}) that can be used in obfuscating the design. In this approach, considering the previous steps, the designer will also find out the total bits of keys needed for the design to be locked. For example, let us assume a for loop copies initial key indices to a temporary variable in an encryption module. Based on the design analysis, locking keys can be added for candidates such as changing array index, adding fake states, skipping states, etc. For each case, the size of the locking key may be different and it can be greater/smaller than the specification. If the number is smaller, then we can move to the next step. But if the required key bits are greater, we need to add a key expansion mechanism or reuse some of the original key combinations.
- **Step 5:** After selecting the number of key bits and locking the design using those keys, it is necessary to make sure locked design leads to correct functionality in case of correct keys and for incorrect keys, correct functionality cannot be achieved for any input. In this step, this validation is performed through simulation or formal methods.
- **Step 6:** In the final step, we need to evaluate the efficacy of the locked design toward various security attacks. To evaluate the security of the locked designs at C/C++, the design will be used by the HLS tool to generate RTL and adding technology library, the RTL can be synthesized to generate the gate-level netlist. The gate-level netlist will be already locked, and we will measure the overhead, i.e., area, timing, and power overheads. The resiliency of the design can be measured using SAT-based [18], key sensitization [31, 32], removal, and machine learning attacks. Finally, we can compare the results of this approach with conventional locking techniques.

The proposed framework is shown in Fig. 6.4. As can be seen from the figure, design specifications are taken as input. Then the intermediate representation and the control data flow graph (CDFG) have of the design are analyzed. After that, possible obfuscation candidates are selected based on the design. Then, the obfuscation key length is determined and if needed, key expansion or reusing the keys will be implemented. Then the design is locked with the locking points and locking candidates at the C/C++ level. Finally, the locked design will be validated and compared with other logic techniques in terms of overhead and resiliency.

The proposed technique is based on three performance parameters attack resiliency, area overhead, and obfuscation key size. Initially for a given design, for each obfuscation point, the design is synthesized to the corresponding gate-level netlist and these parameters are determined for each locking candidate and stored in

6 Security Assessment of High-Level Synthesis

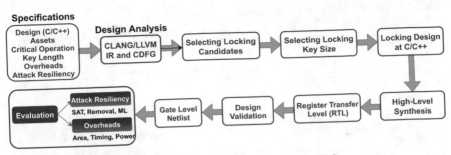

Fig. 6.4 A framework for locking designs at the higher level using HLS

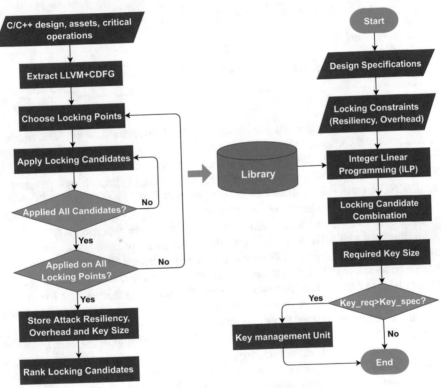

Fig. 6.5 Library generation and meeting performance specifications

a library. For a given specification, the scenario becomes an optimization problem where the target is to achieve maximum resiliency and minimum overhead and key size. To achieve that, we feed the library information to an optimization tool such as Integer Linear Programming (ILP) and for a large array of specifications, the tool provides the list of locking points and candidates needed to meet the specification. The set of locking points and subsequent locking candidates can be translated to the required key size to meet the specifications. The steps can be seen from Fig. 6.5.

Table 6.1 Comparison between different locking techniques

Locking type	Locking key size	Area overhead	Power overhead	SAT resiliency
Random insertion [33]	440 bits	14%	33.21%	CPU time = 3.381 s
Inserting XNOR/XNOR to avoid fault-analysis attack [31]	532 bits	14%	148%	CPU time = 224 s
Maximizing HD using MUX [34]	565 bits	14%	64.21%	CPU time = 2.61 s
Maximizing HD using XOR [34]	436 bits	14%	139%	CPU time = 24.76 s
Minimizing low controllability locations by inserting AND, OR [35]	1012 bits	14%	13.11%	CPU time = 0.325 s
Proposed approach	**16 bits**	**14%**	**7.95%**	**CPU time = 4416 s**

The proposed technique has been tested on the opensource 128 bit AES encryption C code and compared with different locking techniques at the gate-level netlist. For imitating a realistic scenario, we use a constrained area overhead less than 15%. The compared techniques were Random Logic Locking [33], Locking by inserting XOR/XNOR gates at specific locations proposed at DAC'12 [31], maximizing the Hamming Distance between the correct and wrong outputs using MUX and XOR gates at ToC'13 [34], and minimizing low controllability by inserting AND and OR gates proposed at IOLTS'14 [35]. For all the techniques, we keep the area overhead to around 14% and measure SAT resiliency, power overhead, and key size required. As can be seen from Table 6.1, the proposed approach provides SAT resiliency from 15× to 10,000× across techniques using a key size factoring at least 25× lower than other techniques with lower power overhead. It also needs to be mentioned that, the reported results here one of the many solutions available from this Approach. As a result, better resiliency/overhead can be achieved by changing the constraints of the optimization tool. Additionally, our approach provides a wide array of solutions with different combinations of attack resiliency, overheads, and key size. So, the designer can prioritize their requirements and get a possible solution that meets the intended performance requirements.

6.6 HLS Security

Although high-level synthesis is proving to be beneficial to different government and industrial entities. As discussed in the previous sections, HLS can be used as a security measure to efficiently integrate obfuscation in the design flow. However, HLS steps have been focused on optimizing the design to obtain performance gains in terms of throughput, latency, area, or power, with very little efforts towards security. As a result, HLS flow may indeliberately introduce some security vulnerabilities in the generated RTL designs due to its underlying optimization strategies. These vulnerabilities can be classified into four broad categories based on their effects/impacts on the generated RTL design. These categories can be summarized as follows:

- *Information Leakage:* It allows an adversary to spill/leak a security asset to observable points of the design.
- *Control Flow Violation:* It allows a malicious attacker to bypass the internal states of HLS generated control FSM to access a protected state of the design in an unauthorized fashion.
- *Fault Injection:* Certain design specification causes the generated RTL design to be vulnerable to fault injection/glitching attacks that aim the security of the design [36].
- *Side-Channel Leakage:* It causes the HLS generated RTL design to be vulnerable to side-channel leakage attacks.

Depending on the threat and use case of the generated RTL, a single source of vulnerability can be associated with multiple categories.

To provide proof-of-concept that existing HLS flow is vulnerable and proper measures should be in place to verify the security of the generated RTL designs, we assume the following threat model. We assume that the attacker may be an end-user directly interacting with the HLS generated hardware design. The attacker may have physical access to the device or is interacting remotely with the device. Moreover, we assume that security vulnerabilities in the generated RTL if not verified could make their way into the post-silicon device. Accordingly, we classify a data variable as a security asset which the company expects to be inaccessible by an adversary. The security asset could be input to the design controlled by other IP block or something which is stored/generated inside the design. Examples of secure assets are the encryption/decryption keys in the cryptographic systems, configuration bits for true-random number generators, or weights/biases of the trained ML/AI model.

In the following sections, we provide a detailed discussion on various performance optimization strategies currently utilized by existing HLS compilers. We also highlight how some of these strategies could become a source of potential security vulnerabilities in the generated RTL designs. Finally, we provide example cases of the introduced vulnerability using the OpenSSL AES test case and emphasize on the development of verification frameworks on top of the existing flow to make HLS flow security-aware.

6.6.1 HLS Optimizations

Figure 6.6 shows different optimization strategies that could be adopted by the HLS to generate efficient RTL. These strategies could be classified into throughput, latency, area, and power. Details on each strategy are discussed as follows.

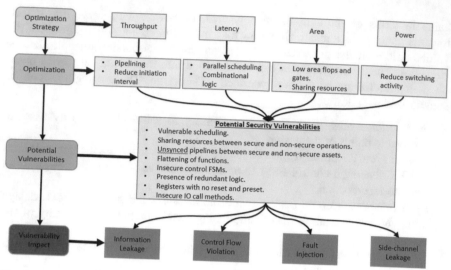

Fig. 6.6 Framework to identify potential security vulnerabilities caused due to HLS optimizations

- *Throughput Optimizations:* HLS could attempt to improve the throughput of the generated RTL by optimizing loops and functions in the C/C++ design specification. For this, HLS could identify such locations and insert registers between operations to generate efficient pipelines. Pipelines allow the design to process new data even before results from the previous iteration is written back. After how many cycles a pipeline can take new data for processing is known as initiation interval (II). Setting II to 1 means that design can take new data as input every clock cycle. The lower the II, the higher the pipeline's throughput.
- *Latency Optimizations:* In order to measure the effectiveness of the RTL, latency is a significant parameter that is defined as the number of clock cycles needed to complete one execution call. By default, the objective of HLS is to decrease the design's overall latency and may take any of the following steps to achieve it.

  - Parallel Scheduling: Schedule operations, functions, or loops in parallel to each other.
  - Partial/Full Loop Unrolling: Schedule multiple iterations of the loop together.
  - Optimize Algorithmic Trees: Identify multicycle adder and multiplication operations and schedule them in a single cycle.
  - Generate Combinational Circuitry: Identify multicycle operations to generate combinational circuit blocks.

- *Area Optimizations:* Another optimization strategy that HLS follows aims to reduce the overall design area. For example, HLS may use gates and flops from the resource pool that have less area compared to others. Table 6.2 provides area comparison of different D-flip flops for a 32 nm library. It can be clearly seen that for a single flop, area difference can range between 8%. This difference could be

6 Security Assessment of High-Level Synthesis

Table 6.2 Area comparison of different types of flip-flop standard cells of saed32hvt_ff0p85v25c library

Type	Standard cell	Area
DFF	DFFX1	6.60774
	DFFX2	7.62432
DFF with reset	DFFARX1	7.116032
	DFFARX2	7.878464
DFF with preset	DFFASX1	7.116032
	DFFASX2	8.132608
DFF with reset and preset	DFFASRX1	7.62432
	DFFASRX1	8.132608

significant for designs with a large number of flip-flops. Similarly, in order to reduce the total number of resources used in the created design, HLS can share resources for operations scheduled for different clock cycles. This could turn out to be very useful for designs with a large number of similar operations.

- *Power Optimizations:* The design's dynamic power consumption directly depends on the design's switching activity, as seen in the following equation.

$$\text{Power} = [(C_{pd} \times f_I \times N_{SW}) + \sum (C_{LN} \times f_{On})] \times V_{CC}^2 \qquad (6.1)$$

where P_T is transient power, C_{pd} is dynamic power dissipation capacitance, V_{CC} is supply voltage, F_I is input frequency, and N_{SW} is number of bits switching.

Therefore, HLS targets to optimize the switching activity of the design in order to control the dynamic power consumption of the design. HLS could schedule operations in such a way that the overall switching activity of the design is within the threshold limits at any given clock cycle. Currently, not all HLS tools are capable to optimize the generated RTL for power.

6.6.2 Potential Security Vulnerabilities Due to HLS Optimizations

In Sect. 6.6.1, we addressed a set of optimizations adopted by the HLS compiler to efficiently translate the C/C++ design into a functionally equivalent RTL design. As discussed, HLS compilers are not security-aware or take security under consideration while optimizing the RTL design different optimization strategies (throughput, latency, area, or power). As a result, translation from C/C++ to RTL may not be secure. In this section, we discuss some such scenarios where certain optimization or HLS steps could potentially become a source of vulnerabilities in the generated RTL.

1. *Non-secure Scheduling:* As previously described, HLS can try to reduce the overall latency of the design by scheduling multiple operations in parallel.

Since HLS is not security-aware, this simple optimization can cause a violation of security rules in certain secure applications. For example, the optimization can overlook the security rule "If X satisfies, then only Y," resulting in pre-computation of operation Y and then discarded them later when is X unsatisfied. For example, in security protocol, "HMAC then decrypt" [37], first the ciphertext is authenticated and only then the decryption should occur.

2. *Non-secure Resource Sharing:* HLS has no notion of secure and non-secure operations in the design. And to reduce the overall resource consumption, HLS can schedule different operations to use the same resources. As a result, secure and non-secure operations in the design may end up using the same resources (adder, DSP, multiplier, etc.). Furthermore, HLS can also share registers between different operations during the register binding process to minimize the overall register consumption of the design [38]. This could allow an adversary to use non-secure operations (assuming it can control them) to leak/flush the previously executed results of the secure operations.

3. *Unbalanced Pipelines:* As previously discussed, to improve the throughput of the design HLS can automatically generate pipelines in the RTL. However, since HLS has no awareness about secure and non-secure assets in the design, it makes no attempts to balance the pipelines between the assets. As a result, the secret asset may arrive early at the processing element (XOR, DSP, multiplier, etc.) while the non-secure asset is still in the pipeline. This could provide opportunities to an adversary to leak/flush out the secret assets from the processing element.

4. *Inlining of Functions:* HLS can optimize the function calls in the C/C++ code while generating RTL design. For example, instead of generating a hardware block for the function and calling it during each function call HLS could inline the function's hardware in the top module. As a result, this could violate the basic assumption that the function's local variables are destroyed after the execution. Due to the inlining of functions, the results/data which should be destroyed after function call may remain in the registers until the entire top module's execution. Depending on the threat model and the attacker's capabilities, the mere presence of secret information in the system registers after an execution call can provide adversary opportunities to cause information leakage.

5. *Registers with No Reset/Preset:* In the generated RTL, HLS uses registers to create efficient pipelines and store intermediate results. In algorithmic complex design, the number of registers required may exponentially increase. As a result, to optimize the area of the RTL, HLS may utilize registers with no reset or preset. Registers with no reset can cause sensitive information to be retained, raising the adversary's likelihood of leaking it. Whereas registers with no presets can cause incorrect initialization in the initial design cycles, leading to faulty results.

6. *Direct Latching of Primary Outputs:* The HLS automatically generates control FSMs which regulates that final computed data is latched to the primary output when the done signal is asserted. But sometimes to save area in the generated design, HLS can remove this output latching control logic. As a result, the registers holding intermediate values are directly connected with the primary outputs and the adversary can visualize all the intermediate values.

For example, Listing 6.1 shows a C code containing a function named "test." The "test" function takes an array "a" as input and returns the sum of all array elements. Providing "test" function to HLS will result in RTL with two primary ports corresponding to input "a" and output "total_sum." The value of "total_sum_t" is computed after each iteration and if the primary output port is directly connected to this register, it can cause the primary output port to reflect those changes after each iteration.

Listing 6.1 Example C code

```
1  int test(int a[10]){
2      int total_sum_t = 0;
3      for( int j=0; j<10; j++){
4          total_sum_t = total_sum_t + a[j];
5      }
6      total_sum = total_sum_t;
7      return total_sum;
8  }
```

For sensitive applications, such as AES, where ciphertext is modified after each round but should only be latched to the output port at the end of the final round, this form of behavior may be unsafe.

7. *Redundant Logic:* The HLS compiler adopts a template-based translation, where it has pre-built FSMs for various control logics like "start," "done," "reset," etc. Users can define which top-level signals it needs in the design or in cases the HLS itself does not use all the control signals when combination RTL is generated. These extra logic remains in the generated RTL but do not contribute towards the functionality of the design. These redundant signals could assist an adversary to orchestrate an attack or insert hardware Trojans in the RTL.

6.6.3 Case Studies

In this section, we present some case studies to demonstrate that potential vulnerabilities discussed in Sect. 6.6.2 do exist in the real-world when secure C applications are provided to HLS compilers without proper checks. For the purpose of demonstrating proof-of-concept we used OpenSSL implementation of AES encryption/decryption [39]. We chose OpenSSL because it is a widely used and most updated crypto library at the time of writing the chapter. The AES design used for this work has been adopted in multiple crypto libraries. For our experiments, we used Xilinx's Vivado HLS [40] which is widely used and commercially available for FPGA prototyping. Although the RTL output of HLS compilers is independent of the target hardware, the Vivado HLS uses Xilinx FPGAs to successfully bind operations to the hardware resources.

6.6.3.1 Unbalanced Pipeline Depths

As discussed in Sect. 6.6, HLS has no notion of secure/non-secure assets and makes no efforts to balance pipelines between them. As a result, the secret asset may reach the processing element early than the non-secret asset, providing opportunities for an adversary to leak the secret asset. In AES, for every round, plaintext undergoes various operations (Sbox substitution, mixcolumn, and shiftrows) before being xored with roundkey. As a result, we hypothesized that during HLS translation, the pipelines between roundkey and plaintext could be unbalanced and that could provide us opportunities to leak the roundkey. To test it, we wrote formal properties to verify if roundkey could be leaked after the first round. As shown in Fig. 6.7, we were able to leak the round key by flushing the pipeline after resetting the registers at a specific time. Figure 6.8 shows the functional validation of the leak in the AES encryption module where part of the first roundkey is directly leaked at the ciphertext output port. The abbreviations *rk*, *pt*, and *ct* in the figure represent the roundkeys, plaintext, and ciphertext, respectively. Whereas, *ap_clk*, *ap_rst* and

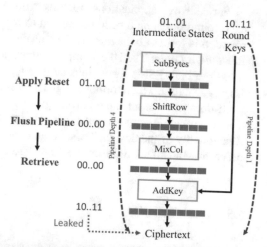

Fig. 6.7 Illustration of different operations in an AES round and unbalanced pipeline between secret and non-secret asset

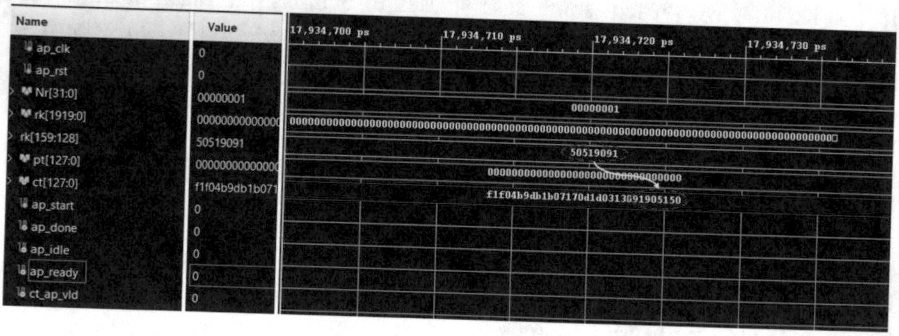

Fig. 6.8 Functional simulation showing leakage of round keys after first round in the AES encryption block

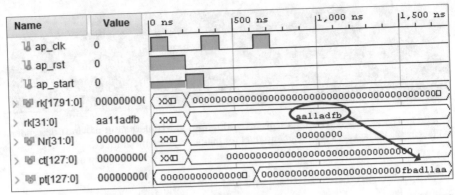

Fig. 6.9 Functional simulation showing leakage of round keys after first round in the AES decryption block

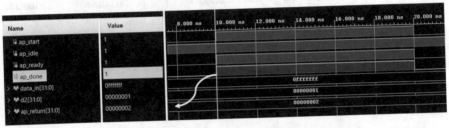

Fig. 6.10 Functional simulation of the RTL generated from conditional statements. The primary output carries the computed result before "ap_done" signal is asserted

ap_start are the top-level control signals introduced by the HLS compiler. We performed the same analysis on the AES decryption module and leaked the first roundkeys as shown in Fig. 6.9.

6.6.3.2 Generation of Combinational Circuits

During analysis, we found that Vivado HLS generates combinational circuits for conditional statements to optimize performance and latency. The HLS generated RTL is also appended with various block-level signals, such as "ap_done," "ap_start," "ap_idle," etc., to provide easy interfacing of the module with other IPs. Due to the generation of the combinational circuits, the computed values from the design get reflected on the primary output even before "ap_done" signal is asserted. Figure 6.10 shows one such simple example. If such HLS generated RTL modules are inserted into the SoC, exposing additional information at the output port when it is not expected to occur, this can become a source of security problems by causing information leakage.

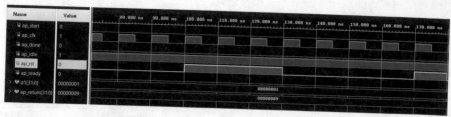

Fig. 6.11 The generated RTL uses registers with no reset. Plaintext from previous decryption cycle remains latched even after applying reset for multiple cycles

6.6.3.3 Uncleared Intermediate and I/O Registers

As discussed previously, HLS can utilize registers with no reset and preset to optimize the overall area of the generated RTL. As a result, the sensitive information may remain in the registers even after applying reset and will be susceptible to leakage by an adversary. We noticed this similar behavior in the Vivado generated RTLs. Figure 6.11 shows the functional simulation of the AES decryption block where the plaintext output port retains the previously decrypted value even after the reset has been asserted. The shown test case reveals the registers connected to primary outputs are not cleared even after applying reset. Similar behavior can be seen in the rest of the design registers.

6.6.4 Automated Verification and Prevention Strategies

The demonstrated test cases have shown that current HLS compilers do not consider security while translating C/C++ design specification to its corresponding RTL. As a result, various security vulnerabilities could be introduced in the generated RTL due to different optimizations performed by the HLS compiler.

Therefore, to prevent and verify for any security vulnerabilities that may exist in the generated RTLs, proper verification strategies and frameworks should be in place. It is possible to incorporate prevention and verification approaches at three distinct levels of the HLS translation.

1. C/C++ Level: This is a pre-translation stage when the developer is incorporating the algorithmic flow of the hardware in C/C++ and provides various specifications (directives and pragmas) to the HLS compiler. Based on these specifications HLS may undergo different types of optimizations. Therefore, strategies and frameworks at this level should target the prevention of security vulnerabilities rather than verification.

   The design developers should be provided with guidelines (Do's and Don'ts) for the secure HLS translation. The developed framework should also be able to automatically verify the design files and prompt warnings to the developer in case it violates any secure translation guidelines.

2. Intermediate/HLS Compiler Level: This is the during-translation stage while the HLS compiler is processing on the C/C++ design files. HLS translation is a multi-step process, and the HLS compiler generates a number of intermediate files after each step before the final RTL is generated. The framework developed at this level establishes a balance between the prevention and verification of security vulnerabilities.

   As previously discussed, the underlying algorithms and HLS optimizations are not security-aware. For example, efforts should be made towards making HLS capable of handling secure and non-secure assets in the design. HLS should also be capable of isolating the secure and non-secure operation of the design. Once the underlying HLS algorithms are security-aware, the verification frameworks could be established that can verify security vulnerabilities in intermediate files before progressing to the next steps of the translation.

3. RTL Level: This is an after-translation stage when HLS has translated the C/C++ design to its corresponding RTL design. The frameworks developed at this stage are focused completely on verification than prevention. The generated RTL design consists of both the algorithmic flow of the design and the control flow which is automatically generated by the HLS compiler. The verification strategies at this level could adopt formal verification methods to automate the verification process against known security weaknesses.

6.7 Need for Secure HLS for Obfuscation

As has been discussed in Sect. 6.6, translation from C/C++ to RTL using HLS is not a secured process by default. On the other hand, design obfuscation using secure HLS is a promising approach to protect IPs against reverse engineering and over-production. A secure HLS has many advantages discussed in Sect. 6.3 compared to conventional flow. On the other hand, HLS flow without a proper guideline to ensure the confidentiality and integrity of the design assets is unacceptable. As a result, to fully extract the best features HLS is capable of providing, in addition to superior performance such as more complexity to conventional locking, the related vulnerabilities need to be addressed and proper coding guidelines should be introduced. These steps can go a long way to not only produce intended performance from HLS but also protect the IPs against reverse engineering, overproduction, malicious functionality insertion, and IP piracy.

6.8 Summary

In this chapter, a locking technique has been discussed for securing the design against reverse engineering, IP piracy, hardware Trojan attacks, overproduction, as well as counterfeiting based on secure HLS. Unlike the conventional locking techniques at the gate-level netlist, this locking technique works on higher levels design abstractions providing the designer access to abstract specifications, design assets, and critical operations. It helps the designer to insert locking keys judiciously in selected locations and achieve optimum output corruptibility to increase SAT resiliency. Also, various steps of HLS and RTL synthesis provide multiple optimizations on the locked design making it difficult to access the locked design using machine learning and removal attacks. However, the core element of this technique is a secure HLS translation which if not ensured can lead to certain vulnerabilities while pushing for the best possible performances. These scenarios have also been discussed in this chapter. To address these issues, HLS needs to have secure programming rules and security analysis tools that can detect and identify these potential weaknesses that can occur during the process. It will not only facilitate better design at the hardware level from high-level abstraction but also pave the way for superior locking mechanisms for IP protection.

References

1. M. Tehranipoor, M. Yilmaz, K. Chakrabarty, Test-pattern selection for screening small-delay defects in very-deep submicrometer integrated circuits. IEEE Trans. Comput. Aided Des. Integr. Circ. Syst. **29**(5), 760–773 (2010)
2. X. Zhang, M. Tehranipoor, H. Salmani, *Integrated Circuit Authentication* (Springer, Cham, 2014)
3. J.C. Li, F. Sijstermans, Working smarter, not harder: Nvidia closes design complexity gap with high-level synthesis. https://go.mentor.com/4N9cP
4. M. Rostami, F. Koushanfar, R. Karri, A primer on hardware security: models, methods, and metrics. Proc. IEEE **102**(8), 1283–1295 (2014)
5. X. Wang, M. Tehranipoor, D. Zhang, M. He, Dynamically obfuscated scan for protecting IPs against scan-based attacks throughout supply chain, in *IEEE 35th VLSI Test Symposium (VTS)* (2017)
6. Q. Shi, G.K. Contreras, M. Tehranipoor, M.T. Rahman, D. Forte, CSST: Preventing distribution of unlicensed and rejected ICs by untrusted foundry and assembly, in *IEEE International Symposium on Defect and Fault Tolerance in VLSI and Nanotechnology Systems (DFT)* (2014), pp. 46–51
7. X. Wang, D. Zhang, M. He, D. Su, M. Tehranipoor, Secure scan and test using obfuscation throughout supply chain. IEEE Trans. Comput. Aided Des. Integr. Circ. Syst. **37**(6), 1867–1880 (2017)
8. M. Tehranipoor, F. Koushanfar, A survey of hardware trojan taxonomy and detection. IEEE Des. Test Comput. **27**(1), 10–25 (2010)
9. An experimental analysis of power and delay signal-to-noise requirements for detecting trojans and methods for achieving the required detection sensitivities.
10. M. Tehranipoor, M. Li, A. Davoodi, A sensor-assisted self-authentication framework for hardware trojan detection, in *Design, Automation & Test in Europe Conference & Exhibition (DATE)* (2012), pp. 1331–1336
11. K. Xiao, D. Forte, M. Tehranipoor, Efficient and secure split manufacturing via obfuscated built-in self-authentication, in *IEEE International Symposium on Hardware Oriented Security and Trust (HOST)* (2015), pp. 14–19
12. H. Salmani, M. Tehranipoor, J. Plusquellic, A layout-aware approach for improving localized switching to detect hardware Trojans in integrated circuits, in *IEEE International Workshop on Information Forensics and Security* (2010)

13. U. Guin, K. Huang, D. DiMase, J.M. Carulli, M. Tehranipoor, Y. Makris, Counterfeit integrated circuits: A rising threat in the global semiconductor supply chain. Proc. IEEE **102**(8), 1207–1228 (2014)

14. X. Zhang, M. Tehranipoor, Design of on-chip lightweight sensors for effective detection of recycled ICs. IEEE Trans. Very Large Scale Integr. (VLSI) Syst. **22**(5), 1016-1029 (2013)

15. U. Guin, X. Zhang, D. Forte, M. Tehranipoor, Low-cost on-chip structures for combating die and IC recycling, in *2014 51st ACM/EDAC/IEEE Design Automation Conference (DAC)* (2014)

16. N. Tuzzio, K. Xiao, X. Zhang, M. Tehranipoor, A zero-overhead IC identification technique using clock sweeping and path delay analysis, in *Proceedings of the Great Lakes Symposium on VLSI*, pp. 95–98 (2012)

17. R. Torrance, D. James, The state-of-the-art in IC reverse engineering, in *Cryptographic Hardware and Embedded Systems - CHES 2009*, ed. by C. Clavier, K. Gaj (Springer, Berlin, 2009), pp. 363–381

18. P. Subramanyan, S. Ray, S. Malik, Evaluating the security of logic encryption algorithms, in *2015 IEEE International Symposium on Hardware Oriented Security and Trust (HOST)* (2015), pp. 137–143

19. M. Yasin, B. Mazumdar, O. Sinanoglu, J. Rajendran, Removal attacks on logic locking and camouflaging techniques. IEEE Trans. Emerg. Top. Comput. (2017)

20. P. Chakraborty, J. Cruz, S. Bhunia, Sail: Machine learning guided structural analysis attack on hardware obfuscation, in *2018 Asian Hardware Oriented Security and Trust Symposium (AsianHOST)* (IEEE, Piscataway, 2018), pp. 56–61

21. anysilicon. ASIC design flow. https://anysilicon.com/asic-design-flow-ultimate-guide/.

22. P. Coussy, D.D Gajski, M. Meredith, A. Takach, An introduction to high-level synthesis. IEEE Des. Test Comput. **26**(4), 8–17 (2009)

23. Y.-W. Lee, N.A. Touba, Improving logic obfuscation via logic cone analysis, in *2015 16th Latin-American Test Symposium (LATS)* (IEEE, Piscataway, 2015), pp. 1–6

24. Y. Xie, A. Srivastava, Anti-sat: Mitigating sat attack on logic locking. IEEE Trans. Comput. Aided Des. Integr. Circ. Syst. **38**(2), 199–207 (2019)

25. P. Chakraborty, J. Cruz, S. Bhunia, Sail: Machine learning guided structural analysis attack on hardware obfuscation, in *2018 Asian Hardware Oriented Security and Trust Symposium (AsianHOST)* (2018), pp. 56–61

26. J. Rajendran, A. Ali, O. Sinanoglu, R. Karri, Belling the cad: toward security-centric electronic system design. IEEE Trans. Comput. Aided Des. Integr. Circ. Syst. **34**(11), 1756–1769 (2015)

27. R.S. Chakraborty, S. Bhunia, RTL hardware IP protection using key-based control and data flow obfuscation, in *2010 23rd International Conference on VLSI Design* (2010), pp. 405–410

28. Y. Lao, K.K. Parhi, Obfuscating DSP circuits via high-level transformations. IEEE Trans. Very Large Scale Integr. (VLSI) systems **23**(5), 819–830 (2014)

29. C. Pilato, F. Regazzoni, R. Karri, S. Garg, Tao: techniques for algorithm-level obfuscation during high-level synthesis, in *Proceedings of the 55th Annual Design Automation Conference* (2018), pp. 1–6

30. P. Chakraborty, J. Cruz, S. Bhunia, Surf: Joint structural functional attack on logic locking, in *2019 IEEE International Symposium on Hardware Oriented Security and Trust (HOST)* (2019), pp. 181–190

31. J. Rajendran, Y. Pino, O. Sinanoglu, R. Karri, Security analysis of logic obfuscation, in *Proceedings of the 49th Annual Design Automation Conference* (2012), pp. 83–89

32. B. Shakya, X. Xu, M. Tehranipoor, D. Forte, Cas-lock: a security-corruptibility trade-off resilient logic locking scheme. IACR Trans. Cryptogr. Hardw. Embed. Syst. **2020**, 175–202 (2020)

33. J.A. Roy, F. Koushanfar, I.L. Markov, Epic: Ending piracy of integrated circuits, in *2008 Design, Automation and Test in Europe*, pp. 1069–1074 (2008)

34. J. Rajendran, H. Zhang, C. Zhang, G.S. Rose, Y. Pino, O. Sinanoglu, R. Karri, Fault analysis-based logic encryption. IEEE Trans. Comput. **64**(2), 410–424 (2013)

35. S. Dupuis, P.-S. Ba, G. Di Natale, M.-L. Flottes, B. Rouzeyre, A novel hardware logic encryption technique for thwarting illegal overproduction and hardware trojans, in *2014 IEEE 20th International On-Line Testing Symposium (IOLTS)* (IEEE, Piscataway, 2014), pp. 49–54
36. K. Yang, Y. Jin, D. Forte, M. Tehranipoor, A. Nahiyan, K. Xiao, AVFSM: a framework for identifying and mitigating vulnerabilities in FSMs, in *Proceedings of the 53rd Annual Design Automation Conference* (2016)
37. J.G. Steiner, B.C. Neuman, J.I. Schiller, Kerberos: an authentication service for open network systems, in *Usenix Winter* (Citeseer, 1988), pp. 191–202
38. Y. Hara-Azumi, T. Matsuba, H. Tomiyama, S. Honda, H. Takada, Impact of resource sharing and register retiming on area and performance of FPGA-based designs. Inf. Media Technol. **9**(1), 26–34 (2014)
39. openssl. https://github.com/openssl/openssl/tree/master/crypto/aes
40. Xilinx. Vivado hls. https://www.xilinx.com/products/design-tools/vivado/integration/esl-design.html

Chapter 7
CAD for Side-Channel Leakage Assessment

Adib Nahiyan, Miao (Tony) He, Jungmin Park, and Mark Tehranipoor

7.1 Introduction

Cryptographic primitives have been widely employed in embedded computing systems, mobile devices and Internet of things (IoT) as well as high assurance electronic systems—military, aerospace, automotive, transportation, financial, and medical. Although the security of cryptographic algorithms has been proven to be mathematically sound, their implementations are often not. It has been demonstrated that the security of a cryptosystem can be broken by exploiting the side-channel information, e.g., power, timing, and electromagnetic signals leaking from its hardware implementation [1, 2]. These attacks pose serious threats because they are non-invasive and are applicable to all cryptographic schemes.

Power side-channel attacks (SCAs) exploit the weaknesses in the hardware implementations of crypto algorithms to leak sensitive information, e.g., the encryption key. A number of SCAs, such as differential power analysis (DPA) [3], correlation power analysis (CPA) [4], mutual information attack (MIA) [5], and partitioning power analysis (PPA) [6], have been proposed and successfully demonstrated over the past two decades. The underlying principle of these attacks is to exploit the relation between intermediate data and power consumption when attempting to extract the secret key [7]. Different countermeasures have also been proposed to defend against SCAs, e.g., "Masking Mechanism" which attempts to randomize the intermediate values of a cryptographic operation in order to mask the dependencies between these values and power consumption [8–10].

Due to rise of power side-channel attacks, power side-channel leakage (PSCL) assessment has become very important. Without an accurate PSCL assessment,

A. Nahiyan (✉) · M. (Tony) He · J. Park · M. Tehranipoor
University of Florida, Gainesville, FL, USA
e-mail: adib1991@ufl.edu; miaohe@uchicago.edu; jungminpark@ufl.edu;
tehranipoor@ece.ufl.edu

© The Author(s), under exclusive license to Springer Nature Switzerland AG 2021
M. Tehranipoor (ed.), *Emerging Topics in Hardware Security*,
https://doi.org/10.1007/978-3-030-64448-2_7

171

security mechanisms of the design can be useless as attackers can bypass them by exploiting side-channel characteristics. It is meant to evaluate the amount of information leakage from crypto hardware. Different metrics, such as signal-to-noise ratio (SNR) [11, 12], t-statistic [13, 14], success rate [15], and information theoretic [16], have been proposed to analyze the vulnerability of crypto hardware to power side-channel. PSCL assessment begins with a security analyst developing an attack model followed by a verification/measurement setup. These processes involve identifying a target function, developing a power model, performing crypto operations for many plaintexts, and collecting their corresponding power traces on a prototype device. The security analyst then evaluates how vulnerable the cryptographic implementation is to SCAs using the aforementioned metrics [17].

It is also important to SCA community to identify side-channel vulnerabilities as early as possible during hardware design and validation phases. Unlike software, there is little to no flexibility in changing or updating the fabricated hardware design. If a vulnerability is discovered after manufacturing while the hardware is fielded, it may cost the company millions of dollars in lost revenues, recalls, and replacement costs. Researchers have proposed techniques suitable for pre-silicon PSCL assessment [18, 19].

7.2 Preliminaries: Power Side-Channel Attacks

In this section, we present a brief background on power side-channel attacks. We primarily focus on DPA and CPA which are considered the most widely adopted power side-channel attacks. These attacks exploit the correlation between the power consumption and some intermediate data related to the key [7]. DPA and CPA attacks rely on the following procedures.

Target Function Identification The first part of the DPA and CPA attack is to identify a specific operation of the encryption algorithm which is a function of the secret information (e.g., the private encryption key) and the variable that an attacker can control (e.g., plaintext) or observe (e.g., ciphertext). We define this operation as a target function, denoted by $T = f(x, k)$. Here, k is the key and x is the controllable or the observable variable to the attacker, i.e., plaintext/ciphertext. An example of a target function is the AES SubBytes or SBox operation. An attacker targets a single bit of T or a subset of T for DPA/CPA attacks.

Power Trace Collection The second step is to measure the power consumption of the cryptographic device while it encrypts or decrypts different data blocks corresponding to the target function. We denote the power trace of data block D_i as $P_i = \{p_{i,1}, \ldots, p_{i,t^*}, \ldots, p_{i,m}\}$, where m denotes the length of the trace and p_{i,t^*} is the power consumption when the target function is performed. Successful DPA and CPA require collecting thousands of P_i traces for different data blocks.

Power Model The third step is to create a hypothetical power consumption model, typically based on Hamming distance (HD) or Hamming weight (HW). Here, the attacker estimates the power consumption by calculating the HD or HW from the output of the target function. The attacker also makes a key hypothesis that enumerates all possible values for the subkey k under attack. Based on the key hypothesis k_j, the attacker applies a plaintext or observes a ciphertext and calculates the hypothetical intermediate value $T_j = f(x_j, k_j)$. In the next step, the hypothetical intermediate values are mapped to the hypothetical power consumption values using either the HD or HW model to estimate the hypothesis power $h_j = HD(T_j)$ or $h_j = HW(T_j)$, respectively.

Key Extraction In the fourth step, the attacker compares the hypothetical power consumption model with the measured power traces to check if the guess key hypothesis is the actual key. In DPA attack, the attacker uses difference-of-means (DOM) δ, i.e., the difference between the average power traces. δ is defined as follows:

$$\delta = E[P_{T=1}] - E[P_{T=0}], \tag{7.1}$$

where $E[P_{T=1}]$ represents the expected value of power traces when the specific bit T (e.g., the least significant bit (LSB)) is 1 and $E[P_{T=0}]$ represents the expected value of power traces when the specific bit T is 0. The DOM δ for the correct key guess should be larger than those for incorrect key guesses, revealing the correct key with higher probability given more traces.

For CPA attack, the attacker computes the Pearson correlation coefficient [4] between the hypothesis power model and the collected power traces. The Pearson correlation coefficient is computed as follows:

$$\rho = \frac{E[H - E[H]][P - E[P]]}{\sqrt{Var[H]Var[P]}}, \tag{7.2}$$

where $E[H - E[h]][P - E[P]]$ represents the covariance between the hypothesis power H and the power traces P, and the $Var[H]$ and $Var[P]$ represent the variance of H and P, respectively. In CPA attack, the attacker exploits the largest ρ value to identify the correct key hypothesis.

7.3 Side-Channel Leakage Assessment

In this section, we present a brief overview of the existing techniques for power side-channel leakage assessment (PSCL). These techniques are used by the semiconductor design house who develops the hardware crypto accelerator for the purpose of identifying potential side-channel leakage vulnerability. We classify these techniques into the following three categories:

7.3.1 Post-Silicon Leakage Assessment

Messerges et al. [11] proposed the signal-to-noise ratio metric (SNR) to evaluate the effectiveness of DPA attacks where a higher SNR value indicates a more effective attack. Gierlichs et al. [5] proposed a distinguisher based on mutual information between the observed power traces and hypothetical power leakage to rank key guesses. Fei et al. [7] proposed a general statistic model based on maximum likelihood estimation that can model and estimate the success rate (SR) of side-channel attacks such as DPA or CPA. Gilbert et al. [13] proposed to use TVLA test to evaluate the side-channel resistance of a cryptographic module. In this technique, the collected power traces are partitioned into two sets based on the intermediate values related to the secret information. Then the t-statistic is calculated to quantify the confidence level of the two sets being statistically different. A t-threshold value of 4.5 was used to test if the device leaks a side-channel information. This technique only provides a pass/fail (not the amount of leakage) test and can produce false positive results [20]. Moradi et al. [20] proposed to use Pearson's χ^2-test and t-test to address some limitations of TVLA.

7.3.2 Simulation Based Leakage Assessment

Veshchikov et al. [21] presented a comprehensive survey of simulators for side-channel analysis, e.g., PINPAS [22], SCARD [23], NCSim [24], OSCAR [25], etc. Among these reported side-channel simulators, only two (SCARD and NCSim) work with hardware crypto modules, whereas the rest deal with software crypto algorithms implemented in microprocessors. These simulators, in general, emulate the execution of a program under analysis and use simple power estimate techniques, e.g., hamming weight or hamming distance. In addition to not being applicable to ASICs and FPGAs, these simulators cannot produce accurate results as their power estimation model is not derived from the actual physical characteristic of the implemented design. SCARD, on the other hand, aims to develop an accurate spice model for power side-channel leakage simulation. However, SCARD requires simulation at spice level for many input patterns which is not feasible as well as cannot be used in pre-silicon when the prototype device is not available. NCSim utilizes commercial power estimation tool to evaluate DPA resistance at the gate-level. However, this technique also requires thousands of plaintexts and therefore, may require prohibitively large assessment time overhead. Bayrak et al. [26–28] propose automated application and verification of power analysis countermeasures on software implementation of crypto algorithms. However, these techniques cannot be applied to hardware PSCL.

Table 7.1 Comparison: pre- and post-silicon PSCL evaluation

	Pre-silicon			Post-silicon
	RTL	Gate-level	Layout	
Time	Medium	High	Very High	Low
Accuracy	Low	Medium	High	Very High
Flexibility	High	Medium	Low	Not feasible (ASIC); difficult (FPGA)

7.3.3 Pre-silicon Leakage Assessment

Huss et al. [17, 29] proposed a framework named AMASIVE for side-channel vulnerability assessment. AMASIVE identifies the hypothesis function for Hamming Weight/Hamming Distance model which is used for side-channel vulnerability assessment. However, this framework can only identify the hypothesis function, and the final vulnerability assessment still needs to be performed on a prototype device requiring thousands of plaintexts.

7.3.4 Pre- vs. Post-Silicon Leakage Assessment

We summarize the assessment time and accuracy of PSCL assessment as well as the flexibility to make design changes at different pre-silicon design stages w.r.t. the post-fabricated device level in Table 7.1. PSCL assessment at different design stages has a trade-off between time/accuracy and flexibility. While the post-silicon PSCL assessment has the highest accuracy and the fastest processing time, flexibility to make design changes to address potential vulnerabilities is the worst. For ASIC, any post-fabrication modification is not feasible, whereas for FPGA post-synthesis bitstream modification is difficult. On the other hand, pre-silicon PSCL evaluation offers better flexibility to address potential vulnerabilities at the cost of poor accuracy and long processing time.

7.4 SCRIPT Framework for Pre-silicon Leakage Assessment

In this section, we describe the SCRIPT framework for PSCL assessment in details. A high-level overview of the SCRIPT is shown in Fig. 7.1. The framework first extracts the underlying properties of the function which cause side-channel leakage. It then utilizes information flow tracking (IFT) engine to identify registers (termed as target registers) which exhibit these properties. The IFT engine works on the gate-level netlist and takes some inputs from the user, e.g., the name of the key and plaintext input ports. SCRIPT also employs formal verification technique

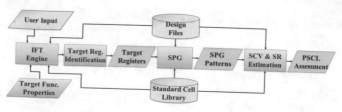

Fig. 7.1 Overview of the SCRIPT framework for power side-channel leakage (PSCL) assessment. PSCL assessment refers to estimating SCA success rate at pre-silicon stage and predicting the number of plaintexts for a successful SCA

to generate specific input patterns (plaintexts) which produce maximum power difference in the target registers. The difference in the power consumption of the target registers can be exploited by the adversary during the side-channel attack. SCRIPT utilizes a metric named SCV to provide a quantitative assessment of how vulnerable the design is to side-channel attack. Before we provide details of the SCRIPT framework, we first present the threat model employed in this framework.

7.4.1 Threat Model

SCRIPT framework is designed to be used by the semiconductor design house who develops the hardware crypto accelerator for the purpose of identifying potential side-channel leakage vulnerability. The verification engineer using SCRIPT has access to the gate-level design of the crypto accelerator along with the standard cell libraries. The verification engineer has the white-box knowledge of the design, i.e., knows the ports corresponding to the key, plaintext, and ciphertext; however, he/she does not have the skill set for performing SCA or analyzing vulnerabilities to SCA. Therefore, he/she will use the SCRIPT framework for performing PSCL assessment.

From a side-channel vulnerability assessment point of view, this framework intends to find the worst-case leakage scenario. Therefore, SCRIPT considers a strong attack model where the attacker performing the side-channel attack has in-depth knowledge of the cryptographic algorithm and has some knowledge about the implementation. For example, the attacker knows at which time instance a specific round operation occurs. SCRIPT also assumes that the attacker has full control over the plaintext and can perform any number of encryption operations on plaintext data. The attacker also has physical access to the power port which allows him to observe the power traces.

7.4.2 Properties of Target Function

A function needs to have specific properties to be targeted for side-channel attacks. SCRIPT utilizes the following properties which are inherently exploited by side-channel attacks.

- **P1: Function of the secret**. The target function should be a function of the secret information, e.g., encryption/ decryption key or an intermediate value which is related to the key. As a simple example, the AES SBox operation is a function of the key.
- **P2: Function of the controllable inputs**. The target function should also be a function of a variable or an input, e.g., plaintext, that an attacker can control. This property allows an attacker to control the input of the target function to create the power hypothesis model and guess the intermediate values of the secret. For example, the AES SBox operation of the first round exhibits this property as it is the function of the plaintext that an attacker can control.
- **P3: Function with confusion property**. This property dictates that the target function needs to possess the confusion property, i.e., one output bit of the target function should depend on more than one key input bit. Confusion property is quantified by the confusion coefficient metric [30] which measures the probability that given two different key guesses the respective power hypothesis model is different. This property ensures that an adversary can isolate the correct key hypothesis from the guessed key space. The AES SBox exhibits this property with high confusion coefficient metric.
- **P4: Function with divide-and-conquer property**. The target function also needs to be a function of a small subset of the secret information. This property allows the adversary to apply divide-and-conquer strategy to focus on one subset of the secret information at a time and extract the subset. The AES SBox operation has also this property since it depends on the 8-bit subkey.

A function having these properties is defined as "Target Function" and denoted by T. The AES SBox operation is used as an example of the target function, though, other crypto functions (e.g., AddRoundKey of AES) or non-crypto functions (e.g., logic obfuscation [31]), having the aforementioned properties can also be used as target functions. SCRIPT formally defines the target function as follows:

$$T = \{f(k', p') | \ k' \subset k \ \& \ size(k') > 1\}, \tag{7.3}$$

where k is the key and p' is the variable that an attacker can control. $k' \subset k$ represents the divide-and-conquer property, whereas $size(k') > 1$ represents the confusion property. These four properties are prerequisites for DPA and CPA as well as template and profiling SCA attacks. However, not all side-channel attacks exploit all four of these properties. For example, simple power analysis attack does not have controllable input requirement (property 2). Therefore, these four properties do not guarantee that they are sufficient for all possible side-channel attacks. However, existing properties could be modified, or additional properties can be added to cover such attacks as well.

Target Register SCRIPT defines the registers that store the output values of the target function (i.e., T) as target registers. The switching characteristic of the target registers embeds the four properties of T that are exploited by side-channel attacks. Therefore, the switching power of these registers and their corresponding fan-in combinational logic gates contribute greatly to power side-channel leakage. This framework autonomously identifies target registers using information flow tracking which is described in the following subsection.

7.4.3 Identifying Target Registers Using IFT

SCRIPT utilizes information flow tracking (IFT) to identify target registers which meet the properties discussed in Sect. 7.4.2. It uses fault propagation to track information propagation [32, 33]. Below, we first describe SCRIPT's IFT engine and then discuss how the IFT is used to identify target registers.

7.4.3.1 IFT Engine

IFT engine is based on the concept of modeling a secret information (e.g., the encryption key) as a stuck-at-0 and stuck-at-1 fault[1] and leveraging the ATPG tool [34] to detect those faults. A successful detection of faults means that the logical value of the key carrying net can be observed through the observe points, i.e., the registers. In other words, there exists an information flow from the fault location to the observe points. ATPG employs path sensitization algorithm [34] which exhaustively analyzes if there exists any information flow path from a key bit to an observe point and automatically derives the input vectors which enables this information propagation. Figure 7.2 shows the overall flow of our IFT engine and its four main steps: Initialization, Analysis, Propagation, and Recursive.

1. **Initialization**: This step takes the name of the key input ports to which IFT will be applied (shown in Fig. 7.2 as *key*), the gate-level netlist of the design, and the technology library (required for ATPG analysis) as inputs. Then, the engine adds scan capability to all the registers/flip-flops (FFs) in the design to make them controllable and observable. Also, it applies masks to all FFs so that it can track key propagation to each FF independently. Applying masks is an important step as it allows controlling fault propagation to one FF at a time.

---

[1]Stuck-at-fault is modeled by assigning a fixed logic value (0 or 1) to a single net or port of a design [34].

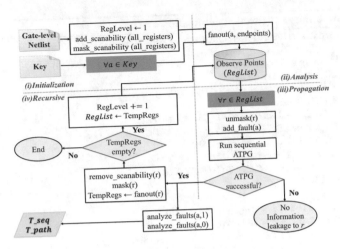

Fig. 7.2 IFT algorithm utilizing partial-scan ATPG to identify the registers where a key bit propagates to

2. **Analysis**: This step utilizes fanout analysis for each key bit $a \in key$ to find the FFs that are in the fanout cone of a. These FFs are potential candidates of target registers as a key bit a can potentially propagate to them. This step generates a list of potential candidates of target registers (called $RegList$). This list is used in the next stage as shown in Fig. 7.2.

3. **Propagation**: This step tracks the propagation of each key bit a to each individual FF. To perform a comprehensive analysis of potential points of the key bit propagation, each FF must be analyzed separately. For each $r \in RegList$ (shown in Fig. 7.2), the applied mask is removed, so the key bit propagation to r can be tracked. The next step adds the key bit a as the only stuck-at fault in the design and runs ATPG algorithm in the sequential mode to find paths to propagate $a = 0$ and $a = 1$ to FF r. If both, $a = 0$ and $a = 1$, can be detected from r, then there exists an information flow from a to r and r is marked as a potential target register. This step also stores the propagation path (T_{path}) as well as the control sequence (T_{seq}) required for the key bit propagation for further analysis. Note that T_{seq} contains the list of input ports and control registers which controls the information propagation from a to r.

4. **Recursive**: This step leverages our partial-scan technique along with sequential ATPG to find propagation paths through all sequential levels until the output or the last-level FFs are reached. Here, the function $removescanability$ (shown in Fig. 7.2) makes the ATPG tool treat r as a non-scan FF for simulation purposes without redoing scan chain insertion. The FF's output ports Q and QN are used to get a new fanout emerging from r to the next level of registers. To find information flow through multiple levels of registers, the scan-ability of all identified registers in $RegList$ is removed incrementally and sequential ATPG

Algorithm 1 Target registers identification

1: **procedure** IDENTIFYING AND GROUPING TARGET REGISTERS
2: **Input:** $RegList, T_{path}$
3: **Output:** Group of Target Registers
4:     **for all** $r \in RegList$ **do**
5:         $TSD_r \leftarrow Extract\ sequential\ depth\ from\ T_{path}$
6:         $TCK_r \leftarrow FanIn(r, TSD_r) \in Key$
7:         $TCPI_r \leftarrow FanIn(r, TSD_r) \in Plaintext$
8:         **if** $((length(TCPI_r) \geq 1)\ \&\ (length(TCK_r) > 1)\ \&$
9:         $(length(TCK_r) < K_{Th}))$ **then**
10:           $r \rightarrow Target\ Register$
11:           $G_r \leftarrow \{r, TCK_r, TCPI_r, TSD_r\}$
12:         **end if**
13:     **end for**
14:     **for** $<i = 1,\ i < length(G_r),\ i++>$ **do**
15:         **for** $<j = i,\ j < length(G_r),\ j++>$ **do**
16:           **if** $G_r(i):TCK_r == G_r(j):TCK_r$ **then**
17:             $G_r(i) = \{G_r(i)||G_r(j)\}$
18:           **end if**
19:         **end for**
20:     **end for**
21: **end procedure**

---

is used to create propagation paths from key bit a to subsequent-level registers. This process continues until the last level of registers.

The output of our IFT engine is a list of registers/FFs where the key propagates to, i.e., key observe points ($RegList$), and the propagation path (T_{path}) along with the stimulus vector (T_{seq}) for asset propagation for each FF, r. In the following subsection, we discuss how these information are utilized to identify target registers.

7.4.3.2 Target Registers Identification

Once SCRIPT has identified the registers $RegList_a$ where a key bit a propagates to, it analyzes the stimulus vector (T_{seqa}) to check if the registers show the properties of the target function (discussed in Sect. 7.4.2). All registers in $RegList_a$ satisfy the first property of *P1: Function of secret information*. The reason is that the key bit a propagates to $RegList_a$. SCRIPT utilizes Algorithm 1 to analyze which registers in $RegList_a$ satisfy the remaining three properties (P2-P4).

Algorithm 1 first takes the $RegList$ and T_{path} as inputs (Line 2). Then for each r in the $RegList$, Algorithm 1 extracts the sequential depth TSD_r from T_{path} (Line 5). Next, the algorithm performs the successive fan-in analysis to find the (1) control key port names TCK_r and (2) control plaintext port names $TCPI_r$ (Lines 6–7). Here, $FanIn(r, TSD_r)$ refers to performing fan-in analysis of r up to the sequential depth TSD_r. Algorithm 1 then checks for the following properties:

- Whether one or more plaintext input ports control key bit propagation to r ($length(TCPI_r) \geq 1$). If yes, then property *P2: Function of the control input* is satisfied.
- Whether more than one key input ports control key bit propagation to r ($length(TCK_r) > 1$). If yes, then property *P3: Function with confusion property* is satisfied.
- Whether the number of key input ports that control key bit propagation to r is less than K_{Th} ($length(TCK_r) < K_{Th}$). If yes, then property *P4: Function with divide-and-conquer property* is satisfied. Here, the K_{Th} value is set to be 32. K_{Th} determines the number of possible key guesses ($2^{K_{Th}}$) for SCAs. Note that $K_{Th} = 32$ does not refer to the key length of an encryption algorithm but refer to the maximum limit considered for the divide-and-conquer property.

If all properties are satisfied (Line 8), Algorithm 1 marks r as a target register (Line 10) and stores $\{r, TCK_r, TCPI_r, TSD_r\}$ in a G_r variable (Line 11). After completing the above analysis for all $r \in RegList$, the algorithm analyzes the G_r variables to place all r with the same control key bits (TCK_r) in the same group (Lines 14–18). It is an important step as it identifies which key bits control which particular set of target registers.

7.4.3.3 Target Registers of AES

In this subsection, a Galois field based AES design is utilized as an example to illustrate how SCRIPT utilizes the IFT engine along with the Algorithm 1 to identify target registers of this AES implementation. The details of the AES design are discussed in Sect. 7.4.5.1.

The analysis starts from a key bit 32 ($KEY[32]$). Here any key bit can be analyzed and IFT engine will find the corresponding target registers, if they exist. Figure 7.3 illustrates the target register identification of the AES implementation. IFT engine first searches for the registers where the $KEY[32]$ bit propagates to. It finds the data register, $Data Reg[32]$ (which stores the intermediate round results) along with some key registers (which stores the intermediate key expansion results). Then SCRIPT framework checks whether these registers satisfy the properties of the target function (discussed in Sect. 7.4.2) using the target register identification algorithm (discussed in Algorithm 1). SCRIPT finds that the key registers are not controlled by plaintext inputs and therefore, property *P2: Function of the control input* is not satisfied. For $Data Reg[32]$, SCRIPT finds that only a key bit controls this register and therefore, *P3: Function with confusion property* is not satisfied. Therefore, the register is not a target register and is marked as red in Fig. 7.3. Then, the SCRIPT framework searches for the second level registers. Now, it finds $Data Reg[0 \text{ to } 31]$ along with some key registers. SCRIPT finds again that key registers do not possess *P2* property and therefore, are not target registers. The $Data Reg[0 \text{ to } 31]$, on the other hand, controlled by 32 key inputs and 32 plaintext inputs and therefore, $Data Reg[0 \text{ to } 31]$ possess all the properties of Algorithm 1

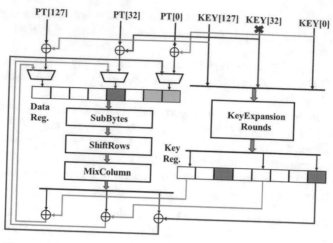

Fig. 7.3 Hardware architecture of the Galois field based AES design. The red marked lines illustrate the IFT process. The green marked registers represent Target Registers

and are the target registers (marked in green in Fig. 7.3). SCRIPT continues to search for subsequent-level registers. However, all the *Data Reg* after the second level violate property *P4: Function with divide-and-conquer property* and therefore, are not target registers. At the end, SCRIPT returns the *Data Reg*[0 *to* 31] along with the 32 controlling key and plaintext input port names and the sequential depth 2.

7.4.4 SCV Metric

SCRIPT uses an IFT-based side-channel vulnerability metric (*SCV*) for PSCL assessment at the pre-silicon design stage. *SCV* is defined as follows:

$$SCV = \frac{P_{signal}}{P_{noise}} = \frac{P_{T.hi} - P_{T.hj}}{P_{noise}}, \qquad (7.4)$$

where the P_{signal} refers to the difference in power consumption during the target function operation and P_{noise} refers to the power consumption of the rest of the design. P_T represents the average power consumed when performing the target function. Let us consider that the target function consumes $P_{T.hi}$ and $P_{T.hj}$ power when the hamming weight (HW) of the output of the target function is $hi = HW(T_i)$ and $hj = HW(T_j)$, respectively, for ith and jth inputs. The difference between $P_{T.hi}$ and $P_{T.hj}$ is exploited during the SCA. Therefore, $P_{signal} = P_{T.hi} - P_{T.hj}$ (Fig. 7.4).

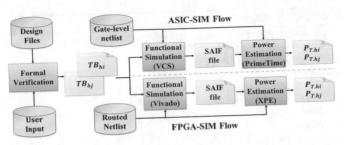

Fig. 7.4 Overall flow for estimating power using SPG. The ASIC-SIM flow shows the gate-level power simulation commonly used for ASIC chips. The FPGA-SIM flow shows the power estimation for Xilinx FPGAs

7.4.4.1 SPG (SCV-Guided Pattern Generation)

SCRIPT utilizes formal verification to generate a few SCV-guided patterns (named, SPG) and use those in functional simulation tool along with power estimation tool to get power corresponding to the patterns. Here, SCV-guided patterns refer to patterns which have following two properties:

- The SCV-guided patterns need to produce our desired hamming weight at the target registers.
- The SCV-guided patterns should only induce switching in the logic related to the target function while muting the switching of the rest of the design.

These properties allow SCRIPT to estimate the power consumption only for producing our desired HW at the target function. The details of the SPG algorithm are presented in Algorithm 2.

Algorithm 2 first takes the list of target registers (r), control plaintext input ports ($TCPI$), and sequential depth (TSD_r) as inputs from the IFT engine (discussed in Sect. 7.4.3.1). It also takes the desired Hamming weight $hi = HW(r)$ and the design files from the user. Then, Algorithm 2 creates some constraints for the formal verification tool. It first applies a fixed input constraint for all key bits and plaintext bits which are not among $TCPI$ (Lines 5–6). Logic 0 is applied for all fixed input constraints. Then, the algorithm applies a bind constraint to $TCPI$ (Line 7) which tells the tool to use the same input plaintext bits in $TCPI$ for all clock cycles. These constraints ensure that switching only occurs in the logic of the target function while muting the switching of the rest of the design. "Reset" constraints are also applied to ensure that every pattern generation process starts from the same initial condition (Lines 8–10).

Next, Algorithm 2 develops the assertions which express the desired behavior of a design under test. In our case, the assertion represents the property when the hamming weight of the r is hi and when the clock count is equal to the sequential depth (TSD_r) (Line 12). The latter part is important because it tells the formal tool to prove the assertion for clock cycles when the key propagates to the "target register." Then, the assertion (with *never* statement) is negated

---

Algorithm 2 SPG algorithm

1: **procedure** FORMAL VERIFICATION BASED PATTERN GENERATION
2: **Input:** Design files, $G_r\{r, TCPI_r, TSD_r\}$, hi
3: **Output:** SCV-guided Pattern (PI_{hi})
4:    **#Constraints**
5:    $Apply\ fixed\ input\ constraint \to K$
6:    $Apply\ fixed\ input\ constraint \to PI \notin TCPI_r$
7:    $Apply\ bind\ input\ constraint \to TCPI_r$
8:    $Apply\ Reset$
9:    Run
10:    $Remove\ Reset$
11:    **#Assertion**
12:    $psl\ A1: assert\ never\ ((HW(r) == hi)\ \&\&\ (clkcount == TSD_r))$
13:    **#Testbench Generation**
14:    $Prove$
15:    $Export\ testbench \to PI_{hi}$
16: **end procedure**

---

and run the formal verification tool to generate a counter example for the given assertion and generate a sequence of input patterns which causes the negation of the assertion to fail. In other words, by forcing the formal tool to verify the negation of the assertion, the tool generates patterns which satisfy this condition: $(HW(r) == hi)\ \&\&\ (clkcount == TSD_r)$. Lastly, the patterns are exported in Verilog testbench format which contains the SCV-guided patterns (PI_{hi}).

7.4.4.2 SPG for AES

In this subsection, a Galois field based AES design is used (discussed in Sect. 7.4.3.3) as an example to illustrate how SCRIPT employs Algorithm 2 to generate SCV-guided patterns. SCRIPT first takes the list of target registers ($r = DataReg[0\ to\ 31]$), control plaintext input ports ($TCPI = PT[0\ to\ 31]$), and sequential depth ($TSD_r = 2$) as inputs from the IFT engine (see Sect. 7.4.3.3). Then, it sets the desired Hamming weight to $hi = 32$ in order to produce the worst-case power side-channel leakage for AES. Next, logic 0 is applied for all fixed input constraints to all key ports ($KEY[0\ to\ 127]$) and plaintext ports ($PT[32\ to\ 127]$) which are not among $TCPI$ as well as apply bind constraint to $TCPI$ ($PT[0\ to\ 31]$). These constraints ensure that switching only occurs in the target function of AES while muting the switching of the rest of the AES design. Next, SCRIPT utilizes the assertion, $assert\ never\ ((HW(r) == hi)\ \&\&\ (clkcount == TSD_r))$, where the hamming weight of the r is hi and when the clock count is equal to the sequential depth TSD_r. Lastly, the formal verification tool is used to generate a counter example for the given assertion which contains the SCV-guided patterns (PI_{hi}).

7.4.4.3 Noise Power Estimation

For evaluating the SCV metric, SCRIPT estimates the average noise power (P_{noise}) by utilizing vector-less power estimation technique. The vector-less power analysis propagates the signal probability and toggle rates from primary inputs to the outputs of internal nodes and repeats the operation until the primary outputs are reached and all nodes are assigned an activity rate. The derived activity rates are then used to compute power consumption numbers [35]. To run the vector-less power estimation technique, the verification engineer needs to define the signal probability and toggle rates of primary input ports. Signal probability is defined as the percentage of the analysis during which the input is driven at a high logic level and toggle rate is defined as the rate at which a net or logic element switches compared to its input(s) [36].

The vector-less power analysis returns the total estimated power P_{total} of the design. To get the P_{noise}, P_{signal} (defined in Eq. (7.4) and derived using SPG technique) is deducted from P_{total}. That is, P_{noise} is given by the following equation:

$$P_{noise} = P_{total} - P_{signal}. \tag{7.5}$$

7.4.5 Experimental Results: SCRIPT

In this section, we present the PSCL assessment of two different implementations of AES algorithm using SCRIPT framework. First, a brief description of the two AES designs: AES Galois Field (AES-GF) and AES look-up table (AES-LUT) is provided. Next, the comparison between SCV (estimated at pre-silicon by SCRIPT) with SNR (evaluated at post-silicon) in terms of correlation coefficient.

7.4.5.1 AES Benchmarks

The SCRIPT framework for PSCL assessment is applied to AES-GF [37] and AES-LUT [38] implementations. Both are open-source designs. Figure 7.3 shows the hardware architecture of the AES-GF encryption module. In this architecture, the AES round operations and the AES key expansion operation occurs in parallel. AES-GF architecture takes 10 clock cycles to encrypt each data block. The main characteristic of this design is that it implements the AES "SubByte" operation using Galois field arithmetic.

AES-LUT design, on the other hand, first performs the key expansion and stores the expanded keys in the key registers. After the key expansion, the round operation starts and takes 10 cycles to perform each encryption. AES-LUT design implements AES "SubByte" operation using a look-up table.

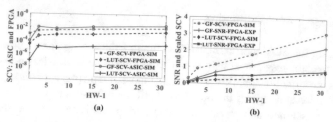

Fig. 7.5 (a) *SCV* metric vs. difference in HW between ASIC-SIM and FPGA-SIM using a base 10 logarithmic scale for the y-axis. (b) Scaled *SCV* value vs. difference in HW in FPGA-SIM and *SNR* metric in FPGA-EXP

7.4.5.2 Results: SCV Estimation and Validation

In this section, the SCV metric calculated by SCRIPT is presented. Figure 7.5a shows $HW - 1$ (hamming weight) vs. SCV values in ASIC-SIM and FPGA-SIM platforms using a base 10 logarithmic scale for the y-axis and a linear scale for the x-axis. According to the SCV metric, the AES-GF implementation is more vulnerable to side-channel leakage as compared to the AES-LUT. Also, note that the SCV metric is greater in ASIC-SIM w.r.t. FPGA-SIM. The reason is that P_{noise} in ASIC-SIM is lower as compared to FPGA-SIM (FPGA has higher power consumption).

Next, the authors validate SCV metric estimated at pre-silicon by calculating the SNR metric experimentally from FPGA. They evaluate the SNR metric in FPGA-EXP for different HWs using the same measurement technique shown in [7, 11]. Figure 7.5b shows the scaled values of SCV metric in FPGA-SIM and SNR metric measured in FPGA-EXP. It shows there exists good correlation between these two metrics. The Pearson correlation coefficient for AES-GF is 0.99, whereas for AES-LUT it is 0.94. This validates that SCV metric can be used to obtain a good estimate of the SNR.

Evaluation Time The SCRIPT framework applied to the gate-level AES designs requires on average 3 min for the IFT analysis and less than 1 min for the SPG. This step is common for both ASIC-SIM and FPGA-SIM flow. The SCV-guided pattern simulation and power estimation on average require less than 1 min for ASIC-SIM and 5 min for FPGA-SIM (per pattern). Therefore, the overall runtime (on average) for SCV estimation requires 6 min for ASIC-SIM and 14 min for FPGA-SIM. The required evaluation time is much smaller than previously proposed pre-silicon-based PSCL assessment [24, 39]. For example, SNR metric evaluation using 10,000 plaintext inputs in the gate-level simulation would take around 31 days to measure it (it takes on average 4.5 min to estimate power for each pattern in our simulation using similar workstation).

Area, Performance vs. Security Table 7.2 shows the comparison between area, performance, and security for AES-GF and AES-LUT designs for both ASIC-SIM and FPGA-SIM flows. Area is represented in terms of the number of gates/LUTs and registers, whereas performance is represented in terms of maximum delay. Security

Table 7.2 Comparison: area (in term of number of gates (**G**), LUTs (**L**), and registers (**R**)), performance (in terms of maximum delay of the design) vs. security (in terms of maximum SCV)

Platform	Design	Area	Timing	SCV
ASIC-SIM	AES-GF	11407 **G**; 796 **R**	89.5 ns	2.1×10^{-2}
	AES-LUT	2695 **G**; 650 **R**	94.3 ns	2.3×10^{-5}
FPGA-SIM	AES-GF	3931 **L**; 796 **R**	5.9 ns	9.4×10^{-3}
	AES-LUT	2480 **L**; 650 **R**	5.0 ns	2.3×10^{-3}

is represented in terms of maximum SCV value. Table 7.2 states that this particular AES-GF implementation requires relatively higher area as compared to the AES-LUT implementation. Also, AES-GF implementation is more vulnerable to power SCA compared to AES-LUT implementation as indicated by the SCV metric.

SCV vs. SNR SCRIPT utilizes formal verification to generate guided patterns. These patterns only introduce switching in the logic related to the target function while muting the switching of the rest of the design and thereby, derive SCV using as few as two patterns. These two patterns produce maximum power difference in the Target function. In contrast, SNR is calculated by performing DPA which requires thousands of plaintexts to make the difference of mean (DOM), i.e., the numerator of the SNR equation measurable [20]. Therefore, evaluating SNR metric at pre-silicon (using 10,000 plaintext) would require 31 days, whereas SCV metric requires 14 min to evaluate using same simulation setup.

7.5 RTL-PSC: Side-Channel Leakage Vulnerability Evaluation Framework

In this section, we present a framework named RTL-PSC [19] which can automatically assess PSCL vulnerability at the earliest pre-silicon design stage, i.e., RTL. RTL-PSC first estimates power profile of a hardware design using functional simulation at RTL. Then it utilizes the evaluation metrics, comprising of KL divergence metric and the success rate (SR) metric based on maximum likelihood estimation to perform power side-channel leakage (PSC) vulnerability assessment at RTL. Next, we present the RTL-PSC framework in details.

7.5.1 RTL-PSC Workflow

Figure 7.6 outlines the side-channel vulnerability evaluation framework. It includes two main parts, RTL Switching Activity Interchange Format (SAIF) file generation shown in the blue box and identification of vulnerable designs and blocks shown in the purple box. Algorithm 3 describes the identification technique for vulnerable

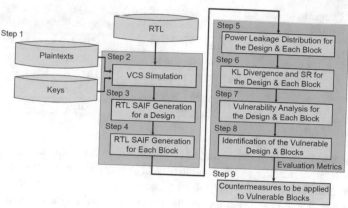

Fig. 7.6 RTL-PSC framework

designs and blocks. Note that, here TC refers to the transition count, $\mathcal{N}$ refers to the Gaussian distribution, f refers to the probability density function (PDF), D_{KL} refers to the KL divergence, and ML refers to the maximum likelihood. Specifically, in Step 1, a group of simulation keys is specified. In Step 2, RTL-PSC utilizes Synopsys VCS to perform functional simulation of the RTL design with the plaintexts and the selected keys as the inputs (key selection process is described in Sect. 7.5.3). In Step 3, once the simulation is complete, the SAIF file for the RTL design is generated. After finishing Step 3, all SAIF files for a group of keys and the applied plaintexts are generated (See Algorithm 3, Lines 4–8). As its name indicates, the SAIF file includes the switching activity information for each net and register in the RTL design. Moreover, the SAIF file generated based on the RTL design has the same hierarchy as the design itself, hence, in Step 4, the SAIF file for each module in the design can be separated for localized vulnerability analysis. Next, the evaluation metrics are applied for leakage assessment. Specifically, in Step 5, the obtained switching activity is exploited to estimate the power leakage distribution for the design and each module within it (Lines 10–11 in Algorithm 3). In Step 6, the Kullback–Leibler (KL) divergence [40] and success rate (SR) based on power leakage distribution are calculated for the design and each block (Lines 12–15 in Algorithm 3). In Step 7, vulnerability analysis is performed for the design and each block. In Step 8, the vulnerable design is identified based on the analysis performed in the previous step. Then the vulnerable blocks in the design that are leaking information the most are identified for further processing. Step 7 and Step 8 correspond to Lines 16–18 in Algorithm 3. Following this, the framework enters into Step 9, where countermeasures only need to be applied to the vulnerable block(s). Note that Step 9 is outside the scope in this paper.

Algorithm 3 Identifying vulnerable blocks

1: **procedure** IDENTIFYING VULNERABLE BLOCKS
2: **Input:** RTL design, $\mathcal{P} = \{Plaintext\}, \{Key_0, Key_1\}$
3: **Output:** $Set_{Vulnerable}$
4:     **for** $Key_j \in \{Key_0, Key_1\}$ **do**
5:         **for** $Plaintext_i \in \mathcal{P}$ **do**
6:             $SAIF_i^{Key_j} \leftarrow VCS(Plaintext_i, Key_j)$
7:         **end for**
8:     **end for**
9:     **for all** $Block_i \in \mathcal{M}$ **do**
10:         $TC_{Block_i}^0 \leftarrow \{SAIF_1^{Key_0}, \ldots, SAIF_n^{Key_0}\}$
11:         $TC_{Block_i}^1 \leftarrow \{SAIF_1^{Key_1}, \ldots, SAIF_n^{Key_1}\}$
12:         $f_i^0 \leftarrow \mathcal{N}(\mu_{TC_{block_i}^0}, \sigma_{TC_{block_i}^0}^2)$
13:         $f_i^1 \leftarrow \mathcal{N}(\mu_{TC_{block_i}^1}, \sigma_{TC_{block_i}^1}^2)$
14:         $KL_i \leftarrow D_{KL}(f_i^0, f_i^1)$
15:         $SR_i \leftarrow ML(f_i^0, f_i^1, n \; Plaintexts)$
16:         **if** $SR_i > SR_{threshold}$ or $KL_i > KL_{norm.th}$ **then**
17:             $Set_{Vulnerable} \leftarrow Block_i$
18:         **end if**
19:     **end for**
20: **end procedure**

7.5.2 Evaluation Metrics

In order to reduce time-to-market and the overall cost of adding security to the design, the RTL power side-channel vulnerability evaluation metrics are proposed to perform a design-time evaluation of PSC vulnerability. The Kullback–Leibler (KL) divergence metric and success rate (SR) metric based on maximum likelihood estimation are developed and combined to evaluate vulnerability of a hardware implementation. The first evaluation metric, i.e., KL divergence metric, estimates statistical distance between two different probability distributions, which is defined as follows [40]:

$$D_{KL}(k_i \| k_j) = \int f_{T|k_i}(t) \log \frac{f_{T|k_i}(t)}{f_{T|k_j}(t)} dt, \qquad (7.6)$$

where $f_{T|k_i}(t)$ and $f_{T|k_j}(t)$ are the probability density functions of the switching activity given keys k_i and k_j, respectively.

For instance, if power leakage probability distributions based on two different keys are distinguishable, KL divergence between these two distributions is high, which provides indication on how vulnerable the implementation is. Hence, KL divergence is suitable for the vulnerability comparison between different implementations. However, the KL divergence value may be difficult to interpret when performing vulnerability analysis for just one implementation. To address this issue, RTL-PSC introduces the second evaluation metric, named success rate (SR) metric based on maximum likelihood estimation. The SR value represents the probability to reveal the correct key and is suitable to evaluate vulnerability for only one implementation. The SR metric is derived as follows: the probability density function of the switching activity T given a key K is assumed to follow a Gaussian distribution, which can be expressed as,

$$f_{T|K}(t) = \frac{1}{\sqrt{2\pi}\sigma_k} e^{-\frac{(t-\mu_k)^2}{2\sigma_k^2}}, \tag{7.7}$$

where μ_k and σ_k^2 are the mean and variance of T, respectively. The likelihood function is defined as $\mathcal{L}(k; t) = \frac{1}{n}\sum_{i=1}^{n}\ln f_{T|K}(t_i)$. Based on the maximum likelihood estimation, an adversary typically selects a guess key $\hat{k}$ as follows:

$$\hat{k} = \arg\max_{k \in K} \mathcal{L}(k; t) = \arg\max_{k \in K} \frac{1}{n}\sum_{i=1}^{n}\ln f_{T|K}(t_i). \tag{7.8}$$

If the guess key ($k_g = \hat{k}$) is equal to the correct key (k^*), the side-channel attack is successful. Thus, the success rate can be defined as follows:

$$SR = \Pr[k_g = k^*] = \Pr[\mathcal{L}(k^*; t) > \mathcal{L}(\langle \bar{k^*} \rangle; t)], \tag{7.9}$$

where $\langle \bar{k^*} \rangle$ denotes all wrong keys, i.e., the correct key k^* is excluded from $\{k_1, k_2, \ldots, k_{n_k-1}\}$.

KL divergence is closely related to SR since the mathematical expectation of $\mathcal{L}(k^*; t) - \mathcal{L}(k_i; t)$ in Eq. (7.9) is equal to KL divergence between $T|k^*$ and $T|k_i$ [41]. Hence, SR increases accordingly as KL divergence increases. Due to the relation between KL divergence and SR based on maximum likelihood estimation, the combination of KL and SR is proposed for leakage assessment.

7.5.3 Selection of a Key Pair

As shown in Algorithm 3, first, a key pair is specified, then the probability distributions of the switching activity based on that key pair can be estimated using Eq. (7.7). The best key pair among all possible pairs is expected to provide the maximum KL divergence for vulnerability evaluation in the worst-case scenario. However, it is impossible and impractical to find the best key pair since the key space is huge, i.e., $\binom{2^{128}}{2}$. Alternatively, an appropriate key pair is able to be chosen, which satisfies the following conditions:

1. Assuming that each set of plaintexts is randomly generated, each key consists of the same subkey, e.g., $Key_0 = \{subkey \ldots subkey\} = 0x151515 \ldots 15$.
2. Hamming distance (HD) between two different subkeys is maximum, i.e., $subkey$ and $\overline{subkey}$.
3. If $D_{KL}(Key_0 || Key_i)$ increases asymptotically as i increases, $i = 1, \ldots n$, Key_0 and Key_n are the appropriate key pair, where Key_i is defined as $[\underbrace{subkey \ldots subkey}_{itimes} \overline{subkey}]$.

These key pairs can be used for the post-silicon validation. While it is difficult to measure the isolated leakage of each block by a random key pair at the post-silicon stage, the leakage of each block can be measured at the pre-silicon stage through applying the key pairs satisfying the above conditions. Moreover, the evaluation metrics would create the worst-case scenario through applying the key pairs with the maximum Hamming distance.

The selected keys applied to an AES RTL design are 16 pairs of keys starting from all $0s$ key until all Fs key. Each key has 128-bit and Hamming distance between Key_i and Key_{i+1} is eight, which is shown in Table 7.3. Also, to take into account not only the key's impact on the power consumption but also the plaintext's impact on power consumption, one thousand random plaintexts are used with the selected key pairs. The AES cipher operation itself is used for the generation of pseudo random plaintext [42]. $Plaintext_0$ is used as seed and then use each ciphertext (j) as the next plaintext $(j + 1)$,[2]

$$Plaintext_0 = 000 \ldots 000 \qquad (7.10)$$
$$Plaintext_{j+1} = AES(Key_i, Plaintext_j), j = 0, 1, \ldots, 999.$$

It can be noted that the key pair Key_0 and Key_{16} in Table 7.3 satisfies the above conditions. Furthermore, it can be seen that Key_0 would create a state similar as the reset state of the design, hence, Key_0 and Key_{16} would create the worst-case scenario.

Table 7.3 Keys used in RTL-PSC framework

Key_0	0x0000_0000_0000_0000_0000_0000_0000_0000
Key_1	0x0000_0000_0000_0000_0000_0000_0000_00FF
......	
Key_{15}	0x00FF_FFFF_FFFF_FFFF_FFFF_FFFF_FFFF_FFFF
Key_{16}	0xFFFF_FFFF_FFFF_FFFF_FFFF_FFFF_FFFF_FFFF

---

[2]This seed is the same as TVLA's setup [43] and 1000 plaintexts are enough to estimate SCA leakage based on our experiments.

7.5.4 Identification of Vulnerable Designs and Blocks

When the appropriate key pair is applied to a design, the same subkey patterns will be propagated to the same blocks, e.g., Sbox blocks in the AES design. The switching activity of each block and the entire design is recorded into SAIF files using VCS functional simulation, which corresponds to power leakage at RTL. The higher the difference between two power leakage distributions is, the higher impact the key has on the power consumption of the design/blocks, the more susceptible the design/blocks are to power analysis attack. Using the KL divergence and SR metrics, the vulnerable design and blocks within the design can be identified. If KL divergence or SR of any design or block is greater than $KL_{threshold}$ or $SR_{threshold}$, the design or the block is considered to be the vulnerable one (Line 16 in Algorithm 3). $SR_{threshold}$ is determined based on the security constraint, e.g., 95% SR with n plaintexts, while $KL_{threshold}$ is determined based on the $SR_{threshold}$ through the relation between KL divergence and SR.

7.5.5 Experiment Results: RTL-PSC

In this section, we present the side-channel vulnerability assessment of two different implementations of AES algorithm using RTL-PSC.

RTL Evaluation Metrics Figure 7.7a shows KL divergence from the second clock cycle to the 11th clock cycle of both AES-GF and AES-LUT RTL implementations, during which 10 round operations for an encryption are performed. At the second clock cycle corresponding to the first round operation, which is mostly exploited by power analysis attack, KL divergence of AES-GF and AES-LUT implementations

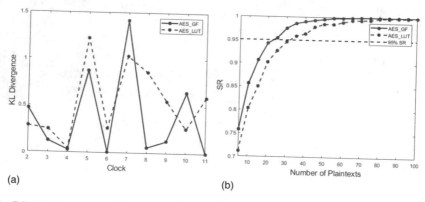

(a)

(b)

Fig. 7.7 KL divergence and SRs for AES-GF and AES-LUT implementations. (a) KL divergence per clock cycle for AES-GF and AES-LUT implementations; (b) SR_{em} corresponding to KL divergence (0.47 and 0.28) for AES-GF and AES-LUT implementations

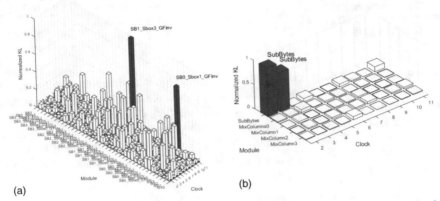

Fig. 7.8 Normalized KL divergence for vulnerable blocks within AES-GF and AES-LUT implementations ($KL_{norm.th} = 0.5$). (**a**) Normalized KL divergence for AES-GF implementation in both time and spatial/modular domains; (**b**) Normalized KL divergence for AES-LUT implementation in both time and spatial/modular domains

are 0.47 and 0.28, respectively. These values correspond to 95% SR_{em}[3] vulnerability level ($SR_{threshold}$) with 25 *plaintexts* and 35 *plaintexts*, respectively, as shown in Fig. 7.7b. Based on KL divergence metric, as shown in Fig. 7.7a, KL divergence of AES-GF implementation at the second clock cycle (i.e., 0.47) is greater than that of AES-LUT implementation (i.e., 0.28). Likewise, based on SR metric, as shown in Fig. 7.7b, the DPA attack success rate of AES-GF implementation is higher than that of AES-LUT implementation with the same number of plaintexts. Similarly, the number of plaintexts required for successful DPA attack on AES-GF implementation is less than that on AES-LUT implementation. Hence, compared with AES-LUT implementation, AES-GF implementation is identified as the more vulnerable design.

Vulnerable Block Identification Once a RTL deign is determined as a vulnerable one, the next step is to identify the vulnerable blocks within the design that contribute to side-channel leakage significantly. Side-channel vulnerability can be evaluated in both time and spatial/modular domains. In other words, the evaluation metrics based on the switching activity of each block within the design are calculated at fine-granularity scale so that vulnerable blocks can be identified per clock cycle.

First, KL divergence in both time and spatial/modular domains is normalized, i.e., KL divergence of each block is divided by the maximum KL divergence of those blocks ($KL_{norm.th} = KL_i/max(KL_i)$). Then, if the normalized KL divergence of any block is greater than $KL_{norm.th} = 0.5$, that block is identified as the vulnerable one and included into the set of vulnerable blocks. Figure 7.8 shows KL

---

[3]The SR_{em} represents the empirical SR based on actual SCA attacks with n plaintexts.

Table 7.4 Correlation coefficient between KL divergence at RTL, GTL, and FPGA silicon level

Benchmark	RTL vs. FPGA silicon level
AES-GF	98.83%
AES-LUT	80.80%

divergence of each block in both time and spatial/modular domains. The identified vulnerable blocks (i.e., KL divergence greater than $KL_{norm.th} = 0.5$) are denoted with blue bars. Specifically, in the AES-GF design, GFinvComp blocks within Sbox0 and Sbox1 blocks are identified as the vulnerable ones; in the AES-LUT design, SubByte blocks are identified as the vulnerable ones. It should be noted that the threshold values ($KL_{threshold}$ and $KL_{norm.th}$) can be adjusted by the SR vulnerability level.

Evaluation Time The evaluation time of RTL-PSC includes VCS functional simulation time of the RTL design as well as the data processing time required for analyzing the SAIF files. The evaluation time of RTL-PSC for AES-GF is 46.3 min and for AES-LUT is 24.03 min. If the same experiments were performed at gate-level designs, it would take around 31 h. Therefore, our RTL-PSC is almost $42\times$ more efficient as compared to similar gate-level assessment. The evaluation time at layout level is going to be even more expensive (more than a month). RTL-PSC also provides the flexibility to make design changes, whereas the post-silicon assessment provides no flexibility.

FPGA Validation For FPGA silicon validation, He et al. [19] first mapped the AES-GF and AES-LUT designs on an FPGA and then, applied the same plaintexts and keys as used at RTL, and measured the power consumption during encryption operation. Following this, the KL divergence metric from the collected power traces were derived. Then, the Pearson correlation coefficient between the KL divergence at RTL and FPGA (as shown in Column 2 of Table 7.4) was calculated. The high correlation coefficient values (>80%) indicate that the PSC vulnerability assessment results generated by RTL-PSC at RTL is almost as accurate as FPGA assessment. In other words, RTL-PSC can accurately analyze PSC vulnerability at RTL.

7.6 Conclusion

CAD for power side-channel leakage assessment is an important component of the recent academic and industrial initiatives for developing CAD frameworks which aim at automating the security vulnerability assessment of hardware designs at design stages [44, 45]. These frameworks would allow the semiconductor industry to systematically and efficiently identify side-channel vulnerabilities before tape-out in order to include proper countermeasures or refine the design to address them.

References

1. P.C. Kocher, J. Jaffe, B. Jun, Differential power analysis, in *Proceedings of the 19th Annual International Cryptology Conference on Advances in Cryptology*, CRYPTO '99, London, UK, 1999 (Springer, Berlin, 1999), pp. 388–397
2. P.C. Kocher, Timing attacks on implementations of Diffie–Hellman, RSA, DSS, and other systems, in *Advances in Cryptology—CRYPTO '96, 16th Annual International Cryptology Conference, Santa Barbara, California, USA, August 18–22, 1996, Proceedings* (1996), pp. 104–113
3. P. Kocher, J. Jaffe, B. Jun, Differential power analysis, in *Annual International Cryptology Conference* (Springer, Berlin, 1999), pp. 388–397
4. E. Brier, C. Clavier, F. Olivier, Correlation power analysis with a leakage model, in *International Workshop on Cryptographic Hardware and Embedded Systems* (Springer, 2004), pp. 16–29
5. B. Gierlichs, L. Batina, P. Tuyls, B. Preneel, Mutual information analysis, in *International Workshop on Cryptographic Hardware and Embedded Systems* (Springer, 2008), pp. 426–442
6. T.-H. Le, J. Clédière, C. Canovas, B. Robisson, C. Servière, J.-L. Lacoume, A proposition for correlation power analysis enhancement, in *International Workshop on Cryptographic Hardware and Embedded Systems* (Springer, Berlin, 2006), pp. 174–186
7. Y. Fei, A. Adam Ding, J. Lao, L. Zhang, A statistics-based fundamental model for side-channel attack analysis. IACR Cryptol. ePrint Arch. **2014**, 152 (2014)
8. A. Moradi, Masking as a side-channel countermeasure in hardware. ISCISC 2016 Tutorial (2006)
9. G. Barthe, S. Belaïd, F. Dupressoir, P.-A. Fouque, B. Grégoire, P.-Y. Strub, Verified proofs of higher-order masking, in *Annual International Conference on the Theory and Applications of Cryptographic Techniques* (Springer, Berlin, 2015), pp. 457–485
10. R. Bloem, H. Gross, R. Iusupov, B. Könighofer, S. Mangard, J. Winter, Formal verification of masked hardware implementations in the presence of glitches, in *Annual International Conference on the Theory and Applications of Cryptographic Techniques* (Springer, Berlin, 2018), pp. 321–353
11. T.S. Messerges, E.A. Dabbish, R.H. Sloan, Examining smart-card security under the threat of power analysis attacks. IEEE Trans. Comput. **51**(5), 541–552 (2002)
12. S. Mangard, Hardware countermeasures against DPA—a statistical analysis of their effectiveness, in *Cryptographers' Track at the RSA Conference* (Springer, Berlin 2004), pp. 222–235
13. B.J. Gilbert Goodwill, J. Jaffe, P. Rohatgi, et al., A testing methodology for side-channel resistance validation, in *NIST Non-Invasive Attack Testing Workshop* (2011)
14. F. Durvaux, F.-X. Standaert, From improved leakage detection to the detection of points of interests in leakage traces, in *Annual International Conference on the Theory and Applications of Cryptographic Techniques* (Springer, Berlin, 2016), pp. 240–262
15. B. Gierlichs, K. Lemke-Rust, C. Paar, Templates vs. stochastic methods, in *International Workshop on Cryptographic Hardware and Embedded Systems* (Springer, Berlin, 2006), pp. 15–29
16. N. Veyrat-Charvillon, F.-X. Standaert, Mutual information analysis: how, when and why? in *Cryptographic Hardware and Embedded Systems-CHES 2009* (Springer, Berlin, 2009), pp. 429–443
17. S.A. Huss, M. Stöttinger, M. Zohner, Amasive: an adaptable and modular autonomous side-channel vulnerability evaluation framework, in *Number Theory and Cryptography* (Springer, Berlin, 2013), pp. 151–165
18. A. Nahiyan, J. Park, M. He, Y. Iskander, F. Farahmandi, D. Forte, M. Tehranipoor, SCRIPT: a CAD framework for power side-channel vulnerability assessment using information flow tracking and pattern generation. ACM Trans. Des. Autom. Electron. Syst. **25**(3), 1–27 (2020)

19. M.T. He, J. Park, A. Nahiyan, A. Vassilev, Y. Jin, M. Tehranipoor. RTL-PSC: automated power side-channel leakage assessment at register-transfer level, in *2019 IEEE 37th VLSI Test Symposium (VTS)* (IEEE, Piscataway, 2019), pp. 1–6

20. A. Moradi, B. Richter, T. Schneider, F.-X. Standaert, Leakage detection with the x2-test. IACR Trans. Cryptogr. Hardware Embed. Syst. **2018**(1), 209–237 (2018)

21. N. Veshchikov, S. Guilley, Use of simulators for side-channel analysis, in *2017 IEEE European Symposium on Security and Privacy Workshops (EuroS&PW)* (IEEE, Piscataway, 2017), pp. 104–112

22. J. den Hartog, J. Verschuren, E. de Vink, J. de Vos, W. Wiersma, Pinpas: a tool for power analysis of smartcards, in *IFIP International Information Security Conference* (Springer, Berlin, 2003), pp. 453–457

23. M. Aigner, S. Mangard, F. Menichelli, R. Menicocci, M. Olivieri, T. Popp, G. Scotti, A. Trifiletti, Side channel analysis resistant design flow, in *Circuits and Systems, 2006. ISCAS 2006. Proceedings. 2006 IEEE International Symposium on* (IEEE, Piscataway, 2006), p. 4

24. Y.A. Durrani, T. Riesgo, Power estimation for intellectual property-based digital systems at the architectural level. J. King Saud Univ. Comput. Inf. Sci. **26**(3), 287–295 (2014)

25. C. Thuillet, P. Andouard, O. Ly, A smart card power analysis simulator, in *International Conference on Computational Science and Engineering, 2009. CSE'09*, vol. 2 (IEEE, Piscataway, 2009), pp. 847–852

26. A.G. Bayrak, F. Regazzoni, P. Brisk, F.-X. Standaert, P. Ienne, A first step towards automatic application of power analysis countermeasures, in *Proceedings of the 48th Design Automation Conference* (ACM, New York, 2011), pp. 230–235

27. A.G. Bayrak, F. Regazzoni, D. Novo, P. Ienne. Sleuth: automated verification of software power analysis countermeasures, in *International Workshop on Cryptographic Hardware and Embedded Systems* (Springer, Berlin, 2013), pp. 293–310

28. A.G. Bayrak, F. Regazzoni, D. Novo, P. Brisk, F.-X. Standaert, P. Ienne, Automatic application of power analysis countermeasures. IEEE Trans. Comput. **64**(2), 329–341 (2013)

29. S.A. Huss, O. Stein, A novel design flow for a security-driven synthesis of side-channel hardened cryptographic modules. J. Low Power Electron. Appl. **7**(1), 4 (2017)

30. Q. Luo, Y. Fei, Algorithmic collision analysis for evaluating cryptographic systems and side-channel attacks, in *2011 IEEE International Symposium on Hardware-Oriented Security and Trust (HOST)* (IEEE, Piscataway, 2011), pp. 75–80

31. M. Yasin, B. Mazumdar, S.S. Ali, O. Sinanoglu, Security analysis of logic encryption against the most effective side-channel attack: DPA, in *2015 IEEE International Symposium on Defect and Fault Tolerance in VLSI and Nanotechnology Systems (DFTS)* (IEEE, Piscataway, 2015), pp. 97–102

32. A. Nahiyan, M. Sadi, R. Vittal, G. Contreras, D. Forte, M. Tehranipoor, Hardware trojan detection through information flow security verification, in *2017 IEEE International Test Conference (ITC)* (IEEE, Piscataway, 2017), pp. 1–10

33. G.K. Contreras, A. Nahiyan, S. Bhunia, D. Forte, M. Tehranipoor, Security vulnerability analysis of design-for-test exploits for asset protection in SoCs, in *2017 22nd Asia and South Pacific Design Automation Conference (ASP-DAC)* (IEEE, Piscataway, 2017), pp. 617–622

34. M. Bushnell, V. Agrawal, *Essentials of Electronic Testing for Digital, Memory and Mixed-Signal VLSI Circuits*, vol. 17 (Springer, Berlin, 2004)

35. Vectorless estimation (2018). https://www.xilinx.com/support/documentation/. Accessed 20 April 2018

36. Power analysis and optimization (2017). https://www.xilinx.com/. Accessed 20 April 2018

37. Galois field based AES Verilog design (2018). http://www.aoki.ecei.tohoku.ac.jp/. Accessed 20 April 2018

38. Lookup table based AES Verilog design. Satoh Laboratory UEC (2018). Accessed 20 April 2018

39. K. Tiri, I. Verbauwhede, Securing encryption algorithms against DPA at the logic level: next generation smart card technology, in *International Workshop on Cryptographic Hardware and Embedded Systems* (Springer, Berlin, 2003), pp. 125–136

40. S. Kullback, R.A. Leibler, On information and sufficiency. Ann. Math. Statist. **22**(1):79–86
41. Y. Fei, A. Adam Ding, J. Lao, L. Zhang, A statistics-based fundamental model for side-channel attack analysis. Cryptology ePrint Archive, Report 2014/152, 2014
42. S.S Keller, NIST-recommended random number generator based on ANSI x9. 31 appendix a. 2.4 using the 3-key triple DES and AES algorithms, in *NIST Information Technology Laboratory-Computer Security Division, National Institute of Standards and Technology* (2005)
43. G. Becker, J. Cooper, E. DeMulder, G. Goodwill, J. Jaffe, G. Kenworthy, T. Kouzminov, A. Leiserson, M. Marson, P. Rohatgi, et al., Test vector leakage assessment (TVLA) methodology in practice, in *International Cryptographic Module Conference*, vol. 1001 (2013), p. 13
44. K. Xiao, A. Nahiyan, M. Tehranipoor, Security rule checking in IC design. Computer **49**(8), 54–61 (2016)
45. A. Nahiyan, K. Xiao, D. Forte, M. Tehranipoor, Security rule check, in *Hardware IP Security and Trust* (Springer, Berlin, 2017), pp. 17–36

Chapter 8
Post-Quantum Hardware Security

Physical Security in Classic vs. Quantum Worlds

Ana Covic, Sreeja Chowdhury, Rabin Yu Acharya, Fatemeh Ganji, and Domenic Forte

8.1 Introduction

The advances in quantum technology have definitely become a milestone in the development of a new computational paradigm. Quantum technologies, more specifically, quantum computers, leverage physical phenomena that classic physics fails to capture. In doing so, the data is stored and processed by the computers in a way that problems, known to be intractable in the classic world, can be efficiently solved [33]. Interestingly enough, the security of various products and services widely used in everyday life relies on certain problems, which are hard to solve using conventional computers. The question of finding an appropriate alternative for them has been answered in the post-quantum cryptography field of study.

Post-Quantum cryptography, or the so-called quantum-safe cryptography, deals with the design and security assessment of cryptosystems, resilient to attacks that can be mounted by an adversary with access to a quantum computer. When implementing such a system, however, the attacker can also launch attacks proven effective in the classic world. Prominent examples of such attacks include ones targeting the *physical* security of the system. In its broadest sense, physical security proposes methods and means used to address the security-related issues

Part of this study has been carried out, when Fatemeh Ganji was with Florida Institute for Cybersecurity Research, University of Florida.

A. Covic · S. Chowdhury · R. Y. Acharya · D. Forte (✉)
Florida Institute for Cybersecurity Research, University of Florida, Gainesville, FL, USA
e-mail: anaswim@ufl.edu; sreejachowdhury@ufl.edu; rabin.acharya@ufl.edu; dforte@ece.ufl.edu

F. Ganji
Worcester Polytechnic Institute, Worcester, MA, USA
e-mail: fganji@wpi.edu

© The Author(s), under exclusive license to Springer Nature Switzerland AG 2021
M. Tehranipoor (ed.), *Emerging Topics in Hardware Security*,
https://doi.org/10.1007/978-3-030-64448-2_8

199

that designers have to face when implementing their cryptosystem in practice. Secure key generation and storage, as well as secure execution of a system, are key objectives of physical security, being fulfilled through true random number generators (TRNGs), physically unclonable functions (PUFs), and measures devised to minimize side-channel leakage.

Numerous studies have been devoted to the challenging task of achieving the above objectives in the quantum world. This chapter attempts to survey these studies by giving background information on quantum computation and its connection to cryptography. Afterward, side-channel analyses performed to assess the security quantum primitives are reviewed. This is followed by discussions on how key generation and storage tasks have been accomplished in the quantum world. Last but not least, lessons learned in this respect are summarized, and future directions helpful to close the gap between physical security and quantum technology are discussed.

8.2 Quantum Computing and Cryptography: A Brief Overview

At the core of quantum technology is the quantum mechanics that describes observations, which could not be attributed to laws defined in the classical physics.[1] More specifically, quantum mechanics explains the behavior of matter and light at the atomic scale, which is unusual from the classical physics point of view [65]. Among fundamental laws governing the quantum world is the one describing the quantum state of a system. According to that, the quantum state is a mathematical representation of the outcome probability of each possible measurement performed on a system. From this definition, the extreme difference between the classic world and the quantum one becomes evident: in contrast to the classic world, solely (at best) the probability density for a particle to be at a given position x at time t can be obtained. This probabilistic nature of the quantum world results in the emergence of a notion of quantum *superposition*, which states that quantum states can be combined together to describe another valid state.

This principle is one of the foundations of quantum computers as it explains the abstract notion of a qubit. A qubit, being analogous to the bit in a conventional computer, is a two-state quantum system that can be seen as any quantum superposition of two physically distinguishable states. This characteristic of qubits makes it possible to compute on both 1 and 0 simultaneously, i.e., being in a superposition state. This can be improved by introducing the entanglement effect between two qubits. For this, the qubits are linked through this effect so that the state of each of

---

[1]Note that this section provides a gentle introduction to the concepts studied in the quantum mechanics and quantum computation. For more formal definitions and discussions, we refer the readers to [65].

those qubits cannot be described independently, regardless of the distance between them [47]. If two qubits exhibit this feature, four calculations can be carried out at the same time.

The above properties contribute to disseminating the idea of employing quantum computers to solve problems known as infeasible-to-solve in the classic world. This has drastic consequences for cryptosystems, whose security has been based in relation to such problems. A prime example is the Rivest–Shamir–Adleman (RSA) system protecting the information exchanged between parties every day. To give a better understanding of how a quantum computer can become a threat for cryptosystems, consider the RSA system, where the public key is a product $N = pq$ with two secret prime numbers p and q. Finding the factors p, q of N assumed to be hard; however, in the 1990s, Shor proposed an algorithm that revolutionized cryptography [75].

Shor's algorithm attempts to factor numbers into their components, which could potentially be prime. Suppose that N fits in n bits. In order to factor N, Shor's algorithm enhanced by Beauregard requires $O(n^3 \log n)$ operations on $2n + 3$ qubits [6]. Consequently, it might be thought that for commonly applied sizes of the keys, such algorithms have to perform billions of operations on thousands of qubits. However, cryptographers have to provide countermeasures against this, irrespective of the probability that an attack can be launched [7].

Besides Shor's algorithm, a range of quantum algorithms has been developed to improve the time-efficiency of searching, collision finding, and Boolean formulae evaluation. Although an exponential speed-up cannot be achieved through these algorithms, they force designers of cryptosystems to consider larger key sizes, even for symmetric key systems, in the hope of stopping attackers [15] (see Fig. 8.1). In line with this effort, a variety of systems have been considered alternatives to ones, whose security could be compromised in the quantum world. For instance, code-based and lattice-based encryption and signature generation, as well as multivariate-quadratic-equation signatures, are among the proposals. Next, we discuss if it is sufficient to stick to measures designed to counterattack the above algorithmic attacks. In particular, the possibility of mounting side-channel analysis on quantum cryptosystems is explored.

8.3 Quantum Computing for Side-Channel Analysis

A side-channel attack (SCA) is a powerful and dangerous process capable of non-invasively extracting secret cryptographic information stored in the chip. SCAs take advantage of the physical flaws of a design implementation that unintentionally leak information. To execute a side-channel attack, attacker needs to transform physical leakage into probability and score vectors [82] used for further key derivation. The key is extracted by the search over sorted information and assembling of the entire key from smaller subsets [82, 90]. The main challenge presented in the side-channel attack execution is presence of noise [82].

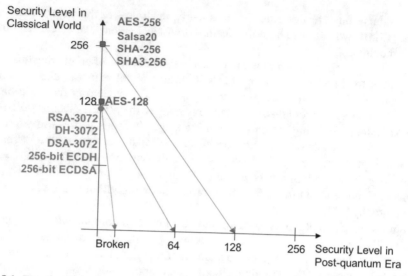

Fig. 8.1 The change in the security level of symmetric and public-key cryptographic schemes (highlighted in blue and orange, respectively). The schemes considered here include encryption, hash functions, key exchange, and signature cryptosystems. This level is obtained with regard to the best attacks demonstrated recently [7]. Security level here refers to the approximate number of operations required by the best attacks, i.e., security level λ means that the best attack needs 2^{λ} operations (see [7], for more details)

8.3.1 SCA in Classic World

Side-channel attacks can be classified as logical and physical attacks. Logical side-channel attacks obtain leaked information from software characteristics, such as data-usage statistics and footprint [23]. An example of logical SCA is a cold boot attack [97]. Contrary, physical side-channel attacks gain leaked information from physical characteristics of the device and its implementation [23], through parameters such as electromagnetic emission or power consumption.

Physical SCA can be classified as non-invasive, semi-invasive, and invasive attacks, as shown in Fig. 8.2. The non-invasive attack observes the system without destroying the IC packaging. Semi-invasive attack decapsulates the IC packaging from the backside, allowing the attacker to execute optical contactless probing [92], photonic analysis [93], or laser stimulation [51]. Finally, the invasive side-channel attacks completely remove the IC packing without affecting its functionality, but its original condition cannot be retrieved.

The non-invasive attacks have been investigated the most extensively. Timing, electromagnetic emissions (EM), and power consumption are the most reliable leakage sources to produce the useful leakage data [64], while less investigated means of side-channel leakage are acoustics [22, 34, 64], and light [81]. Among

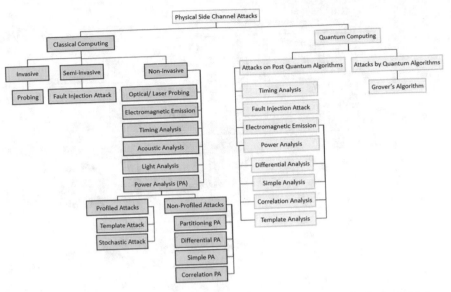

Fig. 8.2 Side-channel taxonomy in classical (blue) and quantum (orange) computing settings

the most investigated side-channel attacks, power analysis (PA) is the first and most reliable method, whose statistical analysis can be applied to the EM leakage data.

Side-channel attacks performed under the classic and quantum world constraints have many similarities, but due to the difference in the nature of the settings, quantum SCAs are explained in further text.

8.3.2 SCA in Quantum World

Side-channel attacks in the quantum world have been investigated on the algorithms which are not susceptible to the quantum computer attacks, as well as using the advantages of the quantum computer, such as speed-up in runtime and space, to launch the side-channel attack [61] on classical computation. A side-channel attack on post-quantum algorithms is a complicated procedure because the sensitive data is not readily available. An attacker needs to put additional effort in finding the relationship between secret data, and the rest of the algorithm, to finally understand the leakage, as shown in Fig. 8.3. The next sections will provide an overview of post-quantum algorithms focusing on code-based post-quantum algorithms, which are the most heavily investigated. It will be followed by side-channel attacks against them.

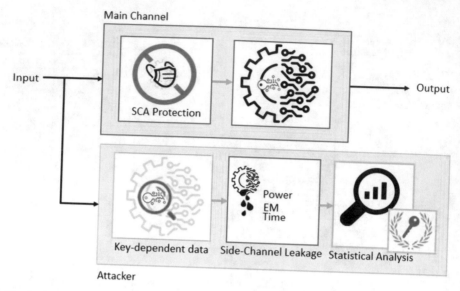

Fig. 8.3 Side-channel attack in post-quantum world

8.3.2.1 Post-Quantum Cryptographic Algorithms

The main cryptographic systems, including public-key encryption, digital signatures, and key exchange, are mainly based on Diffie-Hellman key exchange, RSA (Rivest–Shamir–Adleman) public-key cryptosystem, DSA (Digital Signature Algorithm), and elliptic curve cryptosystems [1]. After creating the large-scale quantum computer, quantum algorithms will be able to break them because the speed-up offered by quantum algorithms ranges from quadratic to exponential [1]. However, exponential speed-up was shown not to be achievable in quantum search, making symmetric encryption and hash functions safe in the post-quantum era [1]. The most promising algorithms against quantum computing are based on the lattice, code, hash, and multivariate public-key problems. These algorithms will be introduced below.

Lattice-Based Algorithms These are one of the most promising because of their simplicity and worst-case security hardness assumptions for security notions. The basis of this algorithm is the lattice shortest vector problem (SVP), and ring learning with error (R-LWE) because it is reducible to SVP, an NP-hard problem.

Code-Based McEliece Scheme This system is based on the hardness of random linear error-correcting code, which is also an NP-hard problem, such as Syndrome Decoding [78]. Optimized McEliece schemes have been created because the original McEliece scheme requires a large key size, and optimization also focuses on encryption speed [63]. The optimized McEliece schemes are quasi-cyclic low- and moderate-density parity code (QC-LDPC and QC-MDPC). The hardness of the problem is based on Syndrome Decoding and Goppa Codes Distinguishing problems. Original McEliece scheme has a secret key created from the parity check matrix, scrambling matrix, and permutation matrix. The vector of errors is used to encrypt the plaintext. To decrypt the ciphertext, a codeword needs to be created by the multiplication of ciphertext with the permutation matrix. Then, Patterson algorithm is used to compute the syndrome of the codeword by multiplying codeword and transpose of the parity check matrix. The multiplication process requires syndrome to be firstly initialized as a vector of zeroes. Then, the algorithm goes through codewords in search of 1s to create the syndrome vector. The syndrome vector is transformed into syndrome polynomial by using the Patterson algorithm and the Extended Euclidean Algorithm (XGCD). To obtain the plaintext, the syndrome and scrambling matrix are multiplied, and the key equation must be solved. The optimized McEliece schemes do not have a permutation matrix but rather a similar matrix Q, which consists of a small number of 1s in each row. The optimized version has a sparse permutation matrix, so the error correction in decryption is significantly sped up [59].

Hash-Based Algorithms These algorithms are based on binary hash tree structure and hash functions. They are primarily used for digital signatures. Binary tree combines one-time signature key pairs which get updated after signing if it is stateful, or unchanged if it is stateless [45].

Multivariate Public-Key Crypto-Algorithms (MPKC) Their hardness is based on problem of solving a set of multivariate quadratic polynomial equations in a finite field. The most reliable and the most efficient ones come from the family of the step-wise triangular systems [99]. Digital signature scheme enTTS is currently the fastest, and it works with 20-byte hashes and 28-byte signatures. The building block of this scheme is small size multivariate polynomials and linear maps. Monomials do not occur more than once, as most coefficients are zeroes. The central linear map contains three layers [18]. Affine transformations, evaluation of polynomials, and solving of the system of linear equations need to performed to generate a signature in the enTTS scheme [99].

8.3.2.2 SCA on Code-Based Post-Quantum Cryptographic Schemes

Side-channel attacks in the post-quantum world are non-invasive attacks that exploit power leakage, timing leakage, and electromagnetic emissions. Fault analysis attack (FA) has been shown as both passive and active attacks, while power analysis (PA) is differential PA (DPA), correlation PA (CPA), and simple PA (SPA). This chapter will focus on side-channel attacks on code-based post-quantum cryptographic algorithms, while a comprehensive analysis of SCAs on all four types of post-quantum cryptographic algorithms can be found in an extended version on this chapter [16].

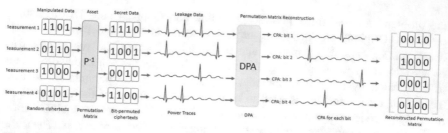

Fig. 8.4 Main steps in differential power analysis attack on permutation matrix, inspired by Fig.3. in [70]

Differential Power Analysis (DPA) This analyzes measured power consumption from traces of the implemented cryptographic algorithm. Traces represent the intermediate variables, which can be expressed as a function of the secret key and known value. To predict the power consumption, different models are used, such as Hamming Weight (HW) or Hamming Distance (HD) models. HD models are used for leakage of switching bits, and HW models are used based on the number of bits in obtained data. To extract the key, the attacker needs to compute the correlation between predictions and obtained power consumption data.

The dependency between secret data and known data is not easily obtainable and recognized in the classical McEliece scheme. At first, it was thought that syndrome needs to be extracted as secret data, but ciphertext does not contain syndrome, so it turned out to be an invaluable asset. DPA attack on code-based scheme original McEliece from [70] is launched on bit permutation of ciphertext, as shown in Fig. 8.4. The Hamming Weight model is applied to individual bits of leakage data. Correlation analysis is then performed for each input bit, which allows for the permutation matrix to be recovered by comparing known ciphertext and permuted ciphertext to the correlation peaks from each measurement. One measurement represents one row of the permutation matrix.

DPA methodology on constant-time multiplication [76], which creates a syndrome, is composed of word rotation and bit rotation to compute the parity matrix. The attack in [76] proposed multiple and single trace attacks to overcome the need for solving linear equations in the case of constant-time multiplication, such as in [72]. The reason for unsuccessful attacks before the one in [76] is that the results of the constant-time multiplication were saved in the same register, once the algorithm is implemented. This creates multiple candidates for secret indices and creates a system of linear equations. Multiple bit shift instructions are attacked by multiple track attack, while single trace attack is performed on the single-bit shift instruction. Multiple-trace DPA procedure consists of partitioning the attack position into two parts. Firstly, masking is performed with all ones or all zeroes. The power at each point is modeled as a summation of data-dependent power and Gaussian noise. Modeling the data-dependent power consumption with Hamming Weight, there is a linear relationship between total power consumption and the modeled value. Pearson correlation coefficient between these two parameters allows for the recovery of the correct indices in the first part. Secondly, the CPA is performed to find the rest of the correct indices by bit rotation.

Comparison to Classic World Algorithms DPA against AES attacks each round of AES and uses the recovered secret as a known variable in the next round, [43]. This attack is direct and does not need an effort to find a relation between a sub-key and known value. Classic world attacks use multi-bit statistical tools to extract the entire key, while post-quantum attacks only use correlation analysis, which makes DPA identical to CPA.

Simple Power Analysis (SPA) Simple power analysis (SPA) visually examines power traces. This attack utilizes key and data dependencies during computation. It is also very dangerous against recognizing specific operations within the algorithm, as each operation has its own power signature [40]. The attack is not successful for higher signal-to-noise (SNR) ratios, and in the post-quantum world, this attack has been performed on code and lattice-based algorithms. SPA performed on original McEliece extracts permutation and parity check matrices if they are computed individually. However, if these two are combined, their recovery is combined as well [40]. The HW value between the original ciphertext and permuted ciphertext is the same because only the position of the ciphertext gets changed. Permutation matrix can be extracted for ciphertext with HW equal to 1. Recognition of summation has been performed in [40, 71]. SPA has also been successful against error vector recovery, through the Euclidean algorithm [60]. By measuring power traces during the decryption, error vector corresponds to the power peaks caused by iteration numbers of the Euclidean algorithm.

Comparison to Classic World Algorithms SPA in the classic world is very prominent, extracting successfully even secret data in RSA and KeeLog. As the post-quantum algorithms do not contain easily distinguishable operations, and since entire matrices need to be extracted, the SPA is not as dangerous.

Timing Attacks (TA) Timing attacks (TA) are a variety of side-channel attacks in which time required for the completion of a logical operation is exploited. The time can variant within the function itself, based on the variety of the inputs processed. These SCAs in the post-quantum world have been successfully launched on code-based and lattice-based algorithms.

Unprotected McEliece algorithm was firstly attacked in [87], in which the decryption execution time was exploited, in addition to the time needed to learn dependencies of decoding algorithm errors and errors in the locator polynomial. The same attack was further expanded in [85] in which the dependencies of error locator polynomial in the Patterson algorithm to the private key are used, as well as the time to solve the key equations which give information about the polynomials. This attack was shown to be unfeasible in [86], where the timing attack in [86] takes advantage of the multiplication of syndrome with a scrambling matrix, more specifically of syndrome inversion. Strenzke [86] elaborates on work in [85] in which the syndrome inversion leakage is combined with the leakage from solving the key equation to create practical analysis that extracts private data: linear and cubic equations, as well as zero-element.

Comparison to Classic World Algorithms Timing attacks on post-quantum algorithms are powerful compared to the classic world in which the countermeasures are remarkably developed, doubting the feasibility of the timing attacks [19]. However, the variety of timing attacks in the classic world is larger, compared to the post-quantum world, which only compromises the error corrections [19].

Fault Attacks (FA) Computation errors can be intentionally created, creating a perfect environment for secret data leakage to occur. Compared to unintentional errors, which can be due to quality, noise, or environment, intentional errors give an opportunity to attacker to control the faulty system and observe its behavior. Comparing faulty behavior to the correct operation, private data can be leaked. These fault attacks (FA) have been performed on post-quantum algorithms, notably code-based algorithms in [14], lattice-based algorithm in [8], MPKC algorithm in [36], and hash-based algorithm in [13].

According to [14] McEliece scheme is intrinsically resistant against fault injection attacks because of underlying error correction code. Theoretical fault analysis in [14] introduces random faults to encryption of the plaintext: multiplication with publicly known key and error addition. This attack showed that denial of service would be achieved if the plaintext is corrupted, while for other scenarios, error correction code corrects the corruption.

Comparison to Classic World Algorithms Fault injection attacks in the post-quantum world are strictly on a software level because many algorithms have not been efficiently implemented yet. Compared to the classic world, these attacks are very underdeveloped, as the classical fault attacks can be categorized into semi-invasive and non-invasive, in which light and radiation are used physically to introduce the fault.

8.3.2.3 SCA by Quantum Computer on Classical Algorithms

Quantum algorithms, based on hardness in computational complexity theory, have significantly been investigated for more than 20 years and provide computational speed-up [1, 61, 62]. In the field of side-channel security, quantum algorithms are used for unstructured search [61]. The unstructured search is an NP-hard problem which evaluates the function $f(x)$ without any prior understanding of the function, and it searches for its solution x if the solution exists. According to [61], the worst-case classically computed solution is found after 2^n evaluations. However, the quantum Grover's algorithm solves the problem in $\sqrt{2^n}$ evaluations, creating quadratic speed-up. SCA consists of two steps. Quantum algorithm would not increase the speed of data acquisition, but of the search for the key. Once data is obtained in the first step and processed into sub-key components, the quantum algorithm analyzes the data and looks for possible private keys. The first step would not benefit from quantum algorithms because of the low complexity of data acquisition. The only way to use the quantum algorithm in the first step of SCA is if the attack has very few queries to obtain data. However, the main issue in the second step is that SCA data is very structured compared to the requirement of Grover's algorithm, which needs unstructured data. To solve this issue, according to [56], research has been done to introduce the advice distribution of the set. However, this is not the only limitation of Grover's algorithm for this application, because independent side-channel leakage data is not sorted and ordered as Grover's algorithm requires. Although there are limitations of Grover's algorithm for search over independent structured data, the work in [56] gives a first step towards using quantum algorithms for SCA on classical algorithms. The work in [56] recognized the limitations based on the knowledge, uses, and applications from classical SCAs. Further research has been done to enhance and optimize the quantum search furthermore, in terms of rank estimation and provable poly-logarithmic time and space complexities, [20], which would not require the sorting and ordering steps of the entire set of leaked data which is currently required for Grove's algorithm.

8.4 Random Number Generators

A wide range of applications offered by random number generators (RNGs) makes them an interesting subject of study for researchers and designers. It also makes them attractive to attackers, mainly attempting to induce bias into the output of these generators to compromise the security of systems built upon them. Such attacks usually target the security of RNGs at the physical level cf. [53]. This section explores these attacks launched against RNGs proposed in the classical and the quantum worlds. It is worth noting here that quantum technology has provided practical assistance in designing and implementing RNGs. In particular, the randomness sources used in quantum RNGs (QRNGs) provide inherent randomness of quantum effects (see, Fig. 8.5 for the modules involved in a QRNG). However, this is beyond the scope of this chapter. For more information regarding the design and implementation of QRNGs, we refer the reader to [16, 39, 84].

8.4.1 Attacks Against TRNGs and QRNGs

In practice, in order to meet the need for uncorrelated and unbiased random bits, the output of a TRNG may be fed into a post-processing module. Regardless, TRNGs could be sensitive to external influences, e.g., a change in their environment. An adversary can take advantage of this by launching attacks with the aim of inducing variations in the setting of the TRNG, by, e.g., manipulating the power supply, emitting electromagnetic signals, changing the temperature outside the nominal operating range, etc. These attacks, broadly speaking, can be classified as invasive

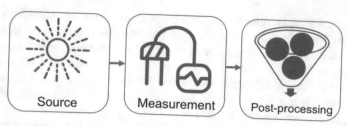

Fig. 8.5 Main blocks included in the architecture of a QRNG. Optical and non-optical randomness sources are connected to a module measuring an unpredictable magnitude of the source. The binary raw bit sequence obtained in this way is then fed into a post-processing unit to achieve a bias-free random sequence

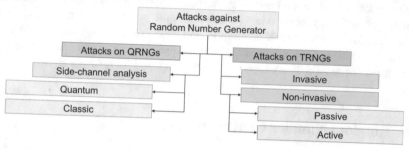

Fig. 8.6 Taxonomy of attacks discussed to assess the security of TRNGs and QRNGs

and non-invasive ones, see Fig. 8.6. Here, the term invasive means that the attack is powerful enough to permanently affect the circuit, e.g., by inserting a pin, burning a hole, etc. [89]. The main focus of this section is non-invasive attacks, being easy-to-mount and effective to introduce bias into the outputs of TRNGs.

From another angle, attacks against TRNGs can be passive or active, where for the former type, the adversary collects some information about the functionality of the TRNG, e.g., frequencies, location, sampling frequency, etc. This information could be helpful to predict the forthcoming outputs of the TRNG, e.g., through passive temperature attacks [11, 55]. On the other hand, adversaries mounting active attacks have the goal of disrupting the functionality of the generator. As an example, active attacks presented in [77] change the temperature of the FPGA, embodying a TRNG to make its output biased. Other examples include frequency injection attacks [54], power signal manipulation [10, 11, 55], injecting electromagnetic (EM) signals [5, 68, 74]. Having these type of attacks in mind, it could be thought that QRNGs could also be vulnerable to them, as discussed in the next section.

8.4.1.1 Attacks Against QRNGs

Compared to TRNGs, QRNGs employ intrinsically random processes; hence, contrary to TRNGs, it is not needed to examine if the source is random and to what extent. With regard to this, QRNGs have become an integral part of quantum cryptosystems, although they could be vulnerable to attacks. A prime example is the attacks against BB84 quantum key distribution (QKD) protocol that have targeted QRNGs (or randomness generator modules) [9, 49].

In the same vein as TRNGs, QRNGs can come under attacks, where the adversary leverages information leaked due to measuring the side-channels or by injecting a fault and then, measuring the side-channels. There are, however, differences between such attacks and their corresponding ones in the classic world. First, for conducting side-channel analysis on QRNGs, the adversary may need access to the device(s). Moreover, it can be possible that the adversary's device is entangled with the devices used by honest parties to extract quantum side-channels, similar to fault injection attacks launched on TRNGs. To study the side-channel attacks against QRNGs more systematically, they are divided into two groups, namely quantum and classic ones, as explained below (see, Fig. 8.6).

Quantum Side-Channel Analysis As mentioned before, side-channel attacks on QKD systems have constituted the first step towards analyzing the security of quantum systems in the face of such attacks. In this regard, it has been demonstrated that due to the time-variant detection efficiency of photon detectors, it is possible to perform side-channel analysis [101]. For this, possible correlations between physical properties of the photons (e.g., temporal properties) with the actual bit values are used to disclose the information exchanged between two parties.

For such attacks, another aspect of the design of QRNGs can be beneficial to the adversary that is arbitrary imperfections in the implementation. For instance, in the attack by Zhao et al. [101], the mismatch between the efficiency of detectors for the bits "0" and "1" plays an important role. Detector blinding attacks introduced in [29, 52] are other examples of such attacks cf. [83]. To launch these attacks, the adversary should control the detectors at the receiver side and perform, e.g., strong illumination so that the randomness in the measurements is eliminated [48].

Classic Side-Channel Analysis Compared to the attacks mentioned above, side-channel analyses that are referred to as classical employ non-trivial information about the QRNGs rather than the quantum-related features. As an example, the information about the seed can be used to mount the attack. This type of attacks targets a specific block in the design of a QRNG, namely the post-processing module, see Fig. 8.5.

Among approaches devised to perform post-processing, randomness expansion techniques are of great importance. A relatively long, random bit sequence can be obtained from a small random seed given to a quantum protocol by employing such methods. This is possible through concatenating a finite number of quantum devices with a certain security level, e.g., [17, 58]. It has been demonstrated that even combining imperfect quantum sources can result in a source with certified randomness of the output [12]. Nevertheless, the security of such randomness expansion methods needs to be assessed under the scenario, where the adversary misuses the side-channel information regarding the seed, see, e.g., [21, 42].

In addition to the above, it has been reported that classical side-channels, namely the classical noise, can provide the adversary with information about the random numbers generated by honest parties [95]. Moreover, in an attempt, Gorbenko et al. have studied whether the EM side-channel attack can be launched against a QRNG [32]. For this purpose, the EM signal leakage from the post-processing module has been considered. Finally, we stress that side-channel analysis of QRNGs is an ongoing study with the aim of understanding different aspects of these quantum systems that make them vulnerable to such attacks.

Other Types of Attacks Against QRNGs The attacks discussed so far are mainly carried out by third parties; however, the risk of an attack launched by the malicious has to be considered and brought under control. In such attacks, the device manufacturer misuses the access to the QRNG to generate a long bit string that is then stored on a memory stick and sold as a QRNG. When using this memory stick mimicking the QRNG, the outputs look random, i.e., pass any statistical test, although the manufacturer can predict the bits in the string. In another variant of this attack, the memory included in a device is exploited to store past measurements, and further use them for future measurements [4].

8.5 Physically Unclonable Functions

Physically unclonable functions (PUFs) produce unique, unbiased, yet reliable fingerprints for anti-counterfeiting, authentication, and cryptographic key generation. PUFs are used for secured storage of sensitive data like cryptographic keys and contribute to a major research topic among security primitives. The idea of PUF originated as "physical one-way functions" described by R. Pappu et al. in [69]. Later, a silicon interpretation was created by utilizing the process variations of a chip by Gassend et al. in [27]. Silicon PUFs generate a unique, unclonable identification key by using the random manufacturing variations within a silicon chip. The output of a PUF is stored in the form of challenge-response pairs (CRPs) obtained by querying the PUF and measuring its response. The state-of-the-art research in silicon PUFs investigates key generation based on metastability, race conditions, etc. in different units present in a silicon chip, including memory (DRAM, SRAM, etc.), ring oscillators (ROs), comparators, latches, current mirrors, etc.

PUFs are categorized depending on their structure, performance, or the number of CRPs generated by them. In this chapter, we divide PUFs into two categories: PUFs in the classical era and PUFs in the post-quantum era. According to [3], classical PUFs generate CRPs involving classical phenomena. Here, the adversary is also limited to only classical interactions with the PUF. In short, the authentication and key-generation process use physical characteristics that can be explained by classical mechanics and do not interact with quantum particles or generate quantum bits. With interesting developments in quantum supercomputers as described in Sect. 8.2, limiting the adversary to only the classical domain may hide potential

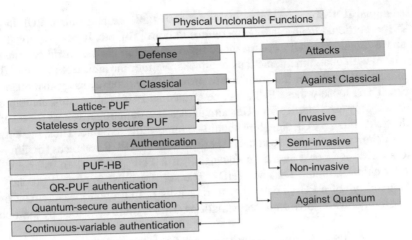

Fig. 8.7 Taxonomy of PUFs in the post-quantum era with research work explored in both defense and attack categories. Note that, especially on the defense side, there are other methods and approaches not categorized in this taxonomy (for more details, see [16])

vulnerabilities within the PUF architecture leading to biased security evaluation. To justify correct performance and security evaluation, PUFs must be studied in the quantum domain where the adversary has access to quantum interactions with the PUF system. This calls for a new perspective on PUFs in the post-quantum era.

Overall, we have categorized the research for PUFs in the quantum domain into two major divisions, (1) *Defense* involves the architectures and design of several PUFs as well as authentication protocols with both quantum and classical architectures built to defend against cloning, piracy, and attacks incorporating quantum phenomena. This can be seen in the taxonomy depicted in Fig. 8.7; and (2) *Attacks* against PUFs in both classical and quantum domains. We describe both the domains in detail in the following sections providing perspectives for current state-of-art research in PUFs.

8.5.1 PUF Preliminaries: Pre-quantum Era

Silicon PUFs generate unique fingerprints from process variations within a chip. Thus, two chips, even with the same design and architecture, provide different CRPs, leading to a unique identifier for every individual chip produced by a company. Silicon PUFs are divided into three types as shown in [100]: (1) *Delay based PUFs* use race conditions in a delay chain producing unique delays due to process variations as seen in Arbiter PUF, RO PUF, etc. [88]; (2) *Memory based PUFs* are made from memory cells like SRAM [46], DRAM [94], etc. and random memory start-up values to generate unique binary strings; (3) *Analog electronic PUFs* belong to a comparatively new category aimed at analog and mixed-signal ICs. Some instances include mono-stable current mirror PUF in [2], PTAT (proportional to absolute temperature) PUF in [50], etc. It must be noted that the above-mentioned PUFs are meant completely for the classic world and are not associated with quantum interactions or attacks. Neither the architecture is based on quantum phenomena nor the architecture provides any robustness against quantum attacks. Thus, loosely these PUFs can be termed as PUFs in the pre-quantum era. The commonly used metrics for PUFs aim to measure their effectiveness for security applications. First, a PUF signature should be impossible to clone. Thus, there must be a sufficient variation between the responses from two PUF instances (in different devices). This is known as the "uniqueness" of a PUF and is often estimated by the inter-chip Hamming Distance (HD). The second most important property is the "reliability" of a PUF, i.e., ability to repeat its output upon measurement over varying temporal and environmental variations. This is often measured by intra-chip HD.

Classical PUFs belong to a mature research field that includes various architectures and authentication schemes. Despite several research works that have been conducted in the classical PUF domain, many of them can be vulnerable to attacks in the post-quantum era. In our future sections, we describe in detail the PUFs which belong to quantum era and the advantages/disadvantages associated with it.

8.5.2 Defense: Innovative PUF Architectures for Anti-Counterfeiting and for Preventing Quantum Attacks

The existence of a quantum adversary can threaten the security of PUFs that have previously been claimed secure against traditional machine learning (ML) attacks. Thus, we require special architectures for PUFs that provide resilience against quantum attacks. Classical quantum-secure PUFs represent those PUF architectures that have already been proven secure against quantum attacks. The availability of quantum systems can completely transform the traditional viewpoint for PUFs. In our next subsections, we try to answer the persisting questions regarding PUFs in the post-quantum world, imploring a new line of research approach for PUFs.

8.5.2.1 Classical Quantum-Secure PUFs

Recent discoveries in quantum technologies have made security primitives vulnerable to quantum attacks. Such attacks against PUFs reduce the complexity of computation and time to brute-force. Thus, PUF designs that are provably secure against quantum attacks are critical for security applications. Only a few instances of classical PUF architectures that are resilient against ML attacks in classical and quantum domains have been proposed so far. The architecture of such PUFs are classical by nature and do not produce CRPs originating from quantum interactions. Nevertheless, the responses provided by these PUFs are mathematically proved to be resistant against quantum attacks, which is an inherent quality of their architectures.

Lattice PUF In [98], the authors propose a strong PUF, which is provably secure against both classical and quantum ML attacks. The PUF called as the lattice PUF entails a physically obfuscated key (POK) and a learning with errors (LWE) decryption function that responds relative to the challenges queried to the PUF. The authors introduce a metric called ML resistance to evaluate the robustness of lattice PUF against attacks. A PUF is said to have k-bit ML resistance if an ML attack needs 2^k attempts to break it. The proposed lattice PUF provides 128-bit ML resistance and produces a CRP size of 2^{136}. The significant advantage of the lattice PUF is that it derives its security from the fact that the decryption function used in the lattice PUF cannot be learned through a PAC (probably approximately correct) framework.

Stateless Cryptographic PUF In [38], a cryptographically secure PUF is described using learning parity with noise (LPN) hard fuzzy extractor. Fuzzy extractors are cryptographic modules capable of eliminating the noise from biometric data (either human or silicon). These modules generate random, uniform keys for security, and cryptographic applications. At first, the authors design a fuzzy extractor which corrects errors (in the order of $O(m)$) within a m-bit response of a POK in polynomial time to decipher an n-bit key. This fuzzy extractor produces a cryptographically secure PUF, which is LPN (learning parity with noise)-hard. This indicates that the PUF cannot be broken by classical as well as quantum attacks. The PUF contains a POK, which entails a series of RO pairs. The frequency of each RO is computed, and a subtractor provides the difference between them. This difference is used as the confidence value to generate the entropy source of the RO. An FPGA implementation of the PUF was provided in [44]. A TRNG creates the challenges for the PUF, and an LPN-hard POK produces the response.

8.5.2.2 Quantum-Secure Authentication of PUFs

Other than PUF architectures, the authentication protocols utilized to ensure secure communication in cryptographic IPs are also at risk against quantum adversaries. An example of a standard authentication protocol using a PUF is provided in Fig. 8.8. Such authentication protocol contains an enrollment phase where the challenge-response pairs (CRPs) are generated and enrolled in a database. The verification phase begins once the protocol is inserted in an IP or any other channel to ensure security to verify a specific set of challenges. New authentication protocols have been proposed to ensure security against quantum attacks. Table 8.1 provides a comprehensive summary of such protocols. Most of these authentication protocols rely on NP-hard problems like LPN and thus can be termed secure against quantum attacks.

PUF-HB The authors use an amalgamation of PUF and the Hopper Blum (HB) function (thus, called PUF-HB) in [35]. HB function was first proposed as an authentication protocol by Hopper and Blum in [41]. These protocols have significant advantages, like the reduction of power consumption for pervasive networks. This protocol also derives its security from the hardness of the LPN problem, making

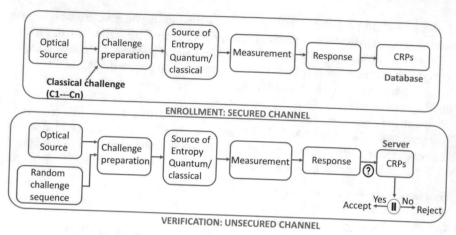

Fig. 8.8 A general structure representing the authentication protocol using PUFs consisting of enrollment and verification stages

PUF-HB theoretically resilient against classical and quantum attacks. The PUF-HB authentication protocol contains two entities, namely the tag and the reader. For a given challenge, the tag utilizes the PUF response along with some additional bit vectors to compute a response, which is sent to the reader. The reader calculates an identical response using the database and verifies the response provided for a series of authentication rounds. After a definite number of rounds is completed, the reader checks the response only when the wrong answers lie within a predetermined level of tolerance. Since the LPN problem is noise parity (NP) hard, the PUF-HB protocol can be proved secure against active ML attacks. The proposed authentication protocol never stores the intermediate responses of the PUF, used to calculate the function. Thus, it is impossible for the attacker to tamper the PUF and attempt to clone the response from the PUF-HB protocol. This makes the PUF tamper resilient as well.

QR-PUF Authentication The significant advantage of quantum PUFs over classical PUFs is that quantum PUFs cannot be cloned. In [30], the authors re-evaluate the uniqueness and robustness metrics of a classical PUF to characterize quantum PUFs like QR-PUFs [79]. In QR-PUF, a generalized authentication scheme is designed, that is applicable to both classical and quantum PUFs like QR-PUFs (described in Sect. 8.5.2). Comparing quantum PUFs against classical PUFs, the authors infer that classical measurements are observed without destabilizing the current configuration of a classical PUF; thus, it is easy to clone the response. After q interactions in case of a classical PUF, an adversary knows exactly q number of CRPs, which are used to create a clone of the classical PUF. In the case of quantum PUFs, the measurements will introduce errors in the quantum response, making it challenging to copy quantum states. The adversary has less than q CRPs to clone the quantum PUF after q interactions. This makes such quantum PUFs superior in

Table 8.1 Comprehensive summary of quantum authentication protocols

Authentication scheme	Type of PUF used	Algorithm used	Advantages/Contribution	Limitations
PUF-HB [35]	Classical PUF	Hopper Blum function	Tamper resilient, unclonable LPN-hard, secure against classical and quantum attacks	Performance against quantum attacks like challenge estimation, quantum emulation not shown
QR-PUF authentication [30]	QR-PUFs	Entity authentication protocols	Characterization of QR-PUF metrics like uniqueness and reliability against classical PUFs	No suggestion provided to improve reliability of QR-PUFs
Quantum-secure authentication [31]	Classical optical PUF	Application of quantum challenges to optical PUF	Resilient against emulation attacks	Reliability and uniqueness not discussed
Continuous variable authentication [66, 67]	Classical optical PUF	Application of quantum challenges using coherent states of light	Resilient against emulation attacks	Reliability and uniqueness not discussed

terms of unclonability. Further, quantum challenges are non-orthogonal in nature. So, a quantum PUF is unable to distinguish between states. This property of non-orthogonality of challenges benefits the no-cloning property but inserts errors in measurement, reducing robustness.

Quantum-Secure Authentication (QSA) The technique proposed in [31] uses a quantum state to challenge the PUF in place of classical states. Traditional authentication protocols using an optical physically unclonable key (PUK) are dependent on classical challenges. Once the attacker has access to the authentication architecture, he may possess sufficient information about the CRP characteristics. It is then easier for the attacker to intercept the classical challenges sent to the verifier. The adversary can mimic such an attack by measuring the challenges from the architecture and providing them to the verifier to obtain correct responses. Thus, the attacker does not need to clone the quantum PUF. Such an attack is called an "emulation attack." However, if the challenges obtained from quantum states are intrinsically unclonable, the attacker will fail to clone the challenges and get a valid response. This protocol authenticates a classical multiple-scattering key by querying it with a light pulse and verifying the spatial structure of the reflected light.

Continuous Variable Authentication Protocol In [67], an authentication protocol using optical PUFs is proposed. The major modification is termed as the utilization of quantum challenges in lieu of classical ones like the QSA technique explained above. The authentication scheme involves quantum challenges obtained from random coherent states of light. The scattering of light emanating from the laser source generates quadrature structures. These are analyzed by a coarse-grained homodyne detector. The homodyne detection helps in obtaining information from the phase and frequency of the quadrature light structures. In [66], the security of the above protocol has been compared against emulation attacks. The authors utilize Helevo's bounds and Fano's inequality to prove that the scheme is robust against emulation attacks.

Concerns for security against quantum attacks have aggravated owing to advancements in the technologies related to the quantum supercomputer. The way such innovations impacts hardware security is still an open-ended question and needs to be investigated to re-evaluate security for PUFs in the quantum world. Quantum superiority has both advantages and disadvantages. It has initiated sufficient progress in security primitives like improvement of unclonability with quantum PUF authentication protocols. The inherent principles of unclonability in quantum PUFs strengthen the PUF performance and increase resilience against tampering and man-in-middle attacks. Whereas, the existence of quantum technology incorporates several vulnerabilities within existing security primitives, compromising the security and privacy of several cryptographic protocols. Quantum technology can be utilized by both the defender and the attacker. In our next section, we discuss several attack strategies against PUFs in the post-quantum era.

8.5.3 Attacks Against PUFs in Post-Quantum Era

Attacks exist against both classical and quantum PUFs. Attack strategies against classical PUFs are well-researched and entail a wide range of strategies, as discussed below. Attacks against quantum PUFs are a comparatively newer topic and still require more research.

8.5.3.1 Attacks Against Classical PUFs

In general, the attacks against classical PUFs are separated into two major subsections (a) classical attacks where the adversary is devoid of any quantum capability and (b) quantum attacks where the attacker possesses quantum competence. We discuss the state-of-art research in both attack types below.

Attacks against classical PUFs are divided into three types, namely (a) invasive attacks, (b) semi-invasive or side-channel attacks, and (c) non-invasive or software/machine learning (ML) attacks. Invasive attacks are capable of changing the physical structure of the PUFs to extract decrypted data by completely altering the PUF configuration. An example attack is demonstrated in [37], where the attackers permanently modify the security fuses of the device by launching a focused-ion-beam (FIB) circuit edit. This enables the attacker to decipher sensitive data using backside micro probing. Semi-invasive attacks on PUFs include the conglomeration of fault injection/side-channel attacks assisted by an ML framework to obtain responsive CRP behavior from the PUFs. In [91], the attackers use laser fault injection attacks along with ML computation to change configurations and undermine the security of XOR-arbiter and RO PUFs in programmable logic devices (180 nm). For non-invasive ML attacks, the attacker uses a small portion of the CRPs to generate a model emulating the challenge-response behavior of the PUF. The first-ever ML attack on PUFs was implemented in [28]. Different types of ML attacks have been investigated against PUFs after that. ML attacks on classical PUFs are divided into two types: empirical attacks and PAC (probably approximately correct) framework. Empirical ML [73] does not have any predefined level of confidence, whereas the PAC framework develops models with definite pre-specified levels of efficiency, as seen in [25, 26].

8.5.3.2 Attacks Against Quantum PUFs

The robustness of quantum readout PUFs against challenge estimation attacks has been provided in [80]. This is a demonstration of a classical attack launched against quantum PUFs. The authors analyze the security of quantum readout PUF against such attacks. Here, the adversary has access to the verification protocol and is capable of decoding the challenge quantum state. The attacker then measures the challenge state and recreates the challenge to emulate PUF behavior. Though such an attack can be executed against quantum-secure authentication protocol (which uses a classical challenge), it fails against protocols that involve quantum states as challenges. In this case, the attacker is unable to measure a quantum state, rendering the attack ineffective. The QSA [31] and continuous variable authentication protocols [67] are secure against such forms of attack. The QR-PUF has also been proved resilient against such attacks, as shown in [80]. The authors mention that the number of photons detected in the protocol is lower by more than a factor of the quantum security parameter $(S)^2$ defined in the paper proving the QR-PUF resilient against such classical attacks.

Another type of attack is based on quantum interactions, where the attacker possesses quantum capabilities. Here, the system that is attacked contains a quantum PUF (QPUF). An example of such an attack is the quantum emulation attack described in [3]. Here the adversary uses quantum emulation (QE) algorithm [57] to attack a QPUF. The authors demonstrate QPUFs as quantum channels with the standard intrinsic quantum qualities of unclonability, robustness, and collision resistance. The authors further add that QPUFs can be modeled by unitary transformations owing to the property of collision resistance. Thus, such PUFs are termed as unitary quantum PUFs or UQPUFs in the paper. The authors explain that quantum emulation attacks can undermine the unclonability property of the UQPUFs. If the adversary can select the target challenge themselves, then quantum primitives will fail to provide security against such attackers. Again, if the attacker has access to the challenge and can query the primitive after deciphering the challenge, then no current quantum primitive can provide protection against such attacks. In short, it is shown that the unclonability property of UQPUFs can be compromised by quantum emulation attacks.

Another critical aspect of emulation attacks has been demonstrated in [24]. In this paper, the authors use a similar framework of intercept-emulation attacks provided in [66] against the continuous variable-protocol. They presume that the attacker has access to a copy of the numerical CRPs and hacks the measurement apparatus to make an informed guess of the other security parameters essential to induce the valid response state of the verifier. The paper provides the resilience of continuous variable authentication protocol against three varieties of emulation attacks and analyzes the security metrics used to ensure the security of the protocol against the above-mentioned attacks. The security against the three varieties of emulation attacks is related to the total number of CRPs enrolled in the data base, which is denoted by N. The authors conclude that the protocol is secured against the attacks if $N > 110$ for a large sample size of M. M is the total number of random queries made to the protocol during verification stage. The value of M should be large enough to ensure a lower value of security threshold (2ϵ) such that, $2\epsilon \leq 3 \times 10^{-4}$.

The metric which increases the robustness of classical PUFs against emulation/modeling attacks is unclonability. It has been mentioned in several papers,

---

[2]$S = \frac{K}{n}$, where K is the order of the information within the challenge and n is the cumulative number of photons.

such as [30, 79], that quantum PUFs are preferable to classical PUFs in terms of unclonability. But with the emergence of new attacks like quantum emulation and classical challenge estimation attacks, the no-cloning property of quantum PUFs can be at risk. With the onset of new attack strategies, new authentication protocols have also been provided that claim to protect against the above attacks, but all these protocols have their own advantages and limitations, as described in Table 8.1. Also, the trait of unclonability adversely affects other metrics like reliability for quantum PUFs. Thus, an optimum trade-off of the above parameters is desirable that can generate an unclonable, as well as a reliable quantum PUF.

8.6 Conclusion

The key goal of this chapter is to draw the attention of both quantum technology and hardware security communities to the mutual impact of quantum technology and physical security. To this end, it is attempted to investigate the state-of-the-art techniques studied in the literature; however, this chapter aims to discuss neither the detailed design of quantum systems nor the attacks mounted on them. Nonetheless, this chapter can be seen as a basis for a systematic and comprehensive study on secure key generation/storage and secure execution of quantum systems.

When it comes to side-channel attacks, it has been demonstrated that post-quantum algorithms are not necessarily resistant against these attacks, while being proven quantum-resistant in some cases. To protect the quantum systems coming under side-channel analysis, more specifically, timing and SPA attacks, countermeasures have been proposed, e.g., in [96]. Compared to their counterpart measures proposed in the classic world, post-quantum countermeasures may not be generic in the sense that the protection against a specific attack is provided while being compromised by others. The situation is even worse if more recent post-quantum algorithms are considered, where their security to side-channel attacks is not well understood and studied.

For secure key generation, thanks to quantum systems studied for many decades, more reliable and technically mature solutions can be found in the literature. Optical QRNGs are examples of such solutions, where light emitted from, e.g., lasers, light-emitting diodes, etc., sources is an easy-to-reach and cost-effective source. In contrast to TRNGs, QRNGs offer a guarantee on intrinsic randomness and a sufficiently high generation rate. Despite these facts, similar to TRNGs, QRNGs must be examined to prevent any defect and interference. This highlights the importance of research on reliable testing methods. Another aspect of the design of QRNGs that requires more attention is the development of high-quality, electronic chip-based quantum sources, which facilitates the implementation of commercial devices. In this regard, the cost and compatibility with off-the-shelf systems, along with the generation rate, should be taken into account.

On the other hand, at least in the classic world, PUFs also seem to be promising candidates for secure key generation and storage. Quantum PUFs are an emerging topic that concerns not only the development of quantum-enhanced PUFs and PUF-based protocols but also the impact of the quantum technology on attacks on PUFs. Yet, due to a lack of systematic approaches, work done so far in this matter can be considered ad-hoc. As a prime example, NP-hard authentication protocols offer the potential for security solutions against quantum attacks. In spite of that, more effort should be put to study their practical feasibility and security in practice. Last but not least, we emphasize the need for a new set of metrics to ensure the security PUFs against quantum attacks.

Acknowledgments The author would like to acknowledge the support of AFOSR under award number FA 9550-14-1-0351. We would like to thank Spencer Dupee, who has contributed to the survey [16], which is the extended version of this chapter.

References

1. G. Alagic, J. Alperin-Sheriff, D. Apon, D. Cooper, Q. Dang, Y.K. Liu, C. Miller, D. Moody, R. Peralta et al.: Status report on the first round of the NIST post-quantum cryptography standardization process. US Department of Commerce, National Institute of Standards and Technology, Maryland (2019)
2. A. Alvarez, W. Zhao, M. Alioto, 14.3 15fj/bit static physically unclonable functions for secure chip identification with $< 2\%$ native bit instability and 140x inter/intra PUF hamming distance separation in 65nm, in *2015 IEEE International Solid-State Circuits Conference - (ISSCC)* (IEEE, Piscataway, 2015), pp. 1–3
3. M. Arapinis, M. Delavar, M. Doosti, E. Kashefi, Quantum physical unclonable functions: possibilities and impossibilities (2019)
4. R. Arnon-Friedman, A. Ta-Shma, Limits of privacy amplification against nonsignaling memory attacks. Phys. Rev. A **86**(6), 062333 (2012)
5. P. Bayon, L. Bossuet, A. Aubert, V. Fischer, F. Poucheret, B. Robisson, P. Maurine, Contactless electromagnetic active attack on ring oscillator based true random number generator, in *International Workshop on Constructive Side-Channel Analysis and Secure Design* (Springer, Berlin, 2012), pp. 151–166
6. S. Beauregard, Circuit for Shor's algorithm using 2n+ 3 qubits. Quantum Inf. Comput. **3**(2), 175–185 (2003)
7. D.J. Bernstein, T. Lange, Post-quantum cryptography. Nature **549**(7671), 188–194 (2017)
8. N. Bindel, J. Buchmann, J. Krämer, Lattice-based signature schemes and their sensitivity to fault attacks, in *2016 Workshop on Fault Diagnosis and Tolerance in Cryptography (FDTC)* (IEEE, Santa Barbara, 2016), pp. 63–77
9. J. Bouda, M. Pivoluska, M. Plesch, C. Wilmott, Weak randomness seriously limits the security of quantum key distribution. Phy. Rev. A **86**(6), 062308 (2012)
10. S. Buchovecká, J. Hlaváč, Frequency injection attack on a random number generator, in *2013 IEEE 16th International Symposium on Design and Diagnostics of Electronic Circuits & Systems (DDECS)* (IEEE, Karlovy Vary, 2013), pp. 128–130
11. Y. Cao, V. Rožić, B. Yang, J. Balasch, I. Verbauwhede, Exploring active manipulation attacks on the TERO random number generator, in *2016 IEEE 59th International Midwest Symposium on Circuits and Systems (MWSCAS)* (IEEE, Abu Dhabi, 2016), pp. 1–4
12. Z. Cao, H. Zhou, X. Yuan, X. Ma, Source-independent quantum random number generation. Phys. Rev. X **6**(1), 011020 (2016)

13. L. Castelnovi, A. Martinelli, T. Prest, Grafting trees: a fault attack against the sphincs framework, in *Proceedings of PQCrypto* (Springer International Publishing, Cham, 2018), pp. 165–184
14. P.L. Cayrel, P. Dusart, McEliece/Niederreiter PKC: sensitivity to fault injection, in *2010 5th International Conference on Future Information Technology* (IEEE, Changsha, 2010), pp. 1–6. https://doi.org/10.1109/FUTURETECH.2010.5482663
15. L. Chen, L. Chen, S. Jordan, Y.K. Liu, D. Moody, R. Peralta, R. Perlner, D. Smith-Tone, Report on post-quantum cryptography, vol. 12. US Department of Commerce, National Institute of Standards and Technology, USA (2016)
16. S. Chowdhury, A. Covic, R.Y. Acharya, S. Dupee, F. Ganji, D. Forte, Physical security in the post-quantum era: a survey on side-channel analysis, random number generators, and physically unclonable functions (2020). Preprint, arXiv:2005.04344
17. M. Coudron, H. Yuen, Infinite randomness expansion with a constant number of devices, in *Proceedings of the Forty-Sixth Annual ACM Symposium on Theory of Computing, STOC 2014* (Association for Computing Machinery, New York, 2014), pp. 427–436
18. P. Czypek, Implementing multivariate quadratic public key signature schemes on embedded devices (2012)
19. J.P. D'Anvers, M. Tiepelt, F. Vercauteren, I. Verbauwhede, Timing attacks on error correcting codes in post-quantum schemes. Cryptology ePrint Archive, Report 2019/292 (2019). https://eprint.iacr.org/2019/292. Accessed 2 May 2020
20. L. David, A. Wool, Poly-logarithmic side channel rank estimation via exponential sampling, in *Topics in Cryptology – CT-RSA 2019*, ed. by M. Matsui (ed.) (Springer International Publishing, Cham, 2019), pp. 330–349
21. A. De, C. Portmann, T. Vidick, R. Renner, Trevisan's extractor in the presence of quantum side information. SIAM J. Comput. **41**(4), 915–940 (2012)
22. G. Deepa, G. SriTeja, S. Venkateswarlu, An overview of acoustic side-channel attack. Int. J. Comput. Sci. Commun. Netw. **3**(1), 15–20 (2013)
23. J. Fan, I. Verbauwhede, *An Updated Survey on Secure ECC Implementations: Attacks, Countermeasures and Cost* (Springer, Berlin, 2012), pp. 265–282
24. L. Fladung, G.M. Nikolopoulos, G. Alber, M. Fischlin, Intercept-resend emulation attacks against a continuous-variable quantum authentication protocol with physical unclonable keys. Cryptography **3**(4), 25 (2019)
25. F. Ganji, S. Tajik, J.P. Seifert, PAC learning of arbiter PUFs. J. Cryptogr. Eng. **6** (2014). https://doi.org/10.1007/s13389-016-0119-4
26. F. Ganji, S. Tajik, J.P. Seifert, Why attackers win: on the learnability of XOR arbiter PUFs, in *Trust and Trustworthy Computing*, ed. by M. Conti, M. Schunter, I. Askoxylakis (Springer International Publishing, Cham, 2015), pp. 22–39
27. B. Gassend, D. Clarke, M. van Dijk, S. Devadas, Silicon physical random functions, in *Proceedings of the 9th ACM Conference on Computer and Communications Security, CCS 2002* (Association for Computing Machinery, New York, 2002), pp. 148–160
28. B. Gassend, D. Lim, D. Clarke, M. van Dijk, S. Devadas, Identification and authentication of integrated circuits. Concurr. Comput. Pract. Exp. **16**(11), 1077–1098 (2004)
29. I. Gerhardt, Q. Liu, A. Lamas-Linares, J. Skaar, C. Kurtsiefer, V. Makarov, Full-field implementation of a perfect eavesdropper on a quantum cryptography system. Nat. Commun. **2**(1), 1–6 (2011)
30. G. Gianfelici, H. Kampermann, D. Bruß, Theoretical framework for physical unclonable functions, including quantum readout. Phys. Rev. A **101**, 042337-1–042337-12 (2020)
31. S.A. Goorden, M. Horstmann, A.P. Mosk, B. Škorić, P.W.H. Pinkse, Quantum-secure authentication of a physical unclonable key. Optica **1**(6), 421–424 (2014)
32. Y. Gorbenko, O. Nariezhnii, M. Krivich, Differential electromagnetic attack on cryptographies modules of a quantum random number generator, in *2017 4th International Scientific-Practical Conference Problems of Infocommunications. Science and Technology (PIC S&T)* (IEEE, Piscataway, 2017), pp. 161–167
33. J. Gruska, *Quantum Computing*, vol. 2005 (McGraw-Hill, London, 1999)

34. H. Gupta, S. Sural, V. Atluri, J. Vaidya, Deciphering text from touchscreen key taps, in *Data and Applications Security and Privacy XXX*, ed. by S. Ranise, V. Swarup (Springer International Publishing, Cham, 2016), pp. 3–18

35. G. Hammouri, B. Sunar, PUF-HB: A tamper-resilient HB based authentication protocol, in *Applied Cryptography and Network Security* (Springer, Berlin, 2008), pp. 346–365

36. Y. Hashimoto, General fault attacks on multivariate public key cryptosystems. IEICE Trans. Fundam. Electron. Commun. Comput. Sci. **E.96-A** (2013). https://doi.org/10.1587/transfun. E96.A.196

37. C. Helfmeier, D. Nedospasov, C. Tarnovsky, J.S. Krissler, C. Boit, J.P. Seifert, Breaking and entering through the silicon, in *Proceedings of the 2013 ACM SIGSAC Conference on Computer and Communications Security* (Association for Computing Machinery, New York, 2013), pp. 733–744

38. C. Herder, L. Ren, M.V. Dijk, M.D. Yu, S. Devadas, Trapdoor computational fuzzy extractors and stateless cryptographically-secure physical unclonable functions. IEEE Trans. on Dependable Secure Comput. **14**(1), 65–82 (2017)

39. M. Herrero-Collantes, J.C. Garcia-Escartin, Quantum random number generators. Rev. Mod. Phys. **89**(1), 015004 (2017)

40. S. Heyse, A. Moradi, C. Paar, Practical power analysis attacks on software implementations of McEliece, in *Post-Quantum Cryptography*, ed. by N. Sendrier (Springer, Berlin, 2010), pp. 108–125

41. N.J. Hopper, M. Blum, Secure human identification protocols, in *Proceedings of the 7th International Conference on the Theory and Application of Cryptology and Information Security: Advances in Cryptology, ASIACRYPT 2001* (Springer, Berlin, 2001), pp. 52–66

42. R. Impagliazzo, D. Zuckerman, How to recycle random bits, in *Proceedings of FOCS*, vol. 30 (IEEE, Piscataway, 1989), pp. 248–253

43. J. Jaffe, A first-order DPA attack against AES in counter mode with unknown initial counter, in *International Workshop on Cryptographic Hardware and Embedded Systems* (Springer, Vienna, 2007), pp. 1–13

44. C. Jin, C. Herder, L. Ren, P. Nguyen, B. Fuller, S. Devadas, M. van Dijk, FPGA implementation of a cryptographically-secure PUF based on learning parity with noise. Cryptography **1**(3), 23 (2017). http://dx.doi.org/10.3390/cryptography1030023

45. M.J. Kannwischer, A. Genêt, D. Butin, J. Krämer, J. Buchmann, Differential power analysis of XMSS and SPHINCS, in *Constructive Side-Channel Analysis and Secure Design*, ed. by J. Fan, B. Gierlichs (Springer International Publishing, Cham, 2018), pp. 168–188

46. P.A. Layman, S. Chaudhry, J.G. Norman, J.R. Thomson, Electronic fingerprinting of semiconductor integrated circuits. U.S. Patent 6 738 294, Sept 2002

47. R.B. Leighton, M.L. Sands, *The Feynman Lectures on Physics: Quantum Mechanics*, vol. 3 (Addison-Wesley, Reading, 1965)

48. H.W. Li, S. Wang, J.Z. Huang, W. Chen, Z.Q. Yin, F.Y. Li, Z. Zhou, D. Liu, Y. Zhang, G.C. Guo, et al.: Attacking a practical quantum-key-distribution system with wavelength-dependent beam-splitter and multiwavelength sources. Phys. Rev. A **84**(6), 062308 (2011)

49. H.W. Li, Z.Q. Yin, S. Wang, Y.J. Qian, W. Chen, G.C. Guo, Z.F. Han, Randomness determines practical security of bb84 quantum key distribution. Sci. Rep. **5**(1), 1–8 (2015)

50. J. Li, M. Seok, Ultra-compact and robust physically unclonable function based on voltage-compensated proportional-to-absolute-temperature voltage generators. IEEE J. Solid-State Circuits **51**(9), 2192–2202 (2016)

51. H. Lohrke, S. Tajik, T. Krachenfels, C. Boit, J.P. Seifert, Key extraction using thermal laser stimulation. IACR Trans. Cryptogr. Hardware Embed. Syst. **4**, 573–595 (2018)

52. L. Lydersen, C. Wiechers, C. Wittmann, D. Elser, J. Skaar, V. Makarov, Hacking commercial quantum cryptography systems by tailored bright illumination. Nat. Photonics **4**(10), 686 (2010)

53. R. Maes, *Physically Unclonable Functions: Constructions, Properties and Applications* (Springer Science & Business Media, Berlin, 2013)

54. A.T. Markettos, S.W. Moore, The frequency injection attack on ring-oscillator-based true random number generators, in *International Workshop on Cryptographic Hardware and Embedded Systems* (Springer, Berlin, 2009), pp. 317–331

55. H. Martin, T. Korak, E. San Millán, M. Hutter, Fault attacks on STRNGs: impact of glitches, temperature, and underpowering on randomness. IEEE Trans. Inf. Forensics Secur. **10**(2), 266–277 (2014)

56. D.P. Martin, A. Montanaro, E. Oswald, D. Shepherd, Quantum key search with side channel advice, in *Selected Areas in Cryptography – SAC 2017*, ed. by C. Adams, J. Camenisch (Springer International Publishing, Cham, 2018), pp. 407–422

57. I. Marvian, S. Lloyd, Universal quantum emulator (2016)

58. C.A. Miller, Y. Shi, Universal security for randomness expansion from the spot-checking protocol. SIAM J. Comput. **46**(4), 1304–1335 (2017)

59. R. Misoczki, J.P. Tillich, N. Sendrier, P.S.L.M. Barreto, MDPC-McEliece: new McEliece variants from moderate density parity-check codes, in *IEEE International Symposium on Information Theory - ISIT 2013* (IEEE, Istanbul, 2013), pp. 2069–2073

60. H.G. Molter, M. Stöttinger, A. Shoufan, F. Strenzke, A simple power analysis attack on a McEliece cryptoprocessor. J. Cryptogr. Eng. **1**(1), 29–36 (2011)

61. A. Montanaro, Quantum algorithms: an overview. NPJ Quantum Inf. **2**(1), 1–8 (2016)

62. M. Mosca, Quantum algorithms (2008)

63. S. Myung, K. Yang, J. Kim, Quasi-cyclic LDPC codes for fast encoding. IEEE Trans. Inf. Theory **51**(8), 2894–2901 (2005)

64. S. Narain, A. Sanatinia, G. Noubir, Single-stroke language-agnostic keylogging using stereo-microphones and domain specific machine learning, in *Proceedings of the 2014 ACM Conference on Security and Privacy in Wireless & Mobile Networks, WiSec 2014* (Association for Computing Machinery, Oxford, 2014), pp. 201–212

65. M.A. Nielsen, I.L. Chuang, I.L. Chuang, *Quantum Computation and Quantum Information*, Chap. 2 (Cambridge University Press, Cambridge, 2000)

66. G.M. Nikolopoulos, Continuous-variable quantum authentication of physical unclonable keys: security against an emulation attack. Phys. Rev. A **97**(1), 012324 (2018)

67. G.M. Nikolopoulos, E. Diamanti, Continuous-variable quantum authentication of physical unclonable keys. Nat. Sci. Rep. **7**, 46047 (2017)

68. S. Ordas, L. Guillaume-Sage, P. Maurine, EM injection: fault model and locality, in *Workshop on Fault Diagnosis and Tolerance in Cryptography (FDTC)* (IEEE, Saint Malo, 2015), pp. 3–13

69. R. Pappu, B. Recht, J. Taylor, N. Gershenfeld, Physical one-way functions. Science **297**(5589), 2026–2030 (2002). https://doi.org/10.1126/science.1074376

70. M. Petrvalsky, T. Richmond, M. Drutarovsky, P.L. Cayrel, V. Fischer, Differential power analysis attack on the secure bit permutation in the McEliece cryptosystem, in *2016 26th International Conference Radioelektronika (RADIOELEKTRONIKA)* (IEEE, Kosice, 2016), pp. 132–137

71. T. Richmond, M. Petrvalsky, M. Drutarovsky, A side-channel attack against the secret permutation on an embedded McEliece cryptosystem (2015). https://hal-ujm.archives-ouvertes.fr/ujm-01186639

72. M. Rossi, M. Hamburg, M. Hutter, M.E. Marson, A side-channel assisted cryptanalytic attack against QcBits, in *Cryptographic Hardware and Embedded Systems – CHES 2017*, ed. by W. Fischer, N. Homma (Springer International Publishing, Cham, 2017), pp. 3–23

73. U. Rührmair, F. Sehnke, J. Sölter, G. Dror, S. Devadas, J. Schmidhuber, Modeling attacks on physical unclonable functions, in *Proceedings of the 17th ACM Conference on Computer and Communications Security, CCS 2010* (Association for Computing Machinery, New York, 2010), pp. 237–249. https://doi.org/10.1145/1866307.1866335

74. J.M. Schmidt, M. Hutter, *Optical and EM Fault-Attacks on CRT-Based RSA: Concrete Results* (Verlag der Technischen Universität Graz, Graz, 2007), pp. 61–67

75. P.W. Shor, Polynomial-time algorithms for prime factorization and discrete logarithms on a quantum computer. SIAM Rev. **41**(2), 303–332 (1999)

76. B.Y. Sim, J. Kwon, K.Y. Choi, J. Cho, A. Park, D.G. Han, Novel side-channel attacks on quasi-cyclic code-based cryptography. IACR Trans. Cryptogr. Hardware Embed. Syst. 2019(4), 180–212 (2019)

77. M. Šimka, P. Komenského, Active non-invasive attack on true random number generator, in 6th PhD Student Conference and Scientific and Technical Competition of Students of FEI TU Košice, Košice, Slovakia. Citeseer, Slovakia (2006), pp. 129–130

78. H. Singh, Code based cryptography: classic McEliece (2019)

79. B. Škorić, Quantum readout of physical unclonable functions. Int. J. Quantum Inf. 10(01), 1250001 (2012)

80. B. Škorić, A.P. Mosk, P.W. Pinkse, Security of quantum-readout PUFs against quadrature-based challenge-estimation attacks. Int. J. Quantum Inf. 11(04), 1350041 (2013)

81. R. Spreitzer, Pin skimming: exploiting the ambient-light sensor in mobile devices. in 4th Annual ACM CCS Workshop on Security and Privacy in Smartphones and Mobile Devices (SPSM) (Association of Computing Machinery, New York, 2014), pp. 51–62. https://doi.org/10.1145/2666620.2666622. In conjunction with the 21st ACM Conference on Computer and Communications Security (CCS)

82. F.X. Standaert, Introduction to Side-Channel Attacks (Springer US, Boston, 2010), pp. 27–42

83. M. Stipčević, Preventing detector blinding attack and other random number generator attacks on quantum cryptography by use of an explicit random number generator (2014)

84. M. Stipčević, Ç.K. Koç, True random number generators, in Open Problems in Mathematics and Computational Science (Springer, Cham, 2014), pp. 275–315

85. F. Strenzke, A timing attack against the secret permutation in the McEliece PKC, in Post-Quantum Cryptography, ed. by N. Sendrier (Springer, Berlin, 2010), pp. 95–107

86. F. Strenzke, Timing attacks against the syndrome inversion in code-based cryptosystems, in Post-Quantum Cryptography, ed. by P. Gaborit (Springer, Berlin, 2013), pp. 217–230

87. F. Strenzke, E. Tews, H.G. Molter, R. Overbeck, A. Shoufan, Side channels in the McEliece PKC, in International Workshop on Post-Quantum Cryptography (Springer, Berlin, 2008), pp. 216–229

88. G.E. Suh, S. Devadas, Physical unclonable functions for device authentication and secret key generation, in Proceedings of the 44th Annual Design Automation Conference, DAC 2007 (Association for Computing Machinery, New York, 2007), pp. 9–14. https://doi.org/10.1145/1278480.1278484

89. B. Sunar, W.J. Martin, D.R. Stinson, A provably secure true random number generator with built-in tolerance to active attacks. IEEE Trans. Comput. 56(1), 109–119 (2006)

90. M. Taha, T. Eisenbarth, Implementation attacks on post-quantum cryptographic schemes. Cryptology ePrint Archive, Report 2015/1083 (2015). https://eprint.iacr.org/2015/1083

91. S. Tajik, H. Lohrke, F. Ganji, J.P. Seifert, C. Boit, Laser fault attack on physically unclonable functions, in 2015 Workshop on Fault Diagnosis and Tolerance in Cryptography (FDTC) (IEEE, Piscataway, 2015), pp. 85–96

92. S. Tajik, E. Dietz, S. Frohmann, H. Dittrich, D. Nedospasov, C. Helfmeier, J.P. Seifert, C. Boit, H.W. Hübers, Photonic side-channel analysis of arbiter PUFs. J. Cryptol. 30(2), 550–571 (2017)

93. S. Tajik, H. Lohrke, J.P. Seifert, C. Boit, On the power of optical contactless probing: attacking bitstream encryption of FPGAs, in Proceedings of the 2017 ACM SIGSAC Conference on Computer and Communications Security (ACM, Dallas, 2017), pp. 1661–1674

94. Q. Tang, C. Zhou, W. Choi, G. Kang, J. Park, K.K. Parhi, C.H. Kim, A dram based physical unclonable function capable of generating> 10 32 challenge response pairs per 1kbit array for secure chip authentication, in 2017 IEEE Custom Integrated Circuits Conference (CICC) (IEEE, Austin, 2017), pp. 1–4

95. J. Thewes, C. Lüders, M. Aßmann, Eavesdropping attack on a trusted continuous-variable quantum random-number generator. Phys. Rev. A 100(5), 052318 (2019)

96. I. von Maurich, T. Güneysu, Towards side-channel resistant implementations of QC-MDPC McEliece encryption on constrained devices, in Post-Quantum Cryptography, ed. by M. Mosca (Springer International Publishing, Cham, 2014), pp. 266–282

97. R. Villanueva-Polanco, A comprehensive study of the key enumeration problem. Entropy **21**(10), 972 (2019)

98. Y. Wang, X. Xi, M. Orshansky, Lattice PUF: a strong physical unclonable function provably secure against machine learning attacks (2019)

99. H. Yi, W. Li, On the importance of checking multivariate public key cryptography for side-channel attacks: the case of enTTS scheme. Comput. J. **60**, 1–13 (2017). https://doi.org/10.1093/comjnl/bxx010

100. J.L. Zhang, G. Qu, Y.Q. Lv, Q. Zhou, A survey on silicon PUFs and recent advances in ring oscillator PUFs. J. Comput. Sci. Technol. **29**(4), 664–678 (2014)

101. Y. Zhao, C.H.F. Fung, B. Qi, C. Chen, H.K. Lo, Quantum hacking: experimental demonstration of time-shift attack against practical quantum-key-distribution systems. Phys. Rev. A **78**(4), 042333 (2008)

Chapter 9
Post-Quantum Cryptographic Hardware and Embedded Systems

Brian Koziel, Mehran Mozaffari Kermani, and Reza Azarderakhsh

9.1 Introduction

PQC Post-quantum cryptography (PQC) refers to the study of cryptosystems that are resistant to attacks by both classical and quantum computers. Virtually all of today's deployed cryptosystems are resistant to classical computers, but vulnerable to more efficient attacks by large-scale quantum computers. It is unknown when large enough quantum computers will be available, but current estimates vary between several years to several decades. Notably, once large scale quantum computers are available, today's public key cryptography schemes will be broken with a quantum computer employing Shor's algorithm for factoring discrete logarithms [1] and symmetric key encryption schemes will have approximately half the security with a quantum computer employing Grover's algorithm for quantum search [2].

NIST Standardization Project Given the time needed to create, evaluate, and deploy new quantum-resistant cryptosystems, the National Institute of Standards and Technology (NIST) has begun publicly evaluating post-quantum cryptosystems in its PQC standardization project that began in 2016 [3]. Currently in its third round of review, NIST has narrowed down the number of standardization candidates from 69 "complete and proper" candidates down to 15 candidates based on (1) security, (2) cost and performance, and (3) algorithm and implementation characteristics. Of these 15 schemes, there are 9 public key encryption (PKE)/key encapsulation

B. Koziel · R. Azarderakhsh
Florida Atlantic University, Boca Raton, FL, USA
e-mail: bkoziel2017@fau.edu; razarderakhsh@fau.edu

M. M. Kermani (✉)
University of South Florida, Tampa, FL, USA
e-mail: mehran2@usf.edu

M. Tehranipoor (ed.), *Emerging Topics in Hardware Security*,
https://doi.org/10.1007/978-3-030-64448-2_9

229

mechanisms (KEMs) and 6 digital signature schemes. Furthermore, 4 PKE/KEM and 3 digital schemes are considered "Third Round Finalists," which will be considered for standardization at the conclusion of the third round. The other candidates are considered "Third Round Alternative Candidates," which NIST expects to have a fourth round of evaluation if they are to be standardized. Although there is no drop-in replacement for today's currently deployed elliptic curve cryptography (ECC) implementations, each of the 9 PKA/KEM schemes features different tradeoffs of security, performance, public key size, implementation size, and secure implementation characteristics.

PKE/KEM Deployment First In the most immediate future, the quantum-resistant PKE/KEM schemes are the top priority to transition to in our digital infrastructure. This is to prevent retroactive decryption, where a malicious party stores encrypted traffic and then later uses a quantum computer to break the private key and retrieve the private data. Given that sensitive secrets such as those related to national security must remain confidential for several years, these quantum-resistant PKE/KEM schemes must be deployed well in advance of the advent of large-scale quantum computers. Quantum-resistant digital signatures are primarily used for authentication. Since retroactively breaking authentication is not a concern, these schemes only need to be enabled once large-scale quantum computers are available, rather than well in advance.

In this work, we primarily examine metric #2 on the cost and performance of these algorithms, in particular how well hardware can be used to accelerate critical computations. Below security, NIST identifies the cost and performance as the second most important criterion when evaluating candidate algorithms. Here, the cost includes the computational efficiency of key generation, key encapsulation, and key decapsulation, the transmission cost for public keys and ciphertexts, and the implementation costs in terms of RAM and gate counts.

Hardware vs. Software Implementations The platform on which a cryptosystem is deployed determines its general computational efficiency and area footprint. A software implementation deploys a cryptographic protocol on some (general-purpose) processor with a high-level programming language such as the C coding language. Depending on performance and code size requirements, this software can be optimized for a particular device using assembly language to directly utilize each processing resource available. In the NIST PQC standardization project, NIST used the low-cost and low-performance ARM Cortex-M4 processor and the high-cost and high-performance Intel Core i7-6700 Skylake processor to compare different software performance targets. Software processors are designed to be general-purpose, capable of doing a multitude of processing tasks depending on the user. On the other hand, a hardware implementation deploys a cryptographic protocol on a Field-Programmable Gate Array (FPGA), Application-Specific Integrated Circuit (ASIC), or many other such devices that implement the connections between transistors. Here, a hardware description language such as VDHL and Verilog is used to describe the basic logic components and blocks that make up computations in the cryptosystem. Although there is less flexibility here, implementing a cryptosystem at the hardware level typically results in huge gains to performance, power, and energy cost. Lastly, there is also the option of a hardware/software (HW/SW) codesign approach where the hardware accelerates expensive low-level computations and the software processor can handle and implement any high-level algorithms. Based on the complexity of the PQC algorithms, the HW/SW codesign route has been the most common implementation style; switching to pure hardware would only produce insignificant speedups.

Hardware Benchmarks In the NIST PQC competition, the Xilinx Artix-7 family of FPGAs was the recommended hardware to compare hardware implementations. This FPGA choice recommends the low-cost and efficient performance point of comparison. With no choice of a high-performance FPGA option, a variety of papers presented their results additionally on Xilinx Virtex-7, Zynq UltraScale+, and Kintex UltraScale+ FPGAs. ASIC results, where the hardware is implemented on a particular technology node, are fairly sparse. One well-reasoned guideline of the hardware community is to use the Xilinx Artix-7 family for lightweight hardware implementations, the Zynq 7000-series for lightweight HW/SW implementations, and the Zynq Xilinx UltraScale+ family for high-performance hardware and high-performance HW/SW implementations [4].

Organization In the following sections, we dive into hardware implementation characteristics of these Round 3 PKE/KEM candidates. First, in Sect. 9.2, we review some implementation features and algorithm costs for each of the PKE/KEM candidates. Then, in Sects. 9.3–9.5, we explore the critical computations among the code-based, lattice-based, and isogeny-based candidates, respectively. In Sect. 9.6, we present and compare the state-of-the-art hardware results for each of these candidates. Finally, we conclude this chapter in Sect. 9.7.

9.2 Overview of Round 3 PKE/KEM Candidates

Round 3 PKE/KEM Candidates There are 9 PKA/KEM schemes that have advanced to the third round of NIST's PQC standardization competition. The 4 finalists are Classic McEliece [5], CRYSTALS-Kyber [6], NTRU [7], and Saber [8]. The 5 alternative candidates are BIKE [9], FrodoKEM [10], HQC [11], NTRU Prime [12], and SIKE [13]. For PKE/KEM schemes, there are three general hard problem families to build these schemes: code-based cryptography, lattice-based cryptography, and isogeny-based cryptography. Classic McEliece, BIKE, and HQC are code-based cryptosystems, CRYSTALS-Kyber, NTRU, Saber, FrodoKEM, and NTRU Prime are lattice-based candidates, and SIKE is the only isogeny-based candidate. Although there are similarities between the security foundation of these cryptosystems in a family, many construction parameters, performance profiles, and security notions are different.

Table 9.1 Overview of NIST Round 3 PKE/KEM candidates. Note that the first four schemes are Round 3 finalists and the next five schemes are Round 3 alternatives. PFS stands for perfect-forward secrecy and DE stands for decryption errors

Scheme	PQC type	PFS?	DE?	NIST security levels				
				1	2	3	4	5
NIST Round 3 PKE/KEM finalists								
Classic McEliece [5]	Code			✓		✓		✓
CRYSTALS-Kyber [6]	Lattice		☹	✓		✓		✓
NTRU [7]	Lattice			✓		✓		✓
Saber [8]	Lattice		☹	✓		✓		✓
NIST Round 3 PKE/KEM alternatives								
BIKE [9]	Code		☹	✓		✓		
FrodoKEM [10]	Lattice		☹	✓		✓		✓
HQC [11]	Code		☹	✓		✓		✓
NTRU prime [12]	Lattice				✓	✓	✓	
SIKE [13]	Isogeny	☺		✓	✓	✓		✓

NIST Security Levels Since these schemes are based on a variety of hard foundational problems with different tunable parameters, NIST asked for specific parameter sets that provide a level of security assurance. These range from a NIST security level 1 to a NIST security level 5. NIST security level 5 is the strongest security level which is considered to be as hard to break as an exhaustive key search attack on AES-256. NIST security level 4 is the next strongest security level and is considered to be as hard as finding a hash collision on SHA-384. NIST security levels 3 and 1 are as hard to break as an exhaustive key search attack on AES-192 and AES-128, respectively. Lastly, NIST security level 2 is as hard as a hash collision attack on SHA-256. Table 9.1 summarizes the supported parameter sets for each Round 3 PKE/KEM candidate. Although there are five security levels, not all NIST PQC submissions chose to include a submission at each of these security levels. The general guidance has been to include a parameter set at NIST security level 1 (low security), securityevel 3 (medium security), and security level 5 (high security).

Security Notion The security model of the PQC scheme has played an essential role. Notably, public key encryption schemes achieve indistinguishability under chosen plaintext attack (IND-CPA) versus key encapsulation mechanisms that achieve the stronger security notion indistinguishability under adaptive chosen ciphertext attack (IND-CCA2). Of the Round 3 PKE/KEM schemes, there are 3 Saber parameter sets that are PKE and thus IND-CPA secure. All other schemes utilize adaptations of the Fujisaki-Okamoto transform [14] for the quantum random oracle model. This allows for the IND-CCA2 stronger security model at the cost of a slower decapsulation. Additionally, KEM constructions will generally use some symmetric cryptosystems such as SHAKE-512 to provide indistinguishability from a random bitstring. Each Round 3 candidate features some standardized hash

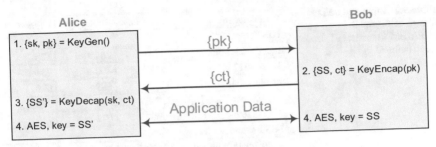

Fig. 9.1 Key Encapsulation Mechanism (KEM) use case

function or AES encryption function to increase the entropy and statistical strength of the shared secret.

Miscellaneous PKE/KEM Features Two additional features of PKE and KEM schemes are perfect-forward secrecy (PFS) and no decryption errors (DE). In a perfect-forward secret cryptosystem, the compromise of one session key does not also compromise all other session keys. SIKE is the only cryptosystem that supports PFS. Decryption errors mean that there is a chance (an extremely small possibility) that some set of public keys or ciphertexts will cause a decryption error with no possibility to recover. This creates some complications for some internet protocols that cannot handle a failed key establishment. Side-channel resistance and ease of implementation are additional nice-to-have features, but these are not discussed here.

Algorithm Storage and Communication Overhead In general, these Round 3 key establishment schemes are KEM constructions which have a private key, public key, ciphertext, and shared secret of various sizes. The KEM interface is shown in Fig. 9.1. Let us consider two parties, Alice and Bob. In the key generation phase, Alice generates a private key and public key. The public key is then broadcast over a public channel. When Bob wants to engage in secure computations with Alice, he performs key encapsulation over Alice's public key to generate a ciphertext and shared secret. Alice then performs key decapsulation by applying some computation with her private key on Bob's ciphertext to also agree on the same shared secret. From there, Alice and Bob use this shared secret to encrypt communications, such as with AES. We compare the implementation overhead for the NIST PQC Round 3 finalists and Round 3 alternatives in Tables 9.2 and 9.3, respectively.

Secret Key The secret key is generally the combination of Alice's private key as well as some public key information. Here, the private key is some random value that must be kept secret throughout the lifetime of the public key. This private key is sampled from some random source and then conditioned with a pseudorandom function such as AES or SHA3 to improve the quality and entropy of the bitstream. Higher private key sizes require more random data from a random source as well as more storage in some protected memory region.

Table 9.2 Input and output storage of Round 3 PKE/KEM finalists. All sizes are in bytes

Parameter set	Secret key	Public key	Ciphertext
NIST security level 1			
mceliece348864 [5]	6452[b]	261,120	128
Kyber-512 [6]	1632	800	736
ntruhps2048509[a][7]	935	699	699
LightSaber-PKE [8]	832	672	736
LightSaber-KEM [8]	1568	672	736
NIST security level 3			
mceliece460896 [5]	13,568[b]	524,160	188
Kyber-768 [6]	2400	1184	1088
ntruhrss701[a][7]	1452	1138	1138
ntruhps2048677[a][7]	1235	931	931
Saber-PKE [8]	1248	992	1088
Saber-KEM [8]	2304	992	1088
NIST security level 5			
mceliece6688128 [5]	13,892[b]	1,044,992	240
mceliece6960119 [5]	13,908[b]	1,047,319	226
mceliece8192128 [5]	14,080[b]	1,357,824	240
Kyber-1024 [6]	3168	1568	1568
ntruhps4096821[a][7]	1592	1230	1230
FireSaber-PKE [8]	1664	1312	1472
FireSaber-KEM [8]	3040	1312	1472

[a] Security level assumes local computational models
[b] Specification only provides the private key size

Public Key The public key is some value that is computed from the secret key. In the KEM model, a public key is broadcast following the key generation phase. For a specific parameter set, the private-public keypair needs to only be generated once. In some cases such as Kyber, it is memory-efficient to regenerate the public key from the private key rather than storing it. Large public key sizes correspond to higher public key storage and cost to transmit the public key over a wire.

Ciphertext The ciphertext is generated from the key encapsulation phase. In the KEM model, Alice will receive the ciphertext from whichever party wants to establish a secure connection. In general, the public key and ciphertext sizes are very similar in size. The primary exceptions are the code-based schemes Classic McEliece and HQC. Although the Classic McEliece scheme suffers from extremely large public keys, the scheme does feature the smallest ciphertexts of any PQC scheme. HQC has the opposite problem where its ciphertexts are approximately double the size of its public keys. Large ciphertext sizes correspond to higher ciphertext storage and cost to transmit the ciphertext over a wire.

Shared Secret In the above tables, the shared secret sizes were not compared since they are generally the same and are significantly smaller than public key or secret key sizes. These shared secrets range from 16 bytes (such as in FrodoKEM-640 or

Table 9.3 Input and output storage of Round 3 PKE/KEM alternatives. All sizes are in bytes

Parameter set	Secret key	Public key	Ciphertext
NIST security level 1			
BIKE level 1 [9]	2244[a]	12,323	12,579
FrodoKEM-640 [10]	19,888	9616	9720
hqc-128 [11]	40[a]	3024	6017
SIKEp434 [13]	374	330	346
SIKEp434_compressed [13]	350	197	236
NIST security level 2			
sntrup653 [12]	1518	994	897
ntrulpr653 [12]	1125	897	1025
SIKEp503 [13]	434	378	402
SIKEp503_compressed [13]	407	225	280
NIST security level 3			
BIKE level 3 [9]	3346[a]	24,659	24,915
FrodoKEM-976 [10]	31,296	15,632	15,744
hqc-192 [11]	40[a]	5690	11,364
sntrup761 [12]	1763	1158	1039
ntrulpr761 [12]	1294	1039	1167
SIKEp610 [13]	524	462	486
SIKEp610_compressed [13]	491	274	336
NIST security level 4			
sntrup857 [12]	1463	1184	1312
ntrulpr857 [12]	1999	1322	1184
NIST security level 5			
FrodoKEM-1344 [10]	43,088	21,520	21,632
hqc-256 [11]	40[a]	8698	17,379
SIKEp751 [13]	644	564	596
SIKEp751_compressed [13]	602	335	410

[a] Specification only provides the private key size

SIKEp434) to 64 bytes (such as in any HQC parameter set). These must be stored. Smaller shared secret sizes do allow the use of smaller encryption functions. For instance, AES-128 can be used with a 16 byte shared secret and AES-256 can be used with a 32 byte shared secret. Overall, the shared secret size is similar across the board.

9.3 Code-Based Cryptosystems

Code-based cryptography relies on the difficulty to decode a general linear code. Since its original analysis of being an NP-complete problem several decades ago [15], no breakthroughs have been made, showing a confidence in its hardness. In

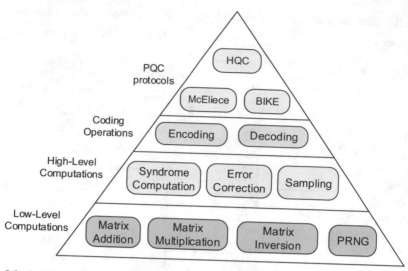

Fig. 9.2 Code-based cryptography computation hierarchy

the NIST PQC competition, the Classic McEliece [5] scheme is the only code-based finalist and the BIKE [9] and HQC [11] schemes are the two code-based alternatives. The choice of code family greatly impacts the public key size and implementation performance. The Classic McEliece scheme utilizes Goppa code with a very large public key, whereas BIKE and HQC utilize Quasi-Cyclic Medium Density Parity Check (QC-MDPC) codes with a much smaller public key. Lastly, BIKE applies some structure to the QC-MDPC code to offer performance optimizations.

State-of-the-Art Overall, the literature for hardware implementations of code-based cryptography is fairly sparse. The work in [16] presents a high-performance implementation of the Classic McEliece scheme, while [17, 18] both focus on optimizations to the BIKE scheme. The BIKE specification [9] features a hardware implementation section. There are currently no existing hardware implementations of HQC in the literature.

Code-Based Computations The critical computations in these three coding-based schemes can be summarized in the hierarchy in Fig. 9.2. The key generation operation involves sampling a random private and public key such that they follow a specific form. For instance, Classic McEliece requires an invertible matrix. Key encapsulation and decapsulation then requires various polynomial or matrix arithmetic to encode and decode errors into the public key and ciphertext, respectively.

Polynomial/Matrix Arithmetic The primary thrust of the code-based literature has been on efficient computation of polynomial multiplication and inversion. The Classic McEliece implementation in [16] uses some optimizations to the additive FFT to perform multiplications by the coefficients within a polynomial (base-2). Meanwhile, the BIKE polynomial multiplication proposed in [17] features a constant-time polynomial multiplication scheme over commutative rings, a key construction of BIKE. The additional structure in the BIKE parameters allow for more efficient polynomial operations. The implementation for polynomial and matrix inversion is then highly dependent on the type of coding scheme. Classic McEliece uses Gaussian elimination to find the matrix inverse. BIKE utilizes Fermat's little theorem ($a^{-1} = a^{p-2}$ for prime p) to compute the inversion over many polynomial multiplications and squarings. Lastly, the HQC specification's software implementation uses the additive Fast Fourier Transform (FFT) algorithm from [19] to compute the polynomial's inverse.

Syndrome Computation The syndrome computations are a key computation in decoding the linear code. For Classic McEliece and BIKE this requires a large matrix multiplication using the additive FFT operation. HQC is more interesting as we compute 2δ syndromes by evaluating the received codeword at various primitives α in $\mathbb{F}_{2^m}$. Here, the additive FFT operation is again the most efficient (and constant-time) method to compute the syndromes. Another option, perhaps for lightweight implementations, could be the use of a linear feedback shift register (LFSR). For high-performance systems, these syndromes can be computed entirely in parallel with enough LFSRs.

Error Correction Error correction is then the next phase of codeword decoding. The standard decoding steps are to find the error-locator polynomial and finding all of the roots (or errors) of the polynomial so that the codeword can be corrected. Each of the three code-based candidates utilize their own method. The Classic McEliece cryptosystem requires a unique codeword that solves the Goppa code. Here, the recommended solution is to use Berlekamp's method [20] to find the codeword. Then, the Berlekamp-Massey algorithm can be used to compute the error-locator polynomial and the Chien search algorithm can be used to find the error-locator roots. Based on BCH codes, the HQC algorithm uses a simplified version of Berlekamp's method [21] to find the error-locator polynomial and polynomial inversion to find the corresponding roots. BIKE uses an entirely separate algorithm for decoding based on the Black-Gray-Flip algorithm from [22] that uses a complex sequence of bit flipping operations.

Sampling Scheme Inputs Code-based cryptography requires a large number of pseudorandom bits (from a seed) to generate keys. For Round 3 code-based candidates, this is generally on the order of several kilobytes. For instance, the Classic McEliece parameter set at NIST Security Level 1 has a key generation that requires 3552 12-bit values, or 42,624 bits (5328 bytes) of uniformly random data. Generally, NIST-standardized hash or encryption functions are used as pseudorandom number generators (PRNGs) to generate these random bits. Here, Classic McEliece uses SHAKE-256, BIKE uses SHA2-384 and AES256-CTR, and HQC uses SHA3-512. Furthermore, BIKE uses AES-CTR with AES-256 to generate pseudorandom bits. Since so many pseudorandom bytes are required for code-based schemes, it is generally recommended to implement a high-performance version of whatever PRNG is specified.

9.4 Lattice-Based Cryptosystems

Lattice-based cryptography relies on the difficulty of computational problems over lattices. The two most popular problems are based on learning with errors (LWE), learning with rounding (LWR), and the shortest vector problem (SVP). Additionally, lattices can be given some structure to speed up their computations. The Ring-LWE (RLWE) and Module-LWR (MLWR) schemes are two such examples. Of the NIST Round 3 finalists, CRYSTALS-Kyber [6] is based on the Module Learning With Errors problem, NTRU [7] is a structured lattice KEM based on the Shortest Vector Problem, and Saber [8] is based on the Module Learning With Rounding problem. For the NIST Round 3 alternatives, FrodoKEM [10] is based on the unstructured Learning With Errors problem and NTRU Prime [12] is based on the Shortest Vector Problem with a slightly different structure than NTRU.

State-of-the-Art The lattice schemes feature the largest implementation literature. In the following, we focus on papers that implement NIST's selection of Round 3 candidates. Since the lattice schemes require a complex control flow, the sampling and matrix multiplication are typically implemented in hardware and most of these implementations are HW/SW. Dang et al. [4] provide an overview of all Round 2 lattice implementations in the literature. On the high-performance side first, [23] benchmark FrodoKEM and Saber, and were later updated in presentation form to include NTRU, and NTRU Prime. Mera et al. [24] and Roy and Basso [25] present high-speed implementations of Saber by optimizing the polynomial multiplier. Farahmand et al. [26] report a high-speed implementation for NTRU. Howe et al. [27] implemented FrodoKEM for high-performance FPGA. One lightweight research approach has been to add ISA extensions to the RISC-V architecture [28–30]. Of these works, Alkim et al. [29] and Xin et al. [28] both supported the CRYSTALS-Kyber algorithm and Ritzmann et al. [28] supported Kyber and Saber. Banerjee et al. [31] implement a (lightweight focused) configurable coprocessor that accelerates various LWE, Ring-LWE, and Module-LWE lattice schemes for key encapsulation and digital signatures. Of note, this implementation supports FrodoKEM and CRYSTALS-Kyber. Lastly, [32] used a high-level synthesis (HLS) approach to synthesize many Round 2 PQC candidates, including Kyber, Saber, NTRU, FrodoKEM, and NTRU Prime.

Lattice-Based Computations The critical computations in the lattice-based cryptosystems can be summarized in the hierarchy in Fig. 9.3. Similar to code-based cryptography, the lattice-based private and public key must be sampled from a PRNG according to a preprescribed form. Then, the key encapsulation and decapsulation operations require a large field multiplication. The method to perform the multiplication depends highly on the structure of the lattice which we will explore in the following paragraphs.

FrodoKEM FrodoKEM requires a large integer matrix multiplication. In hardware, the multiply and accumulate (MAC) approach is an efficient way to break

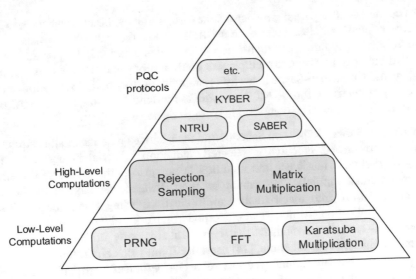

Fig. 9.3 Lattice-based cryptography computation hierarchy

up the matrix multiplication into chunks. The work in [27] utilized the DSP slice to efficiently compute row-column MACs of the matrix multiplication. Likewise, [23] utilized many 4MAC hardware structures in parallel to compute partial products of the matrix in parallel. On the lightweight side, [31] use a single multiplication unit, but must iterate over the i-th row and j-th column of the matrix.

NTRU and NTRU Prime NTRU and NTRU Prime are lattice-based schemes with a Ring-LWE-like structure. This structure allows for much more efficient polynomial multiplication. Here, the general strategy is to use the Discrete Fourier Transform (DFT) to transform the polynomial from the normal domain to the spectral domain, where multiplication is much more efficient. The Number Theoretic Transform (NTT) is an efficient version of the DFT where the twiddle factors are elements of $\mathbb{Z}_q$, where q is the prime in the NTRU system. To convert back from the spectral domain to the normal domain, the Inverse Number Theoretic Transform (INTT) operation is used. The Fast Fourier Transform (FFT) is then an efficient implementation of the NTT and INTT algorithms by using a divide-and-conquer approach where a large NTT/INTT computation is broken into smaller NTT/INTT computations. The smaller NTT/INTT computation here is the butterfly operation that consists of a multiplication by a power of the twiddle factor, an addition, and a subtraction in $\mathbb{Z}_q$. For the NTT and INTT computations, the Cooley-Tueky [33] and Gentlemen- Sande [34] butterflies can be instantiated, respectively.

When implementing NTT/INTT in hardware, the design and interface for each butterfly unit is critical. Furthermore, high-performance implementations have the option to parallelize each of the butterflies over more resources as each round of small NTT/INTT computations are independent. Dang et al. [4] instantiated an LFSR over $\mathbb{Z}_p$ to perform a multiplication each cycle and many MAC units to accumulate the partial products. Likewise, lightweight implementations can utilize a single efficient butterfly unit. At its core, the butterfly consists of $\mathbb{Z}_p$ multiplication, addition, and subtraction. The implementation in [31] opted to use a unified butterfly unit, where multiplexers could easily switch between the Cooley-Tukey and Gentlemen-Sande butterfly operations. Furthermore, [31] optimized the memory architecture for the NTT computation around four single-port SRAMs to allow each butterfly to be computed in a single cycle.

Kyber and Saber Kyber and Saber define their lattices with the module structure. Since a power-of-2 modulus is selected, reduction is trivial in hardware. The only downside is that since the modulus is not prime, the NTT/INTT method for polynomial multiplication cannot be used. Instead, the large multiplication be broken down into much smaller multiplications using Karatsuba or Toom-Cook multiplication method. These methods break a polynomial multiplication inputs into smaller polynomials that can then efficiently utilize sums of small partial products. In the literature, [24] used the Toom-Cook algorithm to accelerate Saber's computations through the use of a 64-coefficient polynomial multiplier with their efficient memory scheduling. On the other hand, the work in [25] utilized schoolbook polynomial multiplication for Saber since parallelization and accumulation in memory were easier to schedule.

Sampling Scheme Inputs Similar to code-based cryptosystems, the lattice-based cryptosystems must sample a large amount of pseudorandom data to generate the private and public keys. For instance, Frodo-640 requires a 640×640 matrix of 16-bit integers during key generation (819,200 bits or 102,400 bytes). These random polynomials are generated through rejection sampling, so there is a chance of rejecting a poorly chosen polynomial. Kyber uses SHAKE-128, SHA3-256, SHA3-512, and SHAKE-256, which are all based on the Keccak sponge function. Saber uses SHAKE-128, SHA3-256, and SHA3-512. NTRU uses SHAKE-256 and SHA3-256. FrodoKEM uses SHAKE-128 for Frodo-640 and SHAKE-256 for Frodo-976 and Frodo-1344. Lastly, NTRU Prime uses SHA2-512. Similar to the code-based cryptosystem recommendation, a high-performance PRNG implementation is recommended for lattice-based cryptography.

9.5 Isogeny-Based Cryptosystems

The Supersingular Isogeny Key Encapsulation (SIKE) [13] scheme is the only isogeny-based cryptosystem in the NIST PQC competition. This scheme is protected by the difficulty to find isogenies, or mappings, between elliptic curves. Although Shor's algorithm [1] will break standard elliptic curve cryptography, this new family of problems with elliptic curves is still hard even for a quantum computer. The SIKE scheme features the smallest public keys of all PKE/KEM submissions, but is generally an order of magnitude slower.

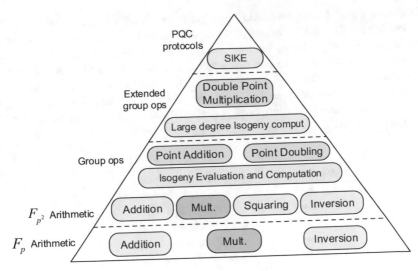

Fig. 9.4 Isogeny-based cryptography computation hierarchy

State-of-the-Art The precursor to SIKE is the supersingular isogeny Diffie-Hellman (SIDH) protocol. In hardware, this was first implemented for high-performance applications in [35–38]. Since isogeny-based cryptography is among the youngest candidates for PQC, various optimizations in elliptic curve formulas, arithmetic, and algorithms resulted in numerous speedups. Koziel et al. [39] and Elkhatib et al. [40] then features a high-performance hardware implementation of SIKE and [41] present an area-efficient HW/SW implementation of SIKE. The implementation in [41] utilizes hardware to accelerate low-level $\mathbb{F}_p$ field arithmetic and Keccak operations, and software to issue all other operations.

Isogeny-Based Computations The critical computations in SIKE can be summarized in the hierarchy in Fig. 9.4. Aside from the SHAKE module, all computations are based on various elliptic curve arithmetic that is defined over various field arithmetic.

Modular Arithmetic At the lowest level, a SIKE implementation requires prime-field addition and prime-field multiplication over $\mathbb{F}_p$ with large prime p, which can be thought of as large modular operations. In the following, we assume $a, b, c \in \mathbb{F}_p$. Prime-field addition, $c = a + b$ is simply an add followed by a conditional subtraction. Multiplication, $c = a \times b$, is much more complex as the modular reduction is expensive. The modular multiplier in [39] utilizes a high-throughput systolic architecture, and [41] utilize a multiply accumulate (MAC) architecture. Both implementations incorporate Montgomery multiplication [42] that trades expensive division operations for bit shifts which are free in hardware. Since there are so few $\mathbb{F}_p$ inversions in SIKE, both implementations use Fermat's little theorem. The inversion can be computed as $a^{-1} = a^{p-2}$.

Extension Field Arithmetic Although the lowest level arithmetic can be defined as simple modular addition or multiplication, SIKE operations are over the quadratic extension field $\mathbb{F}_{p^2}$. Here, we can perform $\mathbb{F}_{p^2}$ operations as a sequence of $\mathbb{F}_p$ operations. To compute extension field arithmetic, we utilize the polynomial reduction $x^2 + 1$. Thus, $x^2 = -1$, and let us assume $i = \sqrt{-1}$. We define $A, B, C \in \mathbb{F}_{p^2}$, where $A = ia_1 + a_0$, $B = ib_1 + b_0$, and $C = ic_1 + c_0$. We define the quadratic extension field operations with the minimum number of $\mathbb{F}_p$ multiplications as

$$
\begin{aligned}
A + B = \quad & c_1 = a_1 + b_1, c_0 = a_0 + b_0 \\
A - B = \quad & c_1 = a_1 - b_1, c_0 = a_0 - b_0 \\
A \times B = \quad & c_1 = a_0 b_1 + a_1 b_0 \\
& c_0 = (a_0 + a_1)(b_0 - b_1) + a_0 b_1 - a_1 b_0 \\
A^2 = \quad & c_1 = 2 a_0 a_1, c_0 = (a_0 + a_1)(a_0 - a_1) \\
A^{-1} = \quad & c_1 = -a_1 \left(a_0^2 + a_1^2\right)^{-1}, c_0 = a_0 \left(a_0^2 + a_1^2\right)^{-1}
\end{aligned}
$$

Group Level Arithmetic The SIKE implementation is defined over Montgomery [43] curves which are known for fast Montgomery ladders. The point multiplication and isogeny formulas have been heavily optimized and the fastest known algorithms can be found in [44] and [13]. These formulas are utilized in all current implementations of SIKE.

Isogeny Algorithms The largest functions in SIKE are the double-point multiplication and large-degree isogeny. The double-point multiplication computes $R = Q + n \times P$, for which the fastest known algorithm was presented in [45]. Surprisingly, this differential ladder uses a single differential point addition and point doubling just like the original Montgomery ladder. The large-degree isogeny is a complex computation that chains a variety of point multiplication, isogeny evaluation, and isogeny computation operations according to a precomputed strategy. The optimal strategy is the one of least cost to compute, which can be computed using the combinatorial method in [46].

Options for Parallelization The SIKE implementations in [39, 40] use a greedy scheduling algorithm with several multipliers to parallelize many of the computations in SIKE. Parallelization can be achieved at multiple levels. For instance, at the extension field arithmetic level, we can parallelize the field multiplications over additions so long as there is not a data dependency. One major parallelization point is also the large-degree isogeny. Typically, isogeny evaluations, or applying an isogeny mapping to a point, are applied serially to multiple stored pivot points. However, there is no data dependency between these pivot points, so the isogeny evaluations can be done entirely in parallel. To achieve a large amount of parallelism, the implementation in [39] stored up to 256 intermediate $\mathbb{F}_p$ values and could support 8 independent field multiplications.

Public Key Compression SIKE's primary advantage over other PQC schemes is that it has the smallest public keys. This is further exemplified by its public key compression schemes that reduce the public key by almost half. Public-key compression was first proposed in [47] and then further optimizations were presented in [48–51]. The public key compression utilizes expensive computations to generate a basis, perform a pairing, and compute a discrete logarithm. Each of these can still be reduced to low-level arithmetic over $\mathbb{F}_p$. However, there are currently no hardware implementations of the SIKE schemes with public key compression as these are complex computations that require a large number of stored elements. Nevertheless, there are many options for parallelization to speed this up in hardware such as computing two bases in parallel, computing the four pairings in parallel, or computing the four discrete logarithms in parallel. In the SIKE specification [13], the software results show that the key compression schemes have almost twice as slow key generation and key encapsulation operations, but the key decapsulation performance is relatively unaffected.

Symmetric Crypto Functions SIKE utilizes the SHAKE-512 hash function to achieve IND-CCA security. Since the isogeny operations take significantly longer than SHAKE, a small or medium-performance SHAKE module can be swapped in to save some hardware gates. This is unlike the code-based and lattice-based implementations. Here, SIKE does not need to use rejection sampling or any complex sampling mechanism to generate the private key.

9.6   PQC HW Implementation Comparisons

Here, we highlight the current state-of-the-art performance results of various PQC HW implementations. We note that this comparison is purely on HW implementation metrics such as area and timing. Other characteristics, such as public key size, security notions, or side-channel protections, are not considered here. The state-of-the-art FPGA implementations for Round 3 PQC candidates are presented in Tables 9.4, 9.5, 9.6, 9.7, and 9.8. The state-of-the-art ASIC implementations for Round 3 PQC candidates are presented in Tables 9.9 and 9.10.

In this survey of implementations, note that not all schemes have a parameter at each of the five NIST security levels. Further, design targets may be high performance or lightweight and can also include support for more than one parameter set or PQC algorithm. Many of these works have various merits and optimizations that are hard to detail through results around. These and many other reasons make a fair comparison between implementations and algorithms difficult.

FPGA Implementations The majority of these implementations are done on FPGA devices. Such devices are designed for fast prototyping. These devices generally have a slower max clock frequency for a given device than ASIC implementations. The general resources on an FPGA vary between family, which

Table 9.4 Hardware FPGA comparison of Round 3 PQC submissions, part 1

Scheme	FPGA	HP/ LW	# FFs	# LUTs	# Slices	# DSPs	# BRAMs	Freq. (MHz)	KeyGen cc × 10³	KeyGen (ms)	Encap cc × 10³	Encap (ms)	Decap cc × 10³	Decap (ms)
NIST security level 1 (as strong as AES-128)														
McEliece348864 [16]	A7	HP	132,190	81,339	–	0	236	106	203	1.92	2.72	0.025	12.7	0.120
McEliece348864 [16]	A7	HP	49,383	25,327	–	0	168	108	1600	14.8	2.72	0.025	18.4	0.169
Kyber-512 [28]	Z7	LW	10,844	23,925	–	21	32	–	150	–	193	–	205	–
Kyber-512 [52]	A7	LW	2539	14,975	4173	11	14	25	75	2.98	132	5.27	142	5.69
Kyber-512 [29]	RV	LW	1634	1842	–	5	34	59	710	12.0	971	16.4	870	14.7
LightSaber [23]	ZUS+	HP	11,288	12,343	1989	256	3.5	322	–	–	–	0.051	–	0.05
LightSaber [28]	Z7	LW	10,844	23,925	–	21	32	–	367	–	526	–	658	–
ntruhps2048677 [23]	ZUS+	HP	19,244	24,328	4972	677	2.5	200	–	–	–	0.386	–	0.114
ntruhrss701 [23]	ZUS+	HP	21,410	27,218	5770	701	2.5	200	–	–	–	0.171	–	0.128
FrodoKEM-640 [23]	ZUS+	HP	6647	7213	1186	32	13.5	402	–	–	–	1.41	–	1.41
FrodoKEM-640a [53]	A7	HP	5335	14,528	4020	16	12.5	149	–	–	–	–	–	1.41
FrodoKEM-640a [53]	A7	LW	2299	10,518	2933	1	12.5	162	–	–	–	–	–	20.4
FrodoKEM-640 [31]	A7	LW	2539	14,975	4173	11	14	25	–	–	–	–	–	481
BIKE [17]	A7	HP	2141	3874	1312	0	10	160	11,454	15.3	11,610	464	12,036	–
BIKE [9]	A7	HP	589	1865	590	0	4	135	2150	13.4	–	–	–	–
SIKEp434 [39]	A7	HP	24,328	21,946	8006	240	26.5	132	7370	54.5	–	–	–	–
SIKEp434 [40]	V7	HP	18,271	12,818	5527	195	32	250	530	4.01	930	7.03	980	7.41
SIKEp434 [41]	A7	HP	11,558	22,595	7491	162	37	145	1474	9.1	2495	15.4	2657	16.4
SIKEp434 [41]	A7	LW	7115	10,976	3512	57	21	109	2188	15.3	3718	26	3947	27.6

a Key decapsulation only

HP High-Performance, LW LightWeight

A7 Artix-7, V7 Virtex-7, ZUS+ Zynq UltraScale+, Z7 Zynq-7000, RV RISC-V RV32IM

Table 9.5 Hardware FPGA comparison of Round 3 PQC submissions, part 2

Scheme	FPGA	HP/LW	Area # FFs	# LUTs	# Slices	# DSPs	# BRAMs	Time Freq. (MHz)	KeyGen $cc \times 10^3$	(ms)	Encap $cc \times 10^3$	(ms)	Decap $cc \times 10^3$	(ms)
NIST security level 2 (as strong as SHA-256)														
sntrup653 [23]	ZUS+	HP	28,143	55,843	8134	0	3	244	–	–	–	0.242	–	0.341
ntrulpr653 [23]	ZUS+	HP	34,050	50,911	7874	0	2	244	–	–	–	1.842	–	1.359
SIKEp503 [39]	A7	HP	27,759	24,610	9186	264	33.5	130	640	4.93	1140	8.78	1200	9.24
SIKEp503 [40]	V7	HP	19,935	13,963	6163	225	34	244	–	–	–	–	–	–
SIKEp503 [41]	A7	HP	11,558	22,595	7491	162	37	145	1733	10.7	2932	18.1	3127	19.3
SIKEp503 [41]	A7	LW	7115	10,976	3512	57	21	109	2603	18.2	4390	30.7	4676	32.7

Table 9.6 Hardware FPGA comparison of Round 3 PQC submissions, part 3

| Scheme | FPGA | HP/LW | Area | | | | | Time | | | | | | |
			# FFs	# LUTs	# Slices	# DSPs	# BRAMs	Freq. (MHz)	KeyGen $cc \times 10^3$	KeyGen (ms)	Encap $cc \times 10^3$	Encap (ms)	Decap $cc \times 10^3$	Decap (ms)
NIST security level 3 (as strong as AES-192)														
McEliece460896 [16]	V-7	HP	168,939	109,484	–	0	446	131	515	3.94	3.36	0.026	17.9	0.137
McEliece460896 [16]	A7	HP	74,858	38,669	–	0	303	107	5002	46,704	3.36	0.031	31	0.29
Kyber-768 [52]	A7	LW	2539	14,975	4173	11	14	25	112	4.46	178	7.1	483	3.08
Saber [28]	ZUS+	HP	11,619	12,566	1993	3.5	256	322	–	–	–	0.067	–	0.069
Saber [24]	Z7	LW	7331	7400	–	28	2	–	–	3.27	–	4.15	–	3.84
Saber [25]	ZUS+	HP	18,705	45,895	–	0	2	250	4.32	0.0173	5.23	0.0209	6.46	0.0258
Saber [25]	ZUS+	HP	10,750	25,079	–	0	2	250	5.44	0.0218	6.62	0.0265	8.03	0.0321
ntruhps4096821 [23]	ZUS+	HP	32,338	29,389	5913	821	2.5	200	–	–	–	0.475	–	0.338
FrodoKEM-976 [23]	ZUS+	HP	6693	7087	1190	17	32	402	–	–	–	2.03	–	2.06
FrodoKEM-976[a] [53]	A7	HP	5087	7213	2042	16	19	157	–	–	–	–	–	3.27
FrodoKEM-976[a] [53]	A7	LW	2153	4888	1390	1	19	162	–	–	–	–	–	47.6
FrodoKEM-976 [52]	A7	LW	2539	14,975	4173	11	14	25	–	–	–	–	–	–
BIKE [9]	A7	HP	557	1884	593	0	5	135	26,005	1040	29,749	1190	30,421	1217
sntrup761 [23]	ZUS+	HP	32,763	62,595	9176	0	3	244	30,450	231	–	–	–	–
ntrulpr761 [23]	ZUS+	HP	39,600	51,295	7978	0	2	244	–	–	–	0.646	–	0.392
SIKEp610 [39]	A7	HP	33,198	29,447	10,843	312	39.5	125	–	–	–	2.09	–	1.53
SIKEp610 [40]	V7	HP	26,757	16,226	7461	270	38.5	239	900	7.18	1810	14.4	1780	14.2
SIKEp610 [41]	A7	HP	11,558	22,595	7491	162	37	145	2916	18	5443	33.6	5508	34
SIKEp610 [41]	A7	LW	7115	10,976	3512	57	21	109	4347	30.4	8108	56.7	8208	57.4

[a] Key decapsulation only

HP High-Performance, LW LightWeight

A7 Artix-7, V7 Virtex-7, ZUS+ Zynq UltraScale+, Z7 Zynq-7000, RV RISC-V RV32IM

Table 9.7 Hardware FPGA comparison of Round 3 PQC submissions, part 4

| Scheme | FPGA | HP/ LW | Area | | | | | Time | | | | | | |
			# FFs	# LUTs	# Slices	# DSPs	# BRAMs	Freq. (MHz)	KeyGen $cc \times 10^3$	(ms)	Encap $cc \times 10^3$	(ms)	Decap $cc \times 10^3$	(ms)
NIST security level 4 (as strong as SHA-384)														
sntrup857 [23]	ZUS+	HP	37,018	70,604	9894	0	3	244	–	–	–	0.727	–	0.437
ntrulpr857 [23]	ZUS+	HP	44,719	58,056	8895	0	2	244	–	–	–	2.36	–	1.71

Table 9.8 Hardware FPGA comparison of Round 3 PQC submissions, part 5

| Scheme | FPGA | HP/LW | Area | | | | | Time | | | | | | | |
			# FFs	# LUTs	# Slices	# DSPs	# BRAMs	Freq. (MHz)	KeyGen cc × 10³	(ms)	Encap cc × 10³	(ms)	Decap cc × 10³	(ms)
NIST security level 5 (as strong as AES-256)														
McEliece6960119 [16]	V7	HP	188,324	116,928	–	0	607	130	974	7.5	5.41	0.0417	25.1	0.194
McEliece6960119 [16]	V7	HP	88,963	44,154	–	0	563	141	11,180	79.6	5.41	0.0385	46.1	0.328
McEliece6688128 [16]	V7	HP	186,194	122,624	–	0	589	137	1046	7.66	5.02	0.0368	29.8	0.218
McEliece6688128 [16]	V7	HP	83,637	44,345	–	0	446	136	12,390	91	5.02	0.0369	52.3	0.412
McEliece8192128 [16]	V7	HP	190,707	123,361	–	0	589	130	1286	9.9	6.53	0.0503	32.8	0.252
McEliece8192128 [16]	V7	HP	88,154	45,150	–	0	525	134	15,185	113	6.53	0.0486	55.3	0.412
Kyber-1024 [28]	Z7	LW	10,844	23,925	–	21	32	–	350	–	405	–	425	–
Kyber-1024 [52]	A7	LW	2539	14,975	4173	11	32	25	149	4.94	223	8.94	241	9.64
Kyber-1024 [29]	RV	LW	1634	1842	–	5	34	59	2203	37.2	2519	44.2	2429	41
FireSaber [23]	ZUS+	HP	11,881	12,555	2341	256	3.5	322	–	–	–	0.094	–	0.086
FireSaber [28]	Z7	LW	10,844	23,925	–	21	32	–	1300	–	1523	–	1898	–
FrodoKEM-1344 [23]	ZUS+	HP	6610	7015	1215	32	17.5	417	–	–	–	–	–	–
FrodoKEM-1344 [31]	A7	LW	2539	14,975	4173	11	14	25	–	–	–	1.977	–	2.608
SIKEp751 [39]	V7	HP	50,079	39,953	15,834	512	43.5	163	67,994	2720	71,501	2860	72,527	2901
SIKEp751 [40]	V7	HP	39,339	20,207	11,136	452	41.5	232.7	1250	7.67	2210	13.5	2340	14.3
SIKEp751 [41]	V7	HP	13,657	21,210	7408	162	38	142	2517	17.7	4166	29.3	4479	31.5
SIKEp751 [41]	V7	LW	7115	10,976	3512	57	21	154	7960	52.3	13,151	86.4	14,186	93.2

Table 9.9 Hardware ASIC area/time comparison of Round 3 PQC submissions. HP = High-Performance, LW = LightWeight

Scheme	Technology	HP/LW	Area		Time	KeyGen		Encap		Decap	
			# kGE	SRAM (KB)	Freq. (MHz)	$cc \times 10^3$	(ms)	$cc \times 10^3$	(ms)	$cc \times 10^3$	(ms)
NIST security level 1 (as strong as AES-128)											
Kyber-512 [30]	TSMC 28 nm	HP	979	12	300	18.6	0.0619	45.9	0.153	80.0	0.267
Kyber-512 [52]	TSMC 40 nm	LW	106	40.25	72	74.5	1.04	132	1.83	142	1.98
Kyber-512 [28]	UMC 65 nm	LW	170	465[a]	45	150	3.32	193	4.27	205	4.53
LightSaber [28]	UMC 65 nm	LW	170	465[a]	45	124	2.74	207	4.58	227	5.01
FrodoKEM-640 [52]	TSMC 40 nm	LW	106	40.25	72	11,454	159	11,610	161	12,036	167
NIST security level 3 (as strong as AES-192)											
Kyber-768 [52]	TSMC 40 nm	LW	106	40.25	72	112	1.55	178	2.47	191	2.65
FrodoKEM-976 [52]	TSMC 40 nm	LW	106	40.25	72	26,005	361	29,749	413	30,421	423
NIST security level 5 (as strong as AES-256)											
Kyber-1024 [30]	TSMC 28 nm	HP	979	12	300	40.0	0.132	81.6	0.272	136	0.455
Kyber-1024 [52]	TSMC 40 nm	LW	106	40.25	72	149	2.06	223	3.10	241	3.37
Kyber-1024 [52]	UMC 65 nm	LW	170	465[a]	45	350	7.73	405	8.96	425	9.38
Kyber-1024 [28]	UMC 65 nm	LW	170	465[a]	45	1300	28.7	1623	35.9	1898	41.9
FireSaber [28]	UMC 65 nm	LW	170	465[a]	45	1300	28.7	1623	35.9	1898	41.9
FrodoKEM-1344 [52]	TSMC 40 nm	LW	106	40.25	72	67,994	944	71,501	99	72,527	1007

[a] SRAM numbers are converted to kGE

Table 9.10 Hardware ASIC power/energy comparison of Round 3 PQC submissions. HP = High-Performance, LW = LightWeight

Scheme	Technology	HP/LW	Area		Freq. (MHz)	Power and energy					
			# kGE	SRAM (KB)		KeyGen		Encap		Decap	
						Power (mW)	Energy (µJ)	Power (mW)	Energy (µJ)	Power (mW)	Energy (µJ)
NIST security level 1 (as strong as AES-128)											
Kyber-512 [30]	TSMC 28 nm	HP	979	12	300	29.26	1.81	23.67	3.62	24.94	6.65
Kyber-512 [52]	TSMC 40 nm	LW	106	40.25	72	5.77	5.97	5.12	9.37	5.69	11.25
Kyber-512 [28]	UMC 65 nm	LW	170	465[a]	45	-	-	-	-	2.58	141.41
LightSaber [28]	UMC 65 nm	LW	170	465[a]	45	-	-	-	-	2.78	431.18
FrodoKEM-640 [52]	TSMC 40 nm	LW	106	40.25	72	6.65	1057.65	7.01	1129.95	6.88	1150.83
NIST security level 3 (as strong as AES-192)											
Kyber-768 [52]	TSMC 40 nm	LW	106	40.25	72	5.28	8.19	5.19	12.80	5.86	15.52
FrodoKEM-976 [52]	TSMC 40 nm	LW	106	40.25	72	6.70	2420.97	7.05	2912.95	6.94	2932.13
NIST security level 5 (as strong as AES-256)											
Kyber-1024 [30]	TSMC 28 nm	HP	979	12	300	35.45	4.69	29.20	7.94	25.57	11.63
Kyber-1024 [52]	TSMC 40 nm	LW	106	40.25	72	5.96	12.27	5.25	16.30	5.91	19.76
Kyber-1024 [28]	UMC 65 nm	LW	170	465[a]	45	-	-	-	-	2.60	307.68
FireSaber [28]	UMC 65 nm	LW	170	465[a]	45	-	-	-	-	2.77	1335.48
FrodoKEM-1344 [52]	TSMC 40 nm	LW	106	40.25	72	6.75	6374.45	7.10	7050.83	7.00	7051.21

[a] SRAM numbers are converted to kGE

makes a fair comparison difficult. A Flip-Flop (FF) represents a single register, but a Look-Up Table (LUT) can be one of several options, such as 6:1 or 4:1 LUTs which map 4 inputs to a single output. The Slice is generally a unit that contains a few LUTs and FFs. The Digital Signal Processor (DSP) is a logic component on the FPGA that accelerates addition, subtraction, multiplication, and MAC operations. Lastly, the Block Random Access Memory (BRAM) is a memory storage unit, for which the size of a single BRAM is dependent on the FPGA family. The recommended devices for benchmarking are the Xilinx Artix-7 and Xilinx Zynq UltraScale+ FPGAs.

ASIC Implementations ASIC implementations instantiate the hardware design using standard cells. ASIC implementations are significantly more expensive to produce than FPGA implementations. The cell library determines the technology node, or size of the basic transistor that is used. This library impacts the maximum frequency, area, memory, and a variety of performance characteristics. The simplest comparison between area is the Gate Equivalent (GE) that represents the size of a NAND gate. Besides strict area, the static random access memory (SRAM) is a typical memory unit on an ASIC device. The ASIC implementations here also provided the power and energy consumption of KEM operations, which are highly dependent on the target frequency and target technology.

Trends Although a fair comparison is difficult between implementations and PQC algorithms, we can still observe trends that these implementation results show. The first Round 3 finalist listed is Classic McEliece for which only a single set of implementations is given. Here, the extremely large public key generation and matrix inversion dominate the key generation phase. McEliece-based key encapsulation and decapsulation, however, is the fastest of all known high-speed PQC implementations, which couples well with the smallest ciphertext size. Both Kyber and Saber, based on Module lattice computations, feature lightweight implementations with a medium area footprint that still have much faster KEM operations than other competitors. The only high-speed implementation of Kyber is in ASIC, which shows an edge over in performance over Saber. The only implementations of NTRU and NTRU prime feature a high-performance optimization target and achieve a medium area footprint with very fast performance relative to other PQC. The BIKE lightweight implementations only targeted key generation and achieved the smallest area of any PQC KEM at the expense of fairly slow key generation. FrodoKEM implementations feature a similar area as Saber and Kyber, but suffer with a slowdown over an order of magnitude. SIKE implementations generally target high-performance targets with a large area and slow relative performance. Lastly, there are no hardware implementations of HQC. We briefly summarize these trends in Table 9.11.

Table 9.11 Overview of NIST Round 3 PKE/KEM candidate trends. These center on the general cost to area and time to implement the PQC algorithm in hardware. Unfortunately, there is still a limited literature of PQC hardware implementations

Scheme	Area overhead	Performance overhead		
		KeyGen	Encap	Decap
NIST Round 3 PKE/KEM finalists				
Classic McEliece [5]	Very large	Very slow	Very fast	Very fast
CRYSTALS-Kyber [6]	Small	Fast	Fast	Fast
NTRU [7]	Large	?	Fast	Fast
Saber [8]	Small	Fast	Fast	Fast
NIST Round 3 PKE/KEM alternatives				
BIKE [9]	Very small	Very slow	?	?
FrodoKEM [10]	Moderate	Slow	Slow	Slow
HQC [11]	?	?	?	?
NTRU prime [12]	Large	?	Fast	Fast
SIKE [13]	Moderate	Very slow	Very slow	Very slow

9.7 Conclusion

Here, we surveyed the hardware computations in each of the NIST PQC Round 3 PKE/KEM schemes. There is no clear winner in these schemes as there are tradeoffs between area, performance, and bandwidth overhead. Throughout the NIST third round of PQC evaluation, further implementation results at the high performance and lightweight levels (especially for HQC where there is no hardware implementation) serve to narrow down which candidate is best suited for deployment. Hardware implementations feature power, energy, and performance optimizations over standard software implementations. Considering that these PKE/KEM schemes will transition to become the norm to establish quantum-secure communications in the future, these hardware benchmarks play a key role in the transition to PQC deployment.

References

1. P.W. Shor, Algorithms for quantum computation: Discrete logarithms and factoring, in *35th Annual Symposium on Foundations of Computer Science, Santa Fe, New Mexico, 20–22 November 1994* (1994), pp. 124–134. https://doi.org/10.1109/SFCS.1994.365700
2. L.K. Grover, A fast quantum mechanical algorithm for database search, in *STOC '96: Proceedings of the twenty-eighth annual ACM symposium on Theory of Computing* (1996)
3. L. Chen, S. Jordan, Report on Post-Quantum Cryptography (2016). NIST IR 8105.
4. V.B. Dang, F. Farahmand, M. Andrzejczak, K. Mohajerani, D.T. Nguyen, K. Gaj, Implementation and benchmarking of round 2 candidates in the NIST post-quantum cryptography standardization process using hardware and software/hardware co-design approaches. Cryptology ePrint Archive, Report 2020/795 (2020). https://eprint.iacr.org/2020/795

5. D.J. Bernstein, T. Chou, T. Lange, I.V. Maurich, R. Misoczki, R. Niederhagen, E. Persichetti, C. Peters, P. Schwabe, N. Sendrier, J. Szefer, W. Wang, M. Albrecht, C. Cid, K.G. Paterson, C.J. Tjhai, M. Tomlinson, Classic McEliece, *NIST Round 3 Submissions* (2020)
6. P. Schwabe, R. Avanzi, J. Bos, L. Ducas, E. Kiltz, T. Lepoint, V. Lyubashevsky, J.M. Schanck, G. Seiler, D. Stehle, CRYSTALS - Cryptographic Suite for Algebraic Lattices: Kyber, *NIST Round 3 Submissions* (2020)
7. C. Chen, O. Danba, J. Hoffstein, A. Hulsing, J. Rivjneveld, J.M. Schanck, P. Schwabe, W. Whyte, Z. Zhang, NTRU, *NIST Round 3 Submissions* (2020)
8. J.-P. D'Anvers, A. Karmakar, S.S. Roy, F. Vercauteren, SABER, *NIST Round 3 Submissions* (2020)
9. N. Aragon, P. Barreto, S. Bettaieb, L. Bidoux, O. Blazy, J.-C. Deneuville, P. Gaborit, S. Gueron, T. Guneysu, C.A. Melchor, R. Misoczki, E. Persichetti, N. Sendrier, J.-P. Tillich, G. Zemor, V. Vasseur, S. Ghosh, BIKE - Bit Flipping Key Encapsulation, *NIST Round 3 Submissions* (2020)
10. M. Naehrig, E. Alkim, J. Bos, L. Ducas, K. Easterbrook, B. LaMacchia, P. Longa, I. Mironov, V. Nikolaenko, C. Peikert, A. Raghunathan, D. Stebila, FrodoKem, *NIST Round 3 Submissions* (2020)
11. C.A. Melchor, N. Aragon, S. Bettaieb, L. Bidoux, O. Blazy, J.-C. Deneuville, P. Gaborit, E. Persichetti, G. Zemor, J. Bos, HQC - Hamming Quasi-Cyclic, *NIST Round 3 Submissions* (2020)
12. D.J. Bernstein, C. Chuengsatiansup, T. Lange, C.v. Vredendaal, NTRU Prime, *NIST Round 3 Submissions* (2020)
13. D. Jao, R. Azarderakhsh, M. Campagna, C. Costello, L. De Feo, B. Hess, A. Jalali, B. Koziel, B. LaMacchia, P. Longa, M. Naehrig, J. Renes, V. Soukharev, D. Urbanik, G. Pereira, SIKE - Supersingular Isogeny Key Encapsulation, *NIST Round 3 Submissions* (2020)
14. E. Fujisaki, T. Okamoto, Secure integration of asymmetric and symmetric encryption schemes, in *Advances in Cryptology - CRYPTO '99, 19th Annual International Cryptology Conference, Santa Barbara, California, August 15–19, 1999, Proceedings* (1999), pp. 537–554. https://doi.org/10.1007/3-540-48405-1_34
15. E. Berlekamp, R. McEliece, H. van Tilborg, On the inherent intractability of certain coding problems (corresp.). IEEE Trans. Inf. Theory 24(3), 384–386 (1978)
16. W. Wang, J. Szefer, R. Niederhagen, FPGA-based Niederreiter cryptosystem using binary Goppa codes, in *Cryptographic Hardware and Embedded Systems—CHES 2017. CHES 2017* 01 (2018), pp. 77–98
17. J. Hu, W. Wang, R.C.C. Cheung, H. Wang, Optimized polynomial multiplier over commutative rings on FPGAs: A case study on bike, in *2019 International Conference on Field-Programmable Technology (ICFPT)* (2019), pp. 231–234
18. A. Reinders, R. Misoczki, S. Ghosh, M. Sastry, Efficient bike hardware design with constant-time decoder, Cryptology ePrint Archive, Report 2020/117 (2020). https://eprint.iacr.org/2020/117
19. S. Gao, T. Mateer, Additive fast Fourier transforms over finite fields. IEEE Trans. Inf. Theory 56(12), 6265–6272 (2010)
20. E.R. Berlekamp, *Algebraic Coding Theory* (World Scientific, Singapore, 2015). https://www.worldscientific.com/doi/abs/10.1142/9407
21. L.L. Joiner, J.J. Komo, Decoding binary bch codes, in *Proceedings IEEE Southeastcon '95. Visualize the Future* (1995), pp. 67–73
22. N. Drucker, S. Gueron, D. Kostic, QC-MDPC decoders with several shades of gray, in *Post-Quantum Cryptography*, ed. by J. Ding, J.-P. Tillich (Springer International Publishing, Cham, 2020), pp. 35–50
23. V.B. Dang, F. Farahmand, M. Andrzejczak, K. Gaj, Implementing and benchmarking three lattice-based post-quantum cryptography algorithms using software/hardware codesign, in *2019 International Conference on Field-Programmable Technology (ICFPT)* (2019), pp. 206–214

24. J.M.B. Mera, F. Turan, A. Karmakar, S.S. Roy, I. Verbauwhede, Compact domain-specific co-processor for accelerating module lattice-based key encapsulation mechanism. Cryptology ePrint Archive, Report 2020/321 (2020). https://eprint.iacr.org/2020/321

25. S.S. Roy, A. Basso, High-speed instruction-set coprocessor for lattice-based key encapsulation mechanism: Saber in hardware. Cryptology ePrint Archive, Report 2020/434 (2020). https://eprint.iacr.org/2020/434

26. F. Farahmand, M.U. Sharif, K. Briggs, K. Gaj, A high-speed constant-time hardware implementation of NTRUEncrypt SVES.' Cryptology ePrint Archive, Report 2019/322 (2019). https://eprint.iacr.org/2019/322

27. J. Howe, T. Oder, M. Krausz, T. Güneysu, Standard lattice-based key encapsulation on embedded devices. IACR Trans. Cryptograp. Hardware Embed. Syst. **2018**(3), 372–393 (2018). https://tches.iacr.org/index.php/TCHES/article/view/7279

28. T. Fritzmann, G. Sigl, J. Sepúlveda, RISQ-V: Tightly coupled RISC-V accelerators for post-quantum cryptography. Cryptology ePrint Archive, Report 2020/446 (2020). https://eprint.iacr.org/2020/446

29. E. Alkim, H. Evkan, N. Lahr, R. Niederhagen, R. Petri, ISA extensions for finite field arithmetic - accelerating Kyber and NewHope on RISC-V. Cryptology ePrint Archive, Report 2020/049 (2020). https://eprint.iacr.org/2020/049

30. G. Xin, J. Han, T. Yin, Y. Zhou, J. Yang, X. Cheng, X. Zeng, VPQC: A domain-specific vector processor for post-quantum cryptography based on RISC-V architecture. IEEE Trans. Circ. Syst. I Regular Papers **67**(8), 2672–2684 (2020)

31. U. Banerjee, T.S. Ukyab, A.P. Chandrakasan, Sapphire: a configurable crypto-processor for post-quantum lattice-based protocols. IACR Trans. Cryptograph. Hardware Embed. Syst. **2019**(4), 17–61 (2019). https://tches.iacr.org/index.php/TCHES/article/view/8344

32. D.T. Nguyen, V.B. Dang, K. Gaj, High-level synthesis in implementing and benchmarking number theoretic transform in lattice-based post-quantum cryptography using software/hardware codesign, in *Applied Reconfigurable Computing. Architectures, Tools, and Applications*, ed. by F. Rincón, J. Barba, H.K.H. So, P. Diniz, J. Caba (Springer International Publishing, Cham, 2020), pp. 247–257

33. J. Cooley, J.W. Tukey, An algorithm for the machine calculation of complex Fourier series. Math. Comput. **19**, 297–301 (1965)

34. W.M. Gentleman, G. Sande, Fast Fourier transforms: For fun and profit, in *Proceedings of the November 7–10, 1966, Fall Joint Computer Conference*, Ser. AFIPS '66 (Fall) (Association for Computing Machinery, New York, 1966), pp. 563–578. https://doi.org/10.1145/1464291.1464352

35. B. Koziel, R. Azarderakhsh, M.M. Kermani, D. Jao, Post-quantum cryptography on FPGA based on isogenies on elliptic curves. IEEE Trans. Circ. Syst. **64-I**(1), 86–99 (2017). https://doi.org/10.1109/TCSI.2016.2611561

36. B. Koziel, R. Azarderakhsh, M.M. Kermani, Fast hardware architectures for supersingular isogeny Diffie-Hellman key exchange on FPGA, in *Progress in Cryptology - INDOCRYPT 2016 - 17th International Conference on Cryptology in India, Kolkata, December 11–14, 2016, Proceedings* (2016), pp. 191–206. https://doi.org/10.1007/978-3-319-49890-4_11

37. B. Koziel, R. Azarderakhsh, D. Jao, Side-channel attacks on quantum-resistant supersingular isogeny Diffie-Hellman, in *Selected Areas in Cryptography - SAC 2017 - 24th International Conference, Ottawa, August 16–18, 2017, Revised Selected Papers* (2017), pp. 64–81. https://doi.org/10.1007/978-3-319-72565-9_4

38. B. Koziel, R. Azarderakhsh, M. Mozaffari-Kermani, A high-performance and scalable hardware architecture for isogeny-based cryptography. IEEE Trans. Comput. Special Sect. Cryptograph. Eng. Post-Quantum World **PP**(99), 1–1 (2018)

39. B. Koziel, A. Ackie, R. El Khatib, R. Azarderakhsh, M.M. Kermani, Sike'd up: Fast hardware architectures for supersingular isogeny key encapsulation. IEEE Trans. Circ. Syst. I Regular Papers **PP**, 1–13 (2020)

40. R. Elkhatib, R. Azarderakhsh, M. Mozaffari-Kermani, Efficient and fast hardware architectures for sike round 2 on FPGA. Cryptology ePrint Archive, Report 2020/611 (2020). https://eprint.iacr.org/2020/611

41. P.M.C. Massolino, P. Longa, J. Renes, L. Batina, A compact and scalable hardware/software co-design of sike. IACR Trans. Cryptogr. Hardware Embed. Syst. **2020**(2), 245–271 (2020). https://tches.iacr.org/index.php/TCHES/article/view/8551

42. P.L. Montgomery, Modular multiplication without trial division. Math. Comput. **44**(170), 519–521 (1985)

43. P.L. Montgomery, Speeding the pollard and elliptic curve methods of factorization. Math. Comput. **48**, 243–264 (1987)

44. C. Costello, H. Hisil, A simple and compact algorithm for SIDH with arbitrary degree isogenies, in *Advances in Cryptology - ASIACRYPT 2017 - 23rd International Conference on the Theory and Applications of Cryptology and Information Security, Hong Kong, December 3–7, 2017, Proceedings, Part II* (2017), pp. 303–329. https://doi.org/10.1007/978-3-319-70697-9_11

45. A. Faz-Hernández, J. López, E. Ochoa-Jiménez, F. Rodríguez-Henríquez, A faster software implementation of the supersingular isogeny Diffie-Hellman key exchange protocol. IEEE Trans. Comput. **PP**(99), 1–1 (2017)

46. L.D. Feo, D. Jao, J. Plût, Towards quantum-resistant cryptosystems from supersingular elliptic curve isogenies. J. Math. Cryptol. **8**(3), 209–247 (2014)

47. R. Azarderakhsh, D. Jao, K. Kalach, B. Koziel, C. Leonardi, Key compression for isogeny-based cryptosystems, in *Proceedings of the 3rd ACM International Workshop on ASIA Public-Key Cryptography, AsiaPKC@AsiaCCS, Xi'an, May 30–June 03, 2016* (2016), pp. 1–10. http://doi.acm.org/10.1145/2898420.2898421

48. C. Costello, D. Jao, P. Longa, M. Naehrig, J. Renes, D. Urbanik, Efficient compression of SIDH public keys, in *Advances in Cryptology - EUROCRYPT 2017 - 36th Annual International Conference on the Theory and Applications of Cryptographic Techniques, Paris, France, April 30–May 4, 2017, Proceedings, Part I* (2017), pp. 679–706. https://doi.org/10.1007/978-3-319-56620-7_24

49. G. Zanon, M.A. Simplicio Jr., G.C.C.F. Pereira, J. Doliskani, P.S.L.M. Barreto, Faster isogeny-based compressed key agreement, in *Post-Quantum Cryptography - 9th International Conference, PQCrypto 2018, Fort Lauderdale, FL, April 9–11, 2018, Proceedings* (2018), pp. 248–268. https://doi.org/10.1007/978-3-319-79063-3_12

50. M. Naehrig, J. Renes, Dual isogenies and their application to public-key compression for isogeny-based cryptography, in *Advances in Cryptology – ASIACRYPT 2019*, ed. by S.D. Galbraith, S. Moriai (Springer International Publishing, Cham, 2019), pp. 243–272

51. G.C.C.F. Pereira, J. Doliskani, D. Jao, X-only point addition formula and faster torsion basis generation in compressed sike. Cryptology ePrint Archive, Report 2020/431 (2020). https://eprint.iacr.org/2020/431

52. U. Banerjee, T.S. Ukyab, A.P. Chandrakasan, Sapphire: A configurable crypto-processor for post-quantum lattice-based protocols (extended version). Cryptology ePrint Archive, Report 2019/1140 (2019). https://eprint.iacr.org/2019/1140

53. J. Howe, M. Martinoli, E. Oswald, F. Regazzoni, Optimised lattice-based key encapsulation in hardware, in *Second NIST Post-Quantum Cryptography Standardization Conference 2019* (2019)

Chapter 10
Neuromorphic Security

Rajesh J. S., Koushik Chakraborty, and Sanghamitra Roy

10.1 Introduction

Neuromorphic computing—emerging bio-inspired hardware architectures truly mark the beginning of a new era in computing system design. The amalgamation of novel non-Von Neumann architectures, new post-CMOS nano-ionic devices, redesigned software stack, and advanced learning algorithms together will act as a tailwind to propel the next wave of artificial intelligence (AI) applications. Currently, the anticipated slowing of growth in compute capacity of conventional architectures encompassing the CPU, GPU, and FPGAs is fast becoming a stumbling block to ubiquitous AI [12]. The end of Dennard scaling, the impending end to Moore's law, and Von-Neumann bottleneck are all contributing factors accelerating the demand for neuromorphic hardware fabricated from post-CMOS nano-ionic devices. These emerging neuromorphic platforms are expected to offer significant performance and power efficiency over existing conventional computing systems.

Neuromorphic chips are expected to soon become the mainstay platform for a diverse set of applications, ranging from day to day rudimentary decision making to safety critical defense and healthcare management. Figure 10.1 illustrates the diverse applications fueled by AI's digital revolution. With the neuromorphic hardware market projected to grow at an annual rate of 20.7% to an estimated $10.81 billion by 2026 [39], *security can no longer take an auxiliary role in system design.*

Rajesh J. S.
AMD, Santa Clara, CA, USA
e-mail: rajesh.js@amd.com

K. Chakraborty (✉) · S. Roy
Utah State University, Logan, UT, USA
e-mail: koushik.chakraborty@usu.edu; sanghamitra.roy@usu.edu

257

Fig. 10.1 Diverse applications due to AI's digital revolution

To understand the security implications, and identify concerns around applications, we consider a few scenarios of compromised neuromorphic systems.

- Homeland security and military aim to deploy neuromorphic hardware to detect and identify radiological threats [7], and target acquisition in autonomous lethal weapons [38]. Unexpected system behavior, mistaken targets, or even faulty trigger can cause unwarranted tensions between nations [8].
- The automotive industry is eager to adopt neuromorphic computing systems to enable faster pattern/object recognition in self-driving vehicles [39]. A compromised system can lead to fatal accidents in self-driving cars [18]. In addition, external hacks to autonomous vehicles can potentially provide anti-social elements new weapons for inciting violence [18].
- In the healthcare industry, neuromorphic chips are expected to play a critical role in disease diagnosis, health risk assessment, therapeutic recommendations, and report generation. But, advances in the domain can be nullified due to incorrect cataloging [8, 18].
- In government and other social applications, security vulnerabilities can be exploited to discriminate against groups, and effectively target political sentiments [8].

The above assortment of grim events is a glimpse into the potential ramifications of neuromorphic security vulnerabilities, underlining the importance of secure neuromorphic hardware design.

Neuromorphic hardware design and adoption is still at a very nascent stage. The complete design and development tool stack, both in hardware and software, are being re-examined and in some cases being built from the ground up. It is imperative to recognize the unique opportunity to incorporate security as a forethought to stay ahead of the threat curve. Reflecting on the last three decades of hardware design, CMOS devices, circuits and architectures experienced an accelerated technological evolution, and security policies struggled to catch up to the advances. We continue to endure the consequences of our design practices where security played an auxiliary role with periodic discovery of security vulnerabilities that affects products across generations [27, 28]. In this chapter, we dissect security of emerging neuromorphic hardware by: (1) examining the scope of neuromorphic domain, and its implications on security, (2) identifying vulnerable surfaces across design layers, and (3) reviewing the state-of-the-art research in this domain.

10.2 Threat Models

Security threats can be broadly classified into two lateral threat models based on the source of security vulnerability. Figure 10.2 outlines the following two threat models, comparing and contrasting key parameters.

- *Untrustworthy design environment*: Insertion of hardware Trojans.
- *External hacks after deployment*: Exploiting implementation-specific vulnerabilities.

To begin with, neuromorphic computing inherits the system-on-chip's (SoC) potentially flawed and untrustworthy design space [6]. Rising design and verification costs, aggressive time-to-market, and complex designs have made the SoC design flow and supply chain vulnerable to malicious Trojans and covert backdoors [55]. CMOS device properties, circuit design, and architectures have been studied for over a decade, yet new vulnerabilities are being discovered frequently [27, 28]. The introduction of a new class of nano-ionic devices, in succession to novel, unfamiliar and intricate brain-inspired architectures raise severe

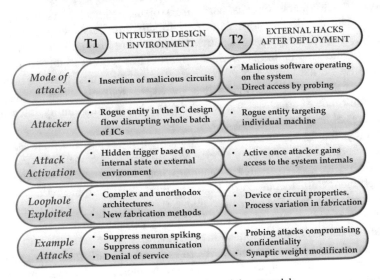

Fig. 10.2 Comparison of key attributes of the two lateral threat models

Fig. 10.3 Neuromorphic design flow illustrating opportunities for hardware Trojan insertion

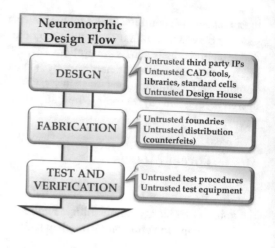

security concerns. Realization of security mechanisms with limited impact on the performance and energy efficiency is dependent on many factors including, but not limited to, device maturity, technology exploration, modeling capabilities, design strategies, fabrication capability, characterization, proven design tool chain, reliable software eco-system, and in-depth comprehension of hardware–software interaction. Research in many of these areas for neuromorphic computing is at a very nascent stage, but the pressure to cater to a growing number of critical applications may force premature adoption of neuromorphic systems.

In this threat model (*T1* in Fig. 10.2), one or more components of the incipient neuromorphic design flow, encompassing third party intellectual property vendors, design automation tool providers, and fabrication houses, are considered to be untrustworthy. Malicious circuits, hardware Trojans, and backdoors can potentially be inserted in the design flow, with their functionality discreetly hidden within the complex and densely connected architectures using innovative rare-event based triggers. Figure 10.3 highlights various opportunities present in the design flow to covertly insert a hardware Trojan into complex architectures, and has been well-studied over the last decade. Once active, the hardware Trojans can be designed to inflict a multitude of attacks, disrupting the *CIA* triad (confidentiality, integrity, and availability) of neuromorphic systems.

Secondly, the development of novel functional materials and devices, such as resistive random-access memory (ReRAM), phase change memory (PCM), and spintronic devices, have incited a quantum leap in computing, and are crucial for an efficient implementation of brain-inspired neuromorphic hardware. These fundamental devices, based on new material physics, are sought after for the properties that mimic the biological brain's synapses. The use of these novel materials in neuromorphic applications requires extensive research on properties such as multi-state behavior, sensitivity to external stimuli, fault tolerance, temperature window, ability to synthesize, electromagnetic interaction, compatibility with existing technology, and side-channel characteristics, among others. Due to the broad range of desirable requirements, most chosen materials often have inherent properties that can expose security vulnerabilities when employed in circuits.

In this threat model (*T2* in Fig. 10.2), the design environment is trustworthy, i.e., the neuromorphic hardware is not intentionally compromised during the design and fabrication stages. However, post-deployment, an external rogue agent targets an individual system, by gaining direct (probing the hardware) or indirect (running malicious software) access to the neuromorphic hardware. The attacker needs to have sufficient information regarding hardware implementation or design internals. On gaining access to the system, the attacker can exploit implementation-specific characteristics arising from device, circuit, architectural, hardware–software inter-action, or even process variations that occur during the fabrication to inflict damage to the system.

Neuromorphic computing has only recently emerged as a hot research area. Although, tremendous progress is being made in circuits, architectures, and systems, currently, there exists a very limited body of work directly addressing security in neuromorphic systems. Hence, it is necessary to examine the scope of neuromorphic domain, potential attack surfaces, its implication on security, and delve into some seminary works on neuromorphic security.

10.3 Scope of Neuromorphic Computing

Realization of efficient neuromorphic systems requires convergence of research from a variety of fields including neuroscience, material science, electrical and computer engineering. Schuman et al. conducted an encyclopedic survey of the domain, and defined five key research areas: neuro-inspired modeling, hardware and devices, supporting systems/IPs, algorithms and learning models, and applications [50]. The article epitomizes over 35 years of research and is an excellent preamble to neuromorphic computing. We align our discussion of neuromorphic security to these key areas, with stronger emphasis on devices, hardware, and supporting IPs given the context of this book. To help drive the discussion on neuromorphic security with relevant examples, this section embeds vulnerability and potency analysis carried out in *BRIDGE lab* research group [44], along with several contemporary research work in this domain.

10.3.1 Bio-inspired Hardware Models

Given that neuromorphic hardware is bio-inspired, fundamental building blocks are derived from abstractions of our understanding of the brain's behavior. Neural building blocks (e.g.,, neuron, synapse, axon and dendrite) dictate functions, behavior and implementation of their hardware counterparts. Neuron is the fundamental computational element in the brain. Its typical behavior is accumulation of charge from incoming signals via input dendrites, and generation of action potential to their axon, which then is communicated to other neurons through synapses. Various classes of neuron models exist differing widely in computational complexity and biological fidelity [50]. At one end of the spectrum is the Hodgkin-Huxley model where the neuron behavior is described using four-dimensional non-linear differential equations (high complexity and high biologic fidelity), and on the other end is rudimentary thresholding functions (artificial neural networks) for spiking neural networks, often trading-off complexity for accuracy. Numerous neuron models have already been implemented in hardware, and there also exists implementations such as SpiNNaker with programmable neuron models [41]. However, research on security implications of these implementations has been rarely considered. In Sect. 10.3.2.2, we have explored one of the simpler neuron implementations known as leaky integrate and fire.

Emerging neuromorphic architectures suggest that another fundamental component—the synapse—will take up the biggest chunk of physical area in neuromorphic hardware, and are key to mimic the brain's learning mechanisms. A variety of synaptic implementations exist for both spiking, as well as, non-spiking neural networks [50]. Implementation of synapse has received special attention with numerous post-CMOS nano-ionic materials being explored to optimize its implementation. Novel materials have distinct physical properties and intrinsic characteristics such as multi-state behavior, sensitivity to external stimuli, noise and fault tolerance, operational conditions, and compatibility with existing circuits. These inherent device/circuit properties can often expose vulnerabilities that can be exploited by external hacks during the system's operational lifetime. Section 10.3.2.1 presents a few exploitable properties, and the sections that follow examine both the beneficial and malicious use of circuit properties in neuromorphic security.

The neuromorphic community continues to deliberate on the degree of biological fidelity that brain-inspired computational systems should emulate. Biological systems have a variety of plasticity mechanisms, where the brain's function is molded by the neural activity and synaptic transmissions. The brain's synaptic plasticity spans over a range of spatio-temporal scales, and impacts learning, memory, and other computations [23]. Neural learning mechanisms and concepts such as spike time dependent plasticity (STDP), long-term potentiation (LTP), and long-term depression (LTD) are key to the first generation of neuromorphic hardware, and mechanisms such as synapto-genesis, neurogenesis, neuro-modulation, and regional restructuring will follow. Novel models and hardware abstractions of biological mechanisms will likely introduce new vulnerabilities that mimic biological disorders possibly simulating effects of progressive degeneration, behavior fluctuation, and soft faults proving hard to identify. There is an imminent need for research on security properties, models, and standards for mindful implementation of bio-inspired hardware models.

10.3.2 Neuromorphic Architecture, Implementation and Devices

A plethora of neuromorphic solutions exist in different design stages: research literature, design phase, prototype, and some operational (e.g., [1, 2, 5, 11, 19, 25, 26, 30, 33, 36, 41, 43, 45, 48, 49, 52, 53, 57, 59, 60]. For example, IBM has developed TrueNorth chip specifically to perform neuromorphic applications [2]. Intel introduced Loihi, a self-learning neuromorphic chip designed and optimized specifically for executing spiking neural network (SNN) algorithms [14]. The implementations vary wildly depending on the material, learning algorithm, and end-user application, and clearly, a single solution does not suffice for all needs. In this section, we briefly look at popular non-standard devices and circuits. Taking note that memristor-based neuromorphic implementations are more prominent than others, we will cover a typical neuromorphic implementation, and look at potential security exploitations.

10.3.2.1 Devices and Circuit Implementation

Development of unconventional neuromorphic systems is reliant on fabrication and integration of novel devices. Memristor, phase change memory (PCM), spintronic devices, and optical/photonic components have garnered interest from the neuromorphic community. These materials display characteristics of differential resistance/capacitance or some form of threshold based switching when an incident control parameter such as current, voltage, magnetic field, or irradiation is varied. Memristors are more prevalent than other components, and they exhibit a property of programmable resistance that can reversibly change with voltage, and subsequently are non-volatile. Memristors have been fabricated using various materials such as metal oxides, chalcogenides, carbon and amorphous silicons. The high interest in memristors for neuromorphic hardware is due to their memory and plasticity properties similar to the biological synapses.

In PCM devices, application of current to a dielectric material results in the change of state between varying levels of crystalline or amorphous state (multistate property), making it desirable for the implementation of a synapse. Spintronic devices are another promising alternative to memristor and PCM. At its fundamental level, a spintronic device uses electron spin to represent and process data. Magnetic Tunnel Junction (MTJ) is an elementary spintronic application, in which the magnetic configuration of two ferromagnetic layers determine the electrical resistance [16, 24]. Another form of spintronics is a magnetic nano-wire that holds multiple bits in terms of magnetic polarity to form a domain wall (DW) [16, 24]. Previously, interest in photonic and optical components for neuromorphic implementation had taken a back seat due to complexity in implementation of memory and logic in optical systems, and lack of necessary manufacturing capabilities. However, recent technological innovations have brought about a resurgence in research activity of

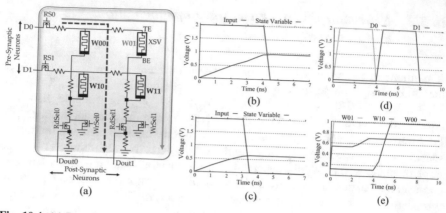

Fig. 10.4 (**a**) Presents a 2 × 2 memristor crossbar array, to demonstrate the inherent properties that can be exploited by external attacks. (**b**) Exhibits the non-volatile property of a memristor (state maintained even after voltage falls to zero). (**b**) and (**c**) Illustrate that the state value of the memristor explicitly depends on the flux injection (change in pulse width changes memristor state value). Finally, (**d**) and **e** show the presence of parasitic leakage paths (red line in (**a**)) that has undesired effects on the memristor state ($W00$)

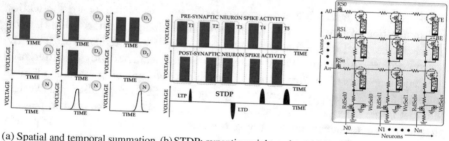

(a) Spatial and temporal summation (b) STDP: synaptic weight update (c) $N \times N$ 1T-1M crossbar array

Fig. 10.5 Properties of memristor-based neuron design, and crossbar structure with sneak path mitigation mechanism. (**a**) Illustrates that the neuron can perform temporal and spatial summation of injected voltage. (**b**) Demonstrates the synaptic update stage following the STDP postulate. (**c**) Shows the schematic of a 1 transistor—1 memristor (1T-1M) based crossbar design to mitigate the sneak paths

photonic neuromorphic systems due to photonics' energy-efficiency, low thermal concerns, low cross-talk, and processing speeds that are far superior to the electronic systems.

Novel nano-ionic devices and new circuits are swiftly consuming the neuromorphic design space, but lack a structured body of knowledge about security vulnerabilities. Devices and circuits realized from these devices all have intrinsic properties that are often exposed to an untrustworthy external user environment. We look at a few properties of memristor-based neuromorphic design implementation using Figs. 10.4 and 10.5 that can potentially be vulnerable to external attacks.

- Figure 10.4b illustrates the non-volatility property of a memristor. The memristor maintains its state even when the supply voltage is removed, leaving it vulnerable to privacy and integrity attacks by physical readout, probing, and fault-injection. However, memristors are densely packed in circuits making them difficult to probe using existing non-invasive mechanisms. In addition, several probing and readout countermeasures exist that aid in securing systems against this form of vulnerability for the moment.

- Figure 10.4b and c together demonstrate that the memristor state is a function of the flux injection (area under the voltage curve defined by pulse width and amplitude). Hence, altering the state by manipulating the voltage amplitude or pulse width can harm system integrity. Flux injection to the circuit is a control parameter that can be manipulated from multiple design layers to carry out the system's desired functions, and can potentially be exposed as an attack surface.

- Figure 10.4a illustrates a 2×2 memristor crossbar array that is typically used to realize a synapse in neuromorphic systems. In Fig. 10.4a, when a voltage is applied across $D0$ (Fig. 10.4d), and $WrSel1$ is enabled, the state of memristor $W01$ changes (Fig. 10.4e). However, we also observe an undesired state change in $W00$, due to the parasitic leakage paths (sneak paths). If left unmitigated, attacks can alter the synaptic weights stored as memristor state value. Memristor sneak path is a fairly well-known vulnerability in memristor crossbar arrays. Although there exist mechanisms to reduce charge through sneak paths, they cannot fully eradicate it due to transistor leakage. Hence, the memristor crossbar is still vulnerable to attacks exploiting sneak paths.

- In SNN, neurons are capable of performing spatial and temporal summation. In Fig. 10.5a, the stimulus from one dendrite (D_1) is insufficient for the neuron to fire. However, the aggregation of inputs from dendrites, D_1 and D_2, either spatially (i.e., stimulus from multiple dendrites) or temporally (i.e., stimulus accumulation in time window) can cause the neuron to fire. Manipulation of spike trains through tampering inputs or address-event-representation (AER) circuits can cause undesired behavior in the neural networks.

- In SNN, the memristor synapse updates its state only when the pre-synaptic and post-synaptic neurons spike at close points in time. Figure 10.5b simulates the STDP process, and shows how voltages with different magnitude, polarity, and pulse width are applied to the synapse. Asymmetric STDP adjusts the synapse weight based on spatial and temporal occurrence of spike trains. In window $T1$, $T4$, and $T5$, the pre-synaptic neuron fires before the post-synaptic neuron resulting in long-term potentiation (LTP), whereas the opposite behavior in $T3$ results in long-term depression (LTD). In window $T2$, the pre-synaptic and post-synaptic neurons fire too far apart to update the memristor state. In a densely connected network, it becomes prohibitive to validate the spiking activity and to identify malicious spiking behavior.

Other devices discussed above have analogous vulnerabilities that can be exploited by an external agent. For example, MTJ and DW are both susceptible to state toggling by the application of spin polarized current or magnetic field

[16, 46]. Further, in DW, the process of *unpinning* a domain from local energy minimum state makes the neighboring domains susceptible to state corruption. These vulnerabilities exacerbate in the presence of process variations and thermal noise [24, 46]. Similar to memristors, spintronic devices are also vulnerable to the pulse width and amplitude modulation attacks on supplied voltage [21].

Given the existence of potential threats to secure neuromorphic systems, we pursue one potential attack surface (*sneak path*) and examine a sample implementation, and impact on application. Sneak paths are associated with problems of increased power consumption, undesired state switching of neighboring memristors, and limited array sizes due to voltage drop away from the driver. In SNN, each of these effects can be exploited by external attacks. Consider a fault-injection attack exploiting a neighboring cell switching due to leakage current flowing through sneak paths.

For this attack vector, the attack surface is exposed primarily due to circuit characteristics of a memristor crossbar array, but can be exacerbated by hardware Trojans in address-event representation (AER) circuit. Two orthogonal attack vectors can exploit this vulnerability. First, an attacker (an external agent/application) can control input spike sequences by selecting appropriate input patterns (images). Second, a hardware Trojan in the AER circuit can manipulate the spike patterns corresponding to the input image to cause undesired state changes in the crossbar array. The stream of input spikes establishes a current flow in the array and continually trains the network by updating the synapses on the path between the pre-synaptic and post-synaptic neurons. However, the current leaks through alternate paths (sneak paths), and continually injects small pulses of charge. When sufficient charge is accumulated sneak paths cause undesired neurons to fire, and hence, alter the states of memristors in the vicinity of the actual path. In addition, the attacker introduces a systematic bias in the neuron synaptic mapping by causing read current variations. Firing an undesired neuron impacts STDP learning rule by strengthening or weakening connections between undesired neuron pairs. These undesired state changes and bias in the mapping will lessen the inference precision.

Contemporary works on memristor crossbar arrays have proposed several sneak path mitigation techniques [61]. Figure 10.5c illustrates one such mechanism, where an additional gating transistor is added to each memristor to remove the alternate paths for current flow. SPICE simulations reveal that the mechanism is able to reduce charge through sneak paths nearly 20% of the original, but cannot fully eradicate it due to the transistor leakage. Hence, the crossbar is still vulnerable to the attack mechanism discussed above.

Since an attack exploiting sneak paths affects synaptic weights stored in the memristor crossbar, exploiting this property leads to significant reduction in recognition rate. Based on one implementation, the recognition rate of SNN can drop by nearly 50%. With an increase in size of the neural network, impact of cross bar sneak paths becomes more localized, thereby reducing the impact of the attack. Figure 10.6a charts out attack induced *malicious misprediction rate* for different MNIST digits. Figure 10.12b reveals a different perspective by presenting the heat map of synaptic weights after network training for an SNN with 100 output neurons.

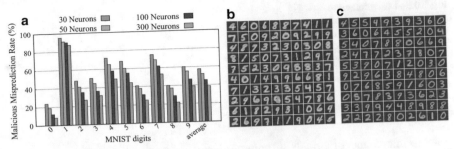

Fig. 10.6 Attack induced misprediction rate signifies the increase in mispredictions caused specifically due to a Trojan attack **exploiting crossbar sneak paths**

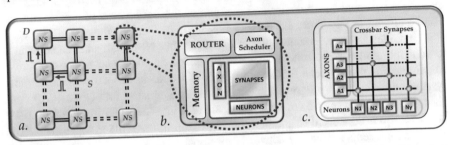

Fig. 10.7 A typical neuromorphic architecture. (**a**) Shows the system level architecture, where many neurosynaptic cores (NS) are connected by a mesh interconnect fabric. It also shows the movement of spike packets from source (S) to destination (D). (**b**) Shows the block diagram of an NS. (**c**) Presents the crossbar synapses, axons, and neurons. When an axon A_2 receives an input spike, the current flows through the highlighted path to the neurons N_3 and N_y. If the membrane potential of either neuron is breached, a spike is generated

Comparing to the Trojan free scenario (Fig. 10.12b), the strength of synaptic weights across the output neurons are weak and hence, MNIST digits are improperly trained due to crossbar sneak paths attack. These results reveal that vulnerabilities arising from device/circuit characteristics can have an undesired effect on the system. In-depth understanding of device and circuit properties from a security standpoint is the need of the hour to reshape secure design and verification practices in emerging neuromorphic systems.

10.3.2.2 Neuromorphic Architecture

Figure 10.7 illustrates a typical spiking neural network (SNN) architecture, at three different levels of abstraction: *system, neurosynaptic core, and crossbar synapse array*. At the system level (Fig. 10.7a), the neuromorphic hardware is composed of many simple distributed processing units (neurosynaptic cores) that are densely connected and operate in parallel. In this example, communication between neurosynaptic cores (NSC) is handled by a two-dimensional mesh network-

Fig. 10.8 Leaky integrate and fire (LIF) neuron model representation and simulation illustrating the neuron spiking

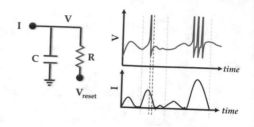

on-chip interconnect fabric. NSC communicates using simple information packets which encompasses spikes (simplest possible temporal message) and the address to route the message over the network. NSC (shown in Fig. 10.7b) mimics the biological neuronal functionality and overcomes the bottlenecks of a conventional Von-Neumann architecture. Each NSC has its own local memory. Further, each NSC includes a router, to route spike packets to other NSC, and an axon scheduler to schedule spike events on selected axons based on incoming packets. From a functional standpoint, each NSC has individually addressable axons, configurable synaptic crossbar, and programmable neurons. The synapse unit (Fig. 10.7c) consists of an $N \times N$ synapse crossbar array, along with other interface circuits. Each input neuron is connected to another output neuron through the synapse array. The crossbar represents a recurrent network topology and can store N^2 synaptic weights among the N neurons. Each neuron element tracks the membrane potential and has spike buffers to store input spikes arriving from other NS, as well as, the output spikes generated when the membrane potential reaches the threshold.

We examine the dynamics of a neuron using a well-known formal spiking model known as the leaky integrate and fire (LIF) as an example [9, 37, 40]. The two basic principles of the LIF model are: (a) the membrane potential of the neuron evolves based on the prevalence of input spikes, and (b) neurons generate a spike event when the membrane potential breaches a set threshold. Figure 10.8 presents the equivalent circuit and the simulation of a single LIF neuron. When the integral of injected charge breaches the threshold, the neuron fires and the membrane potential is reset. When there is no charge injection in a time window, the membrane potential decreases (leaky phase). To model the biological neuron, a refractory period (delay) is considered during which the membrane potential does not increase even when a strong stimulus is provided.

Figure 10.7 illustrates the typical journey of a spike packet. The spike travels from axons (horizontal lines), through active synapses in the crossbar to drive inputs for the neurons (shown in Fig. 10.7c). Axons are activated by spike events arriving from an external stimulus, or other post-synaptic neurons. When a neuron on an NSC spikes, the router packetizes the spike event, destination address of the correlated axon, and the associated delay. The spike packet is injected into the interconnection fabric and directed towards the destination address (Fig. 10.7a). Once the spike packet arrives at the destination NS, the router depacketizes and forwards the event to the axon scheduler, which then schedules it on the specific axon with the associated delay.

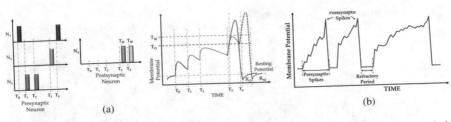

Fig. 10.9 Dynamics of attack vector manipulating the neuronal threshold and refractory period. (a) Impact of threshold voltage manipulation. (b) Refractory period manipulation

Fig. 10.10 Automatic gain control to maintain homeostasis

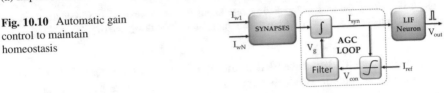

The temporal correlations between the spikes in pre-synaptic axons and post-synaptic neurons adjust the strength of the connections known as spike timing dependent plasticity (STDP). Consistent with Hebb's postulate for unsupervised learning, stronger synaptic connections increase the likelihood of pre-synaptic neuron induced firing of post-synaptic neuron [56]. In short, neuromorphic systems are computationally simple and distributed neurosynaptic cores that are densely connected with sparse communication traffic between them.

In typical neuromorphic architectures, the neuron's overall firing rate is regulated within the operating boundaries, even with undesired changes such as temperature drifts by a process called homeostasis. Homeostasis is implemented using automatic gain control (AGC) mechanisms to scale the synaptic weights and modulate the threshold voltage. Figure 10.10 illustrates the block diagram of an AGC loop based on the work presented in [47] by Rovere et al.. Additionally, the interface circuits allow individual control of the refractory period of each neuron. Programmable biases, and current sources can be used to reset the membrane potential for a fixed period, during which it will not be possible to stimulate the neuron.

Consider a neuron suppression attack by manipulating neuron threshold and refractory period. Figure 10.9 illustrates simplified dynamics of such an attack vector. Figure 10.9a shows that pre-synaptic neuron spikes raise the membrane potential of the neuron. By manipulating the threshold, the embedded Trojan alters the post-synaptic neuron spike response to input spikes. In Fig. 10.9a, the original spike should have occurred at $T_3 P$ in response to the green spike, but the altered threshold forces the neuron to respond to spike at $T4$. After the membrane potential breaches the threshold, the neuron fires and the membrane potential drops to its resting potential. Instead of re-polarization due to input spikes, neuron blocks all inputs during the refractory period and holds the membrane potential steady as seen in Fig. 10.9b. The STDP learning process is compromised as it is dependent on both, the time delta between pre and post-synaptic neurons, as well as, refractory period (Fig. 10.5b).

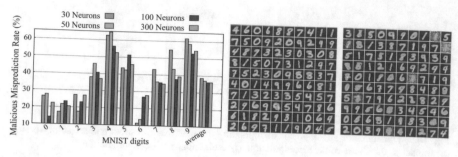

Fig. 10.11 Attack induced misprediction rate signifies the increase in mispredictions caused specifically due to a Trojan attack from manipulating **neuron spiking activity**

The hardware Trojan embedded in the gain control loop can maliciously scale the synaptic currents to produce output spikes at a firing rate proportional to the amplitude of the malicious control voltage. The negative feedback loop, depicted in Fig. 10.10, maintains the proper operating conditions of the neurons by adjusting the neuron current, I_{syn}, close to reference current, I_{ref} [47]. The Trojan in the feedback control manipulates the reference current and change the magnitude of the control voltage which is further fed to the filter to gradually decrease the scaling factor. Thus, the Trojan in the feedback control affects the STDP as illustrated in Fig. 10.9. Further, a minor modification is built into the neuron circuit to explicitly manipulate the refractory period and produce a wide range of spiking behavior resulting in compromised learning.

In one implementation of the attack, the recognition rate of the system under attack drops by more than 30%, across the different sizes of output layers. Figure 10.11a projects an analytical view of the attack potency, as it charts the attack induced *malicious misprediction rate* for different MNIST digits. By suppressing a subset of the neurons from firing, observe that the degree of attack varies significantly across the digits. SNN mapping, in conjunction with neurons suppressed during an attack determine the impact on the malicious misprediction rate for different handwritten digits. Recognition of digits 4, 5, 8, and 9 are severely affected in these experiments, as compared to other digits. Figure 10.11c shows the heat map of synaptic weights after training for a neural network with 100 output neurons. The Trojan suppresses certain neurons from firing correctly, thereby affecting the training of synaptic weights. From Fig. 10.11c, we can observe that there is uncertainty in recognizing handwritten digits 4, 5, and 9.

The introduction of novel unorthodox non-Von Neumann architectures makes it paramount to understand the range of possible threats, the type of attacks, the associated design overheads of embedding hardware Trojans and their subsequent impact on applications. The work discussed above provides a small glimpse into potential attack vectors in emerging neuromorphic hardware. Going forward, well-defined testing, verification, and validation strategies are required to support security assurance in neuromorphic hardware.

10.3.3 Supporting IPs/Systems

Although the focus of novel devices and architectures is firmly on the design of neurons, synapses, and circuits for multiple plasticity mechanisms, neuromorphic systems will not be feasible without infrastructure supporting inter and intra-chip communication, power delivery, and management and sensory functions. Attacks on any of these supporting IPs can have a potent impact on AI applications deployed on neuromorphic hardware. Scalable network-on-chip (NoC) communication fabric is one option that is being examined carefully to connect highly parallel and distributed neurosynaptic cores. NoCs designed for neuromorphic systems present new challenges as they are required to support direct axon addressing, time multiplexing, multicast routing, and have low latency [15, 34]. Multiple network configurations including mesh, tree, toroidal, shared bus, and ring are being explored to support dense connections, and multicast communication is required to extract high performance from neuromorphic systems [15, 34, 50].

Contemporary works on NoC security demonstrate the existence of security vulnerabilities of relevance to neuromorphic systems [3, 10, 51, 58]. We examine the behavioral impact on an application by adopting a previously uncovered attack vector. Selective denial of service attack vector explored in our research group previously [51] can prove to be more potent due to the added design complexity, and the impact on STDP learning. For example, delaying communication packets on an existing Von-Neumann architecture based multi-processor system-on-chip (SoC) maintains functional correctness and affects system performance. However, a similar delay in a neuromorphic system can impact strength of neuron connections, changing the system behavior to inputs based on how the attack is implemented. Adopting the concept presented in [51], we study consequences of a rogue NoC in neuromorphic systems. A malicious circuit can be embedded in the NoC to inflict selective denial of service (DoS) attack to a subset of the neuronal spike communication. The hardware Trojan will classify neuron pairs based on frequent firing rate during the presentation of a defined number of labeling samples. Once, a set of highly correlating spike communication packets are identified, the Trojan will covertly deny/delay the delivery of packets. Deriving from discussion in Fig. 10.5, propagation of spikes from multiple neurons dictates a neuron's membrane potential, spiking activity, synaptic weight tuning, and the strength of connection between the two neurons. Delayed or dropped communication will weaken the connection between the two neurons affecting the STDP.

Attack implementation differences will arise due to differences in traffic characteristics, sensitivity to communication delays, and learning adaption for repeatedly delayed or dropped spike communication. In SNN, spiking activities can be sparse, sporadic, and multicast, and together with time-multiplexed neurons, a new approach may be required for the selection of victim pairs. We observed that on an average, the recognition rate degrades by 48% in a preliminary evaluation of one specific implementation. Figure 10.12a shows a detailed impact of the attack by charting the attack induced *misprediction rate* for different digits of the MNIST dataset. In these simulations, the suppression of spike communication has had a

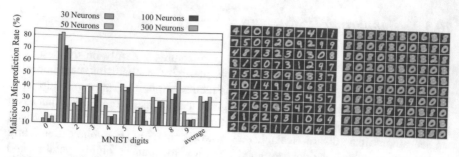

Fig. 10.12 Attack induced misprediction rate signifies the increase in mispredictions caused specifically due to a Trojan attack from manipulating **neuron spike communication**

smaller impact on digits 0 and 6, in contrast to the other digits. Figure 10.12c shows the heat map of synaptic weights after network training for an SNN with 100 output neurons. Comparing the figure with the Trojan free version (Fig. 10.12b), several digits are not trained and do not appear in the heat map. Figure 10.12c represents an extreme case of DoS attack, and different flavors of DoS can be realized.

We would like to reiterate that limited existing research on the security implications of untrustworthy supporting IPs in neuromorphic systems presents open opportunities to advance the state of the art.

10.3.4 Algorithms and Learning Models

Operational data and learning model, are two key components of AI, and the chosen neuron, synapse and network models and implementation depend on the choice of these two components. Neuromorphic systems have piqued researchers interest due to their potential for online learning. Tremendous effort is going into the domain to find optimized implementation of learning models and algorithms tailored for neuromorphic hardware. Once again, the article by Schuman et al. provides a comprehensive overview of different supervised and unsupervised on-chip learning algorithms. From a security standpoint, both data and learning models are vulnerable to security threats, with dire consequences. Attacks against learning models and algorithms can be classified based on three properties described by Huang et al. [22]: (a) *influence of attack*—altering training process vs model/data discovery by probing, (b) *security violation*—confidentiality, integrity, and availability, and (c) *attack specificity*—targeted vs indiscriminate.

Learning models have previously been shown to be vulnerable to adversarial inputs [42], where subtle noise is ingrained into the test data to manipulate learning model behavior. Similarly, SNN trained as classifiers are also vulnerable to such adversarial perturbations, where spikes may be added or removed to cause adversarial behavior [4]. Sensitivity studies and research on training SNNs towards adversarial perturbations are still at a very nascent stage with researchers considering small networks, and datasets with small set of features. Adversarial

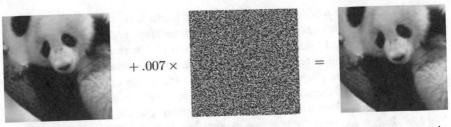

Fig. 10.13 Adversarial example generation from [54]. The original image is classified as a *panda*, but the manipulated image is mis-classified

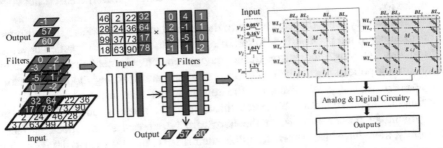

Fig. 10.14 Convolutional operation on memristor-based neuromorphic system used by Liu et al. in [32]

training still suffers from the curse of dimensionality in complex dataset, where the test input can reside far from the distribution of the adversarial training dataset, thus, limiting the impact of adversarial training.

Liu et al. investigated a residual learning based algorithm to build a restoration network reusing computational components of the neuromorphic system to defend against adversarial perturbations [32]. Figure 10.13 sourced from research article by Szegedy et al. illustrates a popular example for adversarial input. The manipulated image is almost indistinguishable from the original to the human eye, but in a neuromorphic system, the noise could carry significant perturbations resulting in classification error [54]. Liu et al. first generate adversarial samples and use the learning framework to develop a residual features map between the original and adversarial samples. The residual features generated are mapped as pixel noise in the same resolution as the adversarial sample to enable quick superposition of one over the other to restore the original sample. To implement this mechanism, they employed convolutional layers (for feature learning), deconvolutional layers (to maintain scaling between adversarial sample and feature map), skip connections (data concatenation between multiple layers to extract multi-level features), leaky rectified linear unit (LRELu) as the activation function, and a batch normalization layer to adjust the batch size. Another key aspect of this work is that restoration network is implemented as part of the neuromorphic system reusing the memristor crossbar array. By applying square to vector transformation, along with filter transformation to vertical tensors, they deploy convolutional and deconvolutional layers on memristor crossbar array as shown in Fig. 10.14.

To evaluate their work, they compared their restoration network performance to techniques such as random cropping, image resizing, and Gaussian blur, and obtained a restoration rate of 38.4%. Restoration rate here is the percentage of adversarial samples that could be restored to defend the classifier, from the total sample set. Since the proposed technique can work in conjunction with Random Crop and Resize techniques, they obtained the best restoration rate by combining these techniques. Their evaluation also highlighted that the restoration network has low design footprint in area and power, but is computationally slow due to large convolutional and deconvolutional layers. However, a key benefit is that the restoration network scales well, and the overheads reduce with increased network complexity.

Model privacy is another concern with the need to protect proprietary algorithms. For instance, in the profit-driven autonomous vehicle industry, companies develop proprietary learning models for traffic navigation, environment processing, and safe driving policies. Leak/theft of proprietary data and learning model can result in severe loss. Liu et al. discussed the feasibility of learning model replication by function approximation [29]. Such threats and their associated challenges rise to gargantuan proportions, when the hardware is exposed as an attack surface. Recently, there have been efforts to utilize memristor characteristics to devise smart countermeasures to tackle the problem of model privacy.

Yang et al. in their work present a detailed study on model replication, and proposed countermeasures leveraging memristor circuit's obsolescence effect, where the resistance varies on applying voltage pulses as depicted in Fig. 10.4. Their approach is learning model agnostic. Furthermore, to establish indifference to choice of learning model, they train a target model using support vector machine (SVM), and three attack models using K-nearest neighbor, SVM, and random-forest models with the same dataset. The experiment revealed that the accuracy of all models approaches 90% after only 1000 I/O pairs. The work exploits memristor circuit's basic principle that applying voltage pulses to inputs can change the conductance of the crossbar, and over time can change the function of the model implemented. When unauthenticated users provide inputs to the system, with the intention to monitor outputs and replicate the model, the system gradually degrades in accuracy preventing model theft. Also, to guarantee system performance to authenticated users, they propose calibrating the crossbar array by refreshing with initial conductance states periodically. To provide additional protection against replication attacks, the authors also evaluated non-linear degradation. Smart design of degradation circuits helps tune trade-off knobs between usability, calibration overhead (reinstating conductance for authenticated inputs), and protection against replication attacks. These model replication attacks fall under the classification of exploratory attacks where the attacker explores and replicates the design by monitoring I/O pairs without changing any parameters of the model. Contrary to these attacks, neural networks can also fall prey to causative attacks, where the training dataset is tampered with to change target model parameters. Although these attack vectors are explored in neural networks, no examples currently exist explicitly for neuromorphic hardware.

Given the lack of research on neuromorphic security, these works serve as a foundation to secure neuromorphic systems. Besides, most existing countermeasures to machine learning privacy attacks are predominantly computation and time intensive encryption mechanisms [31]. Forthcoming works should consider trade-offs such as *ease of use vs. security, cost of security vs cost of failure/recovery*, and *design complexity vs protection offered.*

10.4 Alternative Security Approach

In previous sections, we aligned existing neuromorphic security research to active domains of research in the neuromorphic community. In addition to these major domains, there have been a couple of research works targeting secure neuromorphic systems. He et al. have recently proposed IPLock, an encryption mechanism to counter IP counterfeiting and neuromorphic IP theft [17]. Their threat model considers a man-in-the-middle attack where the system architecture is stolen from an IP vendor in the untrustworthy design environment. They propose a hybrid encryption mechanism combining elliptic curve cryptography (ECC) and SM4 simultaneously. SM4 algorithm is used to encrypt the neuromorphic architecture information and parameters extracted from design modules, and ECC encrypts SM4 key. They employ sub-module level encryption to protect key components such as neuron computing module, synaptic architecture module, and configuration unit module.

Finally, we would like to throw the spotlight on an emerging technology that could disrupt research on an entire class of secure hardware systems: transient electronics. Wu et al. demonstrated physically transient devices fabricated on dissolvable and biodegradable silk fibroin substrates, in line with multiple previous works exploring resistive RAM devices based on natural bio-materials that are dissolvable [13, 20, 35]. These devices, dissolvable in deionized water are sought after for transient memory systems, and implantable medical devices. In their recent work, they fabricated an artificial synapse network based on a dual in-plane-gate Al–Zn–O neuromorphic transistor. To add to the excitement, they showcase excitatory post-synaptic current, paired-pulse-facilitation, and temporal filtering characteristics along with dynamic spatio-temporal learning rules and neuronal arithmetic. These devices are soluble in deionized water in approximately 120 s. There exist other research works that demonstrate robust memristive characteristics such as retention, endurance, and STDP. Transient electronic technology could prove to be an exciting and promising solution for a subset of secure hardware designs.

Acknowledgments This work was supported in part by National Science Foundation grants (CNS-1117425, CAREER-1253024, CCF-1318826, CNS-1421022, CNS-1421068). Any opinions, findings, and conclusions or recommendations expressed in this material are those of the authors and do not necessarily reflect the views of the NSF.

References

1. S. Agarwal, D. Rastogi, A. Singhal, The era of neurosynaptics: neuromorphic chips and architecture. Eur. Sci. J. **11**(10) (2015). https://eujournal.org/index.php/esj/article/view/5716
2. F. Akopyan, J. Sawada, A. Cassidy, R. Alvarez-Icaza, J. Arthur, P. Merolla, N. Imam, Y. Nakamura, P. Datta, G.J. Nam, B. Taba, M. Beakes, B. Brezzo, J.B. Kuang, R. Manohar, W.P. Risk, B. Jackson, D.S. Modha, Truenorth: design and tool flow of a 65 mW one million neuron programmable neurosynaptic chip. IEEE Trans. Comput. Aided Design Integr. Circ. Syst. **34**(10), 1537–1557 (2015). https://doi.org/10.1109/TCAD.2015.2474396
3. D.M. Ancajas, K. Chakraborty, S.Roy, Fort-NoCs: Mitigating the threat of a compromised NoC, in *IEEE/ACM Design Automation Conference (DAC)* (2014), pp. 1–6
4. A. Bagheri, O. Simeone, B. Rajendran, Adversarial training for probabilistic spiking neural networks, in *19th IEEE International Workshop on Signal Processing Advances in Wireless Communications, SPAWC 2018, Kalamata, June 25–28, 2018* (2018), pp. 1–5
5. B.V. Benjamin, P. Gao, E. McQuinn, S. Choudhary, A.R. Chandrasekaran, J.M. Bussat, R. Alvarez-Icaza, J.V. Arthur, P.A. Merolla, K. Boahen, Neurogrid: a mixed-analog-digital multichip system for large-scale neural simulations. Proc. IEEE **102**(5), 699–716 (2014). https://doi.org/10.1109/JPROC.2014.2313565
6. S. Bhunia, M.S. Hsiao, M. Banga, S. Narasimhan, Hardware trojan attacks: threat analysis and countermeasures. Proc. IEEE **102**(8), 1229–1247 (2014)
7. C. Bobin, O. Bichler, V. Lourenço, C. Thiam, M. Thévenin, Real-time radionuclide identification in gamma-emitter mixtures based on spiking neural network. Appl. Radiat. Isot. **109**, 405–409 (2016). *Proceedings of the 20th International Conference on Radionuclide Metrology and its Applications 8–11 June 2015, Vienna*
8. N. Bostrom, Strategic implications of openness in AI development. Global Policy **8**(2), 135–148 (2017)
9. A.N. Burkitt, A review of the integrate-and-fire neuron model: I. homogeneous synaptic input. Biol. Cyb. **95**(1), 1–19 (2006)
10. S.V.R. Chittamuru, I.G. Thakkar, S. Pasricha, S.S. Vatsavai, V. Bhat, Exploiting process variations to secure photonic NoC architectures from snooping attacks. CoRR **abs/2007.10454** (2020). https://arxiv.org/abs/2007.10454
11. M.H. Choi, S. Choi, J. Sim, L.S. Kim, Senin: An energy-efficient sparse neuromorphic system with on-chip learning, in *2017 IEEE/ACM International Symposium on Low Power Electronics and Design (ISLPED* (IEEE, Piscataway, 2017), pp. 1–6
12. P.R. Cohen, E.A. Feigenbaum, *The Handbook of Artificial Intelligence*, vol. 3 (Butterworth-Heinemann, Oxford, 2014)
13. B. Dang, Q. Wu, F. Song, J. Sun, M. Yang, X. Ma, H. Wang, Y. Hao, A bio-inspired physically transient/biodegradable synapse for security neuromorphic computing based on memristors. Nanoscale **10**(43), 20089–20095 (2018)
14. M. Davies, N. Srinivasa, T.H. Lin, G. Chinya, Y. Cao, S.H. Choday, G. Dimou, P. Joshi, N. Imam, S. Jain, et al., Loihi: a neuromorphic manycore processor with on-chip learning. IEEE Micro **38**(1), 82–99 (2018)
15. H. Fang, A. Shrestha, D. Ma, Q. Qiu, Scalable NoC-based neuromorphic hardware learning and inference, in *2018 International Joint Conference on Neural Networks, IJCNN 2018, Rio de Janeiro, Brazil, July 8–13, 2018* (2018), pp. 1–8
16. S. Ghosh, Spintronics and security: prospects, vulnerabilities, attack models, and preventions. Proc. IEEE **104**(10), 1864–1893 (2016)
17. G. He, C. Dong, Y. Liu, X. Fan, IPlock: An effective hybrid encryption for neuromorphic systems IP core protection, in *2020 IEEE 4th Information Technology, Networking, Electronic and Automation Control Conference (ITNEC)*, vol. 1 (2020), pp. 612–616
18. M. Hengstler, E. Enkel, S. Duelli, Applied artificial intelligence and trust—the case of autonomous vehicles and medical assistance devices. Technol. Forecast. Soc. Change **105**, 105–120 (2016)

19. M. Hu, H. Li, Y. Chen, Q. Wu, G.S. Rose, R.W. Linderman, Memristor crossbar-based neuromorphic computing system: a case study. IEEE Trans. Neur. Netw. Learn. Syst. **25**(10), 1864–1878 (2014)

20. W. Hu, J. Jiang, D. Xie, S. Wang, K. Bi, H. Duan, J. Yang, J. He, Transient security transistors self-supported on biodegradable natural-polymer membranes for brain-inspired neuromorphic applications. Nanoscale **10**(31), 14893–14901 (2018)

21. Y. Huai, Spin-transfer torque MRAM (STT-MRAM): challenges and prospects. AAPPS Bulletin **18**(6), 33–40 (2008)

22. L. Huang, A.D. Joseph, B. Nelson, B.I. Rubinstein, J.D. Tygar, Adversarial machine learning, in *Proceedings of the 4th ACM Workshop on Security and Artificial Intelligence* (ACM, New York, 2011), pp. 43–58. https://doi.org/10.1145/2046684.2046692

23. C.D. James, J.B. Aimone, N.E. Miner, C.M. Vineyard, F.H. Rothganger, K.D. Carlson, S.A. Mulder, T.J. Draelos, A. Faust, M.J. Marinella, et al., A historical survey of algorithms and hardware architectures for neural-inspired and neuromorphic computing applications. Biol. Inspir. Cogn. Arc. **19**, 49–64 (2017)

24. J.W. Jang, J. Park, S. Ghosh, S. Bhunia, Self-correcting STTRAM under magnetic field attacks, in *2015 52nd ACM/EDAC/IEEE Design Automation Conference (DAC)* (IEEE, Piscataway, 2015), pp. 1–6

25. N.K. Kasabov, NeuCube: a spiking neural network architecture for mapping, learning and understanding of spatio-temporal brain data. Neur. Netw. **52**, 62–76 (2014)

26. D. Kim, J. Kung, S. Chai, S. Yalamanchili, S. Mukhopadhyay, NeuroCube: a programmable digital neuromorphic architecture with high-density 3d memory, in *2016 ACM/IEEE 43rd Annual International Symposium on Computer Architecture (ISCA)* (IEEE, Piscataway, 2016), pp. 380–392

27. P. Kocher, J. Horn, A. Fogh, D. Genkin, D. Gruss, W. Haas, M. Hamburg, M. Lipp, S. Mangard, T. Prescher, M. Schwarz, Y. Yarom, Spectre attacks: exploiting speculative execution. Commun. ACM **63**(7), 93–101 (2020)

28. M. Lipp, M. Schwarz, D. Gruss, T. Prescher, W. Haas, A. Fogh, J. Horn, S. Mangard, P. Kocher, D. Genkin, Y. Yarom, M. Hamburg, Meltdown: Reading kernel memory from user space, in *27th USENIX Security Symposium, USENIX Security 2018, Baltimore, MD, August 15–17, 2018* (2018), pp. 973–990

29. B. Liu, C. Wu, H. Li, Y. Chen, Q. Wu, M. Barnell, Q. Qiu, Cloning your mind: Security challenges in cognitive system designs and their solutions, in *Proceedings of the 52Nd Annual Design Automation Conference* (ACM, New York, 2015), pp. 95:1–95:5 https://doi.org/10.1145/2744769.2747915

30. X. Liu, M. Mao, B. Liu, H. Li, Y. Chen, B. Li, Y. Wang, H. Jiang, M. Barnell, Q. Wu, et al., Reno: A high-efficient reconfigurable neuromorphic computing accelerator design, in 2015 52nd ACM/EDAC/IEEE Design Automation Conference (DAC) (IEEE, Piscataway, 2015), pp. 1–6

31. B. Liu, C. Yang, H. Li, Y. Chen, Q. Wu, M. Barnell, Security of neuromorphic systems: Challenges and solutions, in *2016 IEEE International Symposium on Circuits and Systems (ISCAS)* (IEEE, Piscataway, 2016), pp. 1326–1329

32. C. Liu, Q. Dong, F. Yu, X. Chen, Rerise: An adversarial example restoration system for neuromorphic computing security, in *2018 IEEE Computer Society Annual Symposium on VLSI, ISVLSI 2018, Hong Kong, July 8–11, 2018* (2018), pp. 470–475

33. C. Luo, Z. Ying, X. Zhu, L. Chen, A mixed-signal spiking neuromorphic architecture for scalable neural network, in *2017 9th International Conference on Intelligent Human-Machine Systems and Cybernetics (IHMSC)*, vol. 1 (IEEE, Piscataway, 2017), pp. 179–182

34. Y. Luo, L. Wan, J. Liu, J. Harkin, L. McDaid, Y. Cao, X. Ding, Low cost interconnected architecture for the hardware spiking neural networks. Front. Neurosci. **12**, 857 (2018)

35. Z.D. Luo, M.M. Yang, M. Alexe, Dissolvable memristors for physically transient neuromorphic computing applications. ACS Appl. Electr. Mater. **2**(2), 310–315 (2019)

36. P. Merolla, J. Arthur, F. Akopyan, N. Imam, R. Manohar, D.S. Modha, A digital neurosynaptic core using embedded crossbar memory with 45pj per spike in 45nm, in *2011 IEEE Custom Integrated Circuits Conference (CICC)* (IEEE, Piscataway, 2011), pp. 1–4
37. Ş. Mihalaş, E. Niebur, A generalized linear integrate-and-fire neural model produces diverse spiking behaviors. Neur. Comput. **21**(3), 704–718 (2009)
38. M.B. Milde, H. Blum, A. Dietmüller, D. Sumislawska, J. Conradt, G. Indiveri, Y. Sandamirskaya, Obstacle avoidance and target acquisition for robot navigation using a mixed signal analog/digital neuromorphic processing system. Front. Neurorob. **11**, 28 (2017). https://doi.org/10.3389/fnbot.2017.00028. https://doi.org/10.5167/uzh-149387
39. C.P. Newswire, Usdollar 1.6 Bn global neuromorphic chip market to witness strong growth in North America (2017). https://www.prnewswire.com/news-releases/us-16-bn-global-neuromorphic-chip-market-to-witness-strong-growth-in-north-america-610928905.html
40. E. Orhan, The leaky integrate-and-fire neuron model. no **3**, 1–6 (2012). [Online]. Available: http://www.cns.nyu.edu/~eorhan/notes/lif-neuron.pdf
41. E. Painkras, L.A. Plana, J. Garside, S. Temple, F. Galluppi, C. Patterson, D.R. Lester, A.D. Brown, S.B. Furber, Spinnaker: A 1-w 18-core system-on-chip for massively-parallel neural network simulation. IEEE J. Solid-State Circ. **48**(8), 1943–1953 (2013)
42. N. Papernot, P.D. McDaniel, I.J. Goodfellow, Transferability in machine learning: from phenomena to black-box attacks using adversarial samples. CoRR **abs/1605.07277** (2016). http://arxiv.org/abs/1605.07277
43. Q. Qiu, Z. Li, K. Ahmed, W. Liu, S.F. Habib, H.H. Li, M. Hu, A neuromorphic architecture for context aware text image recognition. J. Signal Proc. Syst. **84**(3), 355–369 (2016)
44. C. Rajamanikkam, R. JS, S. Roy, Chakraborty, K.: Understanding security threats in emerging neuromorphic computing architecture. https://engineering.usu.edu/ece/faculty-sites/bridge-lab/
45. S.G. Ramasubramanian, R. Venkatesan, M. Sharad, K. Roy, A. Raghunathan, Spindle: Spintronic deep learning engine for large-scale neuromorphic computing, in *Proceedings of the 2014 International Symposium on Low Power Electronics and Design* (ACM, New York, 2014), pp. 15–20
46. N. Rathi, S. Ghosh, A. Iyengar, H. Naeimi, Data privacy in non-volatile cache: Challenges, attack models and solutions, in *2016 21st Asia and South Pacific Design Automation Conference (ASP-DAC)* (IEEE, Piscataway, 2016), pp. 348–353
47. G. Rovere, Q. Ning, C. Bartolozzi, G. Indiveri, Ultra low leakage synaptic scaling circuits for implementing homeostatic plasticity in neuromorphic architectures, in *IEEE International Symposium on Circuits and Systems (ISCAS)* (2014), pp. 2073–2076
48. V. Roychowdhury, D. Janes, S. Bandyopadhyay, X. Wang, Collective computational activity in self-assembled arrays of quantum dots: a novel neuromorphic architecture for nanoelectronics. IEEE Trans. Electr. Dev. **43**(10), 1688–1699 (1996)
49. I.K. Schuller, R. Stevens, R. Pino, M. Pechan, Neuromorphic computing–from materials research to systems architecture roundtable. Technical Reprt, USDOE Office of Science (SC)(United States) (2015)
50. C.D. Schuman, T.E. Potok, R.M. Patton, J.D. Birdwell, M.E. Dean, G.S. Rose, J.S. Plank, A survey of neuromorphic computing and neural networks in hardware. CoRR **abs/1705.06963** (2017). http://arxiv.org/abs/1705.06963
51. R.J. Shridevi, D.M. Ancajas, K. Chakraborty, S. Roy, Runtime detection of a bandwidth denial attack from a rogue network-on-chip, in *ACM/IEEE International Symposium on Networks-on-Chip (NOCS)* (2015), pp. 8:1–8:8
52. D. Soudry, D. Di Castro, A. Gal, A. Kolodny, S. Kvatinsky, Memristor-based multilayer neural networks with online gradient descent training. IEEE Trans. Neur. Netw. Learn. Syst. **26**(10), 2408–2421 (2015)
53. V. Sze, Y.H. Chen, J. Einer, A. Suleiman, Z. Zhang, Hardware for machine learning: challenges and opportunities, in *Custom Integrated Circuits Conference (CICC)* (IEEE, Piscataway, 2017), pp. 1–8

54. C. Szegedy, W. Liu, Y. Jia, P. Sermanet, S.E. Reed, D. Anguelov, D. Erhan, V. Vanhoucke, A. Rabinovich, Going deeper with convolutions, in *IEEE Conference on Computer Vision and Pattern Recognition, CVPR 2015, Boston, MA, June 7–12, 2015* (2015), pp. 1–9
55. M. Tehranipoor, H. Salmani, X. Zhang, Hardware Trojan detection: Untrusted third-party IP cores, in *Integrated Circuit Authentication* (Springer, Berlin, 2014), pp. 19–30
56. M.C. Van Rossum, G.Q. Bi, G.G. Turrigiano, Stable Hebbian learning from spike timing-dependent plasticity. J. Neurosci. **20**(23), 8812–8821 (2000)
57. S. Venkataramani, A. Ranjan, K. Roy, A. Raghunathan, AxNN: Energy-efficient neuromorphic systems using approximate computing, in *Proceedings of the 2014 International Symposium on Low Power Electronics and Design* (ACM, New York, 2014), pp. 27–32
58. K.Wang, H.Zheng, A. Louri, TSA-NoC: learning-based threat detection and mitigation for secure network-on-chip architecture. IEEE Micro **40**, 1–1 (2020)
59. S. Wozniak, A. Pantazi, S. Sidler, N. Papandreou, Y. Leblebici, E. Eleftheriou, Neuromorphic architecture with 1M memristive synapses for detection of weakly correlated inputs. IEEE Trans. Circ. Syst. II: Express Briefs **64**, 1342–1346 (2017)
60. L. Xia, B. Li, T. Tang, P. Gu, P.Y. Chen, S. Yu, Y. Cao, Y. Wang, Xie, Y., H. Yang, MNSIM: Simulation platform for memristor-based neuromorphic computing system. IEEE Trans. Comput. Aided Design Integr. Circ. Syst. **37**, 1009–1022 (2017)
61. M.A. Zidan, H.A.H. Fahmy, M.M. Hussain, K.N. Salama, Memristor-based memory: The sneak paths problem and solutions. Microelectr. J. **44**(2), 176–183 (2013)

Chapter 11
Homomorphic Encryption

Mehdi Sadi

11.1 Introduction

As we enter the era of the connected world of Internet-of-Everything and social media, our dependence on cloud-based facilities for data sharing and storage is at an all-time high. Moreover, to harness the full potentials of Artificial Intelligence and Machine Learning assisted data analytics, individuals as well as businesses are increasingly utilizing cloud-based computing facilities. However, the cloud-based facilities are not necessarily trusted, or secure against malicious hacking. As a result of the tremendous traffic of private data between users and the cloud, the need for secure encryption is more than ever. An encryption scheme for secure and privacy-preserving trusted storage and computation in untrusted locations has long been sought [1, 2]. In this regard, homomorphic encryption which allows processing on private encrypted data (without decryption) has the potential to be the ultimate solution of data privacy in cloud computing and storage [3–7]. Broadly, the term Homomorphism means similarity of form. However, it has been widely used in different areas of science and mathematics and has its own meaning in a particular context, albeit they refer to a somewhat similar concept in all aspects. In abstract algebra, homomorphism refers to the special correspondence between the two systems (sets, or fields) preserving a similar structure, while having different elements and operations. In cryptography, the term homomorphic encryption is an encryption type where the user can analyze and manipulate (in a meaningful way) the encrypted data without the need to decrypt the data. Meaningful manipulation means any computable functions, such as algebraic operation (addition, multiplication), query (search, comparison) operations [3–7]. Contrary to other forms of

M. Sadi (✉)
Auburn University, Auburn, AL, USA
e-mail: mzs0190@auburn.edu

© The Author(s), under exclusive license to Springer Nature Switzerland AG 2021
M. Tehranipoor (ed.), *Emerging Topics in Hardware Security*,
https://doi.org/10.1007/978-3-030-64448-2_11

281

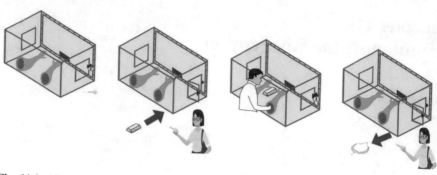

Fig. 11.1 Alice's workers are working within a glovebox

encryption, homomorphic encryption uses an algebraic system to allow functions to be performed on the data while it is still encrypted [1, 2]. Simply stated, encryption is a technique to preserve the privacy and secrecy of sensitive information. "Ordinary" encryption mechanisms, such as (AES), do not let anyone, even the owner of the data, to perform any complex operation on the encrypted data without first decrypting [8, 9]. It seems fine as long as the owner of the data is manipulating it, as he/she has access to the private key, and decrypting his/her data will not hamper any trust issue. However, it becomes a non-trivial issue when the owner of the data is not capable of processing the data, but wants someone else, while still maintaining the secrecy of data, to process it for him/her [10, 11]. The concept and motivation of homomorphic encryption can be simply illustrated with the famous story of Alice's Jewelry store given by Craig Gentry [10], and depicted in Fig. 11.1. Alice owns a jewelry store. She wants her workers to turn precious raw materials—gold, diamond, silver, etc.—into intricate rings or necklaces. But, she does not trust her workers. In other words, she wants her workers to process the raw materials into a complete piece of jewelry without giving them physical access to the raw materials. To deal with such a situation, she came up with the plan of a transparent impenetrable glovebox, those have locks on them for which only she has the keys. She puts the raw materials inside the boxes, locks those with her key, and gives them to the workers. The worker can stick his/her hand inside the gloves and manipulate the raw materials inside and create the rings and necklaces. Alice finally unlocks her box to pull out the finished ring or jewelry. Thus, the worker processed the raw ingredients into a finished product, without ever having true access to the ingredients.

Encryption techniques are designed to preserve the privacy, security, confidentiality, integrity, and authenticity of data. Any encryption scheme is usually constituted of three algorithms: (1) Key Generation, (2) Encryption, and (3) Decryption. Depending on the availability of the public key, encryption methods can be classified broadly into two classes: symmetric encryption and asymmetric encryption schemes. In a symmetric, or secret key, encryption scheme, a single key is generated by the Key Generation algorithm which is then used in both Encryption and Decryption phases. The encryption phase uses the key to map a message to a ciphertext, while the decryption phase uses it to map the ciphertext back to the message. In the case of asymmetric, or public key, encryption scheme, the Key Generation phase produces two keys—a public encryption key pk, which is made available to everyone, and a secret decryption key sk, which is only available to someone who is supposed to have it [5, 8]. From the security perspective, there are two types of encryption schemes, deterministic and probabilistic encryption schemes. Deterministic encryption means for a fixed encryption key, a given plaintext will always be encrypted into the same ciphertext under these systems. This leads to some security problems due to the following reasons [8]: (1) a particular plaintext may be encrypted in a too much structured way. For example, in the case of RSA, messages 0 and 1 are always encrypted as 0 and 1, respectively. (2) In some deterministic schemes, such as RSA, the ciphertext generally leaks single bit information about the plaintext through the Jacobi symbol [12]. (3) It is easy to detect the plaintext when the same message is sent twice while processed with the same key. On the other hand, probabilistic encryptions, free of these limitations, are therefore our preferred choice. In probabilistic encryption schemes, the ciphertexts are larger than plain texts, i.e., plaintexts m are encoded into ciphertext c where $c > m$. This phenomenon is known as an expansion [5, 8].

11.2 Defining Homomorphic Encryption

In addition to basic *KeyGen, Encrypt, Decrypt* algorithms, the homomorphic encryption has an additional algorithm *Evaluate*, which evaluates a permitted set of functions F over the cipher texts [2, 3]. For any functions f in F and cipher text $c_1, c_2, \ldots, c_t$ that was encrypted from $m_1, m_2, \ldots, m_t$ with key pk, the algorithm $Evaluate(pk, f, c_1, \ldots c_t)$ outputs a cipher text c such that $Decrypt(sk, c) = f(m_1, \ldots, m_t)$, sk is the secret key. It should be noted that, for a function f not in F, the above scenario might not hold true, i.e., there is no guarantee that Evaluate will output anything meaningful for the functions f which are not enlisted in the permitted function list F [3–5, 8]. In other words, homomorphic encryption can be interpreted as an encryption scheme which encrypts the messages $m_1, m_2, \ldots, m_t$ into cipher texts $c_1, c_2, \ldots, c_t$ under some key, and given any efficiently computable function f, such that anyone (not just the key holder) can compute a ciphertext (or set of ciphertexts) that encrypts $f(m_1, m_2, \ldots, m_t)$ under that key without leaking any information about $m_1, m_2, \ldots, m_t$ or $f(m_1, m_2, \ldots, m_t)$ [10].

Homomorphic Encryption is often defined in additive and multiplicative categories [3, 10]. Assume a cryptosystem C with encryption function E encrypts plain text m_n into ciphertext c_n (i.e., $E(m_n) = c_n$). Additive and Multiplicative homomorphism are defined as:

Additive Homomorphic: For some operation Δ, C is called additive homomorphic if and only if,

$$\exists \Delta \ E(m_1) \Delta E(m_2) = E(m_1 + m_2) \tag{11.1}$$

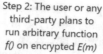

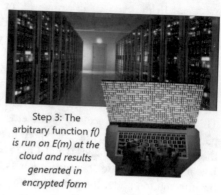

Step 2: The user or any third-party plans to run arbitrary function f() on encrypted E(m)

Step 1: User encrypts the data (m) and sends the encrypted cyphertext E(m) to cloud

Step 4: Encrypted results from cloud is sent back to the user who can decrypt it

Step 3: The arbitrary function f() is run on E(m) at the cloud and results generated in encrypted form

Fig. 11.2 Homomorphic encryption and execution of functions on encrypted ciphertext

Multiplicative Homomorphic: For some operation Δ, C is called multiplicative homomorphic if and only if,

$$\exists \Delta \ E(m_1) \ \Delta \ E(m_2) = E(m_1 \times m_2) \tag{11.2}$$

A homomorphic encryption scheme can either be symmetric or asymmetric, and deterministic or probabilistic. The asymmetric and probabilistic homomorphic encryption schemes are more desirable, as the asymmetric ones are more secured and can allow multiple algebraic functionalities than their symmetric and determin-istic counterpart [2, 8–14]. At high-level, the workflow of homomorphic encryption can be segmentized into 4 steps as illustrated in Fig. 11.2.

Step 1: User encrypts her data using any homomorphic encryption scheme and sends it to cloud. In Alice's jewelry store analogy, this step can be mapped to the step where Alice puts the raw materials inside the box and locks it.

Step 2: If the user or some authorized party wants the cloud server to perform any function $f()$, such as query over her data, the function is sent to the server. This step represents the step where Alice distributes her locked gloveboxes to the workers.

Step 3: Server performs the operation of function $f()$ over the encrypted data through the evaluate algorithm, computes $f(c_1, c_2 \ldots, c_l)$, and sends it back to the user. In this step of Alice's story, the workers are working hard to turn the raw materials into jewelry and return those back to Alice after completing their tasks.

Step 4: User receives the encrypted function-output and decrypts its equivalent plaintext using the secret key, sk. In Alice's jewelry store analogy, finally, Alice receives the gloveboxes finished jewelry inside and unlocks the box to pull those out.

11.3   Types of Homomorphic Encryption

Over the years, the algorithms of homomorphic encryption schemes had been kept improving in terms of supported operation over the encrypted ciphertexts. Homomorphic encryptions can be classified into three types considering the operations (how many and to what level of complexity) it can support [3, 4].

11.3.1   Partially Homomorphic Encryption

Partial homomorphic encryption (PHE) is the earliest form of homomorphic encryption which only allows a single operation (addition or multiplication) over the ciphertexts [1, 15, 16]. Although the researchers have already reached the milestone of Somewhat Homomorphic Encryption (SWHE) and Fully Homomorphic Encryption (FHE) [10], PHEs laid the base of homomorphic encryption schemes. Some of the PHE schemes from the past that made significant contributions in the growth of FHE are discussed in the following.

11.3.1.1   RSA

RSA Encryption Algorithm [17] is the first public-key cryptosystem [18]. It is a deterministic cryptosystem. Interestingly, RSA was not developed as a PHE, rather it was developed as a public-key cryptosystem and later its homomorphism property was realized in [1]. The security of RSA encryption algorithm rests on the part of factoring two large prime numbers [3, 17].

KeyGeneration Algorithm: For large prime numbers p and q, compute $n = p * q$, and $\phi = (p - 1) * (q - 1)$; choose e such that $e = gcd(e, \phi)$; calculate d, where $e * d = 1 \, mod \, \phi$ (d is the multiplicative inverse of e); release (e, n) as public-key pair and keep (d, n) as secret-key pair.

Encryption Algorithm: Encryption algorithm is divided into two steps: first, represent the message as an integer between 0 and $(n - 1)$ using any standard representation. The purpose of this step is to break the long message as a series of blocks and to represent each block as an integer to make it ready to encrypt. Second, encrypt the message by rising it to the eth power of modulo n. Mathematically, $c = E(m) = m^e \, (mod \, (n))$, c represents ciphertext and m represents message.

Decryption Algorithm: Recover the message m from the ciphertext c using the secret-key pair (d, n) as follows:

$$m = D(c) = c^d \, (mod \, (n)) \tag{11.3}$$

Homomorphic Property of RSA: Let us consider, $m_1, m_2 \in M$

$$E(m_1) * E(m_2) = \left(m_1^e(mod\ n)\right) * \left(m_2^e(mod\ n)\right) = (m_1 * m_2)^e\ (mod\ n)$$
$$= E(m_1 * m_2)$$

(11.4)

The above equation implies that RSA is multiplicatively homomorphic which means it can compute the multiplication operation over the ciphertexts.

11.3.1.2 Goldwasser-Micali Scheme

This scheme [19] is considered one of the pioneer of homomorphic encryption scheme as many of the subsequent HE algorithms were inspired by this scheme [8, 9]. It is the first probabilistic public-key cryptosystem. The hardness of *quadratic residuosity problem* decides the security of the GM cryptosystem [20]. Any number a is said to be quadratic residue modulo n if there exists an integer x such that $x^2 \equiv a\ (mod(n))$. The quadratic residuosity problem decides whether a given number q is quadratic modulo n or not [3, 8].

KeyGeneration Algorithm: Compute $n = p * q$ for two large prime numbers p and q; choose x as one of the quadratic nonresidue modulo n values with $(x/n) = 1$; Publish (x, n) as public key and keep (p, q) as secret key.

Encryption Algorithm:

Step 1:   Convert message m into a string of bits; for each bit of message m_i,
          produce a quadratic nonresidue value y_i such that $gcd(y_i, n) = 1$.
Step 2:   Encrypt each bit into c_i as follows:

$$c_i = E(m_i) = y_i^2 x^{m_i}(mod\ n), \forall m_i = \{0, 1\}$$

(11.5)

where $m = m_0, m_1 \ldots m_t$; $c = c_0, c_1 \ldots c_t$, and t is the block size used for the message space and x is picked from $\mathbb{Z}_n^*$ at random for every encryption, where $\mathbb{Z}_n^*$ is the multiplicative subgroup of integers modulo n which includes all the numbers smaller than t and relatively prime to t [3, 19].

Decryption Algorithm: To decrypt ciphertext c_i, check if c_i is a *quadratic residue modulo n* or not; if so, m_i returns 0, otherwise m_i returns 1. (As x is picked from the set $\mathbb{Z}_n^*$ $(1 < x \leq n - 1)$, x is *quadratic residue modulo n* for only $m_i = 0$)

Homomorphic Property: For bits $m_i \in \{0, 1\}$,

$$E(m_1) * E(m_2) = \left(y_1^2 x^{m_1}(mod\ (n)) * y_2^2 x^{m_2}(mod\ (n))\right)$$
$$= (y_1 * y_2)^{m_1+m_2}(mod\ (n)) = E(m_1 + m_2)$$

(11.6)

The above equation shows that it is possible to calculate the encrypted sum $E(m_1 \oplus m_2)$ from the encrypted message $E(m_1)$ and $E(m_2)$. This implies that this scheme is additively homomorphic. As GM only deals with the binary numbers, XOR can be used to represent the addition [21].

11.3.1.3   ElGamal Method

This scheme proposed by ElGamal in [15] is the improvement of original public-key cryptosystem [18] whose security is based on the difficulty of computing discrete logarithms over finite fields [22]. This system was described only over Galois Field $GF(p)$. The size of public key of this system rises with the size of the messages as the public-key system can be easily extended to any $GF(p^m)$ [15].

KeyGeneration Algorithm: Create a cyclic group G with order n from the generator g; Compute $h = g^y$, where y is randomly chosen from $\mathbb{Z}_n^*$; Publish (G, n, g, h) as public key and keep x is the secret key, where x is randomly chosen from the set $\{0, 1, 2, \ldots, n - 1\}$.

Encryption Algorithm: Encrypt the plaintexts m into the ciphertext $c = (c_1, c_2)$ using the following equation:

$$c = E(m) = (g^x, mh^x) = (g^x, mg^{xy}) = (c_1, c_2) \tag{11.7}$$

Decryption Algorithm: Compute $s = c_1^y$ using the secret key, y; compute $c2$ using the following equation:

$$c_2.s^{-1} = mg^{xy}.g^{-xy} = m \tag{11.8}$$

Homomorphic Property: The multiplicative homomorphic property of ElGamal cryptosystem can be realized from the following equation [3]:

$$E(m_1) * E(m_2) = (g^{x_1}, m_1 h^{x_1}) * (g^{x_2}, m_2 h^{x_2})$$
$$= (g^{x_1+x_2}, m_1 * m_2 h^{x_1+x_2}) = E(m_1 * m_2) \tag{11.9}$$

11.3.1.4   Benaloh Scheme

This method proposed by Benaloh is a dense probabilistic encryption model which was similar to GM scheme [23]. However, it had many advantages over GM scheme [19]. This scheme was based on the higher residuosity problem x^n, which is the generalization of quadratic residuosity problem (x^2) used in GM scheme [3, 20].

KeyGeneration Algorithm: Choose block-size r, large prime number p and q such that $gcd(r, (p - 1)/r) = 1$ and $gcd(r, (q - 1)) = 1$; compute $n = pq$ and $\phi = (p - 1)(q - 1)$; $y \in \mathbb{Z}_n^*$ is chosen such that $y^\phi \neq 1 \mod n$, where $\mathbb{Z}_n^*$ is the multiplicative subgroup of integers modulo n that includes all the numbers smaller than r and relatively prime to r [15]. Publish (y,n) as public key, and keep (p,q) as the secret key.

Encryption Algorithm: u is chosen randomly such that $u \in \mathbb{Z}_n^*$, for the message $m \in \mathbb{Z}_r$, where $\mathbb{Z}_r = \{0, 1, \ldots, r - 1\}$; encrypt message m into cyphertext c using the equation [3]:

$$c = E(m) = y^m u^r (mod \; n) \tag{11.10}$$

Decryption Algorithm: m is recovered by an exhaustive search for $i \in \mathbb{Z}_r$ such that

$$(y^{-i}c)^{\phi/r} \equiv 1 \tag{11.11}$$

where $m = i$, i.e., the message corresponds to the value of i.

Homomorphic Property:

$$E(m_1) * E(m_2) = \left(y^{m_1}u_1^r (mod \; n)\right) \times \left(y^{m_2}u_2^r (mod \; n)\right)$$
$$= y^{m_1+m_2}(u1 \times u2)^r (mod \; n) = E(m_1 + m_2 \; (mod \; n)) \tag{11.12}$$

The equation shows that multiplication of the ciphertexts is equivalent to the addition in plaintext, which means, Benaloh scheme is additively homomorphic [3].

11.3.1.5 Okamoto-Uchiyama Scheme

This scheme is also a public-key probabilistic cryptosystem which employs polynomial time algorithms for solving discrete logarithm over a specific finite group [24].

KeyGeneration Algorithm: Choose large prime numbers p and q; compute $n = p^2q$; Choose $g \in (\mathbb{Z}/n\mathbb{Z})^*$ randomly such that the order of $g^{p-1} \neq 1 \, mod \, p^2$. Here, $gcd(p, q - 1) = 1$ and $gcd(q, p - 1) = 1$. Compute $h = g^n \, mod \, n$, where h is a supplementary parameter for improving the efficiency of encryption, since h can be easily calculated from g and n. Publish (n, g, h, k) as public key and store (p, q) as secret key [24].

Encryption Algorithm: For any plaintext m $(0 < m < 2^{k-1})$, select random integer $r \in \mathbb{Z}/n\mathbb{Z}$. Encrypt plaintexts using the following equation:

$$C = g^m h^r \, mod \, n \tag{11.13}$$

Decryption Algorithm: Recover plain texts using:

$$M = \frac{L\left(C^{p-1} \bmod p^2\right)}{L\left(g^{p-1}\right)} \bmod p \tag{11.14}$$

Homomorphic Property: The homomorphic property of this scheme can be realized from the equation: $E(m_0, r_0)E(m_1, r_1) \bmod n = E(m_0 + m_1, r_2)$, if $m_0 + m_1 < p$. Such homomorphic property is useful in electronic voting [24].

11.3.1.6 Paillier Scheme

This probabilistic public-key encryption system [16] was based on composite residuosity problem [3, 25].

KeyGeneration Algorithm: Choose large prime numbers p and q such that $gcd(pq, (p-1)(q-1)) = 1$; compute $n = pq$ and $\lambda = lcm(p-1, q-1)$; select a random integer $g \in \mathbb{Z}_{n^2}^*$ by checking if $gcd(n, L(g^{\lambda \bmod n^2})) = 1$, where L is defined as $L(u) = (u-1)/n$ for every u from the subgroup $\mathbb{Z}_{n^2}^*$ that is a multiplicative subgroup of integers modulo n^2 instead of n as in the Benaloh cryptosystem. In this scheme, the public-key pair is (n, g) and the secret-key pair is (p, q).

Encryption Algorithm: Choose r randomly and encrypt m into c using the following equation [3]:

$$c = E(m) = g^m r^n \left(\bmod n^2\right) \tag{11.15}$$

Decryption Algorithm: For any ciphertext $c < n^2$, recover the message using the equation [3]:

$$D(c) = \frac{L\left(c^\lambda \left(\bmod n^2\right)\right)}{L\left(g^\lambda \left(\bmod n^2\right)\right)} (\bmod n = m) \tag{11.16}$$

Where (p, q) is the private key pair.

Homomorphic Property: For any multiplication operation over the ciphertext, Paillier's encryption scheme is homomorphic over addition [3].

$$E(m_1) * E(m_2) = \left(g^{m_1} r_1^n \left(\bmod n^2\right)\right) * \left(g^{m_2} r_2^n \left(\bmod n^2\right)\right)$$

$$= g^{m_1+m_2}(r_1 * r_2)^n \left(\bmod n^2\right) = E(m_1 + m_2) \tag{11.17}$$

11.3.1.7 Damgard-Jurik Scheme

This [26] is the generalization of above explained Paillier scheme which has better (reduced) expansion factor [27]. The DJ public-key cryptosystem we use computations modulo n^{s+1} where n is an RSA modulus and s is a natural number. In case of Paillier scheme the value of s was 1 [26].

KeyGeneration: Choose two large odd prime number p and q; compute $n = pq$ and $\lambda = lcm(p-1, q-1)$; choose $g \in \mathbb{Z}^*_{n^{s+1}}$ such that $g = (1+n)^j x \bmod n^{s+1}$ for a known j relative prime to n and $x \in \mathbb{H}$, where $\mathbb{H}$ is isomorphic to $\mathbb{Z}^*_n$; choose d such that $d \bmod n \in \mathbb{Z}^*_n$ and $d = 0 \bmod \lambda$. (In case of Paillier scheme, $d = \lambda$); publish (n, g) as public key and store d as secret-key.

Encryption: Any message $m \in \mathbb{Z}_{n^s}$ can be encrypted into c by the following equation: $C = g^m r^{n^s} \bmod n^{s+1}$, where r is randomly chosen from $r \in \mathbb{Z}^*_{n^{s+1}}$

Decryption: For any ciphertext $c \in \mathbb{Z}^*_{n^{s+1}}$. Compute $c^d \bmod n^{s+1}$. Then the recursive version of Paillier Decryption mechanism is applied to compute jmd. As jd is known m can be retrieved by $m = (jmd) \cdot (jd)^{-1} \bmod n^s$.

Homomorphic Property: This scheme is Homomorphic over addition [3].

11.3.1.8 Others PHE Methods

Other researchers also proposed PHE schemes that various performance trade-offs. A review of other PHE schemes is available in [3–5]

11.3.2 Somewhat Homomorphic Encryption (SWHE)

Somewhat homomorphic encryption system is the precursor to the fully homomorphic encryption. From the perspective of developing algorithms, many of the state-of-the-arts FHE begin by constructing an SWHE and are then modified [2].

11.3.2.1 Sander's Method

In 1999 Sander et al. [28] in an attempt to reduce the number of communication rounds between the sender and the receiver of encrypted data designed a protocol for secure evaluation of circuits which is in polynomial time for NC^1 circuits. This protocol was also found to be homomorphic over a semigroup and expanded the range of algebraic structures which can be encrypted homomorphically. The scheme allows the computation of many logical AND operation along with one OR/NOT operation on encrypted data, where the ciphertext size increased by a constant multiplication with each OR/NOT gate. This phenomenon limits the evaluation of circuit depth [28]. In [29] a technique was developed which can result in a one-round secure protocol which provides unconditional security over server. It had a significant improvement over the previous protocols in terms of complexity.

11.3.2.2 Boneh's Method

In 2005, Boneh et al. [30] provided a public-key encryption scheme which can evaluate quadratic multivariate polynomials on ciphertext providing the resulting values falls within a small set. It was based on finite groups of composite order that supports a bilinear map. Its similarity with the Paillier scheme makes it additively homomorphic [3]. Additionally, the bilinear map property supports for one multiplication on encrypted values. Thus, this system supports one multiplication followed by arbitrary additions on encrypted data resulting in the evaluation of multivariate polynomials of total degree 2 on encrypted values. The subgroup decision problem decides the security of the scheme—given an element of a group of composite order $n = q_1 q_2$, it is infeasible to decide whether it belongs to a subgroup of order q_1 [30].

KeyGen Algorithm: Public key is (n, G, G_1, e, g, h).
     Where e is a bilinear map so that $e : G \times G \rightarrow G_1$, where G, G_1 are groups of order $n = q_1 q_2$, g and u are the generators of G and set $h = u^{q_2}$ and h is the generator of G with order q_1. Keep q_1 as secret-key.

Encryption Algorithm: Choose r randomly from set $\{0, 1, 2, \ldots, n-1\}$; encrypt the message as follows:

$$c = E(m) = g^m h^r \bmod n \tag{11.18}$$

Decryption Algorithm: Compute $c' = c^{q_1}$ and $g' = g^{q_1}$; Retrieve the message as follows:

$$m = D(c) = log_{g'} c' \tag{11.19}$$

It should be noted that computation time of decryption algorithm increases with the size of the message.

Homomorphic Property:

$$(E(m_1) * E(m_2))h^r = \left(g^{m_1} h^{r_1}\right) \left(g^{m_2} h^{r_2}\right) h^r = g^{m_1 + m_2} h^{r'} \tag{11.20}$$

where $r = r_1 + r_2 + r$. Above equation reveals that it is homomorphic over arbitrary number of additions. To perform multiplication over the ciphertext, consider g_1 with order n and h_1 with order q_1; set $g_1 = e(g, g)$, $h_1 = e(g, h)$, and $h = g^{\alpha q_2}$. The homomorphic multiplication of messages m_1 and m_2 using the ciphertexts $c_1 = E(m_1)$ and $c_2 = E(m_2)$ are computed as follows:

$$c = e(c_1, c_2)h_1^r \tag{11.21}$$

$$= e\left(g^{m_1}h^{r_1}, g^{m_2}h^{r_2}\right)h_1^r \tag{11.22}$$

$$= g_1^{m_1 m_2} h_1^{m_1 r_2 + r_2 m_1 + \alpha q_2 r_1 r_2 + r} \tag{11.23}$$

$$= g_1^{m_1 m_2} h_1^{r'} \tag{11.24}$$

From the equation it is seen that now c belongs to the group of $\mathbb{G}_1$ instead of $\mathbb{G}$, which implies that no more homomorphic multiplication is allowed in $\mathbb{G}_1$, as there is now pairing from the set $\mathbb{G}_1$. Thus, it has only one homomorphic multiplication function over the ciphertexts. However, this does not affect the multiple homomorphic additions over the ciphertexts [3].

11.3.2.3 Polly Cracker type Cryptosystem

In [31] a polynomial-based public-key cryptosystem, known as Polly Cracker, was proposed. It was one of first proposed SWHE system which allowed both multiplications and additions over ciphertext [3]. However, it was not efficient in terms of the length of ciphertexts. As it is a polynomial-based system, computing an encryption of, $E(m_1 * m_2)$, the product of two messages m_1 and m_2 by multiplying the corresponding ciphertext polynomials $E(m_1)$ and $E(m_2)$ leads to an exponential blow-up in the total number of monomials. Further research revealed that they are either insecure since they are vulnerable to certain attacks, being too inefficient to be practically implemented, or they lose their algebraic homomorphic property [3]. The above discussion suggests that ciphertext size is constant only in case of Boneh scheme [30], while in others it grows either exponentially or linearly. This property of Boneh scheme is considered a cornerstone of designing FHE schemes [3].

11.3.3 Fully Homomorphic Encryption

The ultimate purpose of homomorphic encryption—to be able to run arbitrary functions on encrypted ciphertext—is basically served by Fully Homomorphic Encryption (FHE) schemes. In FHE, any efficiently computable function can operate on the ciphertext while still ensuring security and privacy. After 30 years of the introduction of the idea of privacy homomorphism [1], the first convincing homomorphic encryption scheme, which was able to perform any arbitrary operations over the ciphertexts, was proposed in 2009 by Craig Gentry in his seminal PhD thesis [2, 10]. Although this work laid the foundation of FHE, it was not enough computationally efficient to be practically implemented in the real-world. Following his general framework for achieving FHE, multiple other researchers have contributed to this domain by improving the efficiency and security of FHE [4, 5]. The field is improving gradually; however, the real-world compatible protocol is yet to be designed. Before we jump into Gentry's FHE scheme, it is good to have some understanding and history about ideal lattice [3].

11.3.3.1   Ideal Lattice and Lattice-Based cryptosystems

A lattice can be defined as the linear combination of basis vectors $b_1, b_2, \ldots, b_n$, which can be formulated as:

$$L = \sum_{i=1}^{n} \mathbf{b_i} * v_i, \quad v_i \in \mathbb{Z} \qquad (11.25)$$

where $b_1, b_2, \ldots, b_i$ is a basis of the lattice L. A lattice contains infinitely many bases. A basis is good if the basis vectors are almost orthogonal, while the basis is bad for the rest. Lattice theory, proposed by Minkowski [32], has become a very attractive foundation for cryptography over the last two decades [3, 33]. The appeal of lattice-based schemes stems from the fact that their security is based on the assumption of worst-case hardness, as a result, they may remain secure against quantum computers [33]. Two important problems related to lattice-based cryptosystems—Closest Vector Problem (CVP) and Shortest Vector Problem (SVP)—were suggested in [34]. In [35] a public-key encryption scheme was proposed, whose security was based on the hardness of lattice reduction problem [35]. In this scheme, the public key and secret key are selected from the good and bad bases of the lattice, respectively. Lattice reduction tries to find a good basis for a given lattice, because CVP and SVP problems can easily be solved in polynomial time for the lattices with the known good bases [3]. After Gentry's pioneering work on FHE [2]. Recently, Learning with error (LWE)—also a lattice-related problem—gained popularity, especially after being used as a base to build an FHE scheme [36]. An FHE approach over integers developed on the Approximate-GCD problems was presented in [37]. The conceptual simplicity of this scheme was the driving factor. Afterward, another FHE scheme whose hardness is based on Ring Learning with Error (RLWE) problems was suggested in [38]. The proposed scheme promises some efficiency features. Lastly, an NTRU-like FHE was presented for its excellent efficiency and standardization properties [3, 39]. Although the homomorphic properties of NTRU-Encrypt were recently realized, it is a well-known and strongly standardized lattice-based encryption method. In summary, research attempts on FHE can be categorized in four families [3]: (1) ideal lattice based [2], (2) over integers [37], (3) (R)LWE type [38], and (4) NTRU-like [39].

11.3.3.2 Gentry's FHE Scheme

The encryption algorithms of typical SWHE schemes introduced some noise parameters with each ciphertext. The subsequent decryption algorithm works fine if the noise is less than a certain threshold. The multiplication and addition operation works on the ciphertexts at the cost of adding and multiplying the noise parameter as well. This issue caused SWHE to support a limited number of operations over the ciphertexts. Gentry addressed this limitation of SWHE by a Recrypt algorithm which takes a ciphertext $E(a)$ with noise $N' < N$ and returns the fresh ciphertext $E(a)$ which has a noise parameter smaller than $\sqrt{N}$. Recrypt algorithm can be obtained by making the SHWE bootstrappable where squashing makes the design bootstrappable. Gentry started with an SWHE based on Ideal Lattices. Ideal lattices allowed him to construct a bootstrappable encryption scheme by reducing the multiplicative depth [10]. The properties of Ideal lattices which help make a scheme bootstrappable are: (1) the circuit complexity of decryption algorithms of typical lattice-based schemes is very low and (2) ideal lattices correspond to ideals in polynomial rings and thus inherit natural *ADD* and *Mult* operations from the ring. The primary construction of Gentry's SWHE can be described in terms of rings and ideals, where an ideal is a property-preserving subset of the ring, later the ideals were represented by the lattice. Gentry's SWHE scheme is described below [3]:

KeyGeneration: *IdealGen*(R, B_I) algorithm generates the pair of (B_J^{sk}, B_J^{pk}), where R is ring and B_I is basis of ideal I. The output of *IdealGen()* are relatively prime, public and the secret key bases of the ideal lattice with basis B_I such that $I + J = R$. This step also uses a sample algorithm *Samp()* to sample from the provided coset of the ideal, where a coset is obtained by shifting an ideal by a certain amount. Public-key set is $(R, B_I, B_J^{pk}, Samp())$ while only B_J^{sk} is kept as secret-key.

Encryption: For randomly chosen vectors $\mathbf{r}$ and $\mathbf{g}$, using the public key (basis) B_pk chosen from one of the "bad" bases of the ideal lattice L, the message $\mathbf{m} \in \{0, 1\}^n$ is encrypted by

$$\mathbf{c} = E(\mathbf{m}) = \mathbf{m} + \mathbf{r} \cdot B_I + \mathbf{g} \cdot B_J^{pk} \tag{11.26}$$

where B_I is the basis of the ideal lattice L. Here, $\mathbf{m} + r\dot{B}_I$ is called a "noise" parameter.

Decryption: The ciphertexts can be decrypted using the following equation:

$$\mathbf{m} = \mathbf{c} - B_j^{sk} \cdot \left\lfloor \left(B_j^{sk}\right)^{-1} \cdot \mathbf{c} \right\rceil mod \ B_I \tag{11.27}$$

where function $\lfloor \cdot \rceil$ returns the nearest integers for the coefficients of the vector.

Homomorphic Property: For plaintext vectors $\mathbf{m_1}, \mathbf{m_2} \in \{0, 1\}^n$, the additive homomorphic operation of this scheme can be realized by the following equation [3]:

$$\mathbf{c_1} + \mathbf{c_2} = E(\mathbf{m_1}) + E(\mathbf{m_2}) = \mathbf{m_1} + \mathbf{m_2} + (\mathbf{r_1} + \mathbf{r_2}) \cdot B_I + (\mathbf{g_1} + \mathbf{g_2}) \cdot B_j^{pk} \quad (11.28)$$

To demonstrate the multiplicative homomorphic property $\mathbf{e} = \mathbf{m} + \mathbf{r} \cdot B_I$ should be set. Then the following equation shows the multiplicative homomorphic property:

$$\mathbf{c_1} \times \mathbf{c_2} = E(\mathbf{m_1}) \times E(\mathbf{m_2}) \quad (11.29)$$

$$= \mathbf{e_1} \times \mathbf{e_2} + (\mathbf{e_1} \times \mathbf{g_2} + \mathbf{e_2} \times \mathbf{g_1} + \mathbf{g_1} \times \mathbf{g_2}) \cdot B_j^{pk} \quad (11.30)$$

If the noise $|\mathbf{e_1} \times \mathbf{e_2}|$ is small enough, the multiplication of plain texts $\mathbf{m_1} \times \mathbf{m_2}$ can be successfully recovered from the ciphertexts $\mathbf{c_1} \times \mathbf{c_2}$. The above described SWHE scheme was then converted to FHE by bootstrapping, a technique introduced by Gentry [10]. However, the bootstrapping technique can be applied to only bootstrappable ciphertexts which have small circuit depth. This is known as leveled fully homomorphic scheme where the homomorphism depends on the circuit depth. An encryption scheme (ϵ) is bootstrappable if it has the self-referential property of being able to evaluate its own (augmented) decryption circuit. To make a homomorphic scheme which is independent of circuit depth, Gentry introduced squashing, which reduces the circuit depth to the extent that the decryption can handle properly [10]. To have a clear understanding of how to modify an SWHE into a bootstrappable scheme, further into FHE, interested readers are encouraged to read chapter 1, 8, and 10 of [10].

11.3.3.3 Integer-Based FHE Schemes

In [37], a new and simple FHE scheme was proposed which also started with an SWHE scheme like Gentry's scheme [2]. This SWHE scheme used addition and multiplication over the integers rather than working with ideal lattices over a polynomial ring. The hardness of the scheme was based on the Approximate-Greatest Common Divisor (AGCD) problems [40]. Details of integer-based FHE are available in [37].

11.3.3.4 LWE Based FHE Schemes

Learning with Error (LWE) was introduced in [41] as an extension of the "learning from parity with error" problem, and is considered one of the hardest problems to solve in practical time for even post-quantum algorithms [3–5]. In this approach, the hardness of worst-case lattice problems like SVP is reduced to LWE problems, this implies that if an algorithm can solve the SVP problem in an efficient time, the same algorithm will also solve the SVP problem in an efficient time [3]. Thus, it has become one of the most promising and attractive topics for post-quantum cryptology with its relatively small ciphertext size [3–5]. In [33] the LWE was modified by introducing the ring-LWE (RLWE) problem. RLWE offers better security as well as efficient practical applications. Details are available in [3, 5].

11.3.3.5 NTRU-Like FHE Schemes

NTRU, proposed by Hoffstein et al. [42], is one of the primitive cryptosystems based on the lattice problem, but its FHE property was realized more recently [3]. NTRU significantly improves the efficiency of FHE both in hardware and software. However, its efficiency came along at the cost of security. The security of the scheme was improved by modifying the key generation algorithm [3, 43] to reduce the security vulnerability to the level of standard worst-case problems over ideal lattices. As the security of the scheme was improved, its efficiency, easy implementation, and standardization issues caught researchers' interest again [3–5]. The fully homomorphic properties of the NTRU encryption were analyzed in [39] and [13] to achieve a practical FHE scheme.

11.4 Applications of Homomorphic Encryption

This section is divided into two parts. The first part gives the examples of real-word applications, while the second part deals with the constructions whose underlying building blocks are Homomorphic encryptions schemes.

11.4.1 Consumer Privacy in Advertising

Today's business-focused but free web-based services such as social media sites, search engines, apps, and web browsers are constantly sending the users advertisements for different products and services. Although in the majority of the cases these advertisements are a source of annoyance to the user, sometimes advertising can be helpful for the user of these services when the contents of the advertisement are custom-fitted to users' tastes and necessities. A machine-learning-based recommender system can easily generate targeted advertisement content. However, for a recommender system to create content that the targeted user will be interested in, it must be trained with private data from the user such as email and text messages exchanged, watched videos, browsed sites, and places visited (for location-based advertising). However, many providers (e.g., Facebook, Google, etc.) of these type of services are concerned about the privacy of their user data. Even though they (server) want to show ads based on the contents of their users' data, they do not necessarily want to know the content exactly. Here homomorphic encryption schemes over the users' confidential data can serve the purpose. There have been several approaches that address similar types of scenarios. In [44], a similar scenario was presented where a user wants recommendations for a product. The scenario considered social networking platform where customized recommendations were created by analyzing the preferences and choices of the user's friends while ensuring strict confidentiality. This system applied homomorphic encryption to allow a user to obtain recommendations from friends without revealing the identity of the recommender. In [45] a recommender system was built, upon a very simple but highly efficient homomorphic encryption dedicated for this specific purpose, where a user gets encrypted recommendations while the system is oblivious of the content. The HE scheme allows the function to be computed on the encrypted user data and then chooses the advertisement for each user while keeping the advertisement encrypted. In another approach to targeted advertising, the provider utilized the user's mobile location and prior browsing history to send customized ads, such as discount vouchers for nearby shops [45]. By using homomorphic encryption, the user's habits and preferences can be hidden from the provider while still providing useful ads and service recommendations to the customer.

11.4.2 Medical Application

Consider the scenario, a medical clinic has record of its patients, such as age, height, weight, blood sugar level, insulin level, stored in its database. A researcher wants to do some analysis on the patients' data to come to a general conclusion about the diabetic patients of that demography. However, it is unethical for the hospital to reveal its patients' information without the patients' consent. Here they can encrypt the data using any HE schemes. In [46], a similar scenario was proposed, while [47] presented an actual implementation of a heart attack prediction by Microsoft [45].

11.4.3 Data Mining

Mining from large data sets is sometimes of great value, but the cost for this is the user's privacy [45]. In [48], functional encryption was used to mine data while preserving privacy. However, they suggested that applying homomorphic encryption is the best solution possible.

11.4.4 Forensic Image Recognition

There are many potential applications of homomorphic encryption in the forensic domain. For example, law enforcement agencies often maintain a database of pictures and fingerprints of citizens for various purposes (e.g., foreign travel). Generally, third-party software is used for quick search or any manipulation on this huge database. Homomorphic encryption can ensure such operations on the data while still ensuring security and privacy [45, 49].

11.4.5 Financial Privacy

Homomorphic encryption has been of great use in the financial sector, because in financial cases privacy is of utmost priority while it needs a lot of numerical operations to be operated on. In fact, the first idea of privacy homomorphism was explained with an example of a financial loan company's data bank by Rivest et al. [1]. Where the loan company uses a time-sharing service to store its customers' financial information. The loan company wants a time-sharing server to analyze their customers' financial data. At the same time, the company is also aware of the fact that the information protection technique of the server is not adequate. So, they decided to encrypt their customers' information and sent it to the server, where the system programmer can analyze the data sent by the company in an encrypted form. Thus financial companies will be a great consumer of the real-world efficient HE.

11.4.6 Voting System

Another widespread use of HE is in electronic voting system where the winner can be identified by calculating the total number of votes without knowledge about who voted who. A protocol can be designed where voting of the users is encrypted in binary (i.e., 0 or 1), and then all the ciphertexts are summed in a homomorphic manner. Recently, Microsoft released a free open-source software tool—ElectionGuard—that uses homomorphic encryption to ensure secure and end-to-end verifiable voting systems [50]. The method works by encrypting each voter's choice, then produces a paper ballot to submit, a paper ballot confirmation and a tracking code. The voter can then enter that tracking number online and confirm that their vote was counted correctly in the final tally. If their vote had been tampered or altered with, it can be identified. Moreover, homomorphic encryption ensures that secrecy and privacy on how a person voted is maintained. Even this method does not allow the voter to use the tracking code to prove to anyone else how they voted. They will only be able to prove that their vote was not changed. It is also possible to add up encrypted data so that only the final tally can be decrypted without revealing

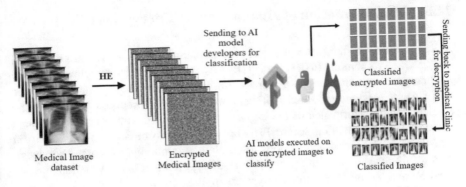

Fig. 11.3 Homomorphic encryption in AI/ML. AI models are executed on encrypted data

any information about the individual votes. This method will find special attention in vote-by-mail cases.

11.4.7 Authentication in Cloud Computing

One of the primary sources of revenue for cloud-based service providers is from the subscription of their extensive computation power that it can offer to the user, such as Amazon Web Services (AWS) form Amazon [51], AutoML from Google [52], etc. However, malicious intervention or malfunction can occur during the data transfer between the cloud and user as well as in the computing platform in the cloud. As a result, the customer requires proof that the computation was done correctly. Homomorphic encryption has been successfully used in such authentications [45]. In this authentication method, during information processing at the cloud a verification tag is generated homomorphically and the holder of the private key can verify it (Fig. 11.3).

11.4.8 Artificial Intelligence (AI) and Machine Learning (ML) Applications

In this era of data analytics, more and more users are depending on Machine Learning and Artificial Intelligence. However, processing ML and AI workload require extensive computing resources which may not be available for all users. As a result, cloud-based AI services such as Google's AutoML, Amazon's AWS Inferentia are gaining a lot of users. To provide end-to-end security and privacy in ML and AI tasks, Homomorphic Encryption has been recently embedded with PyTorch and TensorFlow [53]. The TenSEAL a library is available for doing homomorphic encryption operations on tensors in AI/ML [53].

11.5 Implementation of Homomorphic Encryptions

In this section, the hardware and software implementation of FHE are discussed briefly. Gentry's seminal FHE scheme [10] was computationally complex and costly for practical implementations as far as the general users were concerned. Since then various new and modified FHE schemes have been proposed to increase the efficiency and performance of the implementations [3, 6]. However, the overhead and cost of implementing secure FHE are still high to be applied seamlessly in user-focused real-world applications (e.g., cloud computing and storage services). As a result, research is being pursued in both algorithm and hardware domains for low-cost and fast homomorphic encryption applications.

11.5.1 Hardware Implementations

The major challenges of practical implementation of FHE stem from the time and resources required in key generation and evaluation phases. Out of the four steps—KeyGen, Encrypt, Decrypt, Recrypt—the majority of the time is spent in KeyGen and Recrypt stages [6]. A survey of run times of recent FHE schemes can be found in [3]. There exists a trade-off between the offered security-level and computational complexity of FHEs. In the implementation of [13], it was demonstrated that to obtain strong security in FHE, a large parameter setting (e.g., lattice dimension 2^{15}) is necessary. But this was impractical as it took 2.2 h for public-key generation and 31 min for ciphertext recryption. In the FHE implementation of [14], the homomorphic evaluation of AES took 36 h. In addition to the requirements of significant computational power, the practical deployment of FHEs also demands ample memory to store the large ciphertext and public keys. These large files ensure the security-robustness of the FHE scheme against lattice-based attacks [6]. In [13], to implement a secure FHE, a large public key requiring 2.25 GB storage was necessary. Most of the costly computation and memory requirements occur in the bootstrapping phase to manage the noise associated with the homomorphic computations on the ciphertexts. Computationally simple FHE schemes can be implemented with integer variants of FHE [37], and by applying compression on the key [54]. However, the security of the simpler approach of [54] was much lower [3] than [13]. In addition to optimizing the components of the FHE algorithm for efficient hardware implementation, researchers have also focused on identifying the most suitable computational platform—ASIC, GPU, or FPGA—for the given homomorphic encryption task [3]. Depending on the type of application, a detailed trade-off analysis of the robustness of the FHE against lattice-based attacks and the corresponding hardware implementation complexity is needed [6]. The typical crypto-primitives, such as modular reduction and large multiplications, can be accelerated with GPU [6]. The Fast Fourier Transform (FFT) algorithm has been utilized to optimize the modular multiplication of very large numbers, which is generally the bottleneck of the lattice-based approaches. Parallel implementation of the FFT algorithm and the concurrent execution at different cores of a GPU can further increase the execution speed of FHE schemes [55]. The reported speedup achieved from GPU ranged from 6X to 174X depending on the type of GPU and the FHE scheme [6]. For integer-based FHE methods [37], the hardware performance can be further improved by using batching techniques and parallel execution where encryption of multiple plaintext bits happens on a single ciphertext. In [56], a custom ASIC implementation of Gentry-Halevi FHE scheme was presented. The design achieved speed up factors of 1.24, 99.44 and 10.32 for decryption, encryption and recryption operations, respectively, when compared to the software implementation [6]. For efficient hardware implementation of LWE and RLWE schemes, intelligent use of Fast Fourier Transform (FFT) and FPGA was demonstrated in [57].

11.5.2 Software Implementation

An open-source software implementation of FHE, called $Hcrypt$, is available online from [58]. An online library for homomorphic encryption, $HElib$ is available in [59] that offers a speedup of 12X for FHE implementation [6]. Another recent open-source tool is Microsoft SEAL [60]. Powered by open-source homomorphic encryption technology, Microsoft SEAL supplies a set of encryption libraries which allow computations to be executed directly on the encrypted data. This enables engineers to build end-to-end secure and encrypted data storage and computation services where the customer never needs to share their key with the service provider. Recently, IBM has revealed its HE toolkit for iOS and MacOS development [61]. In summary, both optimization at the algorithmic level and customized hardware designs are necessary to reduce the computational cost of FHE, and this will enable the widespread application of FHE in encryption to ensure robust privacy and security of user data. Some of the SWHE approaches (leveled-FHE) [14] are computationally low-cost to get closer to practical implementations. But, for the FHE schemes, more research is needed to improve the performance of bootstrapping techniques and to reduce the hardware cost of homomorphic multiplications.

11.6 Future Research Directions on Full Homomorphic Encryptions

Although the challenge of FHE issued by Rivest et al. [1] has been conclusively answered by Gentry in theory in 2009, the promise of a truly practical homomorphic encryption scheme continues to intrigue cryptographers in academia, industry, and government. SWHE, which offers fast execution time and requires low memory, has shown success in real-world applications in the medical, financial, and the advertising domains [3, 5]. However, SWHE only supports a limited number of operations on encrypted ciphertext. The major challenges of practical implementation of FHE stem from its, (1) excessive runtime and computational overhead, and (2) high memory requirements. Gentry projects that it'll be another 5 years or so before the system is ready for enterprise adoption [14]. In addition to improving these two key performance limiters, future research on FHE will also focus on multi-key scenarios, AI applications and quantum computing.

11.6.1 Improvement of the Execution Speed of FHE

The applications which require running very complex and large algorithms homomorphically face different types of complexities. Operations on the encrypted ciphertext in FHE require a significant amount of computations to the unencrypted version. These computations are related to large polynomial in size that demands large runtimes even on server-class CPUs. As a result, homomorphic computation of complex functions on the encrypted ciphertext is often impractical. In the survey paper [3], the execution time required for several real-world FHE implementations are presented. For FHE implementation of AES, the execution time reached up to 113 h. The only way to make FHE practical is to limit its runtime within several minutes [4, 5]. Extensive research on adopting high-performance and parallel computing techniques in FHE, as well as improvement of the FHE algorithm, are essential to achieve this runtime target.

11.6.2 Memory Usage

The size of the public key of some of the secure FHE schemes can be as large as 2.25 GB, which can be challenge to store [3]. A trade-off exists between the key size and robustness of the security and privacy offered by FHE [3, 4, 6]. Apart from these, recently there is a lot of interest beyond the vanilla FHE.

11.6.3 Multi-key FHE

The regular FHE methods only allow computation on ciphertext encrypted under a single key. In [39], the concept of multi-key FHE was presented. In essence, a multi-key FHE scheme extends the FHE functionality to allow homomorphic computation on ciphertexts encrypted under different, independent keys. All of the corresponding secret keys are required in the decryption of the result. The need for multi-key FHE arises in many multiparty computation (MPC) platforms. In a recent paper [62], researchers from Microsoft proposed an efficient—in terms of both asymptotic and concrete complexity—multi-key FHE which can evaluate a binary gate on ciphertexts—encrypted under different keys—followed by the bootstrapping. Further research is needed in this domain to increase the execution speed of the implementation.

11.6.4 Evaluating Quantum Circuits

Private delegation of computation in the cloud implies the computational power of the evaluator is qualitatively superior to that of the client. A special scenario arises when the client needs to perform quantum computation, but only a few institutions (e.g., IBM Quantum Hardware) in the world has access to quantum computing hardware. In this case, the client has to encrypt the data, send it to the service provider and remotely execute the task on the quantum hardware. It was recently shown that this can be achieved under similar assumptions to those required from classical FHE [63]. The quantum variant of FHE allows applying homomorphic functionalities over the quantum computation. In [63] it is suggested that Quantum Fully homomorphic encryption (QFHE) is as secure as classical FHE [64]. They designed a QFHE with classical key generation, encryption, and decryption algorithm. The security is dictated by the hardness of the learning with errors (LWE) method with polynomial modulus. In [63, 64], the authors demonstrated a link between the functionality of evaluating quantum gates and the circuit privacy and security property of classical homomorphic encryption. These findings pave the way towards using classical FHE schemes with polynomial modulus for constructing QFHE with the same modulus [45]. The secure and hardware efficient design of the quantum variants of FHE is also considered an open research topic as well. In summary, FHE is a challenging and open research area with a great potential. The recent rapid advances imply that we are not very far from its real-word commercial implementation.

11.7 Conclusion

As we enter the era of the connected world, our dependence on cloud-based computing and services, data mining, e-commerce, pervasive social media, etc. are only going to escalate. Security and privacy of users' data is of utmost importance as these sensitive data can move across untrusted third-party service providers. Any breach in the security and privacy of the data can severely damage the user financially, psychologically, and socially. Regular encryption techniques are not suitable when it comes to third-party processing on users' confidential data, because these encryption methods require decrypting the data with the secret key before any manipulation on the data can be performed, which severely jeopardizes the privacy of the user. Homomorphic Encryption (HE) can eliminate this problem by allowing processing on the encrypted data by anyone without needing the secret key or decryption. Since the inception of the concept of HE, it took 30 years to demonstrate the Fully Homomorphic Encryption (FHE). Whereas Partial HE (PHE) and Somewhat HE (SWHE) can perform limited operations on the encrypted data, a truly FHE scheme opens up a new horizon of cloud computing and cloud-based services as it allows virtually all possible operations on the encrypted data. However, the FHE technology is yet to reach a point where it can be economically and transparently deployed across the edge-to-cloud spectrum for general as well as corporate users. Further research is needed in both the software and hardware domains of FHE to enhance its execution speed, and reduce the storage requirements of the large keys, while still maintaining robust security. The development of a computationally efficient and secure fully homomorphic cryptosystem—the holy grail of cryptography—would have phenomenal implications on shaping the future of computation in all aspects from storage, e-voting, pervasive AI/ML, e-commerce, medical services, quantum computing, etc.

References

1. R. Rivest, L. Adleman, M. Dertouzos, On data banks and privacy homomorphisms, in *Foundations of Secure Computation* (1978), pp. 169–180
2. C. Gentry, Computing Arbitrary functions of encrypted data. Commun. ACM **53**(3) (2010)
3. A. Acar, H. Aksu, A. Uluagac, M. Conti, A survey on homomorphic encryption schemes: theory and implementation. ACM Comput. Surv. **51**(4), Article 79 (2018)
4. P. Martins, L. Sousa, A. Mariano, A survey on fully homomorphic encryption: an engineering perspective. ACM Comput. Surv. **50**(6), Article 83 (2017)
5. B. Zvika, Fundamentals of fully homomorphic encryption: a survey, in *Electronic Colloquium on Computational Complexity*, Report No. 125 (2018)
6. C. Moore, M. O'Neill, E. O'Sullivan, Y. Doröz, B. Sunar, Practical homomorphic encryption: a survey, in *IEEE International Symposium on Circuits and Systems (ISCAS), Melbourne VIC* (2014), pp. 2792–2795
7. C. Fontaine, F. Galand, A survey of homomorphic encryption for nonspecialists. EURASIP J. Inf. Security **2007**(January 2007), Article ID 15 (2007). Hindawi Publishing Corporation, New York, NY, USA
8. J. Sen, *Homomorphic Encryption: Theory and Applications* (2013)
9. F. Armknecht, S. Katzenbeisser, A. Peter, Group homomorphic encryption: characterizations, impossibility results, and applications. Des. Codes Cryptogr. **67**, 209–232 (2013)
10. C. Gentry, Fully homomorphic encryption using ideal lattices, in *Symposium on the Theory of Computing (STOC)* (2009), pp. 169–178
11. C. Gentry, Computing on the edge of chaos: structure and randomness in encrypted computation, in *Electronic Colloquium on Computational Complexity (ECCC)* (2014)
12. C. Aguilar-Melchor, S. Fau, C. Fontaine, G. Gogniat, R. Sirdey, Recent advances in homomorphic encryption: a possible future for signal processing in the encrypted domain. IEEE Signal Process. Mag. **30**(2), 108–117 (2013)
13. C. Gentry, S. Halevi, Implementing Gentry's Fully-Homomorphic Encryption Scheme, in *Advances in Cryptology—EUROCRYPT 2011 (EUROCRYPT 2011)*, ed. by K.G. Paterson. Lecture Notes in Computer Science, vol. 6632 (Springer, Berlin, 2011)
14. C. Gentry, S. Halevi, N. Smart, Homomorphic evaluation of the AES circuit, in *IACR Cryptology* (2012)

15. T. ElGamal, A public key cryptosystem and a signature scheme based on discrete logarithms, in *Advances in Cryptology* (Springer, Berlin, 1985), pp. 10–18
16. P. Paillier, Public-key cryptosystems based on composite degree residuosity classes, in *Advances in Cryptology—EUROCRYPT '99 (EUROCRYPT 1999)* ed. by J. Stern. Lecture Notes in Computer Science, vol. 1592 (Springer, Berlin, 1999)
17. R. Rivest, A. Shamir, L. Adleman, A method for obtaining digital signatures and public-key cryptosystems. Commun. ACM **21**(2), 120–126 (1978)
18. W. Diffie, M. Hellman, New directions in cryptography. IEEE Trans. Inf. Theory **22**(6), 644–654 (1976)
19. S. Goldwasser, S. Micali, Probabilistic encryption and how to play mental poker keeping secret all partial information, in *Proceedings of the 14th Annual ACM Symposium on Theory of Computing* (ACM, New York, 1982), pp. 365–377
20. B. Kaliski, *Quadratic Residuosity Problem* (Springer US, Boston, 2005), pp. 493–493
21. https://mathworld.wolfram.com/XOR.html. Cited 29 Aug 2020
22. S. Kevin, The discrete logarithm problem, in *Cryptology and Computational Number Theory*, vol. 42 (1990)
23. J. Benaloh, Dense probabilistic encryption, in *Proceedings of the Workshop on Selected Areas of Cryptography* (1994), pp. 120–128
24. T. Okamoto, S. Uchiyama, A new public-key cryptosystem as secure as factoring, in *Advances in Cryptology (EUROCRYPT'98)* (Springer, Berlin, 1998), pp. 308–318
25. T. Jager, The generic composite residuosity problem, in *Black-Box Models of Computation in Cryptology* (Vieweg+Teubner, New York, 2012)
26. I. Damgård, M. Jurik, A generalisation, a simplification and some applications of Paillier's probabilistic public-key system, in *Public Key Cryptography* (Springer, Berlin, 2001), pp. 119–136
27. S. Pohfig, M. Hellman, An improved algorithm for computing logarithms over GF(p) and its cryptographic significance, in *IEEE Transactions on information Theory*, vol. IT-241 (1978), pp.106–110
28. T. Sander, A. Young, M. Yung, Non-interactive cryptocomputing for NC1, in *Proceedings of the 40th Annual Symposium on Foundations of Computer Science* (1999), pp. 554–566
29. Y. Ishai, A. Paskin, Evaluating branching programs on encrypted data, in *Theory of Cryptography (TCC 2007)* ed. by S.P. Vadhan. Lecture Notes in Computer Science, vol. 4392 (Springer, Berlin, 2007)
30. D. Boneh, E.J. Goh, K. Nissim, Evaluating 2-DNF formulas on ciphertexts, in *Theory of Cryptography (TCC 2005)*, ed. by J. Kilian. Lecture Notes in Computer Science, vol. 3378 (Springer, Berlin, 2005)
31. M. Fellows, N. Koblitz, Combinatorial cryptosystems galore!, in *Contemporary Mathematics*, vol. 168 (1994), pp. 51–51
32. M. Hermann, *Geometrie Der Zahlen*, vol. 40 (1968)
33. V. Lyubashevsky, C. Peikert, O. Regev, On ideal lattices and learning with errors over Rings, in *Advances in Cryptology—EUROCRYPT 2010 (EUROCRYPT 2010)*, ed. by H. Gilbert. Lecture Notes in Computer Science, vol. 6110 (Springer, Berlin, 2010)
34. M. Ajtai, Generating hard instances of lattice problems, in *Proceedings of the 28th Annual ACM Symposium on Theory of Computing* (ACM, New York, 1996), pp. 99–108
35. O. Goldreich, S. Goldwasser, S. Halevi, Public-key cryptosystems from lattice reduction problems, in *Advances in Cryptology—CRYPTO '97 (CRYPTO 1997)*, ed. by B.S. Kaliski. Lecture Notes in Computer Science, vol. 1294 (Springer, Berlin, 1997)
36. Z. Zhang, Revisiting fully homomorphic encryption schemes and their cryptographic primitives, in PhD thesis (University of Wollongong, Wollongong, 2014)
37. M. van Dijk, C. Gentry, S. Halevi, V. Vaikuntanathan, Fully homomorphic encryption over the integers, in *Advances in Cryptology—EUROCRYPT 2010 (EUROCRYPT 2010)*, ed. by H. Gilbert. Lecture Notes in Computer Science, vol. 6110 (Springer, Berlin, 2010)

38. Z. Brakerski, V. Vaikuntanathan, Fully homomorphic encryption from ring-LWE and security for key dependent messages, in *Advances in Cryptology—CRYPTO 2011 (CRYPTO 2011)*, ed. by P. Rogaway. Lecture Notes in Computer Science, vol. 6841 (Springer, Berlin, 2011)

39. A. López-Alt, E. Tromer, V. Vaikuntanathan, On-the-fly multiparty computation on the cloud via multikey fully homomorphic encryption, in *Proceedings of the Forty-Fourth Annual ACM Symposium on Theory of Computing (STOC '12)* (Association for Computing Machinery, New York, 2012), pp. 1219–1234

40. S. Galbraith, S. Gebregiyorgis, S. Murphy, Algorithms for the approximate common divisor problem. LMS J. Comput. Math. **19**(A), 58–72 (2016)

41. D. Micciancio, O. Regev, Lattice-based cryptography, in *Post-Quantum Cryptography*, ed. by D.J. Bernstein, J. Buchmann, E. Dahmen (Springer, Berlin, 2009)

42. J. Hoffstein, J. Pipher, J.H. Silverman, NTRU: a ring-based public key cryptosystem, in *Algorithmic Number Theory (ANTS 1998)*, ed. by J.P. Buhler. Lecture Notes in Computer Science, vol. 1423 (Springer, Berlin, 1998)

43. D. Stehlé, R. Steinfeld, Making NTRU as secure as worst-case problems over ideal lattices, in *Advances in Cryptology—EUROCRYPT 2011 (EUROCRYPT 2011)*, ed. by K.G. Paterson. Lecture Notes in Computer Science, vol. 6632 (Springer, Berlin, 2011)

44. A. Jeckmans, A. Peter, P. Hartel, Efficient privacy-enhanced familiarity-based recommender system, in *Computer Security—ESORICS 2013 (ESORICS 2013)*, ed. by J. Crampton, S. Jajodia, K. Mayes. Lecture Notes in Computer Science, vol. 8134 (Springer, Berlin, 2013)

45. F. Armknecht, C. Boyd, C. Carr et al., A guide to fully homomorphic encryption, in *IACR Cryptology ePrint Architecture*, vol. 2015 (2015)

46. M. Naehrig, K. Lauter, V. Vaikuntanathan, Can homomorphic encryption be practical? in *Proceedings of the 3rd ACM Workshop on Cloud Computing Security Workshop (CCSW '11)* (Association for Computing Machinery, New York, 2011), pp. 113–124

47. L. Kristin, *Practical Applications of Homomorphic Encryption* (2015)

48. Z. Yang et al., Privacy-preserving classification of customer data without loss of accuracy, in *Proceedings of the SIAM International Conference on Data Mining* (2005), pp. 92–102

49. C. Bösch et al., SOFIR: securely outsourced forensic image recognition, in *IEEE International Conference on Acoustics, Speech and Signal Processing (ICASSP), Florence, 2014* (2014), pp. 2694–2698

50. Microsoft ElectionGuard (2020). https://github.com/microsoft/electionguard. Cited 29 Aug 2020

51. https://aws.amazon.com/machine-learning/inferentia/. Cited 29 Aug 2020

52. https://cloud.google.com/automl. Cited 29 Aug 2020

53. Homomorphic Encryption in PySyft with Seal and PyTorch (2020). https://blog.openmined.org/ckks-homomorphic-encryption-pytorch-pysyft-seal/. Cited 29 Aug 2020

54. J.S. Coron, D. Naccache, M. Tibouchi, Public key compression and modulus switching for fully homomorphic encryption over the integers, in *Advances in Cryptology—EUROCRYPT 2012 (EUROCRYPT 2012)*, ed. by D. Pointcheval, T. Johansson. Lecture Notes in Computer Science, vol. 7237 (Springer, Berlin, 2012)

55. W. Wang, Y. Hu, L. Chen, X. Huang, B. Sunar, Accelerating fully homomorphic encryption using GPU, in *IEEE Conference on High Performance Extreme Computing, Waltham, MA* (2012)

56. Y. Doröz, E. Öztürk, B. Sunar, Accelerating fully homomorphic encryption in hardware. IEEE Trans. Comput. **64**(6), 1509–1521 (2015)

57. T. Pöppelmann, T. Güneysu, Towards practical lattice-based public-key encryption on reconfigurable hardware, in *Selected Areas in Cryptography—SAC 2013 (SAC 2013)*, ed. by T. Lange, K. Lauter, P. Lisoněk. Lecture Notes in Computer Science, vol. 8282 (Springer, Berlin, 2014)

58. H. Perl, M. Brenner, M. Smith, *HCRYPT* (2011). http://www.hcrypt.com/scarab-library/. Cited 29 Aug 2020

59. S. Halevi, V. Shoup, *HElib, Homomorphic Encryption Library* (2012). https://github.com/shaih/HElib. Cited 29 Aug 2020

60. Microsoft SEAL (2020). https://www.microsoft.com/en-us/research/project/microsoft-seal/. Cited 29 Aug 2020
61. IBM Homomorphic Toolkit (2020). https://www.ibm.com/blogs/research/2020/06/ibm-releases-fully-homomorphic-encryption-toolkit-for-macos-and-ios-linux-and-android-coming-soon/. Cited 29 Aug 2020
62. H. Chen, I. Chillotti, Y. Song, Multi-key homomorphic encryption from TFHE, in *Advances in Cryptology—ASIACRYPT 2019 (ASIACRYPT 2019)*, ed. by S. Galbraith, S. Moriai. Lecture Notes in Computer Science, vol. 11922 (Springer, Cham, 2019)
63. Z. Brakerski, H. Yuen, *Quantum Garbled Circuits 2020* (2020). https://arxiv.org/abs/2006.01085. Cited 29 Aug 2020
64. Z. Brakerski, Quantum FHE (almost) as secure as classical, in *Advances in Cryptology—CRYPTO 2018 (CRYPTO 2018)*, ed. by H. Shacham, A. Boldyreva. Lecture Notes in Computer Science, vol. 10993 (Springer, Cham, 2018)

Chapter 12
Software Security with Hardware in Mind

Muhammad Monir Hossain, Fahim Rahman, Farimah Farahmandi, and Mark Tehranipoor

12.1 Introduction

Software becomes a necessary part of our daily life. It has significant contributions to the automation of industries. Nowadays, the software has been an influential tool as a medium of transferring information. People store and transmit sensitive information using software that is connected to the Internet. The sensitive data might range from personal and commercial data (e.g., account information of financial transactions, online banking, social networking, etc.) to the national defense secret. Exposure of this sensitive information to the connected world through the Internet becomes the center of attraction for the adversaries. The adversaries intend to explore the hardware and software covert channels in the systems and exploit the sensitive information. The deployment of developed software without considering the security may dangerously risk our lives and properties. Therefore, nowadays, software security becomes a critical security aspect of the modern computing system.

Software security aims to understand the security prospects of application during the architecture design and implementation phase, thus protecting them from malicious adversaries based on the threat model. Assurance of application security resembles that code is secure at the development stage and less prone to potential vulnerabilities. The comprehensive security of software is achieved through vulnerability testing, application scanning from the development stage to the end of the software release. When the vulnerabilities are not identified in the development or testing phases, these may assist in leaking sensitive information

M. M. Hossain (✉) · F. Rahman · F. Farahmandi · M. Tehranipoor
Department of ECE, University of Florida, Gainesville, FL, USA
e-mail: hossainm@ufl.edu; fahimrahman@ece.ufl.edu; farimah@ece.ufl.edu;
tehranipoor@ece.ufl.edu

© The Author(s), under exclusive license to Springer Nature Switzerland AG 2021
M. Tehranipoor (ed.), *Emerging Topics in Hardware Security*,
https://doi.org/10.1007/978-3-030-64448-2_12

illegitimately during run-time. The adversaries can also exploit those vulnerabilities to execute unauthentic malicious programs.

The primary goal of software security is to develop it in such a way so that it can resist, tolerate, and recover from various malicious attacks [1]. This goal can be achieved by emphasizing on typical high-level features of secure software during the development, such as predictability of execution, trustworthiness, and conformance. The software should be capable of eliminating the attacker's ability to tamper the execution or the result as much as possible, ensuring the program execution's predictability. The program should contain the least number of exploitable vulnerabilities to achieve the trustworthiness of the software. Lastly, software features, components, and system architecture should conform to the specified software requirements.

However, the increased complexity of software challenges the enhancements of security. Additionally, the attackers' domain knowledge and methodologies are also becoming more robust than any previous time. As a result, it becomes challenging to secure the software. As the software runs on the hardware, security from the hardware perspective can also provide more protection to the applications. Hence, the integration of hardware security and software countermeasures can be a practical approach for comprehensive protection.

In this chapter, we discuss various attributes of secure software in Sect. 12.2 and hardware-assisted software security in Sect. 12.3. Different kinds of software and hardware vulnerabilities and corresponding countermeasures are described in Sect. 12.4. Finally, the chapter concludes in Sect. 12.5.

12.2 Attributes of Secure Software

There are several core attributes of a typical secure software. Developers need to be very cautious about ensuring these attributes. Here, we discuss them in brief.

- Confidentiality: The software needs to assure the users that the sensitive data, functional behavior, characteristics, contents, etc. are not leaked to the adversaries or unauthorized entities due to the execution of the software.
- Integrity: Software must be resilient against tampering. Suppose the software comes with protection based on authentication or password. In that case, the attacker should not be capable of running the modified software that bypasses the protection, as mentioned earlier.
- Availability: The software should be available to the owner or authentic consumers anytime. On the contrary, the software should neither run nor produce correct results, while the unauthentic users want to access it.

12.3 Hardware-Assisted Software Security

Hardware-assisted software security utilizes hardware components to enhance the software's security, including BIOS, operating systems, or any other consumer-level application. There is a significant difference between hardware security and hardware-assisted security. Hardware security assists in protecting the hardware devices only from the adversarial attacks. On the other hand, hardware-assisted security approaches protect the software running on the hardware computing systems.

Major device manufacturers such as Intel, IBM, and AMD have offered many hardware-assisted approaches to enhance software security against various adversarial attacks. Hardware-assisted security approaches are widely used for Trusted Computing (TC), primarily to protect data usage. For example, several hardware-assisted approaches are developed for the Trusted Platform Module, such as Intel's TXT [2]. Additionally, Intel offers SGX [3] as Trusted Execution Environment (TEE), whereas ARM has TrustZone [4] and AMD's SEV [5]. ARM's TrustZone separates the hardware and software resources of a System-on-Chip based on the execution of a secure world and non-secure world. Intel proposed Threat Detection Technology (TDT) [6] for malware detection, which scans the memory-based malware from CPU to GPU. Intel also developed hardware-based AES encryption AES-NI [7], for faster and increased security. AES-NI reduces the timing and cache side-channel leakage.

As discussed above, state-of-the-art research proves the effectiveness of integrating hardware approaches with the software protection schemes to maximize software security for the applications. Overall, hardware-assisted approaches prevent the vulnerabilities and accelerate the overall performance, where only software patches usually incur severe lags and degrade the performance significantly due to inherent overhead.

12.4 Vulnerabilities, Attacks, and Countermeasures

Securing software can be achieved through a better understanding of the software and hardware vulnerabilities and taking necessary steps as countermeasures while developing the software. Hence, we analyze various exploitable vulnerabilities in this chapter. We discuss cache-based side channels, Spectre, Meltdown, Ransomware, buffer and integer overflow-based vulnerabilities. We explain how adversaries exploit these vulnerabilities and can cause severe threats to the consumer application. We also discuss various countermeasures against those adversarial attacks.

12.4.1 Cache-Based Side Channel

Side channels are some ways to leak out information that are not supposed in ideal systems. These leaks out of secret may come in several forms, such as timing, power consumption, data traffic, etc. Among all of these side channels, cache-based timing channels are notoriously bad, which can ease the path for exploitation of theoretically secure cryptographic algorithms, such as RSA [8], AES [9, 10], ElGamal [11, 12], etc. These kinds of attacks become possible due to timing differences in executing different kinds of memory access patterns, while the attackers share the same low-level cache L1 with the victim. Cache-based timing attacks intend to leak out the secret information to the malicious parties based on the program's cache behaviors in run-time, which does not require physical access to confidential computation. In Listing 12.1, there are two types of cache-based side channels. One is a secret-dependent array accesses, and another is secret-dependent branches. In Listing 12.1, line 3 represents the secret-dependent array access as the index for *Sbox* is part of the round key, which may reveal the decryption key [13, 14]. Line 6 has a secret-dependent branch condition as it depends on the *i*th bit of *key*. When *i*th bit of *key* is set, the arithmetic operations mentioned in lines 7 and 8 occur, and create timing difference in terms of execution as compared to the case when *i*th bit of *key* is not set. Thus, the adversaries may predict the *key bits* observing the timing difference during the execution of the program.

Listing 12.1 Simplified code of modular exponentiation algorithm

```
1    void Test(){
2      ... // computing RK[0..3] from key
3      RK[4] = RK[0] ^ Sbox[(RK[3] >> 8) & 0xFF];
4      res = res * res;
5      res = res % mod;
6      if(bit_set_at_i(key[0], i)){
7          res = base * res;
8          res = res % mod;
9      }
10   }
```

Developers of the common cipher algorithm manually intend to find the side channels and patch the software with the countermeasures. However, the manual process cannot detect many critical side channels, which allows the invention of new side channel, as Wang et al. mentioned in [15]. Again, the released high-level patches against side-channel attacks are tough to elucidate for the non-specialists. Wang et al. and Doychev et al. proposed CacheD [15] and CacheAudit [16, 17], respectively, for the automatic detection of cache-based side channels. Although these proposed solutions can detect side channels in real-world software, these techniques still have some limitations. Brotzman et al. proposed a symbolic execution-based technique in [18], which explored only a few dynamic paths; existing techniques mostly lack comprehensive code coverage. These techniques cannot detect side channels in static time without executing the programs and in case of conditional branches with confidential information. Some techniques are based on abstract interpretation. These techniques provide comprehensive coverage but cannot provide the locations where the side channels exist and justify why they are the side channels. However, there are some hardware-assisted architectures in preventing cache-based side channels that we describe below.

12.4.1.1 Countermeasures

The root cause of cache-based side channels is creating a timing difference, i.e., cache interference while executing some instructions to perform a particular task in the program. The best way to eliminate side-channel attacks is to prevent inference of cache line evictions, i.e., cache interference. Two approaches can remove cache interference. One approach does not allow sharing of cache lines (cache partitioning), where the other technique does allow but randomizes the cache interference. In this section, we discuss both approaches.

12.4.1.2 Partition-Locked Cache

The cache partitioning technique suggests allocating cache space for various applications according to their memory demands and consumption. Cache partitioning is a promising methodology that ensures shared cache's capacity benefits while maintaining private caches' isolation. On the contrary, if the cache partitioning is not implemented efficiently, it may significantly degrade the performance, such as statically partitioned cache. When a software process requires more cache lines, the running process cannot utilize unused cache lines of other partitions; it reduces the performance. Hence, a statically partitioned cache is not much efficient, although it can prevent sharing. Here, we discuss about an efficient technique, Partition-Locked Cache (PLcache) [19].

Overview of Technique: PLcache creates a private partition and locks the sensitive cache lines. The locked cache lines cannot be evicted during the cache accesses, not relevant to the corresponding private partition. Thus, PLcache prevents cache interference.

PLcache involves both hardware and software enhancements. In order to implement PLcache, an additional hardware module and a system interface are required to define and control which cache lines are needed to be locked in the cache. The implementation requires two new tags for each cache line; one is for a lock–unlock bit, and the other represents an ID for the cache line owner as shown in Fig. 12.1.

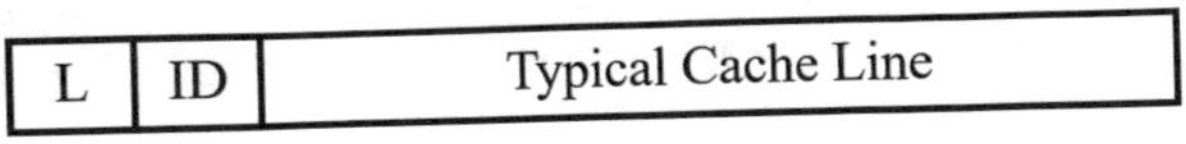

Fig. 12.1 Cache line in PLcache implementation

L	ID	Typical Cache Line

On the other hand, there are two methods to implement the control interface, which provides the platform to lock a cache line.

1. ISA Extension: The base Instruction Set Architecture (ISA) can include a lock–unlock sub-opcode with a new set of load/store instructions. Using these new instructions, the developers can control what kind of data should be locked while compiling the program.
2. Segment/Page-Based Protection: The memory region, which contains AES or RSA relevant table, can be locked in the cache memory by setting the lock–unlock bit. Accessing these memory regions should lock the corresponding cache line. This can be implemented by using an additional bit (LL bit), which remains for each Translation Look-aside Buffer (TLB) entry, page-table entry, or segment descriptor. LL bit indicates whether access to a page or a segment should lock the corresponding cache line. The implementation scheme requires two functions to lock or unlock a particular memory region during the program's compilation. The example functions may be as follows:

int lock_cache_region(unsigned long address_start, unsigned long length);
int unlock_cache_region(int region_id);

In order to lock a memory region, the function *lock_cache_region* is called, which also sets the corresponding LL bit. On the other hand, the function *unlock_cache_region* is used to unlock a memory segment, clearing the corresponding LL bit and invalidating the locked cache lines.

In PLCache, the hardware can ensure the locking of cache lines, while software needs to ensure the correct use of locking.

12.4.1.3 Random Permutation Cache (RPcache)

This approach permits cache sharing, unlike PLcache [19]. It randomizes the cache interference such that no information can be gathered in which cache line was evicted.

Overview of Technique: The main technique for RPcache is the permutation of the main memory to cache mapping. The logical view of the RPcache implementation is shown in Fig. 12.2. As similar to PLcache, a P bit and ID field are added to each cache line to represent the lock status and owner of the cache line, respectively. In the RPcache implementation, a permutation table stores the main memory to cache mapping. The number of entries of this table equals the number of sets in the cache memory. Each table entry includes a different M-bit number for representing the new set. M set bits of an effective address are used as the Permutation Table index to obtain the new set bits representing the index for cache set array. The randomization of the main memory to cache mapping can be obtained by swapping

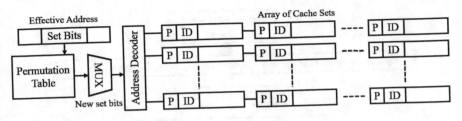

Fig. 12.2 Implementation of RPcache

two table entries. Assume that ith and jth table entries are mapped to S and S' in the cache. After swapping that two entries, the new mapping for them becomes $i \rightarrow S'$ and $j \rightarrow S$.

In the RPcache, there are several permutation tables. One or more processes sharing a cache can use a single table or separate table for each individual process. Sensitive processes, such as AES encryption, can use one table, while other typical processes can share a single table. However, the implementation of many permutation tables for a computer system might be costly. Hence, only a few tables should be sufficient, depending on the type of applications. The non-critical applications can proceed through normal mapping, where only critical processes utilize the main memory to cache remapping.

12.4.2 Spectre

Today's modern processor significantly increases the performance by using branch predictor and speculative execution. When the destination of a branch instruction depends on a memory value that the processor still works on, the CPU guesses the destination and starts execution. When the memory value is computed, the CPU commits if it meets the branch condition; otherwise, it discards the speculative computation it does. This speculative execution opens the door for an attack known as Spectre. Spectre attack induces a victim to perform speculative operations that would not happen in sequential instructions. The adversaries can exploit this attack to leak out secret information of the victim process.

12.4.2.1 Spectre Attacks and Variants

In order to perform Spectre attack, initially, the adversary mistrains the processor by performing some operations. As a result, the processor makes a wrong prediction later, which might be exploited. After completing the initial training, the processor starts execution speculatively, which opens up a covert channel that may leak out

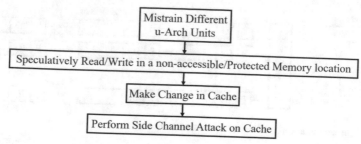

Fig. 12.3 Generic steps for a Spectre attack

sensitive information from the victim process. When the attacker requests the victim process to perform a specific task through a system call, a socket connection, or file access, the attack might be triggered. The speculative execution assists in reading a memory value at the target address for an attacker. Then, the attacker process performs memory operations to modify the cache state to such a state that reveals the sensitive value. In the final state, the sensitive data is recovered through various attacks such as Flush+Reload or Evict+Reload.

There are several variants of Spectre attacks. All variants follow the same direction to perform an attack, as we described before in this section (also shown in Fig. 12.3.

We describe all Spectre variants briefly in Table 12.1. Different variants of Spectre may exploit direct or indirect branch and Return Stack Buffer (RSB). The adversaries may get access to both read/write from/to inaccessible memory through Spectre attacks.

In this section, we explain the Spectre Variant 1 with an example. A tiny code is shown in Listing 12.2 to explain the attack. Assume that the value of the variable y is provided by the attacker. To ensure that there is no out-of-bound access to $array1$, the if statement has a boundary condition. This example explains how the attacker can achieve the data from the process's address space bypassing the if conditional statement. In the first step of the attack, the attacker mistrains the CPU's branch predictor by providing valid inputs in the program mentioned in Listing 12.2.

Listing 12.2 Spectre variant: Conditional branch exploitation

```
1    if ( y < size ( array1 ))
2        z = array2 [ array1 [y] * 4096];
```

Once the branch predictor is trained, the attacker then invokes the code with a value outside of the range of $array1$ for variable y. As the branch predictor is already mistrained, it predicts that the branch condition would be $TRUE$. Hence, the CPU starts executing speculatively to get the result for $array2[array1[y] * 4096]$ with a malicious value of y without waiting for the actual branch result. The CPU loads the data for $array2$ into the cache at the address that depends on $array1[y]$. When the result for branch condition is computed, the CPU finds

Table 12.1 Variants of Spectre attacks

Attack name	Type	Entry point	Exit point	Exploited Unit	Action
Variant 1	Bound check bypass	Branch target buffer (BTB) in BPU[a]	Cache	Exploits BPU[a] by training it to make a misprediction	Read from a non-accessible memory
Variant 1.1	Bound check bypass	Branch target buffer in BPU[a]	Cache	Exploits branch predictor unit by training it to make a misprediction	Write in a non-accessible memory
Variant 1.2	Bound check bypass	Branch target buffer in BPU[a]	Cache	Exploits branch predictor unit by training it to make a misprediction	Write in a user-defined read-only memory
Variant 2	Branch target buffer poisoning	Indirect branch target buffer (iBTB) in BPU[a]	Cache	Exploits BTB by poisoning it with frequently used target location	iBTB gets misled to the target location
Spectre-return stack buffer	Return stack buffer (RSB) poisoning	RSB in BPU	Cache	Exploits RSB by poisoning it to make a misprediction	Changes the return address of a function

[a] BPU: branch predictor unit

the error and reverts all changes that are made due to speculative execution. Nevertheless, the CPU does not revert the cache state that is changed due to the wrong prediction. As a result, the attacker can perform a side-channel attack to retrieve the secret value from the cache, which it would not get access in a sequential processor as the memory access for *array2* is the out-of-bound address for the victim memory.

12.4.2.2 Countermeasures

Several countermeasures have been proposed for mitigating Spectre attacks. Each approach addresses some of the features upon which the attack depends. In this section, we discuss different types of countermeasures and how these can prevent the attacks effectively.

Speculative execution is the root of Spectre attacks. According to the ascertained control flow, the execution of the program can eliminate the Spectre attacks. The modification of the microcode can lead to the elimination of speculative execution. However, this would not be an efficient approach. Speculative execution improves the performance of modern CPUs significantly. Again, this solution is applicable only for future processors, not applicable to the already manufactured processor. However, to provide a generic solution against the Spectre attacks, Intel and AMD suggest using speculative blocking instructions, such as *lfence* [20, 21]. As mentioned above, the elimination of speculative instructions would be a devastating performance for the processors. Instead, it might be better to reduce the number of usage of some critical speculative instructions. Static analysis can be of great assistance in identifying the most vulnerable speculative instructions.

Some countermeasures are based on preventing speculatively executed instructions from accessing the secret data. For instance, in the Google Chrome web browser, each website runs in an individual process [22]. Spectre attacks exploit the victim's permission. As a result, a process running on a different website cannot be exploited by the adversaries. On the other hand, WebKit [23] applies two strategies to restrict access to sensitive data while executing instructions speculatively. Firstly, when accessing an array, WebKit applies a bitmask instead of boundary checking of the array to ensure that the index cannot be much bigger than the array size. This may detect out of bound access for the array (input is greater than the array size) and restrict the attackers to access arbitrary memory.

Tracking data caused by speculative operation might be helpful to detect potential Sptectre attacks. Modification of the computer architecture and providing built-in hardware modules are required to track the data, whether it results from speculative execution. If the data results from speculative execution, preventing that data for use in the subsequent operations can efficiently prevent the leak out of data and prevent Spectre attacks.

12.4.3 Meltdown

Meltdown attack results from the side effect of the out-of-order execution of the modern processors. The adversaries can read arbitrary kernel-memory locations for leaking out personal data or passwords through this attack. Today, many processors such as Intel are prone to Meltdown attacks. Hardware is the root cause of this attack. Hence, a Meltdown attack does not depend on the operating system or any software vulnerabilities. Meltdown provides the adversaries to read data from other processes or virtual machines in the cloud illegitimately. Today software security largely depends on the memory's isolation, separating the kernel region and user space. The kernel space is protected from the users, which is ensured by an operating system. As a result, today's computing system can run multiple applications on a single cloud machine. There is a fundamental difference between the side-channel attack and Meltdown. The adversaries can leak the victim application's secret information through side-channel attacks. Side-channel attacks require only information about the victim or target application. On the contrary, Meltdown opens up the door for the adversaries to read the entire kernel address space.

Example of Attack: Here, we describe how a Meltdown attack occurs with an example code mentioned in Listing 12.3. The code describes that array access is supposed to occur after an exception is raised. During the program's execution, when there is an exception, the array access should never happen; the control flow jumps to the corresponding exception handler in the OS. In this case, the exception might arise due to memory access (e.g., invalid memory access), CPU exception (e.g., divide by zero).

Listing 12.3 Example of Meltdown attack

```
1    raise_exception ();
2    // below line is never reached
3    access ( probe_array [ data * 4096]);
```

According to the program flow, the statement relevant to the array access cannot take place. As the processor executes instructions out of order, the CPU might execute the array access as this does not depend on the exception. The control flow for the example is shown in Fig. 12.4. This out-of-order execution does not bring any visual architectural side effects. Nevertheless, the data from the referenced memory is fetched to a register and also in the cache. If the CPU decides later to discard the results caused due to the out-of-order execution, it does not commit the register and memory contents. Nonetheless, the memory contents (results) remain in the cache. The adversaries then utilize a micro-architectural side-channel attack, for example, Flush+Reload [24], Prime+Probe [10, 25], Evict+Reload [26], and Flush+Flush [27] to know whether a particular memory location is cached.

A different cache line is accessed when executing the array accessing instruction in out of order depending on the value of *data* in Listing 12.3. In the example, the variable *data* is multiplied by 4096, which indicates that *probe_array* is being

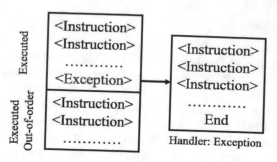

Fig. 12.4 In case of an exception during the execution of a program, the control flow diverts to the exception handler. In the meantime, the subsequent instructions of the exception can be executed due to out-of-order architecture, which is eventually discarded by the CPU

accessed with a 4 KB (assuming page size) distance over the cache. There is an implicit mapping between victim data and the memory page. Each value of *data* represents a different page in the memory. After flashing the memory pages by the attacker (performing Flush+Reload attack), if the victim process accesses a memory page and it is cached, the attacker can perceive the data by reading all cache lines and the corresponding time. The attacker is able to read the cache line in a short time (cache hit) when the victim recently used this particular cache line. Thus, the sensitive data from victim memory space gets exposed to the adversaries.

The entire Meltdown attack consists of three steps in summary.

- Step 1: Reading Secret Data: The content of the victim's memory, which cannot be accessed by the attacker, is loaded into a register.
- Step 2: Transmitting Secret Data: Some transient instructions access the cache line based on the register's secret content.
- Step 3: Receiving Secret Data: The adversary utilizes a cache-based side-channel attack to find out the victim's accessed cache line and retrieves the secret data stored in the selected memory location.

The adversary can retrieve all data from the entire memory space iterating all relevant memory addresses by repeating the steps mentioned above. As almost all operating systems map the physical memory to the kernel address space for every process, the attacker can retrieve all the victim computing device's physical memory data.

12.4.3.1 Countermeasures

The hardware itself is the root of the cause for the Meltdown attack. Enhancements of the hardware design can protect against this attack. As Meltdown happens due to out-of-order execution, eliminating this feature would diminish Meltdown completely. However, this would not allow parallelism of the modern computer systems and hence degrade the performance significantly. Therefore, this is not a feasible solution.

A plausible solution might be to introduce a hard split of user space and kernel space [28]. In order to implement, the kernel may be using a new hard split bit in a CPU control register. If the hard split is enabled, the upper half of the address space would be used for the kernel, where the lower half for user space. A memory fetch instantly can identify whether a memory fetch of the destination address violates the boundary condition of the security as the privileged level can be determined from the virtual address without using any other lookup table. Hence, the performance impact would be minimal while implementing the hard split. The backward compatibility is also achieved as the hard split bit is not set by default. The kernel sets the hard split bit.

On the other hand, it is usually challenging to provide a hardware solution. Hence, there are some software-based solutions against the Meltdown attacks, such as KAISER [29]. KAISER modifies the kernel not to be mapped in the user space to prevent the side-channel attacks.

12.4.4 Ransomware

Many security vulnerabilities are associated with modern computing systems, given rise to numerous cyber attacks and malware that can cause privacy breach, data loss, and financial damages and compromise national and critical infrastructures. According to McAfee quarterly threat report, 176 new cyber threats are emerging every minute [30]. Among such cyber threats and attacks, Ransomware has gained much attention due to its malicious nature and alarming effect [31–34]. Multiple Ransomware variants are growing in number with the capabilities of evasion from many anti-viruses and software-only malware detection schemes that rely on static execution signatures. The rise of Ransomware can be attributed to the appearance of multiple extremely successful variants. This success has been used as a template by later variants, which results in today's mass proliferation.

Ransomware, comprising the words "ransom" and "malware," is a malware class that asks for ransom/money from the victim via anonymous payment mechanisms by holding the system/files as a hostage, in exchange for restoring the hijacked functionality. Although the first Ransomware relied on encrypting information on the victim's computer to demand payment for the key or software to decrypt the data, today, these malware programs have diversified in the way they extort money from the victim. Most of the Ransomware is initiated through user actions such as— clicking on a malicious link in a spam e-mail or visiting a malicious or compromised website, which is fundamentally different from malware. In a few scarce instances, cyber threat actors specifically target a victim. This may occur after the actors realize that a sensitive entity has been infected or because of specific infection attempts.

12.4.4.1 Types of Ransomware

Depending on the behavior, Ransomware can be classified into different categories. In this section, we will discuss those categories elaborately.

- Encrypting ransomware: During execution, this type of Ransomware silently searches for the valuable files stored in the victim's machine and encrypts the files. When the first step is complete, the user can see a message, asking for ransom for the hidden files. Detailed information is provided to the user to pay the ransom. After the ransom is paid, the victim will get a key or code to decrypt the files. CryptoWall, CryptoLocker, WannaCry, Locky, etc. are some common examples of encrypting Ransomware.
- Non-encrypting Ransomware: Non-encryption Ransomware targets the locking mechanism of victim's machine and asks for user action that ends up costing money to unlock. Sometimes the user needs to pay the ransom in advance. Examples of common locker Ransomware include Winlocker and Reveton.
- Leakware: This type of Ransomware does not block access to the victim machine like others. However, it collects sensitive information from the machine silently and uses it as a weapon to blackmail the user. The attackers store the information on servers or other infected machines and threaten the victim to publish it if the ransom is not paid.
- Mobile Ransomware: Since mobile data is easy to store and restore to and from the cloud, mobile devices are easy target to Ransomware. Unlike others, mobile Ransomware relies on a mobile device rather than the data stored on it. When this malware is installed or ran on the victim machine, some attempt to display a blocking message on top of the UI, while others use a form of click high-jacking to cause the user to allow it higher privileges.

12.4.4.2 Countermeasures

Since most of the ransomware families implement naive locking or encryption techniques, different detection methods have been proposed over time. Kharraz et al. [35, 36] proposed a scheme called UNVEIL to continuously monitor ransomware infection by automatically generating data in artificial user environments, where these are created for any suspicious activities. However, deployment and monitoring of artificial users put significant overhead on execution time and resources without considering kernel applications. Scaife et al. [37] proposed an early warning scheme that analyzes variable type changes, similarity measurements, and entropy of user data. This method requires significant data analysis and is not suitable for early detection. Sgandurra et al. [38] presented a dynamic analysis of Ransomware for higher detection accuracy using signature matching and monitoring dominant features such as API calls, which may exhibit limitations for obfuscated sub-routines in the malware.

All the schemes, as mentioned earlier, implement software-only schemes and suffer from intrinsic limitations. Therefore, it is apparent that the existing software-only techniques are not adequate to thwart Ransomware attacks. Another effective approach can be a hardware-assisted dynamic Ransomware detection scheme by utilizing a commodity computing platform. Hardware performance counters (HPCs) in the performance monitoring units (PMUs), commonly available in recent generation processors [39, 40], monitor events at micro-architectural level. HPCs can be utilized to acquire hardware-level activity information for quickly and accurately detect Ransomware. Although originally designed for performance monitoring, HPCs (and PMUs) can be intelligently used for security by analyzing whether a run-time event profile is malicious or not.

12.4.5 Buffer Overflow

Buffer overflow is the most dangerous software vulnerability. The exploitation of buffer overflow may cause severe security issues for a program. It may leak out sensitive information [41] or allow arbitrary code execution to perform code injection, unauthorized memory R/W, denial of service (DoS), etc. Since the invention of this vulnerability, buffer overflow is still prevalent in most software applications [42]. In this section, we discuss buffer overflow, how exploitation happens, and the countermeasures.

The primary objective of buffer overflow attacks is to possess the control flow of the program. Buffer overflow attacks are possible when an array or pointer allows out-of-boundary memory access in the memory, i.e., when a lack of boundary checking for an array or pointer access causes arbitrary read or write in the stack or heap memory.

An example of buffer overflow is shown in Listing 12.4. In the example, an array $buffer$ is declared with the size of 100. The index, i, equals to 100, which is being used to access the buffer. As $i \notin [0, 99]$ for the declared array, this memory access is out of bound, and hence this causes buffer overflow.

Listing 12.4 Basic example of buffer overflow

```
1    int  buffer[100];
2    int  i = 100;
3    buffer[i] = 12;
```

The above example shows the simplistic manner of buffer overflow. The above-mentioned buffer overflow case can be exploited when the index can be controlled by the adversaries. If we rewrite the code as in Listing 12.5. In this example, the index of the array comes from the user input. Hence, the adversary can read some arbitrary data from memory.

Listing 12.5 Basic example of buffer overflow

```
1    int  buffer[100];
```

```
2      int i;
3      scanf("%d", &i);
4      buffer[i] = 12;
```

On the other hand, there are many built-in APIs that are vulnerable to buffer overflows, such as *scanf* and *strcpy* in C programming. Those vulnerable APIs do not check boundaries while manipulating an array or pointer. Hence, attackers can exploit these APIs and cause severe threats to the program.

12.4.5.1 Exploit of Buffer Overflow

In order to explain how buffer overflow can be exploited, a stack smashing attack has been discussed. The typical memory layout of a C program is shown in Fig. 12.5. While executing a function in a program, all variables and functions and relevant data are stored in the stack. Before jumping to another function for execution, the processor stores the return address of the caller function in the stack. The attackers intend to modify this return address in the stack, so that after executing the callee function, it jumps to the adversary's desired instruction address. The replacement

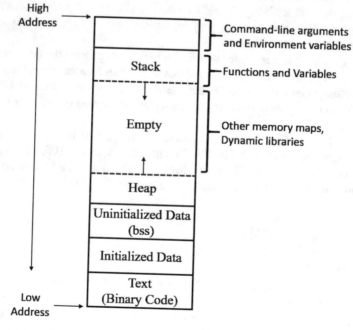

Fig. 12.5 A typical memory layout of a C program. Stack address goes down while heap address goes up

of the return address becomes possible through the buffer overflow. An example is
shown in Listing 12.6 to illustrate the attack.

Listing 12.6 An example of buffer overflow exploitation

```
1       void  secret (){
2           //PASSWORD
3           . . . . . . . . .
4       }
5       void  callee (){
6           char  buffer [30];
7           scanf ("%s", buffer );
8           . . . . . . . . .
9       }
10      void  caller (){
11          callee ();
12      }
```

The stack memory content for the example code is shown in Fig. 12.6. The
attacker initially analyzes the program's binary file by dissembling it and extracts
the necessary information such as the *secret* function's return address and the
address of return of the *caller* function. Then, the attacker defines the payload as
the input for *scanf* API. For instance, assume that for a 32-bit program, the address
of *secret* function is $0x080484A6$. Analyzing the offset between the *buffer* and
return address of *caller* function, the payload would be as follows to exploit the
above buffer overflow program:

$$Payload : [42(38+4) \ bytes \ of \ garbage \ data, \ 0x080484A6] \qquad (12.1)$$

Once the caller return address is modified, the *callee* function returns to the
secret function instead of *caller* function. Thus, the attacker can execute an
unauthorized function by exploiting the buffer overflow. Similarly, the attacker can
also execute arbitrary programs exploiting the function in the C program.

Fig. 12.6 Stake memory
status. The attacker's primary
target is to replace the address
of return with the address of
secret () function

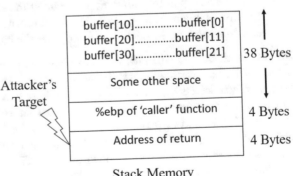

Stack Memory

12.4.5.2 Countermeasures

There are both software and hardware approaches for the protection of software against the buffer overflow attacks. The techniques for buffer overflow detection can be divided into four categories.

- Static Analysis: Several tools, such as Frama-C and KLEE [43], have been developed for detecting buffer overflow vulnerabilities. Static analyzers inspect the code in static time and search for the use cases of arrays and pointers in the user code and verify the boundary condition of the buffer overflow traversing as many control paths as it can. There are many static approaches for detecting buffer overflow, such as [44–47].
- Compiler Modifications: When the source code is available to the users, it might be convenient to use a modified compiler to insert the buffer overflow detection codes automatically, such as StakeGuard [48], ProPolice, StackShield, RAD, etc. A detailed description is provided about the StackGuard approach in the latter part of this section. The ProPolice [49] compiler reorders the variables of the program. Arrays are in the front of pointers to prevent indirect attacks, where local variables are after the pointers from arguments to ease the detection of buffer overflows. RAD [50] comes as patch for GCC compiler. It inserts the detection code in the epilogues and prologues of the function calls. Similarly, as StackSchield, it stores the return addresses in storage.
- OS Modifications: OpenBSD [51] is one of the most comprehensive sets of changes for detecting and preventing buffer overflows. The developer modifies the binary by combining stack-gap randomization with the ProPolice compiler to make the attacks more difficult. Moreover, the developer modifies the OS allocated memory segments to remove the executing permissions as much as possible. Libsafe [52] enforces safe functions for all function calls. The safe functions always check the boundary condition of a buffer.
- Hardware Modifications: SmashGuard [53] proposed to modify the micro-coded instructions for *CALL* and *RET* opcodes. This approach crests a secondary stack on the memory of the processor. The modified *CALL* instruction stores a copy of the return address in the secondary stack while keeping this in the primary stack. Upon returning after executing the *CALL* instruction, the *RET* instruction compares the return address stored in the primary and secondary stacks. Mismatch in the comparison indicates the buffer overflow.

StackGuard: StackGuard [48] is a compiler extension that can prevent the stack smashing buffer overflow attack. It can circumvent the attack in two approaches. One approach detects the change of return address before returning from the callee function. Another approach prevents writing the return address. The first approach is more effective and portable. Here, we discuss the first approach. To detect the change of the return address, a randomly generated canary word is inserted before the return address in the stack, as shown in Fig. 12.7.

When the callee function returns, it checks whether the canary word has been tampered. As the canary word is not known to the attackers, they modify the canary

Fig. 12.7 Canary-based
buffer overflow protection

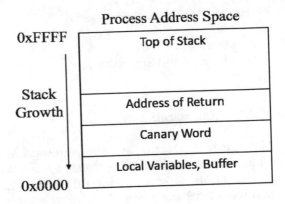

word while replacing the return address during buffer overflow attack. Thus, the change of the canary word can detect the exploitable buffer overflow attack.

On the other hand, MemGuard [54] protects the return address from tampering.

12.4.6 Integer Overflow

Many critical integer overflow vulnerabilities are reported, such as in OpenSSH [55], Mozilla Firefox [56], etc. In those real attacks, arbitrary code was executed, exploiting the vulnerabilities. Several reasons cause integer overflow in a program. The typical reason is a lack of concern about signed vs. unsigned types and incorrect usages of these types. Also, wrong typecasting may lead to integer overflow vulnerabilities. However, the detection of integer overflow bugs is challenging as all overflows are not bugs.

Integer overflow happens when the resultant due to an arithmetic operation at run-time exceeds the maximum value that the resultant variable can hold. An integer overflow scenario is shown in Listing 12.7. Assume that the size of declared unsigned int variables is 8-bit. At line 3, the maximum value for z would be 65025 $((2^8 - 1)^2)$, which cannot be presented for an unsigned integer type of variable. Hence, an integer overflow occurs in this case.

Listing 12.7 An example of integer overflow

```
1        unsigned int  x;
2        unsigned int  y;
3        unsigned int  z = x * y;
```

An integer overflow may lead to a buffer overflow in some cases and can be exploitable. For example, due to an integer overflow, if the overflowed value becomes a very large number, which is used to allocate the dynamic memory as the code in Listing 12.8.

Listing 12.8 An example of integer overflow

```
1
2        size = a − b;
3        ptr =   malloc(size);
```

12.4.6.1 Countermeasures

Many kinds of research have been performed in order to detect integer overflow vulnerabilities. Most of the approaches are software-based. Brumley et al. [57] present a static approach to detect the integer relevant vulnerabilities. There are also some other static analysis-based techniques, such as EXE [58] and KLEE [43]. IntFlow [59] leverages static information flow tracking and dynamic program analysis. This approach mainly focuses on potentially exploitable vulnerabilities. IntFlow improved arithmetic error detection accuracy utilizing information flow tracking, but it lacks implicit information-flow support. In this section, we discuss two techniques for the detection of integer overflow.

Taint Analysis-Based Integer Overflow Detection [60]: Most of the integer overflow cause due to the untrusted data received as a user input source, such as input files or command line. However, the severity caused by integer overflow depends on how the overflowed value is being used in the program. When integer overflow occurs in some sensitive points, which are called sinks, it might lead to severe vulnerabilities. Typically, the sink points might be memory allocation (e.g., malloc in C), array access (overflowed value as index or pointer offset), branch statement, etc. Most of the integer overflow vulnerabilities occur due to the misuse of overflowed values in sinks. Hence, tracking the tainted data and checking whether a path has sufficient checks for integer overflow can prevent integer overflow. Instead of checking all cases of integer overflows, it might be efficient to check whether arithmetic operation causes overflow and tainted data is in the sink. An overview of the technique is shown in Fig. 12.8. At first, the binary program is decompiled to generate an intermediate representation (e.g., Panda) and generate the control flow graphs (CFG). The technique marks the external input sources as tainted data and monitors how this untrusted data affects the program's other data. A symbolic memory environment is used to execute the binary program symbolically. Traversing all feasible paths in the program utilizing a symbolic execution engine, for each sink where tainted data propagates, integer overflow checking is performed. Thus, this technique narrows down the search space for integer overflow vulnerabilities.

Tool—Integer Overflow Checker (IOC): Dietz et al. [61] proposed a tool, IOC that elaborates Integer Overflow Checker, which is a dynamic tool for finding integer overflow in C and C++ codes. This tool is integrated with Clang-LLVM as a pass from the LLVM version of 3.1. The architecture of the tool, IOC is shown in Fig. 12.9. It has mainly two parts, compile-time instrumentation transformation and run-time handler. The transformation inserts numerical error-checking logic through

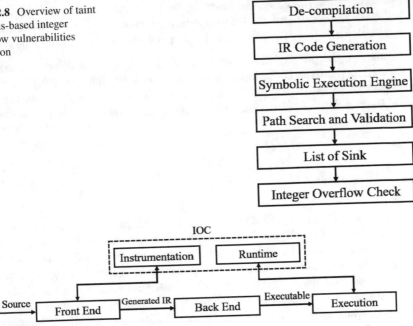

Fig. 12.8 Overview of taint analysis-based integer overflow vulnerabilities detection

Fig. 12.9 Architecture of Integer Overflow Checker (IOC)

a compiler pass. The transformation happens on the Abstract Syntax Tree (AST), which is generated from the user C or C++ code. The Clang LLVM compiler is used to generate the AST. During run-time of the program, if an integer overflow-checking logic fails, it indicates the integer overflow. Detected integer overflow cases are recorded for future developer evaluation.

Integer overflow detection is trickier. In case of n-bit addition or subtraction types of arithmetic operations, $(n + 1)$ bits of precision are required, while $2n$ bits are required for n-bit multiplication; when the result exceeds this specified length in bits, integer overflow occurs. For example, IOC detects integer overflow due to an operation on two signed integers s_1 and s_2 in three ways, which are described below:

- Pre-condition Test: Without executing the instructions in real time, an overflow can be detected through the static analysis when the following expression becomes $TRUE$ for an arithmetic operation of between s_1 and s_2:

$$((s_1 > 0) \wedge (s_2 > 0) \wedge (s_1 > (INT_{MAX} - s_2))) \vee$$
$$((s_1 < 0) \wedge (s_2 < 0) \wedge (s_1 < (INT_{MIN} - s_2)))$$

- CPU Flag-Based Post-condition Test: Many processors have hardware support to detect the arithmetic overflow. IOC utilizes LLVM intrinsic functions that can provide the overflow flag. For example, an addition operation returns a structure of both the result and the corresponding overflow flag. The LLVM backend generates necessary code to check the proper CPU flag for a specific processor.

However, the primary disadvantage of IOC is that it can detect vulnerability only during run-time. If the input vector fails to trigger the bug, integer overflow will go undetected. Again, IOC tools incur significant overhead in terms of code size. Despite these disadvantages, IOC is pretty successful in detecting the integer overflows.

12.5 Summary

Software is an integral demand in our daily life nowadays. The software developed without security in mind can lead to severe damages to our lives and properties. Hence, strengthening the software and making it resilient against attacks are the prerequisites for using the software. Understanding the critical software vulnerabilities can help the developers to avoid those vulnerabilities while developing the software. The developers should act as the attackers while developing the software, which eases to find potentially exploitable vulnerabilities. Therefore, we discuss different kinds of software and hardware vulnerabilities in this chapter. We discuss how the adversaries can exploit those vulnerabilities and how the vulnerabilities can be prevented. However, software-based protection scheme does not always provide an efficient solution. Only software-based approaches may incur significant overhead to the implementation of the system and degrade the overall performance. Therefore, hardware-assisted countermeasures should be more emphasized in preventing software vulnerabilities.

References

1. G. McGraw, Software security. IEEE Secur. Privacy **2**(2), 80–83 (2004)
2. https://www.intel.com/content/www/us/en/support/articles/000025873/technologies.html
3. https://software.intel.com/content/www/us/en/develop/topics/software-guard-extensions.html
4. T. Alves, D. Felton, *Trustzone: Integrated Hardware and Software Security-Enabling Trusted Computing in Embedded Systems (July 2004)* (2014)
5. https://developer.amd.com/sev/
6. https://www.intel.com/content/dam/www/public/us/en/documents/product-briefs/tdt-product-brief.pdf
7. S. Gueron, Intel advanced encryption standard (AES) instructions set. Intel. White Paper Rev. 3, 1–94 (2010)
8. C. Percival, *Cache Missing for Fun and Profit* (2005)
9. D.J. Bernstein, *Cache-Timing Attacks on AES* (2005)
10. D.A. Osvik, A. Shamir, E. Tromer, Cache attacks and countermeasures: the case of AES, in *Cryptographers' Track at the RSA Conference* (Springer, New York, 2006), pp. 1–20

11. Y. Zhang, A. Juels, M.K. Reiter, T. Ristenpart, Cross-vm side channels and their use to extract private keys, in *Proceedings of the 2012 ACM Conference on Computer and Communications Security* (2012), pp. 305–316

12. Y. Zhang, M.K. Reiter, Düppel: retrofitting commodity operating systems to mitigate cache side channels in the cloud, in *Proceedings of the 2013 ACM SIGSAC Conference on Computer and Communications Security* (2013), pp. 827–838

13. D. Gullasch, E. Bangerter, S. Krenn, Cache games–bringing access-based cache attacks on aes to practice, in *Proceedings of the 2011 IEEE Symposium on Security and Privacy* (IEEE, New York, 2011), pp. 490–505

14. E. Tromer, D.A. Osvik, A. Shamir, Efficient cache attacks on AES, and countermeasures. J. Cryptol. **23**(1), 37–71 (2010)

15. S. Wang, P. Wang, X. Liu, D. Zhang, D. Wu, Cached: identifying cache-based timing channels in production software, in *Proceedings of the 26th {USENIX} Security Symposium ({USENIX} Security 17)* (2017), pp. 235–252

16. G. Doychev, B. Köpf, L. Mauborgne, J. Reineke, Cacheaudit: a tool for the static analysis of cache side channels. ACM Trans. Inf. Syst. Secur. (TISSEC) **18**(1), 1–32 (2015)

17. G. Doychev, B. Köpf, Rigorous analysis of software countermeasures against cache attacks, in *Proceedings of the 38th ACM SIGPLAN Conference on Programming Language Design and Implementation* (2017), pp. 406–421

18. R. Brotzman, S. Liu, D. Zhang, G. Tan, M. Kandemir, CASYM: Cache aware symbolic execution for side channel detection and mitigation, in *Proceedings of the 2019 IEEE Symposium on Security and Privacy (SP)* (IEEE, New York, 2019), pp. 505–521

19. Z. Wang, R.B. Lee, New cache designs for thwarting software cache-based side channel attacks, in *Proceedings of the 34th Annual International Symposium on Computer Architecture* (2007), pp. 494–505

20. https://newsroom.intel.com/wp-content/uploads/sites/11/2018/01/Intel-Analysis-of-Speculative-Execution-Side-Channels.pdf

21. https://developer.amd.com/wp-content/resources/Managing-Speculation-on-AMD-Processors.pdf

22. http://www.chromium.org/Home/chromium-security/site-isolation

23. https://webkit.org/blog/8048/what-spectre-and-meltdown-mean-for-webkit/

24. Y. Yarom, K. Falkner, Flush+ reload: a high resolution, low noise, l3 cache side-channel attack, in *Proceedings of the 23rd {USENIX} Security Symposium ({USENIX} Security 14)* (2014), pp. 719–732

25. F. Liu, Y. Yarom, Q. Ge, G. Heiser, R.B. Lee, Last-level cache side-channel attacks are practical, in *Proceedings of the 2015 IEEE Symposium on Security and Privacy* (IEEE, New York, 2015), pp. 605–622

26. M. Lipp, D. Gruss, R. Spreitzer, C. Maurice, S. Mangard, Armageddon: cache attacks on mobile devices, in *Proceedings of the 25th {USENIX} Security Symposium ({USENIX} Security 16)* (2016), pp. 549–564

27. D. Gruss, C. Maurice, K. Wagner, S. Mangard, Flush+ flush: a fast and stealthy cache attack, in *International Conference on Detection of Intrusions and Malware, and Vulnerability Assessment* (Springer, Berlin, 2016), pp. 279–299

28. M. Lipp, M. Schwarz, D. Gruss, T. Prescher, W. Haas, A. Fogh, J. Horn, S. Mangard, P. Kocher, D. Genkin et al., Meltdown: reading kernel memory from user space, in *Proceedings of the 27th {USENIX} Security Symposium ({USENIX} Security 18)* (2018), pp. 973–990

29. D. Gruss, M. Lipp, M. Schwarz, R. Fellner, C. Maurice, S. Mangard, Kaslr is dead: long live kaslr, in *International Symposium on Engineering Secure Software and Systems* (Springer, Berlin, 2017), pp. 161–176

30. C. Beek, D. Dinkar, Y. Gund, G. Lancioni, N. Minihane, F. Moreno, E. Peterson, T. Roccia, C. Schmugar, R. Simon et al., Mcafee labs threats report, in *McAfee, Santa Clara, CA, USA, Technical Report* (2017)

31. K. hutcherson, *Ransomware Reigns Supreme in 2018, as Phishing Attacks Continue to Trick Employees.* https://www.cnn.com/2018/03/27/us/atlanta-ransomware-computers/index.html

32. R. Vamosi, *Wannacry Ransomware Attack Takes the World by Storm.* https://www.synopsys.com/blogs/software-security/how-to-prevent-ransomware-attacks-2019/

33. G. O'Gorman, G. McDonald, Ransomware: a growing menace, in *Symantec Corporation* (2012)

34. *The Wannacry Ransomware Attack has Spread to 150 Countries.* https://www.theverge.com/2017/5/14/15637888/authorities-wannacry-ransomware-attack-spread-150-countries

35. E. Kirda, Unveil: a large-scale, automated approach to detecting ransomware (keynote), in *Proceedings of the 2017 IEEE 24th International Conference on Software Analysis, Evolution and Reengineering (SANER)* (IEEE, New York, 2017), pp. 1–1

36. A. Kharraz, W. Robertson, D. Balzarotti, L. Bilge, E. Kirda, Cutting the Gordian knot: a look under the hood of ransomware attacks, in *International Conference on Detection of Intrusions and Malware, and Vulnerability Assessment* (Springer, Berlin, 2015), pp. 3–24

37. N. Scaife, H. Carter, P. Traynor, K.R. Butler, Cryptolock (and drop it): stopping ransomware attacks on user data, in *Proceeding of the 2016 IEEE 36th International Conference on Distributed Computing Systems (ICDCS)* (IEEE, New York, 2016), pp. 303–312

38. D. Sgandurra, L. Muñoz-González, R. Mohsen, E.C. Lupu, Automated dynamic analysis of ransomware: Benefits, limitations and use for detection. arXiv preprint: 1609.03020 (2016)

39. *Intel® 64 and ia-32 Architectures Software Developer's Manua.* https://software.intel.com/sites/default/files/managed/a4/60/325384-sdm-vol-3abcd.pdf

40. *Arm Cortexa9 Technical Reference Manual—Chapter 11 Performance Monitoring Unit.* https://developer.arm.com/documentation/ddi0433/c/performance-monitoring-unit

41. L. Szekeres, M. Payer, T. Wei, D. Song, Sok: eternal war in memory, in *Proceedings of the 2013 IEEE Symposium on Security and Privacy* (IEEE, New York, 2013), pp. 48–62

42. https://nvd.nist.gov/vuln/search/results?adv_search=true&form_type=advanced&results_type=overview&query=buffer+overflow

43. C. Cadar, D. Dunbar, D. R. Engler et al., Klee: unassisted and automatic generation of high-coverage tests for complex systems programs, in *OSDI*, vol. 8 (2008), pp. 209–224

44. P. Akritidis, M. Costa, M. Castro, S. Hand, Baggy bounds checking: an efficient and backwards-compatible defense against out-of-bounds errors, in *USENIX Security Symposium* (2009), pp. 51–66

45. N. Hasabnis, A. Misra, R. Sekar, Light-weight bounds checking, in *Proceedings of the Tenth International Symposium on Code Generation and Optimization* (2012), pp. 135–144

46. F.C. Eigler, Mudflap: pointer use checking for c/c+, in *GCC Developers Summit* (Citeseer, New York, 2003), p. 57

47. K. Serebryany, D. Bruening, A. Potapenko, D. Vyukov, Addresssanitizer: a fast address sanity checker, in *Presented as part of the 2012 {USENIX} Annual Technical Conference ({USENIX}{ATC} 12)* (2012), pp. 309–318

48. C. Cowan, C. Pu, D. Maier, J. Walpole, P. Bakke, S. Beattie, A. Grier, P. Wagle, Q. Zhang, H. Hinton, Stackguard: automatic adaptive detection and prevention of buffer-overflow attacks, in *USENIX Security Symposium, San Antonio, TX*, vol. 98 (1998), pp. 63–78

49. www.research.ibm.com/trl/projects/security/

50. T.-C. Chiueh, F.-H. Hsu, Rad: a compile-time solution to buffer overflow attacks, in *Proceedings 21st International Conference on Distributed Computing Systems* (IEEE, New York, 2001), pp. 409–417

51. https://www.openbsd.org/

52. A. Baratloo, N. Singh, T.K. Tsai et al., Transparent run-time defense against stack-smashing attacks, in *USENIX Annual Technical Conference, General Track* (2000), pp. 251–262

53. M. Prasad, T.-c. Chiueh, A binary rewriting defense against stack based buffer overflow attacks, in *USENIX Annual Technical Conference, General Track* (2003), pp. 211–224

54. C. Cowan, D. McNamee, A.P. Black, C. Pu, J. Walpole, C. Krasic, P. Wagle, Q. Zhang, *A Toolkit for Specializing Production Operating System Code* (1997)

55. M. Corporation, *Cve-2002-0639: Integer Overflow in SSHD in Openssh* (2002). http://cve.mitre.org/cgi-bin/cvename.cgi?name=CVE-2002-0639

56. *Cve-2010-2753: Integer Overflow in Mozilla Firefox, Thunderbird and Seamonkey* (2010). http://cve.mitre.org/cgi-bin/cvename.cgi?name=CVE-2002-0639
57. D. Brumley, T.-c. Chiueh, R. Johnson, H. Lin, D. Song, *Rich: Automatically Protecting Against Integer-Based Vulnerabilities* (2007)
58. C. Cadar, V. Ganesh, P.M. Pawlowski, D.L. Dill, D.R. Engler, Exe: automatically generating inputs of death. ACM Trans. Inform. System Security (TISSEC) **12**(2), 1–38 (2008)
59. M. Pomonis, T. Petsios, K. Jee, M. Polychronakis, A.D. Keromytis, Intflow: improving the accuracy of arithmetic error detection using information flow tracking, in *Proceedings of the 30th Annual Computer Security Applications Conference* (2014), pp. 416–425
60. T. Wang, T. Wei, Z. Lin, W. Zou, Intscope: automatically detecting integer overflow vulnerability in x86 binary using symbolic execution, in *NDSS* (Citeseer, New York, 2009)
61. W. Dietz, P. Li, J. Regehr, V. Adve, Understanding integer overflow in C/C++. ACM Trans. Softw. Eng. Methodology (TOSEM) **25**(1), 1–29 (2015)

Chapter 13
Firmware Protection

Muhammad Monir Hossain, Fahim Rahman, Farimah Farahmandi, and Mark Tehranipoor

13.1 Introduction

Firmware is the core asset of an embedded system. It defines the functionalities and controls of an embedded application, such as the Internet of Things (IoT). Due to the increased demands for automation in our domestic life, industries, medical sectors, and national defense, embedded devices' usage increases exponentially. In the meantime, security has become a significant concern as devices are being used in sensitive applications, where any incautious activity can cause severe damages to human civilization. Many attacks for tampering the firmware are reported in recent years. Zhang et al. [1], LeMay et al. [2], and Maskiewicz et al. [3] mentioned the firmware tampering for network devices, where Morais et al. [4] mentioned the usage of tampered firmware in the electronic game console. Denial et al. in [5] have shown that the adversaries can insert malware into the printers while updating the firmware remotely. Garcia Jr et al. [6] reported the firmware attack on the car control system.

Various studies have shown several approaches to protect the firmware. J. Nagra and C. Collberg in [7], S. Schrittwieser and S. Katzenbeisser in [8], and Junod et al. in [9] proposed some software-based approaches. Unfortunately, software-based approaches incur significant performance overhead. Additionally, these approaches do not bind the firmware with the specified hardware system. As a result, the adversaries can utilize the same firmware for developing counterfeit hardware solutions. On the contrary, hardware-assisted approaches can provide an enhanced performance without degrading the execution speedup. They can bind the firmware

M. M. Hossain (✉) · F. Rahman · F. Farahmandi · M. Tehranipoor
Department of ECE, University of Florida, Gainesville, FL, USA
e-mail: hossainm@ufl.edu; fahimrahman@ece.ufl.edu; farimah@ece.ufl.edu;
tehranipoor@ece.ufl.edu

© The Author(s), under exclusive license to Springer Nature Switzerland AG 2021
M. Tehranipoor (ed.), *Emerging Topics in Hardware Security*,
https://doi.org/10.1007/978-3-030-64448-2_13

with the hardware to prevent the firmware tampering, cloning, and also reverse engineering.

In this chapter, we discuss the motivation of firmware protection in Sect. 13.2 and firmware assets and corresponding attacks in Sect. 13.3. We discuss encryption and obfuscation-based firmware protection and their pros and cons in Sects. 13.4 and 13.5, respectively. We present different obfuscation techniques in Sect. 13.6. Finally, we conclude the chapter in Sect. 13.7.

13.2 Motivation of Firmware Protection

A firmware development process goes through significant efforts and investments. Figure 13.1 illustrates how a firmware of an embedded system evolves and becomes mature gradually. Initially, the requirements are analyzed for a specific embedded application. The entire embedded system usually divides into several modules. For each module, a modular firmware is developed to observe the feasibility of implementation, i.e., developing the prototype. Prototype testing is required to perform with much care. Once all prototype testing pass successfully, the developers start to implement the entire firmware for the embedded system. Rigorous testing is then required to evaluate the functionalities and security of the whole application. There are lots of back and forth efforts between the testing and development phases. Once all of the issues relevant to functionalities and security are resolved, the product gets released. Although a product goes to the market through lots of tests, the customer may find some issues. Therefore, the developers require investigating

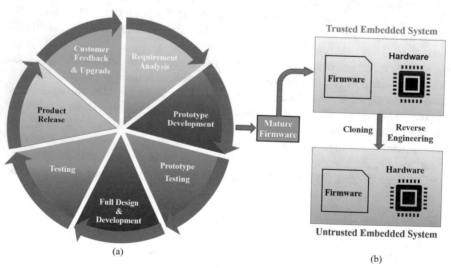

(a)

(b)

Fig. 13.1 (a) Firmware development process. (b) Illegitimate usage of firmware in untrusted embedded system

the issues and upgrading the firmware. Thus, the firmware becomes mature and runs without any issues on the embedded platform.

A company achieves mature firmware for an embedded solution through lots of monetary investments, time, and researches. When the adversaries perform cloning or reverse engineering of the hardware system and firmware at a low cost, they can make more profit in the market. Regretfully, many customers are not aware of the products' genuineness, so the counterfeit systems go undetected. As a result, the original company loses vast profits and brand values as the counterfeit company may sell lower-grade products to the customers. Moreover, the counterfeit entities may tamper sensitive applications, such as medical devices, resulting in serious injury to human beings. Hence, the assurance of the trusted firmware protection is obligatory.

13.3 Firmware Assets and Attacks

The control flow skeleton is the major asset of a firmware. Firmware may also possess secret information, such as keys or experimental data, e.g., calibration coefficients for a sensor module, which are achieved through rigorous experiments. Usually, the firmware is stored in an external memory, which is always vulnerable. The adversaries can extract the plaintext firmware from memory, which can be readily deployed in cloned hardware. Even if the processor has lock bits for protecting the firmware, the adversaries can perform a side-channel attack, such as optical probing, to disable the lock bits and read the firmware [10]. Again, the attacker can snoop the bus, while the processor downloads the firmware from the cloud. The layout of the firmware is also a valuable asset. When the firmware is not obfuscated, the attacker can obtain the sequences of the executing instructions by snooping on the bus. Moreover, recurrence leakage may also happen, which exposes the loops and branches of the firmware. As the firmware has some valuable asset, it should not be kept as plaintext in the program memory. The firmware must be obfuscated before storing it in the program memory to protect its assets from malicious attacks.

13.4 Encryption-Based Protection

Primarily, firmware protection is based on encryption and obfuscation. For the encryption-based technique, the security depends on the encryption scheme itself and the length of the key. There are several standard encryption techniques, such as AES, DES, etc., which are used to encrypt the firmware. During run-time, the processor of an embedded system decrypts the whole firmware and executes the program. However, there are severe performance issues on the embedded system that implements the encryption-based firmware protection scheme. The encryption-based scheme requires decrypting the firmware each time for execution, which

Fig. 13.2 Control flow leak
out from address accesses:
Snooping on the bus provides
the information to the
attackers about the sequence
of instructions that are being
executed. After the execution
of the instructions I_0 and I_1,
program counter jumps to
either I_2 or I_3, which
indicates a conditional
instruction. Executing I_4, I_5
and I_0 instructions in order
indicates a loop

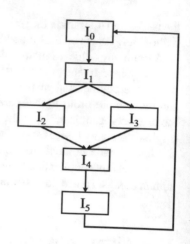

demands a significant decryption time and computational power, and increases memory consumption. Although, hardware-based encryption/decryption engines can improve the performance a little bit, the encryption-based firmware protection is not feasible for most of the low-cost embedded systems. On the other hand, encryption can protect the instructions but fails to keep the control flow unrevealed. An example is illustrated in Fig. 13.2 to show how control flow information gets revealed even if the instructions are encrypted. The leak out of a program's control flow sometimes eases the adversaries' path to attack it. Due to extreme time-to-market pressure, firmware developers often use legacy code-base and modules from the public domain. A study shows that about up to 70% of the code is taken from existing code [11]. A firmware falls under attack when it uses some legacy algorithmic implementation, and the attacker identifies it by analyzing the control flow (comparing the binaries). The attack becomes severe when the data-dependent control flow leaks out the secret information to the attackers. For a conditional branch instruction, some conditional data are compared, and it directs the program to choose which path to execute. As the firmware developers use the typical implementation for cryptographic algorithms, which are most possibly known to the attackers, the critical data may be revealed due to a lack of control flow protection.

For example, Diffie–Hellman and RSA private-key operations [12] compute $R = y^x \bmod n$, where x is the secret key. Consider that the developer implements the simple modular exponentiation algorithm for the computation mentioned above. As this algorithm is ubiquitous, assume that the attacker identifies this in the firmware through CFG matching. The code for the above algorithm and corresponding control flow is shown in Fig. 13.3. Inside of the loop, *IF-branch* (B_2) is executed if a current bit of x is logic 1, otherwise *ELSE-branch* (B_3) is taken. The attacker can identify the respective bits of x by monitoring the address bus from which addresses the instructions are being fetched. If the execution sequence is $B1 \rightarrow B2 \rightarrow B4$, the value of a particular bit of x is 1. On the other hand, for the execution sequence of

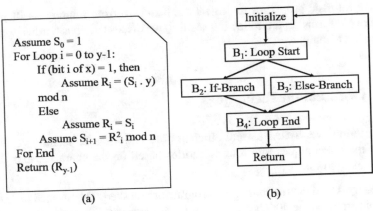

Fig. 13.3 Secret information leak out from modular exponentiation algorithm. (**a**) presents the code and (**b**) reflects the control flow

$B_1 \rightarrow B_3 \rightarrow B_4$, the value of the corresponding bit of x is 0. Suppose the attacker can identify the conditional branch. In that case, the direction of the execution path can be detected by analyzing the instructions' sequence, which eventually provides the branch's comparison results. Again, it is apparent that the path through *IF-branch* takes significant execution time compared to *ELSE-branch*, as it has to perform a multiplication. Hence, a timing channel arises in this case, which can be exploited by the attacker [12].

Although encryption-based firmware protection schemes can protect the plain-text instructions, it fails to protect the firmware's control flow and hidden assets. Hence, the developers must be meticulous while utilizing an encryption-based protection methodology.

13.5 Obfuscation-Based Protection

Obfuscation is the classical approach to hide the control flow of a firmware. The obfuscation technique scrambles the control flow. The objective of obfuscation is to convert the original program into a functionally and semantically equivalent form that is more difficult to understand and perform reverse engineering of the firmware [13]. The obfuscation technique primarily involves software- and hardware-assisted techniques. Software obfuscations are based on layout obfus-cation, control flow obfuscation, or data obfuscation [14]. Software obfuscation incurs significant memory overhead as usually bogus or dead code is inserted in the firmware [14]. Although software obfuscation can hide the control flow, it significantly slows down the program execution. If the hardware information can be utilized in the obfuscation process, it would prevent the running of illegitimate firmware on the authentic hardware. Hence, many techniques evolve leveraging hardware intrinsic properties to protect the firmware from running on illegitimate hardware platforms. In the following section, we discuss various hardware-assisted obfuscation techniques.

13.6 Obfuscation Techniques

In this section, we discuss several efficient obfuscation-based techniques. We can classify these techniques into four categories based on the characteristics that we describe as follows:

1. Category-1: All instructions in the program are masked. The masked instructions remain in the same addresses in the program memory. The techniques utilize a hardware signature to bind the hardware with its firmware. Techniques 13.6.1 and 13.6.2 fall under this category.
2. Category-2: Only some instructions of the program are masked and dislocated into different locations. This kind of technique also leverages hardware intrinsic property for binding the hardware with its firmware. Techniques 13.6.3 and 13.6.4 are examples for this category.
3. Category-3: Program memory blocks are obfuscated dynamically during the execution of the program. This technique does not bind the hardware with the firmware. Technique 13.6.5 falls under this category
4. Category-4: The techniques are based on some key validation. Techniques also do not allow binding the hardware with the firmware. Technique 13.6.6 falls under this category

13.6.1 Instruction-Level Opcode Elimination-Based Obfuscation

In this technique [15], the machine code and the embedded processor authenticate each other for executing each instruction using a Physically Unclonable Function (PUF). The system developer initially prepares the obfuscated firmware from the plaintext instructions and the PUF responses, which is then burned into the processor's program memory. While running the embedded system, the processor recodes each machine instruction.

13.6.1.1 Overview of Technique

The technique allows executing an instruction upon verifying two security aspects. At first, the firmware must recognize the hardware environment as an authentic or trusted platform. Secondly, the processor needs to verify whether the machine

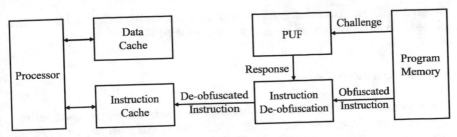

Fig. 13.4 Overview of instruction-level authentication

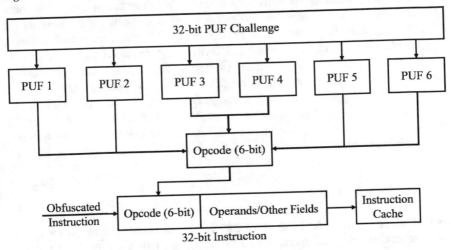

Fig. 13.5 De-obfuscation of instruction

instructions stored in the program memory are authorized to execute. When the requirements, as mentioned earlier, are fulfilled, an instruction can be executed in the proposed method. In the technique, the opcode of each instruction is either missing or garbage. Thus, all instructions without original opcodes are stored in the program memory along with the corresponding PUF challenges. The corresponding challenge is read from the main memory and provided to the PUF device for executing an instruction. The PUF response is then combined with the instruction for the de-obfuscation of the opcode. The de-obfuscated instruction is written into the instruction cache, where the processor fetches and executes the instruction. Figure 13.4 describes the overall technique.

The processor does two kinds of *READ* operations for executing a single instruction. For a requested instruction, the recoding logic fetches the corresponding PUF challenge word from memory. Then, the processor fetches the instruction word, which is in obfuscated format. The de-obfuscation process of the opcode is described in Fig. 13.5 in detail. For the scheme's implementation, OpenRISC OR1200 processor, which has 32 bit of PUF challenge, can be used [15]. Six delay PUFs can be utilized in parallel to de-obfuscate 6 bits of an opcode with the generated response. Finally, the processor can recode the obfuscated instruction and execute it upon receiving the response from the PUF.

13.6.1.2 Overhead Analysis

The technique has a large memory overhead. For each instruction in 32-bit Instruction Set Architecture, four bytes of PUF challenge words must be stored. Hence, memory consumption becomes double as the program size. Moreover, the proposed technique has a significant performance overhead. Each instruction requires some additional cycles to de-obfuscate the opcode as its de-obfuscation depends on the PUF response. For a sequential processor, the proposed technique incurs a significant slowdown of the execution. However, the proposed technique can be an effective solution for the pipe-lined processor. Performing the de-obfuscation process outside of the instruction cache can minimize the performance overhead. On the contrary, firmware managing cost might be increased due to the method. The system integrator needs to compile obfuscated firmware for individual embedded systems. It may require storing them in the database for loading to the systems, which might not be cost-effective.

13.6.1.3 Security Analysis

As PUF response varies from one device to another, it is challenging for the adversaries to find the original instruction. The significant advantage of this technique is that it does not require any key for the authentication and de-obfuscation of the instructions. As a result, the adversaries cannot obtain the original instructions and cannot clone without the firmware's source code.

13.6.2 Instruction-Level Obfuscation Based on XOR'ing with PUF Response

This technique [16] binds the software with the intended hardware platform leveraging hardware intrinsic security properties. The technique ensures that illegitimate hardware cannot run legitimate firmware or vice versa. As the technique obfuscates the firmware using hardware ID, which is unique from one hardware system to another, running the firmware on another hardware platform generates incorrect results. Personalizing the firmware for a specific hardware system makes it very difficult for the adversaries for tampering and cloning.

13.6.2.1 Overview of Technique

In this technique, all instructions of firmware are obfuscated with a PUF response. A masking technique is used to obfuscate the instructions. The masked instruction is determined based on the following equation:

$$I'_i = I_i \oplus R(C_i) \tag{13.1}$$

where C_i is the challenge for an instruction I_i, and R is the challenge/response function for PUF device.

An overview diagram of the technique is shown in Fig. 13.6. Initially, a challenge is applied to the PUF device. The corresponding instruction is then XOR'ed with the obtained response from the PUF device. Similarly, all instructions are XOR'ed with the corresponding PUF responses. Thus, finally, we get the obfuscated version of the original program. As the technique requires n challenges for n number of instructions, it consumes a significant memory. However, memory consumption can be reduced in two ways. An obfuscated instruction can be used as the challenge for the next instruction, as the following equation:

$$C_i = I'_{i-1} \tag{13.2}$$

Choosing the obfuscated instructions as the challenges can get rid of huge memory consumption. Only the challenge (C_0) for the first instruction is required to store in memory. Otherwise, the challenge (C_0) can also be generated by a secure Pseudo-Random Number Generator (PRNG). The function R can be a hash or block cipher.

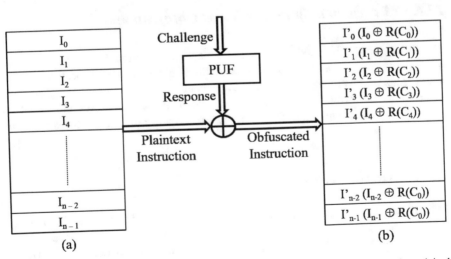

Fig. 13.6 Overview of obfuscation technique based on PUF. Table (**a**) shows the original instructions and Table (**b**) shows the obfuscated version of the firmware

13.6.2.2 Overhead Analysis

The technique does not slow down the execution of sequential instructions because the previous obfuscated instruction, which is the challenge for the current instruction, is already in the CPU memory for the de-obfuscation. In the case of $RETURN$, $JUMP$, or $CALL$ instruction, the program counter jumps from one address to another, requiring pulling the previous obfuscated instruction from memory before executing the current instruction. As a result, the execution takes some extra cycles. Overall, the percentage of instructions for which program counter jumps is significantly low compared to arithmetic or logical instructions for a program. Hence, the performance degradation is not noticeable for a reason, as mentioned earlier. On the other hand, significant overhead might incur if the hardware platform takes much clock cycles to generate the response ($R(C_n)$).

13.6.2.3 Security Analysis

The security of the technique depends on the secrecy of any two values among instructions (I_n), response function ($R(C_n)$), and obfuscated instructions (I'_n). The adversaries may access the obfuscated instructions as these are usually stored in the off-chip program memory. As the instructions are de-obfuscated within the CPU, the adversaries cannot obtain them. Until the attackers reveal the PUF responses, they cannot estimate the original instructions as the obfuscation process involves XOR operation between plaintext instruction and PUF response.

13.6.3 Instruction Swapping-Based Obfuscation

This technique [17] is based on swapping a subset of all instructions in the firmware without any encryption. Swapping of instructions hides the actual locations of the instructions to the adversaries. During the execution of the program, the processor reconstructs the original instructions from the memory that stores the swapped instructions. The technique utilizes a hardware signature to bind the firmware with specific hardware.

13.6.3.1 Overview of Technique

Initially, some instructions are chosen for swapping. In this technique, the security depends on how it is difficult for the adversaries to find out the swapped instructions. Therefore, the process for selecting instructions for swap should meet certain criteria or rules for creating a maximum illusion to the adversaries. Here, some rules are discussed to find out the best candidates for swapping.

- Branch Instructions: A branch instruction should not be swapped. If the branch instruction is swapped, the adversaries can identify this by monitoring in run-time. When the adversaries notice the instruction being executed from an unexpected location, it indicates the instruction swap.
- Function Headers and Footers: Some instructions, such as HLT or RETURN, act as headers and footers for a function. Removing these instructions provide an easy guess for the adversaries about the swaps. Therefore, these instructions should not be chosen for swapping.

Once suitable instructions are selected in order to swap, it is also important to make some pairs in such a way the adversaries cannot find while analyzing the sequences of instructions. Here, some rules are discussed for making possible swaps.

- Equivalent Instructions: If two instructions perform the same operation and also have the same operands, these instructions should not be paired.
- Register Initialization: When a swapped instruction is initialized somewhere else, not in few immediate instructions, the adversaries can easily acknowledge this as swapped instruction.
- Register Utilization: If the destination register of instruction is never used, the adversaries can consider this as a swapped instruction. Hence, this needs to be considered while making a pair.
- Redundancy Operation: Instructions should not be swapped into the location when the swapped instruction is just bogus or redundant.
- Index Distinction: If the processor has an instruction cache implementation, the index of the cache mapping should not be the same for two candidate instructions for swapping. It might create a collision in the cache.

An obfuscation key (K_O) is formed based on the addresses of the swapped instructions. Assuming that the obfuscation process has m number of swaps, then K_O can be expressed as the below equation:

$$S = (I_1 \Longleftrightarrow I_2), (I_3 \Longleftrightarrow I_4) \ldots \ldots (I_{2m-1}) \Longleftrightarrow I_{2m}) \tag{13.3}$$

$$K_O = [Addr_1, Addr_2, \ldots \ldots, Addr_{2m-1}, Addr_{2m}] \tag{13.4}$$

where S represents the set for pairs of swapped instructions, and $Addr_{2m}$ represents the memory address of last instruction in m^{th} pair. As for each swap, the memory address of two instructions belongs to the key, and the length of the obfuscation key is as follows:

$$L_{KO} = N \times m \times 2 \tag{13.5}$$

where N represents the length of instruction and, m represents the number of swaps.

When the obfuscation key is loaded into the embedded devices, the adversaries can clone the key and eventually would be able to de-obfuscate the firmware. In order to protect the firmware from cloning, a unique hardware signature can be utilized. If a unique key for a specific system is derived as the following equation, the system can be secure.

$$K = ID_{HW} \oplus K_O \tag{13.6}$$

ID_{HW} can be generated from the PUF response, which is obtained during run-time. This ID can also be used to maintain the database records for future usage of a hardware system.

To de-obfuscate the program during run-time, an on-chip memory device such as cache memory is required. During boot-up, the processor generates the PUF response and calculates the obfuscation key. The processor retrieves the original instructions by XOR'ing the key and the obfuscated instructions and, stores them in a reorder cache memory. When the processor requests an instruction from a specific memory location and if the instruction is swapped (i.e., instruction is available in the reorder cache), it should be fetched from the reorder cache memory for the program execution.

13.6.3.2 Overhead Analysis

The technique does not store any extra bytes in the program memory for the firmware's obfuscated version. Although boot-loader code needs much memory as this has to perform hardware ID generation, retrieve the swapped instructions, and store them in the recorder cache memory for the execution. On the other hand, the system's SRAM can be used for generating hardware ID (PUF); therefore additional hardware requirements will be reduced for the ID generation.

13.6.3.3 Security Analysis

The technique utilizes hardware fingerprint while obfuscating the firmware, which is unique from one hardware to another. As a result, it becomes challenging for the adversaries to obtain the original version of the firmware. Besides, the attacker's effort to obtain the firmware can be considered as a measure of how robust the obfuscation method is. The attacker's effort can be defined as how many trials or arrangements an attacker requires breaking an encrypted or obfuscated system.

To find out L swaps in the firmware, it is required to calculate the number of ways the attacker can choose L disjoint edges. The attacker chooses 2 instructions out of N, multiplied by the number of ways the attacker chooses 2 instruction out of the remaining $(N-2)$ instructions, and so on. As the order of swaps needs not to be considered, the term should be divided by $L!$. So, the expression of the attacker's effort for the worst case becomes as below:

$$AE_{\text{worst}} = \left(\frac{1}{L!}\right)\binom{N}{2}\binom{N-2}{2}\cdots\binom{N-2L+2}{2}$$

$$= \frac{N(N-1)\ldots(N-2L+1)}{2^L L!} \tag{13.7}$$

In the best situation for the attackers, every swappable instruction can only be swapped with only one other instruction. The adversary requires to choose L edges from $N/2$ probable swaps. Hence, in this case, the efforts to find the remaining swaps are as follows:

$$AE_{\text{best}} = \binom{N/2}{L} = \frac{N(N-2)\ldots(N-2L+2)}{2^L (L)!} \tag{13.8}$$

From the above attacker's effort equations, we can conclude that even for a small program, the number of trials is exceptionally high, which is difficult for the adversaries to obtain the original firmware. Thus, the technique can protect the firmware from the adversaries from being cloned. Again, tampered firmware cannot run as the firmware binds the hardware with the system ID (SID).

13.6.4 Instruction Relocation-Based Obfuscation

This technique [18] allows mutual authentication where hardware authenticates the firmware and vice versa using a PUF based hardware system ID (SID). In the technique, some instructions are displaced to a separate location in the program memory.

13.6.4.1 Overview of Technique

Initially, a system ID is generated from the chips and PCBs available in the embedded system [18]. The SID is then used to obtain obfuscated instructions by XOR'ing the plaintext instructions with it. The architecture for the constructions of machine code is shown in Fig. 13.7. The number of obfuscated instructions depends on the length of SID and Instruction Set Architecture. For 32 bits of ISA, such as MIPS32, 32 instructions can be obfuscated when the SID has 1024 bits in length. In this case, every 32 bits of the SID segment is XOR'ed with an instruction to generate the obfuscated instruction. The system integrator generates an obfuscation key based on the obfuscated instructions (OI),

$$K_i = SID_i \oplus OI_i \tag{13.9}$$

where $i = [0, 31]$.

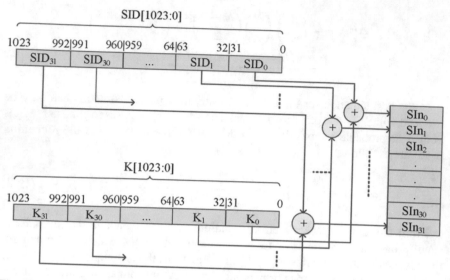

Fig. 13.7 Architecture for generating machine instructions for obfuscation [18]

The key has to be stored in a secure memory, one-time programmable memory [19]. During run-time, the processor retrieves the original instruction by XOR'ing the SID and the key.

However, the detailed process flow for obfuscation in design time and de-obfuscation in run-time is shown in Fig. 13.8. Before starting the obfuscation process, Directed Acyclic Graph (DAG) is generated using a Depth-First Search algorithm [20] to gather all control flow paths. Then, some instructions are removed greedily in such a way so that the execution of the program generates incorrect results. The obfuscated firmware is loaded into the embedded system's non-volatile memory, which the adversaries can access. On the other hand, the removed instructions and their relative addresses are stored in on-chip non-volatile memory (NVM) so that the adversaries cannot access them. As a result, the adversaries cannot construct the original instruction sequences and execute the firmware on illegitimate hardware. However, during run-time, the processor needs to retrieve the removed instructions during boot-up and insert them to their corresponding locations in the obfuscated firmware after de-obfuscating the displaced instructions.

13.6.4.2 Overhead Analysis

The main overhead comes due to the boot memory section. As the technique requires de-obfuscating and reconstructing the original firmware from the obfuscated version and writing page by page of the program memory, the system requires much memory and execution time depending on the size of the memory page and firmware. Moreover, the system requires an on-chip NVM to store

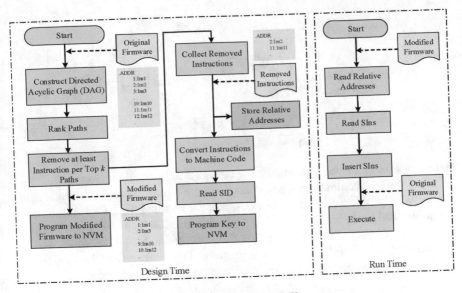

Fig. 13.8 Firmware obfuscation and de-obfuscation flow [18]

the removed instructions and corresponding relative addresses, which incurs an additional overhead. The technique also requires a one-time pad (OTP) for the de-obfuscation. The size and overhead depend on the size of the SID. Finally, a one-time programmable memory is required to store the obfuscation key.

13.6.4.3 Security Analysis

The attacker's effort to break the obfuscated firmware is estimated here to evaluate the technique's security. Let us assume that there are N instructions in firmware, and I number of instructions is removed for dislocation. There are m paths in total, of which k number of paths is modified in the firmware for obfuscation. At first, the attacker requires to find the k path, for which the attacker's efforts will be $\binom{m}{k}$. Assuming that there are r instructions (average) in every path, at least an instruction has been removed for obfuscation from each path. The lengths of m paths are $L_1, L_2, \ldots, L_m$. For a specific path P_1, the attacker's effort will be as follows:

$$E_{P_1} = \sum_{i=1}^{r} \binom{L_1}{i} \times C \times M \tag{13.10}$$

where M represents the types of instructions for an ISA, and C is the average number of combinations for the operands of an instruction.

The total attacker's effort for all paths becomes as follows:

$$E_P = \binom{m}{k}\{E_{P_1} + E_{P_2} + \ldots + E_{P_k}\}$$

$$= \binom{m}{k}\left\{\sum_{i=1}^{r}\binom{L_1}{i}C \times M + \ldots + \sum_{i=1}^{r}\binom{L_k}{i}C \times M\right\}$$

$$= \binom{m}{k}\left\{\sum_{j=1}^{k}\sum_{i=1}^{r}\binom{L_j}{i}C \times M\right\} \qquad (13.11)$$

Here, $k \times r = I$, I is the total number of constructed instructions from SID.

When the adversaries do not know the firmware's functionality, it is challenging to correlate the obfuscated instructions. Again, the required trials for reconstructing the obfuscated firmware are incredibly high, as estimated from the above equation. Hence, the system is much secure under this obfuscation technique. Moreover, the technique utilizes PUF-based hardware intrinsic ID in obfuscation. As a result, any tampered firmware cannot run on the authentic hardware or vice versa.

13.6.5 Program Memory Block Shuffling-Based Obfuscation

This technique [21] obfuscates the firmware's control flow by dynamic shuffle, which protects the firmware from leaking out layout and recurrence leakage.

13.6.5.1 Overview of Technique

The technique is based on shuffling the program memory blocks each time they are written out by the processor in the program memory. The same instructions are loaded from different locations in the memory. As a result, the adversaries face a tremendous challenge in tracking the program execution pattern, i.e., the instructions' layout. An overview block diagram of the technique is shown in Fig. 13.9. The technique utilizes a shuffle buffer, which is inside of the processor. Shuffle buffer assists in relocating the blocks in different locations, which evokes the block address table (mapping of current address to the original address of the program blocks) to update the current location. In order to access the block address table faster, a cache is considered. When the processor requests an instruction, the controller initially looks for the instruction address in the block address table cache. If the controller does not get the address there, it fetches the current address from the block address table and updates the corresponding cache. Then, the controller fetches the corresponding instruction from the shuffle buffer. If the instruction is not available in the shuffle buffer, the controller evicts a block from it when it is full and fetches a new block from the program memory into the shuffle buffer. In the technique, the crucial part is shuffling. An example has been shown in Fig. 13.10 to

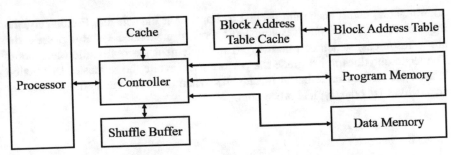

Fig. 13.9 Overview of program memory block shuffling-based obfuscation technique

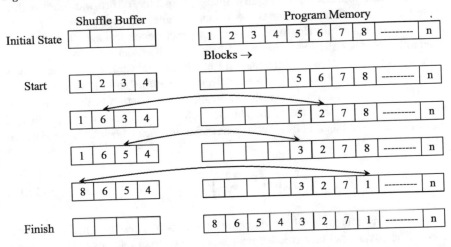

Fig. 13.10 Shuffling technique of program memory blocks

illustrate how the shuffling technique works. Assume that there are n program blocks and the shuffle buffer can contain 4 blocks. Initially, the shuffle buffer is empty, and so the controller fills up the shuffle buffer by fetching the first 4 blocks from the program memory. The first four blocks of the program memory remain empty during the program's entire execution. When block 6 is accessed, the controller evicts a block randomly from the shuffle buffer and places this into the incoming block. In this case, randomly evicted block is 2. The newly fetched block 6 takes the place of the recently evicted block in the shuffle buffer. Similarly, the shuffling of blocks continues until the program ends. At the end of execution, all blocks from the shuffle buffers are written into the first 4 empty blocks in the program memory.

13.6.5.2 Overhead Analysis

In order to implement this obfuscation scheme, several hardware modules are required. A Pseudo-Random Number Generator (PRNG), a shuffle buffer, a block address table, and a cache are needed to implement the scheme. Fortunately, the implementation is not much costly. Most modern embedded systems possess the above resources, which can be mapped to implement the obfuscation technique. The technique does not degrade the performance, while the requested instruction is available in the shuffle buffer (cache hit). When there is a cache miss, a penalty occurs likewise other typical processors.

13.6.5.3 Security Analysis

This is a robust obfuscation technique from the perspective of firmware control flow protection. The technique prevents the layout leakage, tampering, and clone type of counterfeiting, while also protecting against recurrence leakage. The attackers need to know when the previous read-in block is written out back to the program memory to obtain recurrence information (recurrence leakage), which is extremely challenging as the processor evicts a block from the shuffle buffer randomly. Hence, the probability of selecting a shuffle buffer block for eviction is the same for all blocks. Assume that the size of the shuffle buffer is M. Then, the probability of previously read-in block getting written out after k times of evictions becomes as below:

$$P_k = \left(1 - \frac{1}{M}\right)^{k-1} \times \frac{1}{M} \qquad (13.12)$$

The probability increases with a decrease of k and reaches the maximum when $k = 1$. The adversarial best case happens when all of their guesses about evictions are correct. In this case, the probability becomes as follows for n recurrences of an instruction:

$$P_{max} = \frac{1}{M^n} \qquad (13.13)$$

From the above equation, it is apparent that the recurrence leakage reduces exponentially with the increase of shuffle buffer size for multi-recurrence. Hence, this technique is a great choice to prevent recurrent leakage also.

13.6.6 Key-Based Control Flow Obfuscation

This section discusses a key-based firmware protection scheme against software piracy, reverse engineering, and tampering for embedded systems. The technique [20] involves a non-cryptographic and key-based approach for control flow obfuscation, which requires minimal hardware.

13.6.6.1   Overview of Technique

The technique generates a sequence of keys while executing the program and compares the keys with some golden stored values in the memory. The values of keys and the sequences depend on the firmware control flow and input values. For different input values, the key values are changed. Suppose the key values are not validated with the corresponding golden values. In that case, the program execution leads to the wrong path and generates incorrect results. However, the memory locations for the golden values are determined in the program; they are not hard-coded in the firmware. The technique generates a control-flow graph accumulating all possible firmware paths using a node traversing algorithm such as Depth-First Search (DFS) and relevant branch dependencies on the input values. The program is modeled as a Directed Acyclic Graph (DAG), which eliminates the loops. A node of the generated graph represents an instruction with one or two children nodes (one of them is the next instruction always). To enhance the protection, it is required to find out the optimal locations to modify the firmware (i.e., inserting instructions to derive the required memory addresses of golden key values) to achieve maximum impact on multiple paths for a single modification of the control flow. Before modification, the graph's edges are ranked in descending order according to the number of paths. In order to make efficient modifications, the following rules can be considered:

- Any two modifications should not be very close to one another to make it difficult for the adversaries.
- First modification should not be between the root node and the node corresponding to one of the edges that connect the root node. Otherwise, the adversary can easily identify the first modification.
- To increase the security level, the instruction-level executions for deriving and comparing the keys' sequence should not appear consecutively.

13.6.6.2   Overhead Analysis

In this technique, some additional instructions are inserted to generate addresses of some keys, which are compared to golden values. These additional instructions increase the memory consumption a little bit and also the execution time.

13.6.6.3   Security Analysis

The security depends on how efficiently the scheme has been developed to modify the code. If the modification seems very realistic to the attackers, it would be extremely challenging to break the obfuscation.

13.7 Summary

In the modern era of technology, embedded devices ease our lives with enormous autonomous solutions. In the meantime, it also raises a security concern. We need to ensure the security of an intended embedded device, which mostly belongs to the firmware. The firmware should not be tampered or cloned; otherwise, it may cause a considerable loss to our civilization. Hence, we discuss the necessity of the firmware protection in this chapter. We also discuss encryption usage in firmware protection and why encryption-based techniques are not much efficient for protecting the core assets of a firmware. We then discuss different obfuscation techniques for firmware protection. We present the overhead and security analysis for each of the presented techniques.

References

1. F. Zhang, H. Wang, K. Leach, A. Stavrou, A framework to secure peripherals at runtime, in *European Symposium on Research in Computer Security* (Springer, Cham, 2014), pp. 219–238.
2. M. LeMay, C.A. Gunter, Cumulative attestation kernels for embedded systems, in *European Symposium on Research in Computer Security* (Springer, Cham, 2009), pp. 655–670
3. J. Maskiewicz, B. Ellis, J. Mouradian, H. Shacham, Mouse trap: Exploiting firmware updates in {USB} peripherals, in *8th {USENIX} Workshop on Offensive Technologies ({WOOT} 14)* (2014)
4. D. Morais, J. Lange, D.R. Simon, L.T. Chen, J.D. Benaloh, Use of hashing in a secure boot loader, Jun. 14 2005, US Patent 6,907,522
5. D. Peck, D. Peterson, Leveraging Ethernet card vulnerabilities in field devices, in *SCADA Security Scientific Symposium* (2009), pp. 1–19
6. A.M. Garcia Jr, Firmware modification analysis in programmable logic controllers, Air Force Institute of Technology Wright-Patterson AFB OH Graduate School of . . . , Technical Report, 2014
7. J. Nagra, C. Collberg, *Surreptitious Software: Obfuscation, Watermarking, and Tamperproofing for Software Protection: Obfuscation, Watermarking, and Tamperproofing for Software Protection* (Pearson Education, Upper Saddle River, 2009)
8. S. Schrittwieser, S. Katzenbeisser, Code obfuscation against static and dynamic reverse engineering, in *International Workshop on Information Hiding* (Springer, 2011), pp. 270–284
9. P. Junod, J. Rinaldini, J. Wehrli, J. Michielin, Obfuscator-LLVM – software protection for the masses, in *2015 IEEE/ACM 1st International Workshop on Software Protection, May* (2015), pp. 3–9
10. J. Obermaier, S. Tatschner, Shedding too much light on a microcontroller's firmware protection, in *11th {USENIX} Workshop on Offensive Technologies ({WOOT} 17)* (2017)
11. C. McClure, Software reuse planning by way of domain analysis. Technical Paper, Extended Intelligence, Inc. http://www.reusability.com
12. P.C. Kocher, Timing attacks on implementations of Diffie-Hellman, RSA, DSA, and other systems, in *Advances in Cryptology| Crypto*, vol. 96 (1996), p. 104113
13. C.S. Collberg, C. Thomborson, Watermarking, tamper-proofing, and obfuscation-tools for software protection. IEEE Trans. Softw. Eng. **28**(8), 735–746 (2002)
14. C. Collberg, C. Thomborson, D. Low, A taxonomy of obfuscating transformations, Technical Report 148, Department of Computer Science, University of Auckland, July 1997

15. J.X. Zheng, D. Li, M. Potkonjak, A secure and unclonable embedded system using instruction-level PUF authentication, in *2014 24th International Conference on Field Programmable Logic and Applications (FPL)* (IEEE, Piscataway, 2014), pp. 1–4

16. R.P. Lee, K. Markantonakis, R.N. Akram, Binding hardware and software to prevent firmware modification and device counterfeiting, in *Proceedings of the 2nd ACM international workshop on cyber-physical system security* (2016), pp. 70–81

17. B. Cyr, J. Mahmod, U. Guin, Low-cost and secure firmware obfuscation method for protecting electronic systems from cloning. IEEE Internet Things J. **6**(2), 3700–3711 (2019)

18. U. Guin, S. Bhunia, D. Forte, M.M. Tehranipoor, SMA: A system-level mutual authentication for protecting electronic hardware and firmware. IEEE Trans. Depend. Sec. Comput. **14**(3), 265–278 (2016)

19. B. Stamme, Anti-fuse memory provides robust, secure NVM option, in *EE Times* (2012)

20. R.S. Chakraborty, S. Narasimhan, S. Bhunia, Embedded software security through key-based control flow obfuscation, in *International Conference on Security Aspects in Information Technology* (Springer, Berlin, 2011), pp. 30–44

21. X. Zhuang, T. Zhang, H.-H.S. Lee, S. Pande, Hardware assisted control flow obfuscation for embedded processors, in *Proceedings of the 2004 International Conference on Compilers, Architecture, and Synthesis for Embedded Systems* (2004), pp. 292–302

Chapter 14
Security of Emerging Memory Chips

Farah Ferdaus and Md Tauhidur Rahman

14.1 Vulnerabilities of Emerging Non-volatile Memory (NVM) Chips

Continual scaling down in technology introduces an enormous challenge for the existing memory chips. Current mainstream volatile memory chips, i.e., static RAM (SRAM) and dynamic RAM (DRAM), suffer from scalability, density, memory persistency, and leakage issues. In contrast, existing non-volatile memory (NVM) chips (e.g., Flash) suffer from performance and endurance problems and the requirement of significantly high current during the write operation. Due to the limitations mentioned above, existing memory chips are incapable of delivering ever-increasing demands of energy-efficient, compact, and high-performance systems [1]. Emerging NVM chips such as Phase-Change Memory (PCM) [2], Resistive RAM (ReRAM) [3], Magneto-resistive RAM (MRAM) [4], Spin-Transfer Torque MRAM (STT-MRAM) [5], Ferroelectric RAM (FRAM)) [6] etc. are being considered as promising alternatives to mainstream memory chips in high-performance computing systems. These emerging NVMs promise a new set of architectures to replace or augment the current computing system potentially. However, these memory devices' and architecture's inherent characteristics can open a new set of vulnerabilities, which can be exploited to leak sensitive information or allow remote access and endanger the integrity, confidentiality, privacy, and safety of a system by executing different attack methodologies. Therefore, it is essential to study the vulnerabilities associated with emerging memory architecture and enforce proper mitigation/prevention techniques.

F. Ferdaus · M. T. Rahman (✉)
Florida International University, Miami, FL, USA
e-mail: fferd006@fiu.edu; mdtrahma@fiu.edu

© The Author(s), under exclusive license to Springer Nature Switzerland AG 2021
M. Tehranipoor (ed.), *Emerging Topics in Hardware Security*,
https://doi.org/10.1007/978-3-030-64448-2_14

357

The rest of the section is organized as follows. Sections 14.1.1, 14.1.2, 14.1.3, and 14.1.4 highlight the vulnerabilities of MRAM, ReRAM, PCM, and FRAM memory chips, respectively. Finally, Sect. 14.1.5 shows a comparative analysis of different memory technologies.

14.1.1 MRAM Architecture and Vulnerabilities

MRAM has considerable potential to become a universal memory technology because of its various advantageous features such as high density, non-volatility, scalability, thermal robustness, very high endurance, high speed, and fast read access, ultralow-power operation, CMOS compatibility, and radiation hardness. However, certain features and device characteristics of MRAM chips make them vulnerable, as discussed below.

The organization of the rest of the section is as follows. Section 14.1.1.1 highlights the vulnerabilities associated with the magnetic field; Sects. 14.1.1.2, 14.1.1.3, and 14.1.1.4 describe the vulnerabilities associated with prolonged read-/write latency, read/write current asymmetry, and parallel read/write operation, respectively, and finally, Sects. 14.1.1.5 and 14.1.1.6 discuss the vulnerabilities associated with ambient temperature and data retention property.

14.1.1.1 Susceptibility to the Magnetic Field

MRAM chip is vulnerable to the external magnetic field (M-Field). Magnetic tunnel junction (MTJ) is the core element of MRAM/STT-MRAM that exploits the spin torque transfer property, manipulating electrons' spin with the polarized current to store the data. The bit cell of MRAM/STT-MRAM is composed of two ferromagnetic layers separated by an oxide layer (shown in Fig. 14.1a). One layer's magnetic orientation is always fixed, which is known as the reference (or fixed) magnetic layer (RML). Another layer's magnetization can freely be oriented, depending on the M-Field known as the free magnetic layer (FML). Storing bits in the memory array is determined by the resistance states. When both the FML and RML are aligned in the same direction (current passed from SelectLine (SL) to BitLine (BL)), the MTJ produces low electrical resistance (state "0"). On the other hand, when their M-Field orientation is opposite, the MTJ exhibits high electrical resistance (state "1").

The significant difference between STT-MRAM and toggle MRAM is that STT-MRAM is current driven while the magnetic field drives toggle MRAM. Writing bits in the memory array requires the passing of a high write current (I_w) to change FML's magnetic orientation [7]. The I_w directly affects the FML orientation in STT-MRAM, but for toggle MRAM, the I_w creates an auxiliary magnetic field that changes FML direction. During the reading cycle, a voltage is applied across the MRAM cell. Depending on parallel (low resistance) or anti-parallel (high resistance)

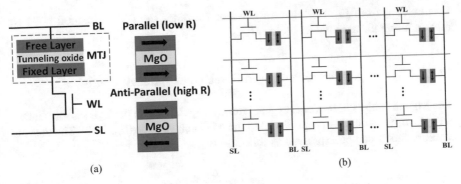

Fig. 14.1 (a) MRAM cell structure with MTJ and (b) MRAM array organization

orientation, a current sensing circuitry (attached with the MRAM cell) experiences different current and latches the appropriate logic ("0" or "1").

Due to the inherent properties (see Sect. 14.1.1.6), sensitive information could be retrieved from non-volatile MRAM. The direction of FML can be altered by manipulating external M-Field. However, the RML is strongly coupled with an antiferromagnet, and therefore it is not impacted by the external M-Field [8]. The FML direction switching follows the Landau Liftshiftz Gilbert (LLG) equation [8, 9] shown in Eq. 14.1:

$$\frac{\partial \hat{m}}{\partial t} = T_{STT} + T_{Field} \tag{14.1}$$

where

$$T_{STT} = (1 + \alpha \hat{m}) \times (-\gamma \hat{m} \times H_{Tot})$$

$$T_{Field} = \frac{\mathcal{I}_s \hbar G(\psi)}{2e} \hat{m} \times (\hat{m} \times \hat{e}_p)$$

$$H_{Tot} = H_{eff} + h_{st}$$

$$H_{eff} = H_{app} + H_k + H_{demag} + H_{ex}$$

$$H_k = H_k^C - 4\pi M_s^2$$

Here, $\hat{m}$ and $\hat{e}_p$ are the unit vector of local magnetic moment and RML magnetization direction. $\mathcal{I}_s$, $\hbar$, $G(\psi)$, e, α, γ, and h_{st} represent the spin current, reduced Planck's constant, transmission coefficient, charge on electrons, Gilbert damping parameter, gyromagnetic ratio, and field due to stochastic noise, respectively. The effective field (H_{eff}) is composed of applied (H_{app}), effective anisotropy (H_k), demagnetization (H_{demag}), and exchange (H_{ex}) field, respectively. Finally, H_k is a function of the magnetocrystalline anisotropy (H_k^C) and demagnetization field.

It is evident from Eq. 14.1 that an attacker can alter the FML direction $(\frac{\partial \hat{m}}{\partial t})$ by manipulating the external M-Field (H_{app}).

14.1.1.2 Susceptibility to Prolonged Read/Write Latency

The MRAM chip's access energy and latency depend on the thermal stability factor (Δ_{TS}), which is directly proportional to the MTJ cell volume (Eq. 14.2). The thermal stability factor can be expressed as follows [10]:

$$\Delta_{TS} \propto \frac{E_b}{T} \tag{14.2}$$

where,

$$Energy\ Barrier,\ E_b \propto H_k M_s At$$

Here, H_k, M_s, A, t, and T represent uniaxial anisotropy, saturation magnetization, area of MTJ, FML thickness, and ambient temperature (Kelvin), respectively. Thus, write latency increases with increasing Δ_{TS}. In MRAM/STT-MRAM, a thickened MTJ barrier requires higher current for a prolonged period to ensure proper switching of M-Field direction. However, the thickness, area, and magnetic resistance value of the MTJ vary with the manufacturing process. Therefore, the process variations in MRAM/STT-MRAM introduce disparity in the magnetization property in different memory cells, which affects the write current and switching delay significantly.

Furthermore, MRAM's data retention period highly depends on Δ_{TS}; an immense value of Δ_{TS} ensures more extended retention period [10]. Hence, the write latency and energy become higher to retain data for a long duration. Therefore, reduced access latency and energy increase the retention failure rate, write failure rate, and read disturbance. Moreover, the inherent random thermal fluctuations can change the magnetic orientation of the MTJ cell (regardless of memory access) [11], which can cause write error and retention failure. The retention time (t_{RT}) with (Eq. 14.3) and without (Eq. 14.4) the presence of external M-Field can be expressed as follows [12–15]:

$$t_{RT} = \tau e^{\Delta_{TS}} \tag{14.3}$$

$$t_{RT} = \tau e^{\Delta_{TS}(1-\frac{H_{eff}}{H_k})^2} \tag{14.4}$$

where

$$\tau = \tau_0\, e^{\frac{E_b}{k_B T}} \approx 1ns$$

Here, τ_0 and k_B are operating frequency and Boltzmann constant, respectively. Besides, the probability of retention failure for a given period (t) can be expressed as follows [14, 15]:

$$P_{RF} = 1 - e^{-\frac{t}{\tau_{RT}}} \qquad (14.5)$$

For both read and write operations, the current flows through the same path, and as a result, the read current can unintentionally change the magnetic orientation of the MTJ cell. This unintentional change produces read disturbance. The probability of such read disturbance error for a particular read operation time (t_{rd}) can be expressed as follows [14, 15]:

$$P_{RD} = 1 - e^{-\frac{t_{rd}}{\tau \, e^{(1-I_r)} \Delta_{TS}}} \qquad (14.6)$$

where

$$I_r = \frac{Read\ Current,\ I_{r0}}{Critical\ Current,\ I_c}$$

$$I_c = \frac{4ek_BT}{h} \cdot \frac{\alpha}{\eta} \cdot \Delta_{TS} \cdot (1 + \frac{4\pi H_{demag}}{2H_K})$$

Here, h, α, η, and ($4\pi H_{demag}$) are Plank's constant, LLG damping constant, STT–MRAM efficiency parameter, and effective demagnetization field, respectively. The disturb current reduces the thermal barrier of an MRAM cell. This reduction increases the probability of the read disturb during the read operation.

On the other hand, the stochastic switching nature of an MTJ cell can cause write errors. An increase in critical current can slow down the alteration of the MTJ cells and thus increases the switching time (t_w) of MTJ. The probability of keeping a similar magnetic orientation within a given write duration (t_w) is known as the write error rate (WER) of an MRAM cell. The probability of proper alteration of the magnetic orientation can be enhanced by escalating the write duration. Therefore, the smaller Δ_{TS} value has a positive impact on the WER, evident from Eq. 14.7 [9].

$$WER_{tw} = 1 - e^{\frac{\pi^2 \Delta_{TS}(I_c-1)}{4(1-I_c e^{Ct_w(I_c-1)})}} \qquad (14.7)$$

where

$$I_w = \frac{Required\ Write\ Current,\ I_{w0}}{Critical\ Current,\ I_c}$$

$$C = \frac{2\alpha\gamma H_k}{1+\alpha^2}$$

MRAM is sensitive to process variations because C depends on the technology-dependent parameters. The MRAM cell of larger size mostly experiences incredibly high read/write latencies due to the random increase in Δ_{TS}. The asymmetric long write latency (due to the random process variation) and high write current (described in Sect. 14.1.1.3) can be a useful tool to initiate a side-channel attack and jeopardize the data privacy [16, 17].

Moreover, the write latency of writing state "0" is distinguishably higher than writing state "1" [18]. Thus, writing "1" is known as fast-writes. Similarly, writing "0" is known as slow-writes [19, 20]. Besides, the transistor's voltage degradation is also responsible for write current reductions, which enhance the writing latency even further.

Similar to MRAM, the asymmetry between the read and write latencies is also present in PCM.

14.1.1.3 Asymmetry in Read/Write Current

As described in Sect. 14.1.1.2, the write current has a direct dependency on Δ_{TS}, retention time, and previous state of the MRAM cell (parallel/anti-parallel configuration). Thus, the write current is considerably higher to ensure proper switching of magnetic orientation [14]. During a read operation, the sense amplifier defines the MRAM cell's resistance state, passing a low current through BL to SL. Hence, to make the read operation robust, it is necessary to maintain a noteworthy amplitude difference between two read currents. Otherwise, the adversary can differentiate the read/write operation by merely observing the current waveform and hence capable of leaking sensitive information.

Moreover, the read/write current's magnitude varies based on the state ("0" or "1") to be written and the previous state. The current waveform is relatively invariant for writing "0"→"0" and "1"→"1" since they maintain the same states. On the other hand, for both read and write operations, the current waveform characteristics of toggling to "0"→"1" are noticeably distinguishable from "1"→"0" [18, 21]. Therefore, by observing the differences mentioned above, the attacker can presume the data's state and extract sensitive information stored in the MRAM.

Taking both Sects. 14.1.1.2 and 14.1.1.3 into account, the asymmetric high read/write current and long write latency are responsible for supply noise [22]. The interconnection resistance between the power supply source and destination is also a source of supply noise. The write current depends on the data pattern to be written in memory. Therefore, a row-hammer attack (described in Sect. 14.2.3) can be initiated by precisely controlling supply noise for specific data patterns. The deterministic supply noise can be generated by writing specified data patterns (such as specified number of 1's and 0's in the data to be written). Using the generated supply noise, the adversary can initiate the fault injection attack by inducing read/write and retention failure to the victim's memory space in the shared network system. Hence, sensitive information such as encryption keys, passwords, PINs, etc. can be leaked.

> The read/write operation's asymmetry is a significant source of side-channel attacks in emerging NVM chips. MRAM, ReRAM, and PCM exhibit asymmetry in read/write operation/current. FRAM exhibits asymmetry in the read operation only.

14.1.1.4 Parallel Read/Write Operation

Multiple clock cycles are required to complete a full read/write operation due to the long write latency. Usually, the required clock cycles for completing a write operation are higher than a read operation. However, the read/write latency's impact can be reduced by introducing parallelism to improve the overall throughput. The parallel read/write access can be performed as follows [23]:

1. Multiple read operations can be initiated along with the write operation in separate banks of the memory to achieve high throughput. Although the read operations are performed in separate banks of the memory, reading multiple banks at the same time requires higher current than reading from a single bank. This high driving current might introduce a ground bounce effect and alter the memory cells connected to the neighbor WordLine (WL) [23].
2. Similarly, multiple write operations can be initiated in parallel along with the read operation in several independent banks. These consecutive write operations, along with a read operation, draw excessive current from the power supply, which can increase noise drastically. As a result, this parallel write operation might cause retention failure, read disturb, read failure, and write failure. Due to parallel access, increasing supply noise introduces faster retention failure.

14.1.1.5 Susceptible to Temperature

As described in Sect. 14.1.1.2, the ambient temperature is inversely related to Δ_{TS} [14]. Besides, larger Δ_{TS} values increase the thermal barrier of MRAM cells. The increase in thermal barrier increases the write latency and current. Thus, the temperature can be used as an essential tool to manipulate MRAM's data retention period to reveal the stored MRAM data.

The resistance of the conductor of ReRAM, PCM, and FRAM is also a function of temperature. An adversary can tamper the stored data in these chips by changing the temperature [24, 25].

14.1.1.6 Data Retention

Unlike DRAM, NVM chips do not require a refresh scheme to retain the stored data, and they can resume instantaneously from the idle mode. Therefore, the NVM chips' data persistence property makes the stored data more vulnerable to various malicious attacks (e.g., micro-probing, optical probing, radiation imprinting, and cross-sectional imaging from the physically acquired memory chips). With the escalating use of the portable devices, such as smartphones, laptops, tablets, etc., these sorts of security issues are becoming more severe since device portability provides more opportunities for the adversaries to get physical access to systems and perform different physical attacks.

14.1.2 ReRAM Architecture and Vulnerabilities

Like MRAM, ReRAM also possesses some advantageous features such as architectural simplicity, non-linearity, very low energy operation, CMOS compatibility, higher resistance value, and better scalability (< 10 nm). The bit cell of ReRAM is composed of two electrodes ($Electrode_{Top}$ and $Electrode_{Bottom}$) separated by an oxide material (shown in Fig. 14.2a). Due to the applied voltage, the electrons pass through the oxygen vacancies depending on the oxide breakdown (deforming) and re-oxidation (forming) process, which defines the resistance states. The state "0" is known as low resistance state, R_{LS}, and the state "1" is known as high resistance state, R_{HS}. Switching from R_{HS} (R_{LS}) to R_{LS} (R_{HS}) is known as set (reset). The applied voltage creates an electric field (E-Field) that is responsible for creating oxygen vacancies.

A small voltage is applied across the ReRAM memory cell during a read operation, and the output current is measured to determine the stored bit. During a write operation, the write voltage polarity, amplitude, and duration are manipulated to ensure the resistance state's switching. Like other NVM chips, ReRAM does not require any access transistor [26, 27]. However, ReRAM requires specially designed peripheral circuit and memory organization. This ReRAM structure can create a new set of vulnerabilities, which are discussed below.

The following sections discuss the vulnerability of ReRAM (exclusively presents in ReRAM), a.k.a., crossbar memory, cross-point resistive memory, etc. The organization of the rest of the section is as follows. Sections 14.1.2.1 and 14.1.2.2

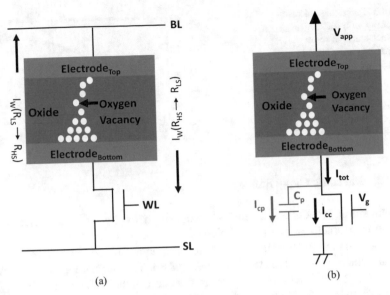

Fig. 14.2 (a) ReRAM cell structure and (b) cell in series with allowed parallel parasitic capacitance

describe the vulnerabilities associated with sneak current and parasitic capacitance, respectively.

14.1.2.1   Sneak Current

During the write operation, the selected bit cells experience full voltage when activating the WL and BL. However, the partially selected cells[1] also experience partial write voltage. Therefore, a leak current is passed through those partially selected cells due to the voltage difference across them known as sneak current [26]. According to [26, 28], due to the sneak current, the write latency of a bit cell is considerably influenced by the state of the cells that are in the close vicinity of the memory array connected to the same row and column(s) of the selected cell(s). An attacker can leverage this latency information of the write operation to extract sensitive information to introduce different security and privacy attacks.

---

[1]The cells that are connected to the same row and column(s) of the selected cells during a write operation are known as partially selected cells.

Fig. 14.3 Percentage of reset probability with different R_{LS} resistance states at room temperature (regenerated from [31])

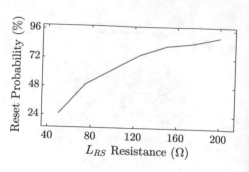

14.1.2.2 Parasitic Capacitance

The deposition process and chemical composition of the oxide material control the unique characteristics of the ReRAM devices. Usually, the applied voltage across the two electrodes of the ReRAM forms a defect-based filament in the oxide material, which controls the switching mechanism. The morphology and geometry of the filament is a function of the current flow across the ReRAM cell. Therefore, the performance of ReRAM is a function of switching voltages and compliance current. Hence, maneuvering the compliance current, an adversary can introduce reliability issues in ReRAM devices.

For a successful operation, the switching current can be controlled using external (1) current compliance circuit or (2) transistor [29]. Inserting an external transistor introduces unintentional parasitic capacitance (shown in Fig. 14.2b). Therefore, during the set process, a transient current flows through the capacitor due to the voltage at the capacitor's connected node, which increases with time [30]. During normal operating conditions, the impact of the current generated by the capacitor is insignificant. However, the higher capacitance value requires a high read/write power due to the filament overgrowth. Moreover, during the reset process, immense voltage is needed to reset due to the higher current level. Furthermore, a longer time is required to reset due to the lower resistance value of the R_{LS}, which can increase the probability of an unsuccessful reset process. Therefore, it is required to maintain a low set/reset voltage and low read/write/erase power and fast switching for successful operation. Higher local parasitic capacitance can introduce an adverse impact on the qualities mentioned above, which in turn makes ReRAM-based memories vulnerable to different kinds of attacks. Figure 14.3 illustrates that the reset probability increases with an increase in R_{LS} resistance, which can introduce failure in the reset process of ReRAM-based devices.

14.1.3 PCM Architecture and Vulnerabilities

Like other NVM chips, PCM also possesses advantageous features such as high density, low standby power, and high scalability. The bit cell of PCM comprises

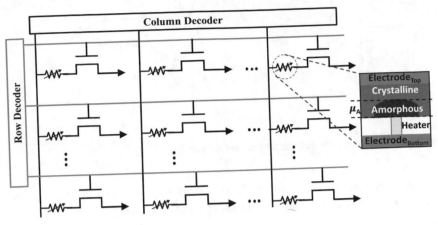

Fig. 14.4 PCM array organization with PCM cell structure

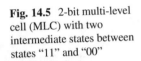 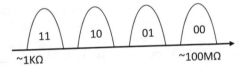

Fig. 14.5 2-bit multi-level cell (MLC) with two intermediate states between states "11" and "00"

an alloy of phase-change (chalcogenide) material such as $Ge_2Sb_2Te_5$ (GST), $In_3Sb_1Te_2$, etc. [32–35]. The chalcogenide material of the PCM cell generally resides between two electrodes (shown in Fig. 14.4). During the write operation, the applied current through the PCM memory cell produces localized heat that forces the PCM cell to be changed the states in between crystalline (also known as set state) and amorphous (also known as reset state) states. The crystalline state (atoms are in a regular pattern) is also known as a low resistance state, and the amorphous state (atoms are in the disordered pattern) is known as a high resistance state. The resistance difference between the crystalline and amorphous states is comparatively significant, allowing us to store multiple bits in a single PCM cell (shown in Fig. 14.5).

Two different types of programming pulses (set and reset) are used to write an appropriate data state (Fig. 14.6). During the reset process (writing "0"), short latency and high-powered reset pulse are applied over the PCM cell, which raises the memory cell's temperature over the melting temperature ($T_{melt} =\sim 660°$) and quickly quenches the heat in a shortened period. On the other hand, during the set process (writing "1"), long latency and moderate-powered set pulse are applied over the PCM cell for a long enough time to raise the memory cell's temperature over the crystalline temperature ($T_{cryst} =\sim 350°$). Due to the mentioned set/reset process, the internal resistance can maintain distinguishably (typically at least one order of magnitude higher or more) two different resistance states [36]. The reset process consumes comparatively more energy than the set process. A small voltage is applied across the PCM cell during the reading cycle, and the sensing circuitry

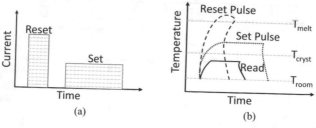

Fig. 14.6 Write (set/reset) and read operations (redrawn from [24])

determines the stored data based on the current differences for different resistance states. However, the read latency and energy are low enough compared to write operation.

Furthermore, the PCM supports MLCs to store multiple bits of data in a single PCM cell to improve the density and capacity. Therefore, MLC PCM is a suitable candidate for storage devices, while single-level cell (SLC) PCM is ideal for the main memory. The MLC PCM uses a program-and-verify approach to ensure proper write operation. This methodology requires precise control over set and reset processes to ensure proper resistance state. Because of the less noise margin, the MLC PCM suffers from (1) very high access latency, (2) very high read failure due to unintentional write, (3) minimal memory lifetime and capacity due to lower write endurance, and (4) ample time for precise write operation due to multiple intermediate states.

PCM chip suffers from a limited lifetime due to wear-out. Besides, the read (write) latency of PCM is 2–3 (20–50) times slower than DRAM [24]. Therefore, a hybrid memory system has been proposed where both high capacity PCM and low latency DRAM are used as main memory and cache, respectively [2, 37–40]. This hybrid model can reduce the number of writes to PCM, which enhances the PCM lifetime. The following section discusses the vulnerability in the PCM chip.

The organization of the rest of the section is as follows. Sections 14.1.3.1 and 14.1.3.2 describe the vulnerabilities associated with limited write endurance and resistance drift.

14.1.3.1 Limited Write Endurance

The write endurance of the PCM cell is a maximum of 10^7 to 10^9 [41]. Since average programs access the entire memory space non-uniformly, the densely written lines' failure rate is faster than other lines. Therefore, PCM cells' lifetime reduces to ~5% of its general expected lifetime, which is less than 1 year [37]. The write operation of PCM is destructive, and the low write endurance is the main bottleneck of PCM technology. This low write endurance makes a PCM chip prone to selective attack [42]. An attacker writes to a few target PCM cells repeatedly to wear them out before the expected endurance in this attack. The programs that do not maintain the application locality appropriately are more vulnerable to such attacks. Recent studies show that repeated writing into a particular address can introduce failure in that memory cell within 32 s [38, 43]. However, secure wear-leveling techniques are used to remap the memory lines randomly to prevent such attacks [38].

14.1.3.2   Resistance Drift

It is possible to set the PCM memory resistance value at a desirable level (from several available levels). However, the resistance value is not constant and gradually drifts with time. Therefore, an increase in cell resistance can corrupt the stored value. Like the charge leakage phenomenon of DRAM, the sensing circuit of PCM senses different values other than the stored one during the read operation due to resistance drift. The following equation (14.8) expresses cell resistance with time [44]:

$$R(t) = R_0 (\frac{t}{t_0})^v \tag{14.8}$$

where

$$t > t_0$$

$$v < 1$$

Here, $R(t)$ is the resistance of the phase-change material at the time, t, R_0 is the initial resistance of the material measured at a reference time, t_0, and v is the drift coefficient, which varies with process variation of the PCM cells. The value of v also depends on the temperature, readout current, mechanical stress, and size of the amorphous materials within the PCM cell's active region. The cells with higher v are more vulnerable to drift errors. Besides, the intermediate states of the PCM cell usually experience the drift error.

Moreover, the programming latency of MLC PCM is comparatively longer; therefore, the cell resistance growth is slower, which deteriorates the memory performance drastically. Hence, resistance drift[2] can introduce transient errors in MLC PCM, which is a severe security threat. Therefore, to overcome the challenges produced by drift errors, the practical MLC PCM design requires the implementation of optimization technique over the conventional design, such as a three-level cell is proposed that is capable of reducing drift error rate substantially [45], optimal state mapping, and smart cell encoding.

---

[2]The slower growth rate of the cell resistance with time is known as resistance drift.

14.1.4 FRAM Architecture and Vulnerabilities

Ferroelectric RAM (FRAM) is relatively larger than other NVM chips and possesses destructive read operation [46]. The single-cell structure of a conventional 1T-1C (single transistor and single capacitor) FRAM cell is similar to that of DRAM, except for a ferroelectric capacitor in place of a dielectric capacitor and a plate line (PL) in place of ground connected at the positive end of the capacitor terminal. The simplified version of the FRAM cell structure is depicted in Fig.14.7a [6, 47]. The primary storage element of a FRAM is the ferroelectric capacitor. A ferroelectric capacitor is a capacitor that is based on ferroelectric material, such as lead zirconate titanate (PZT), instead of dielectric materials used in traditional capacitors.

PZT has a perovskite crystal structure. The central atom of PZT has two equal and stable energy states. These states determine the polarization of the atom. If an electric field is applied in the proper direction, the atom will be polarized in one energy state and aligned according to the applied electric field's direction. When the applied electric field's direction is reversed, the atom will polarize from one energy state to another and consequently move in the opposite direction. Since both the energy states are stable, there remains a remnant polarization even when the applied electric field is removed. After that, if the applied electric field is increased in the opposite direction, the remnant polarization will diminish at a point, and then, the polarization will be reversed with the increment of an applied electric field. Similar behavior is observed in the opposite direction as well.

The polarization vs. applied electric field forms a hysteresis loop (shown in Fig. 14.7b) [6]. The remnant charge, Q_r, the saturation charge, Q_s, and the coercive voltage, V_C are three significant parameters of a hysteresis loop. When the voltage across the capacitor is $0V$, the capacitor assumes one of the two stable states expressed by "0" or "1". The total charge stored on the capacitor is Q_r, for a "0", or $-Q_r$, for a "1". Both Q_r and $-Q_r$ are bound charges that cannot be released for sensing until a voltage pulse is applied to the ferroelectric capacitor. When

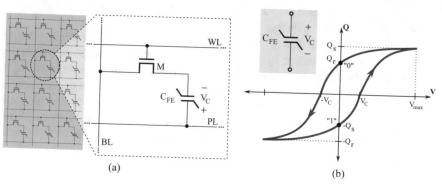

(a)   (b)

Fig. 14.7 (a) Simplified cell structure of 1T-1C FRAM and (b) hysteresis characteristic of the ferroelectric capacitor [47]

voltage is applied to the capacitor's terminal, its polarization may switch depending on the stored data. If it switches, the change in the total charge in the capacitor is $\Delta Q_{switched}$, and if it does not, the change in charge will be $\Delta Q_{unswitched}$. Since

$$\Delta Q_{switched} > \Delta Q_{unswitched} \tag{14.9}$$

where

$$\Delta Q_{switched} = (2Q_r + Q_s)$$
$$\Delta Q_{unswitched} = (Q_s - Q_r)$$

As a result, the capacitance observed is different for switched and unswitched cases since, by definition, $C = dQ/dV$.

$$C_{swithced} > C_{unswitched} \tag{14.10}$$

FRAM stores data by exploiting the spontaneous polarization of the ferroelectric material of the ferroelectric capacitor. In order to write a "1" in a bit cell, the corresponding BL is raised to V_{DD}, and WL is raised to $(V_{DD} + V_T)$, where V_T is the threshold voltage of the access transistor. Then, PL is pulsed. This allows a full $-V_{DD}$ to appear across the ferroelectric capacitor for some duration since the voltage convention is measured as delineated in Fig. 14.7a. In order to write "0" in the bit cell, BL is pulled down to the ground (GND) instead of raising to V_{DD}, and thus, the voltage applied across the capacitor is V_{DD} [6].

On the other hand, the read operation of a FRAM is destructive. Read access begins by precharging the BL to $0V$, followed by activating the WL and raising PL to V_{DD}. This establishes a capacitor divider consisting of C_{FE} and C_{BL} between the PL and the ground, where C_{FE} represents the ferroelectric capacitor's capacitance, and C_{BL} represents the parasitic capacitance of the BL. Depending on the bit stored, the ferroelectric capacitor's capacitance can be approximated by C_0 and C_1. If the stored bit is "1", the polarization will switch when the PL is high. Therefore, C_1 will be greater than C_0 (Eq. 14.10). The voltage developed on the BL (V_x) can be one of the following two values:

$$V_x = \begin{cases} V_0 = \frac{C_0}{C_0 + C_{BL}} V_{DD} \\ V_1 = \frac{C_1}{C_1 + C_{BL}} V_{DD} \end{cases} \tag{14.11}$$

The sense amplifier is activated to drive the BL to full V_{DD} if the voltage developed on the BL is V_1 or to $0V$ if the voltage on the BL is V_0. The WL is kept activated until the sensed voltage on the BL restores the original data back into the memory cell, and the BL is precharged back to $0V$.

Table 14.1 Comparison of emerging NVM technologies with existing mainstream memory technologies (adopted from [16, 24, 48])

Memories	Non-volatile	Endurance	Read/write latency	Read/write voltage (V)	Program energy/bit	Cell area (F^2)	Feature size (nm)	Retention
DRAM	No	10^{16}	10/10 ns	1.8/2.5	Low	6	36	64 ms
SRAM	No	10^{19}	<1 ns	1.65–2.2	Low	50–120	65	0 ns
HDD	Yes	10^6	8.5/9.5 ms	–	High	2/3	–	>10 y
NAND (Flash)	Yes	10^6	25/200 us	1.8/15	High	4	22	10 y
MRAM (Toggle)	Yes	10^{16}	10/20 ns	1.8/1.8	High	20	65	>10 y
STT-MRAM	Yes	10^{16}	1/10 ns	1.8/1.8	Low	6	65	>10 y
ReRAM	Yes	10^{12}	50/100 ns	<1	Low	4,8	<65	>10 y
PCM	Yes	10^9	20/100 ns	1.2/3	High	6	45	>10 y
FRAM	Yes	10^{15}	65/65	1.3–3.3	Low	22	180	10 y

Table 14.2 Summary of the unique features of emerging NVM technologies (adopted from [16])

Unique features	MRAM	ReRAM	PCM
Data retention	Yes	Yes	Yes
Thermal susceptibility	Yes	Yes	Yes
Magnetic susceptibility	Yes	No	No
Humid susceptibility	No	Yes	No
Read/write latency variation	Yes	Yes	Yes
Asymmetric read/write latency	Yes	Yes	Yes
Asymmetric read/write current	Yes	Yes	Yes
High write current	Yes	Yes	Yes
Supply voltage droop	Yes	Yes	Yes
Ground bounce	Yes	Yes	Yes
Endurance	Yes	Yes	Yes

14.1.5 Comparative Analysis between Different Memory Technologies

The comparison between emerging memory chips such as toggle MRAM, STT-MRAM, ReRAM, PCM, and FRAM and the existing mainstream memory technologies such as SRAM, DRAM, NAND Flash, and hard disc drive (HDD) is summarized in Table 14.1. Furthermore, the unique features of emerging NVM are summarized in Table 14.2.

Table 14.3 Different kinds of attacks against emerging NVMs (adopted from [16])

Memories		MRAM	ReRAM	PCM	FRAM
Side-channel attack		Yes	Yes	Yes	Yes
Row-hammer attack		Yes	Yes	Yes	No
DoS attack	Thermal attack	Yes	Yes	Yes	Yes
	M-Field attack	Yes	No	No	No
	E-Field attack	No	No	No	Yes
Fault injection attack		Yes	Yes	Yes	No
Information leakage attack		Yes	Yes	Yes	No
Physical attack		Yes	Yes	Yes	Yes

14.2   Attacks and Countermeasures

The vulnerabilities of emerging NVM technologies can be used to mount various attacks. Table 14.3 summarizes the potential attack methodologies in emerging NVM. The side-channel attack and the row-hammer attack methodologies can be used to jeopardize data privacy [16, 21, 23]. On the other hand, denial-of-service (DoS) attacks can be executed by manipulating the thermal and external (magnetic/electric) fields. In the following sections, the data security and privacy attacks on emerging NVM memories and corresponding countermeasure techniques are explained. This section discusses different kinds of attacks that exploit the vulnerabilities of emerging NVM technologies described in Sect. 14.1. This section also focuses on the countermeasure techniques against these attacks.

The rest of the section describes different security threats and privacy attacks and existing mitigation/prevention techniques and is organized as follows. Sections 14.2.1, 14.2.3, and 14.2.5 present side-channel attacks, row-hammer attack, and DoS attack on MRAM, and Sects. 14.2.2, 14.2.4, and 14.2.6 present the state-of-the-art research on the existing prevention of these attacks. Sections 14.2.7 and 14.2.9 explain the fault injection attack and information leakage attack on ReRAM. Sections 14.2.8 and 14.2.10 present their possible countermeasures, respectively. Finally, Sect. 14.2.11 describes other possible attacks on NVM memory chips along with mitigation/prevention techniques.

14.2.1   Side-Channel Attack

The asymmetric nature of the read/write current and latency introduces a side-channel attack that disrupts MRAM cache's data privacy during memory updates (read/write operation) in the compromised operating system (OS) [21, 49]. Recent studies show that the Advanced Encryption Standard (AES) algorithm, used for encryption/decryption, can be broken by maneuvering the write current vulnerability [21]. The MRAM cache's current (drawing from the power supply for the read/write operations) waveform is monitored and analyzed by inserting a small resistor in series with the supply voltage or ground in order to recover the AES encryption key while executing the side-channel attack.

In MRAM, the power consumption reveals essential information of the data to be written and existing data. The power consumed is negligble when the stored data state and the newly written data state are the same. In contrast, a remarkable change in power consumption is observed during transition state ("1"→"0" or "0"→"1") of the input. According to the attack model [21], the information can be leaked by computing the hypothetical values from known and guessed information using the divide-and-conquer approach. In this approach, sensitive information can be revealed statistically (using Pearson correlation [50]) from the hypothetical leaked information and side-channel traces (e.g., time, power, and electromagnetic emission) based on the highest correlation value. The attack model is summarized as follows [21]:

- As attacker observes physical characteristics (such as power consumption during different input transitions) of underlying computation of the target implementation that he/she intends to attack (i.e., leaking sensitive information).
- An attacker can capture the changes in MRAM cells by counting the number of bit transitions.
- The attacker tests the statistical dependency between the observed and hypothetical leakage power traces.
- The adversary reconstructs the full sensitive information using the divide-and-conquer approach from the small portion that he/she can recover.
- Finally, the attacker measures the statistical dependence to reveal the information successfully.

There are different ways to collect and characterize power traces to leak sensitive information (e.g., encryption keys). Simple power analysis (SPA) and differential power analysis (DPA) are the two most popular tools used to recover cryptographic keys and operations. In SPA, an attacker interprets power consumption during cryptographic operations directly to reveal key or operations. On the other hand, in the DPA attack, an attacker identifies data-dependent correlations from the power trace [21]. Several write traces are generated using the same key with different plaintexts for all AES rounds in the DPA attack [21].

However, the signal-to-noise ratio (SNR) of STT-MRAM is minimal, and therefore pre-processing is required for a successful attack [21]. During AES encryption, XOR operation is performed between the round key and ciphertext. In this circumstance, the leaked information can be divided into three segments: (i) HW[3] of the round key, (ii) HW of ciphertext, and (iii) HW of the XOR output. The write current consumed in each segment is proportional to the data state for segments (i) and (ii). For segment (iii), the write current depends on the switching from the old state to the new state. In the pre-processing stage, the new power

---

[3]Hamming Weight (HW) counts the number of 1's in a string of bits [51].

Table 14.4 Minimum traces required for successful recovery of the first byte of the AES key for different emerging NVM types [16, 21]

Memories		STT-MRAM	Toggle MRAM	ReRAM	PCM
#Traces required	Write	300	–	900	200
	Read	40	15	200	200

traces are computed from the differences of the write current values in different time stamps. These new power traces increase the attack window and escalate the side-channel attack.

If an attacker wants to avoid pre-processing, he/she might need more traces to reveal the full keys. In STT-MRAM, 2000 traces can recover 8 out of 16 bytes of the AES key without any pre-processing [16, 21]. Table 14.4 presents the minimum traces required for the successful recovery of the first byte of the AES key [16]. The results show that the read (write) operation of MRAM (PCM) is more vulnerable to side-channel attacks compared with other emerging NVM types. Hence, a robust countermeasure technique is vital. Performing both the read and write operations on the same key during the AES rounds can help the adversary steal the key more effectively [21]. The key extraction process's speed and accuracy can be further enhanced by performing a cross-correlation between the asymmetric read/write currents.

Most of the existing NVM chips are susceptible to side-channel attacks. Like MRAM, the read current of ReRAM and PCM is also asymmetric [16]. Moreover, ReRAMs exhibit read/write current asymmetry. If the stored data is "0", the average current drawn by the memory cell is significantly higher than its opposite counterpart, valid for both ReRAM and PCM.

Furthermore, a compromised OS of the PCM memory design can deduce the shuffled pattern during the wear-leveling process by manipulating the malicious process and exploiting the leaked side-channel information. Hence, the wear-out process becomes accelerated, and the adversary can track and pinpoint the weared out memory cell's location [52].

14.2.2 Mitigation/Prevention of Side-Channel Attacks

The side-channel attack that exploits vulnerabilities in emerging NVM chips needs to be mitigated/prevented before deploying them into the systems. The following techniques can mitigate/prevent the side-channel attack:

- **Encoding of MSB bits.** Encoding MSB bits can be used to mitigate side-channel attacks [16, 53]. This technique complicates and merges the states in the supply current signature, making it very difficult for an attacker to hamper the privacy. In the encoding scheme, the first few MSB bits are encoded to the opposite polarity. For example, data bit "0" is considered as "1" and vice versa. Although

Table 14.5 Key extraction from ReRAM after implementing the encoding technique [16]

#Encoded first MSB bits		8	16	32	64
#Traces required	Write	800	950	600	600
	Read	–	–	350	200

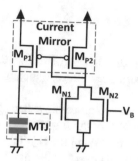

Fig. 14.8 Circuit implementation for constant current supply. Two PMOSs (M_{P1} and M_{P2}) form the current mirror. The parallel configuration of M_{N1} and M_{N2} controls the mirror current depending on the MTJ resistance. M_{N1} and M_{N2} act as a voltage-controlled current source (in saturation) and constant DC bias current. V_B controls the current magnitude (redrawn from [57])

this encoding technique cannot prevent the system from side-channel attacks, Tables 14.4 and 14.5 show that the encoding of more bits makes it complicated by a certain level for an attacker to reveal sensitive information. Like encoding, mitigation techniques such as noise addition, signal reduction, scrambling, and shuffling are incapable of entirely preventing the side-channel attack [16, 54, 55]. Therefore, a strong mitigation technique is essential to prevent the system from side-channel attacks.

- **Constant read/write current.** Asymmetry in the read/write operation is the primary source of the side-channel attack. Therefore, we can design emerging memory chips to consume the same amount of current while writing "1" or "0". A symmetrical structure in the STT-MRAM cell is proposed to prevent DPA attacks by minimizing the asymmetric behavior in power consumption [56]. In another approach, memory controller circuitry can be redesigned to ensure that the write current is constant. Figure 14.8 illustrates the circuitry to produce constant write current. The write latency of the states ("1" and "0") becomes different for constant write operation; hence, a minor modification is required for the WL driver's design. Write current is a function of word size only. Hence, the worst-case write current is selected to ensure proper functionality, which in turn introduces power wastage. Lowering the constant write current increases WER, and hence a trade-off is required. Therefore, the power overhead can be minimized either by increasing the acceptable write latency or by accepting WER up to some acceptable bound.

The property of all-spin logic devices (ASLDs) can also be used to prevent side-channel attacks. ASLD can switch the nano-magnet direction without any additional resources [58], and this property can be used for identical power

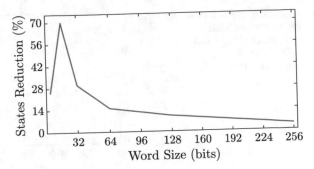

Fig. 14.9 Percentage of state reduction using 1-bit even parity for different word sizes (regenerated from [60])

dissipation during state transition. Both techniques provide excellent means to resist against side-channel attack [58, 59].

- **Reduction of retention time.** Reduction in MRAM's retention time improves the write latency and current (see Sect. 14.1.1.2). Hence, this retention time reduction is capable of mitigating the side-channel attack [60]. Lowering the write current reduces the difference between the current states. Therefore, the adversary is incapable of distinguishing the different current states. In contrast, the retention time increases drastically at low temperatures [60]. Therefore, a trade-off is required.

- **Inclusion of parity bit.** Merging multiple supply currents with the side-channel currents makes the speculation process (accurate prediction of the current state) difficult for the attacker. Appending a parity bit along with the actual data can accomplish this goal. Adding a 1-bit even parity with 4-bit data can reduce the five states (i.e., 1111, 0111, 0011, 0001, and 0000) of data into three states (1111/0 or 0111/1, 0011/0 or 0001/1, and 0000/0) [60]. Consequently, the total read/write current depends on the total number of 0's and 1's rather than the order of their positions. Adding of parity bit makes the current waveform constant, and therefore it makes the system resilient to side-channel attack.

  Figure 14.9 illustrates that 1-bit parity can reduce the number of states significantly for 16- to 32-bit word sizes. However, the effect of 1-bit parity becomes insignificant for larger word sizes (> 32-bits). Furthermore, for much smaller word sizes (< 16-bits), the reduction rate (i.e., reduction of the number of states) also becomes insignificant due to the availability of a few numbers of states. We can use the parity encoding feature of the existing error correction code (ECC) scheme.

- **Randomization of side-channel signature.** Due to numerous current states, adding 1-bit parity is incapable of providing acceptable accomplishment for words of larger size. Adding random bits in different locations in a long word can be a useful tool to mitigate side-channel attacks. The added random bits obfuscate the side-channel signature by merging different combinations of states of supply current. Figure 14.10 illustrates that the higher number of random bits can significantly reduce the number of states to mitigate the side-channel attack.

Fig. 14.10 Percentage of state reduction adding multiple random bits for different word sizes (regenerated from [60])

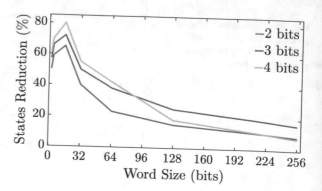

- **Increasing of word size.** The supply current is a function of word size. Due to process variation, an increase in the word size of memory reduces the prediction accuracy. Therefore, a long word is more resistant against side-channel attacks [60].
- **Obfuscating the address information.** Simply hiding the memory address is incapable of refraining the side-channel attack entirely. Therefore, obfuscation of address information is required to defend side-channel attacks [61, 62].

14.2.3 Row-Hammer Attack

In a row-hammer attack, an attacker repeatedly reads from the same address to corrupt nearby addresses' memory contents. Besides, repeated writing to a specific address in a shared memory space produces high ground bounce due to high write current. As a result, retention failure and read disturb are experienced in the nearby bit cells when the originated bounce propagates through WL and SL (for writing "0" → "1") or BL (for writing "1" → "0") drivers. Furthermore, write (read) failure happens in the nearby bit cells when the originated bounce propagates through SL/BL (SL) drivers. The adversary can manipulate the parallel read/write access (described in Sect. 14.1.1.4) to initiate the row-hammer attack. In emerging memory, the SL/BL drivers of both selected and nearby unselected bits share the same supply rails. Therefore, bounce happens in WL and SL/BL jointly, and because of physical distance, the farthest WL drivers receive the bounce later compared to the closer ones.

With an access transistor weakly ON ($V_{GS} > 0$V) for a short period of time, a disturb current is generated that flows through the nearby unselected cells. This enhanced ground bounce decreases the retention time. If the reduced retention time is shorter than the duration of the disturb current flow, data ("0" or "1") is written according to the disturb current's direction.

Temperature and process variations impact the disturbance errors in emerging memory chips. At high temperature, the access transistor's threshold voltage and the retention time decrease drastically [60]. Therefore, the MTJ experiences high current at high temperature. This high temperature makes data corruption faster. The process variation also impacts the row-hammer attack. With process variations, weaker cells experience low thermal stability. These types of cells are more vulnerable to the row-hammer attack.

Read current can influence disturbance errors in emerging NVMs. An increase in read current increases the sense margin but at the cost of read disturbance. A conscientious choice of read current can reduce the read disturbance and provide a better sense margin. However, the disturb current originated from the ground bounce reduces the thermal barrier and increases the read disturbance due to the lower thermal barrier. Besides, the read disturbances increase at high temperatures.

Longer write latency can cause write failures. Different write patterns have different vulnerabilities to write errors. Write failure during "1" → "0" requires higher ground bounce compared to write failure during "0" → "1". However, a parallel write operation is incapable of providing such huge ground bounce to originate write failure. Therefore, write failure during "1" → "0" is only vulnerable to a row-hammer attack.

PCM and ReRAM chips also experience read/write failure from high supply noise during a parallel write operation. PCM is also vulnerable to resistance drift (see Sect. 14.1.3.2) [16] that can be used for a row-hammer attack.

14.2.4 Mitigation/Prevention Techniques Against Row-Hammer Attack

The row hammer can be used in various attacks such as crashing or corrupting a system, breaking memory-protection schemes, hijacking the control of the system or mobile devices, taking over a remote server, gaining random access in a web browser, and compromising co-hosted virtual machine. Below are the few techniques that can be used to prevent or mitigate the row-hammer attack:

- **Stall insertion.** Stalls can be inserted during a write operation to reduce the average disturb current significantly and the disturbance errors [16].
- **Sequential access.** Non-pipelined (sequential) read/write access is capable of refraining the adversary to generate supply noise [22]. However, it hampers system performance significantly.
- **Row-hammer resistant design strategies.** Considering ReRAM as a test case, the generated ground bounce is at the highest (lowest) while writing "0" → "0" ("1" → "1"). The ground bounce generated by other patterns falls in between this

Fig. 14.11 Overview of the
DoS attack (redrawn
from [22])

range. Assume that the victim is writing "1" → "1" in one bank, and the attacker
is writing "0" → "0" in another bank. The victim suffers more ground bounce if
the value of R_{int}[4] is low. Therefore, a higher R_{int} value will make it difficult for
an attacker to create noise bounce in the victim's memory cells. A good quality
grid (power/ground) with scaled read/write current can also help reduce the path
resistance between the intermediate metal layers to reduce the amount of supply
noise. Separate power rails can also be used to minimize the power bounce.

14.2.5 Denial-of-Service (DoS) Attack

The MRAM chip's stored data can be tampered or corrupted by changing the
temperature or applying external M-Field because the free magnetic layer (FML)
of MRAM is sensitive to both temperature and M-Field. An attacker can use these
properties to execute a denial-of-service (DoS) attack [17]. Figure 14.11 illustrates
the overview of the DoS attack in emerging NVM chips. Writing different patterns
into memory creates a different amount of supply noise. Writing all zeros produces
the highest amount of supply noise. An attacker can deliberately create high enough
supply noise by writing a predefined specific data pattern, which can be used to
launch DoS attacks [22].

An attacker can manipulate the M-Field to perform DoS attacks. The direct
dependency on M-Field on MRAM chips makes them vulnerable to both DC and
AC M-Fields [8, 63, 64]. The impact of AC M-Field is more severe than DC M-Field
due to the bipolar effect of the AC field. On the other hand, the external DC M-Field
introduces retention failures if FML and the external DC M-Field direction are not
aligned [65]. However, in 3D space (i.e., in any direction), both AC and DC M-
Fields are almost equally effective in launching attacks. An attacker can manipulate
the M-Field by changing the AC M-Field's frequency, angle, or magnitude.

---

[4]Internal resistance (R_{int}) is modeled as the equivalent resistance between the local ground of one
address of one bank and the local ground of another address of another bank.

In the MRAM chip, the direction of the current flow for the read operation is unipolar. Therefore, the read operation experiences the most adverse impact from external current/M-Field due to (i) the presence of the read disturb current and (ii) the frequency of the read current is comparatively higher than the write current. However, the external M-Field can also affect the write operation and data retention [17] as the current direction for the write operation is altered. This alteration depends on the type of data ("0" or "1") to be written into the memory chip. Applying opposite external M-Field/current to the write ones can accelerate the DoS attack [17].

Like M-Field manipulation, a DoS attack can be performed by altering temperature as well since the M-Field is a function of operating temperature. The retention time can be prolonged by adjusting the operating temperature, which increases the data persistence of the NVM chips. Therefore, through unauthorized access, those persistent bits can be compromised while the device is turned ON.

> The external M-Field can be used to launch DoS attacks since the applied M-Field increases the total number of bit cells that suffer from bit-flip and increase in writing latency. Furthermore, the external M-Field degrades the sense margin of the sense amplifier.
>
> On the other hand, FRAM is sensitive to the external electric field (see Sect. 14.1.4) because of the presence of ferroelectric material in the capacitor. This external electric field may flip the bit polarization and corrupt the data.

14.2.6 Mitigation/Prevention Techniques against DoS Attack

Emerging NVM chips can be one of the primary means of DoS attacks. The following techniques can be used to prevent or mitigate DoS attacks:

- **Increasing retention time.** The total amount of corrupted bits in the presence of external M-Field can be reduced by enhancing the retention time of STT-MRAM. Equations 14.3 and 14.4 show that the retention time changes exponentially with temperature or Δ_{TS}. Moreover, we need to increase the volume of the FML to increase Δ_{TS} (Eq. 14.2). In contrast, higher retention time requires higher write energy; thus, a trade-off is required [66]. Furthermore, as described in Sect. 14.2.2, the side-channel attack can be mitigated by reducing the retention time. In contrast, the retention time enhancement is vital to resist M-Field attacks. Therefore, a trade-off is required at the design phase.
- **Using error correction code (ECC).** ECC scheme can fix a certain amount of errors. However, ECC adds latency, energy, and area overheads [66].

14.2.7 Fault Injection Attack

Injection of fault in memory cells can leak sensitive system information (e.g., keys of encryption technique). An attacker can inject fault in a single memory cell or multiple cells. Figure 14.12 illustrates that the key is revealed as a ciphertext if an attacker can set all the plaintext bits to "0" by injecting fault. In emerging NVM chips, the deterministic supply noise can create a fault in a memory cell (see Sect. 14.1.1.3).

The attacker can launch a specified polarity write failure by generating a specified amount of noise through writing a particular data pattern. Recent studies show that 10.4 mA of write current can originate 130 mV of supply noise [22]. Moreover, the generated supply noise while writing "0" → "0" ("1" → "1") is around 352.46 mV (51.42 mV). The supply noise originated from other patterns falls in between the above mentioned bound.

During the write operation, switching to state "0" → "1" ("0" → "0") requires $80\mu A$ ($100\mu A$) of current [22]. After flashing an address, the adversary can introduce "0" → "1" write failure by generating 130mV of noise through parallelly writing a specific data pattern such as 0x0000FFFFFFFFFFFFFFFFFFFFFFFFFFFF in the victim's Last-Level Cache (LLC) memory space. In ReRAM, during the parallel write operation, generated supply noise from other banks can incur "0" → "1" write failure, whereas writing "1" → "0" does not experience any failure. As mentioned above, leveraging the information along with the side-channel information can boost the process of extracting the cryptographic key. Furthermore, the weak cells are more vulnerable to suchlike attack under temperature and process variations. The fault can be injected by introducing failure by manipulating ambient conditions, read/write sequences, memory locations, and careful selection of patterns.

Fig. 14.12 Outline of the fault injection attack, launched using specified polarity (redrawn from [22]). The attacker can regulate the polarity of the error during the write operation

Asymmetric and high read/write current and long write latency of MRAM produce supply noise. Therefore, like ReRAM, both MRAM and PCM chips are also vulnerable to the fault injection attack.

14.2.7.1 Leveraging Write Failure

The supply noise generated during parallel read/write operation increases the write latency of switching state "0" → "1" very much rapidly than altering state "1" → "0". During the write operation, toggling to state "1" → "0" can tolerate ∼120 mV of supply noise [22]. On the other hand, write failure happens while switching to state "0" → "1" when the supply noise is only 10 mV. Although the supply noise (∼50 mV) reduces the resistance value ∼24% while switching to state "0" → "1", this is not considered a write failure as the sense amplifier (shown in Fig. 14.13) can distinguish between the resistance states during the read operation. Therefore, supply noise (in between 50 and 120 mV) can launch write fails while switching to state "0" → "1" and can introduce successful write while altering to state "1" → "0". Hence, manipulating the range of the supply noise skillfully, the adversary can introduce the fault injection attack, revealing full or partial of sensitive information.

Identifying the memory address location, which the victim uses to perform the write operation, can amplify the attack severity. The adversary can store all 0's patterns, which produce the highest supply noise, in his/her different memory locations. Observing the read failure by frequent reading from those memories can determine that the victim is initiated write operation parallelly in nearby memory locations. The parallel read operation is incapable of producing sufficient supply noise to incur read failure; therefore, it defines that the victim has initiated write operation. Due to long write latency, write failure cannot be conducive to this attack model.

Fig. 14.13 Schematic of the ReRAM read circuitry (redrawn from [16, 22, 67, 68]). For a successful read operation, the final resistance value must be higher than the R_{ref}

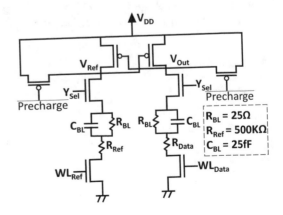

14.2.7.2 Leveraging Read Failure

Supply noise can be generated by performing either both read and write operations or only the write operation by the adversary. The amount of supply noise also varies on the data pattern to be written. Supply noise can reduce the sense margin of the read circuitry by lowering the discharge current and V_{GS} value of the access transistor; thus, read failure is introduced. The read operation can successfully determine the state "0" when the supply noise is 350 mV (high enough) [16, 22]. Therefore, state "0" is not vulnerable to read failure.

14.2.8 Mitigation/Prevention Techniques Against Fault Injection Attack

The mitigation techniques mentioned in Sect. 14.2.4 are also applicable to resist fault injection attacks. Moreover, read/write current scaling can also be implemented to mitigate fault injection attacks. However, current scaling is incapable of eliminating the attack.

14.2.9 Information Leakage Attack

The supply noise generated through writing specific data patterns by the victim propagates to the attacker's memory space. Therefore, the attacker experiences a read failure while reading the previously stored known data pattern from his/her memory space. Hence, the attacker can determine the HW of the victim's written (sensitive) data by analyzing the generated supply noise. Figure 14.14 illustrates the overview of the information leakage attack concept.

As described in Sect. 14.2.7, more than 120 mV of supply noise can introduce write failure while toggling to both "1" → "0" and "0" → "1". If the supply noise is between 50 and 120 mV, only "0" → "1" experiences write failure. Furthermore,

Fig. 14.14 Concept of the information leakage attack (redrawn from [16])

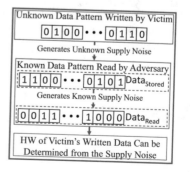

less than 50 mV of supply noise is incapable of introducing write failure. Besides, as described in Sect. 14.2.1, the write current of ReRAM is a function of data pattern to be written and the previously stored data in the memory. Therefore, analyzing the generated supply noise and the current waveforms diligently, the adversary can calculate the HW of the victim's written data. The information extraction process through the read/write operation from the addresses in the vicinity of the read failed addresses is summarized as follows:

1. If the adversary experiences read failure while reading a specific pattern (produces the lowest noise), it ensures that the victim's written data provide the supply noise that is higher than 150 mV. Therefore, the HW of the data written by the victim is higher than 66.77% [16, 67].
2. If the adversary experiences successful read operation while reading a specific pattern (produces the highest noise), which ensures that the HW of the data written by the victim is lower than 67.76% [16, 67]. In short, the range of the HW is (66.77, 67.76%)
3. A read operation is occurred right after a write operation to verify a successful write by the adversary. In this case, if the adversary experiences write failure while writing both "1" → "0" and "0" → "1", which ensures that the supply noise generated by the victim is (120, 150 mV). Thus, the range of the HW is (67.76, 75.37%) [16, 67].
4. Suppose the adversary experiences write failure while writing "0" → "1", which ensures that the supply noise generated by the victim is (50, 120 mV). Thus, the range of the HW is (75.37, 93.23%) [16, 67].

In reality, the boundary conditions might be different as the adversary has very little or no knowledge about the victim's writing location. Furthermore, this leaked information can also enhance the side-channel attack by reducing the search space considerably.

> MRAM chips are also vulnerable to information leakage attacks because they are sensitive to supply noise that causes disturbance errors.

14.2.10   Mitigation/Prevention Techniques Against Information Leakage Attack

The mitigation techniques mentioned in Sect. 14.2.4 are also applicable to resist information leakage attacks. Moreover, read/write current scaling can also be implemented to mitigate information leakage attacks. However, like fault injection attacks, current scaling is incapable of eliminating the attack.

14.2.11   Other Attacks and Their Mitigation/Prevention

Non-volatility of NVM chips makes them more vulnerable to physical attacks since the data stored in the NVM main memory can be retained when the system is OFF. An attacker can physically remove the NVM chip and extract sensitive information [69–71]. Mobile devices, like smartphones, tablets, and laptops, are more vulnerable to such attacks. Encryption schemes such as AES-based and counter-mode XOR-based can be used to prevent such attacks [70]. However, traditional encryption schemes suffer from high performance, resource, and latency overhead. Besides, using encryption on NVM MLC memory chips such as PCM requires immense write energy, which significantly degrades the memory cells' writing endurance. Researchers have proposed several wear-leveling schemes over the last few years, which are suitable for these NVM chips to solve this problem [69]. These wear-leveling techniques can reduce the complexity of encryption design and improve the write endurance [38]. Furthermore, access-control memory (ACM) is a low-cost and high-performance substitute for encryption suitable for NVM chips with limited endurance [69].

As discussed in Sect. 14.1.3.1, the write attack leverages the limited write endurance of the PCM cell. Depending on the side-channel information, the write attack can be classified into several types. Figure 14.15 illustrates the canonical form overview of the repeat address attack (RAA) and slightly modified RAAs. These attacks can bypass almost all levels of modern processor caches with ease. RAA performs an extensive write operation to a single cache line to make the memory cell wear out. In the birthday paradox attack (BPA), the working set is usually altered after several millions of write operations. Finally, stealth mode attack (SMA) performs only on a single line; however, it keeps disguised in other (n-1) cache lines. Usually, the attack density, Δ,[5] is considered 1 for secure wear-leveling techniques to defend the attack. Besides, the following five properties need to be ensured as well [62, 72, 74–76]:

- **Secure mapping.** The user should be completely unaware of the address mapping mechanisms, and therefore, robust and random mapping techniques are preferable.
- **Efficiency.** If the system experiences attack in the logical address, the remapping scheme should quickly alter the mapped physical address of the logical address to mitigate the attack's severity.
- **Sufficiency.** During remapping, to distribute the wear sufficiently, the potential physical address space should be large enough for holding the logical address.
- **Low overhead.** The wear-leveling overhead should be minimal.
- **Error recovery.** Repeated writing introduces mechanical stress to the scaled-down memory cell; hence, permanent stuck-at faults increase drastically. Using

---

[5]Attack density, Δ, is defined as the ratio of the number of writes to the most frequently written line to the total number of writes within a given period [72, 73].

Fig. 14.15 Typical four types of write attacks: (i) RAA, (ii) generalized RAA (GRAA), (iii) BPA, and (iv) SMA (redrawn from [72])

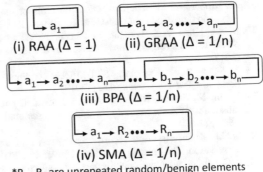

(i) RAA (Δ = 1)   (ii) GRAA (Δ = 1/n)

(iii) BPA (Δ = 1/n)

(iv) SMA (Δ = 1/n)

*R_2 - R_n are unrepeated random/benign elements

lightweight error recovery techniques with the existing wear-leveling algorithms can correct those multiple stuck-at faults [74]. A hardware-efficient multi-bit stuck-at fault error recovery technique (SAFER) is proposed where data blocks are partitioned dynamically for ensuring at most one fail bit in each partition [75].

References

1. J.A. Mandelman, et al., Challenges and future directions for the scaling of dynamic random-access memory (DRAM). IBM J. Res. Dev. **46**(2–3), 187–212 (2002)
2. P. Zhou et al., A durable and energy efficient main memory using phase change memory technology. ACM SIGARCH Comput. Archit. News **37**(3), 14–23 (2009)
3. Y. Zhang et al., A novel ReRAM-based main memory structure for optimizing access latency and reliability, in *Proceedings of the 54th Annual Design Automation Conference 2017* (2017)
4. B. Dieny, R.B. Goldfarb, K.J. Lee, *Introduction to Magnetic Random-Access Memory* (Wiley, Hoboken, 2016). ISBN: 978-1-11907-935-4
5. H. Ohno, et al., Spin-transfer-torque magnetoresistive random access memory (STT-MRAM) technology, in *Advances in Non-Volatile Memory and Storage Technology* (2014), pp. 455–494
6. H. Ishiwara, M. Okyama, Y. Arimoto, Operation principle and circuit design issues, in *Ferroelectric Random Access Memories Fundamentals and Applications*. Topics in Applied Physics, vol. 93 (Springer, Berlin 2004)
7. Y. Nishi, M.-K. Blanka, *Advances in Non-volatile Memory and Storage Technology* (Woodhead Publishing, Duxford, 2019)
8. J.-W. Jang, et al., Self-correcting STTRAM under magnetic field attacks, in *Proceedings of the 52nd Annual Design Automation Conference on - DAC'15* (2015)
9. K. Munira, W.H. Butler, A.W. Ghosh, A quasi-analytical model for energy-delay-reliability tradeoff studies during write operations in a perpendicular STT-RAM cell. IEEE Trans. Electron Dev. **59**(8), 2221–2226 (2012)
10. C.W. Smullen et al., Relaxing non-volatility for fast and energy-efficient STT-RAM caches, in *2011 IEEE 17th International Symposium on High Performance Computer Architecture* (2011)
11. A.F. Vincent, et al., Analytical macrospin modeling of the stochastic switching time of spin-transfer torque devices. IEEE Trans. Electron Dev. **62**(1), 164–170 (2015)

12. W. Rippard, et al., Thermal relaxation rates of magnetic nanoparticles in the presence of magnetic fields and spin-transfer effects. Phys. Rev. B **84**(6), 064439 (2011)
13. M.N.I. Khan, A.S. Iyengar, S. Ghosh, Novel magnetic burn-in for retention testing of STTRAM, in *Design, Automation &; Test in Europe Conference &; Exhibition (DATE), 2017* (2017)
14. N. Sayed, et al., Exploiting STT-MRAM for approximate computing, in *2017 22nd IEEE European Test Symposium (ETS)* (2017)
15. Y. Jin, M. Shihab, M. Jung, Area, power, and latency considerations of STT-MRAM to substitute for main memory, in *2014 International Symposium on Computer Architecture (ISCA)* (2014)
16. M.N.I. Khan, Assuring security and privacy of emerging non-volatile memories. Ph.D. Dissertation, The Pennsylvania State University (2019)
17. S. Ghosh, et al., Security and privacy threats to on-chip non-volatile memories and counter-measures, in *Proceedings of the 35th International Conference on Computer-Aided Design* (2016)
18. R. Bishnoi, et al., Improving write performance for STT-MRAM. IEEE Trans. Magn. **52**(8), 1–11 (2016)
19. X. Fong, S.H. Choday, K. Roy, Bit-cell level optimization for non-volatile memories using magnetic tunnel junctions and spin-transfer torque switching. IEEE Trans. Nanotechnol. **11**(1), 172–181 (2012)
20. Y. Zhang, et al., Asymmetry of MTJ switching and its implication to STT-RAM designs, in *2012 Design, Automation & Test in Europe Conference & Exhibition (DATE)* (2012)
21. M.N.I. Khan, et al., Side-channel attack on STTRAM based cache for cryptographic application, in *2017 IEEE International Conference on Computer Design (ICCD)* (2017)
22. M.N.I. Khan, S. Ghosh, Fault injection attacks on emerging non-volatile memory and countermeasures, in *Proceedings of the 7th International Workshop on Hardware and Architectural Support for Security and Privacy* (2018)
23. M.N.I. Khan, S. Ghosh, Analysis of row hammer attack on STTRAM, in *2018 IEEE 36th International Conference on Computer Design (ICCD)* (2018)
24. S. Rashidi, M. Jalili, H. Sarbazi-Azad, A survey on PCM lifetime enhancement schemes. ACM Comput. Surv. **52**(4), 1–38 (2019)
25. G.W. Burr, et al., Recent progress in phase-change memory technology. IEEE J. Emerg. Sel. Topics Circ. Syst. **6**(2), 146–162 (2016)
26. C. Xu, et al., Overcoming the challenges of crossbar resistive memory architectures, in *2015 IEEE 21st International Symposium on High Performance Computer Architecture (HPCA)* (2015)
27. A. Kawahara, et al., An 8Mb multi-layered cross-point ReRAM macro with 443MB/s write throughput, in *2012 IEEE International Solid-State Circuits Conference* (2012)
28. V.R. Kommareddy, et al., Are crossbar memories secure? New security vulnerabilities in crossbar memories. IEEE Comput. Archit. Lett. **18**(2), 174–177 (2019)
29. W. Lu, et al., Self-current limiting MgO ReRAM devices for low-power non-volatile memory applications. IEEE Electron. Dev. Lett. **6**(2), 163–170 (2016)
30. B. Long, et al., Effects of Mg-doping on HfO2-based ReRAM device switching characteristics. IEEE Electron. Dev. Lett. **34**(10), 1247–1249 (2013)
31. T. Schultz, et al., Vulnerabilities and reliability of ReRAM based PUFs and memory logic. IEEE Trans. Reliab. **69**(2), 690–698 (2020)
32. U. Russo, et al., Modeling of Programming and Read Performance in Phase-Change Memories—Part I: cell optimization and scaling. IEEE Trans. Electron Dev. **55**, 506–514 (2008)
33. F. Pellizzer, et al., Novel /spl mu/trench phase-change memory cell for embedded and stand-alone non-volatile memory applications, in *2004 Symposium on VLSI Technology, Digest of Technical Papers* (2004), pp. 18–19
34. E.T. Kim, J.Y. Lee, Y.T. Kim, Investigation of electrical characteristics of the In3Sb1Re2 ternary alloy for application in phase-change memory, physica status solidi (RRL). Rapid Res. Lett. **3**(4), 103–105 (2009)

35. J.-K. Ahn, et al., Metalorganic chemical vapor deposition of non-GST chalcogenide materials for phase change memory applications. J. Mater. Chem. **20**, 1751–1754 (2010)
36. M. Boniardi, et al., Optimization metrics for Phase change memory (PCM) cell architectures, in *2014 IEEE International Electron Devices Meeting* (2014), pp. 29.1.1–29.1.4
37. M.K. Qureshi, V. Srinivasan, J.A. Rivers, Scalable high performance main memory system using phase-change memory technology. ACM SIGARCH Comput. Archit. News **37**(3), 24–33 (2009)
38. M.K. Qureshi, et al., Enhancing lifetime and security of PCM-based main memory with start-gap wear leveling, in *Proceedings of the 42nd Annual IEEE/ACM International Symposium on Microarchitecture - Micro-42* (2009)
39. W. Zhang, T. Li, Exploring phase change memory and 3D die-stacking for power/thermal friendly, fast and durable memory architectures, in *2009 18th International Conference on Parallel Architectures and Compilation Techniques* (2009)
40. L.E. Ramos, E. Gorbatov, R. Bianchini, Page placement in hybrid memory systems, in *Proceedings of the International Conference on Supercomputing - ICS'11* (2011)
41. G. Wuet, et al., CAR: securing PCM main memory system with cache address remapping, in *2012 IEEE 18th International Conference on Parallel and Distributed Systems* (2012)
42. N.H. Seong, D.H. Woo, H.-H.S. Lee, Security refresh: prevent malicious wear-out and increase durability for phase-change memory with dynamically randomized address mapping. ACM SIGARCH Comput. Archit. News **38**(3), 383–394 (2010)
43. K. Shamsi, Y. Jin, Security of emerging non-volatile memories: attacks and defenses, in *2016 IEEE 34th VLSI Test Symposium (VTS)* (2016)
44. D. Ielmini, et al., Reliability impact of chalcogenide-structure relaxation in phase-change memory (PCM) cellspart I: experimental study. IEEE Trans. Electron Dev. **56**(5), 1070–1077 (2009)
45. D.H. Yoon, et al., Practical nonvolatile multilevel-cell phase change memory, in *Proceedings of the International Conference for High Performance Computing, Networking, Storage and Analysis on - SC'13* (2013)
46. Y.M. Kang, S.Y. Lee, The challenges and directions for the massproduction of highly-reliable, high-density 1t1c FRAM, in *2008 17th IEEE International Symposium on the Applications of Ferroelectrics*, vol. 1 (2008), pp. 1–2
47. M.I. Rashid, F. Ferdaus, B.M.S.B. Talukder, P. Henny, A. Beal, M.T. Rahman, True random number generation using latency variations of FRAM, in *IEEE Transactions on Very Large Scale Integration (VLSI) Systems*, pp. 1–10, https://doi.org/10.1109/TVLSI.2020.3018998
48. E. Jabarov, et al., PCR*-tree: PCM-aware R*-tree. J. Inf. Sci. Eng. **33**, 1359–1374 (2017)
49. N. Rathi, et al., Data privacy in non-volatile cache: challenges, attack models and solutions, in *2016 21st Asia and South Pacific Design Automation Conference (ASP-DAC)* (2016)
50. E. Brier, C. Clavier, F. Olivier, Correlation power analysis with a leakage model, in *Cryptographic Hardware and Embedded Systems - CHES 2004*. Lecture Notes in Computer Science (2004), pp. 16–29
51. P. Rauzy, S. Guilley, Z. Najm, Formally proved security of assembly code against power analysis. J. Cryptograph. Eng. **6**(3), 201–216 (2015)
52. Z. Wang, R.B. Lee, New cache designs for thwarting software cache-based side channel attacks. ACM SIGARCH Comput. Archit. News **35**(2), 494–505 (2007)
53. H. Maghrebi, V. Servant, J. Bringer, *There Is Wisdom in Harnessing the Strengths of Your Enemy: Customized Encoding to Thwart Side-Channel Attacks*, Fast Software Encryption Lecture Notes in Computer Science (2016), pp. 223–243
54. T. Güneysu, A. Moradi, Generic side-channel countermeasures for reconfigurable devices, in *Cryptographic Hardware and Embedded Systems – CHES 2011*. Lecture Notes in Computer Science (2011), pp. 33–48
55. N. Veyrat-Charvillon, et al., Shuffling against side-channel attacks: a comprehensive study with cautionary note, in *Advances in Cryptology – ASIACRYPT 2012*. Lecture Notes in Computer Science (2012), pp. 740–757

56. S. Ben Dodo, R. Bishnoi, M.B. Tahoori, Secure STT-MRAM bit-cell design resilient to differential power analysis attacks. IEEE Trans. Very Large Scale Integr. Syst. **28**(1), 263–272 (2020)

57. H. David, Effects of silicon variation on nano-scale solid-state memories. Ph.D. Dissertation, University of Toronto (2011)

58. Q. Alasad, J. Yuan, J. Lin, Resilient AES against side-channel attack using all-spin logic, in *Proceedings of the 2018 on Great Lakes Symposium on VLSI* (2018)

59. B. Behin-Aein, et al., Proposal for an all-spin logic device with built-in memory. Nat. Nanotechnol. **5**(4), 266–270 (2010)

60. A. Iyengar, et al., Side channel attacks on STTRAM and low-overhead countermeasures, in *2016 IEEE International Symposium on Defect and Fault Tolerance in VLSI and Nanotechnology Systems (DFT)* (2016)

61. S. Bhatkar, D.C. DuVarney, R. Sekar, Address obfuscation: an efficient approach to combat a broad range of memory error exploits, in *12th USENIX Security Symposium, 2003, Washington*

62. N.H. Seong, D.H. Woo, H.-H.S. Lee, Security refresh: prevent malicious wear-out and increase durability for phase-change memory with dynamically randomized address mapping, in *International Symposium on Computer Architecture ISCA-37* (2010)

63. A. Holst, J.-W. Jang, S. Ghosh, Investigation of magnetic field attacks on commercial Magneto-Resistive Random Access Memory, in *2017 18th International Symposium on Quality Electronic Design (ISQED)* (2017)

64. J.-W. Jang, S. Ghosh, Performance impact of magnetic and thermal attack on STTRAM and low-overhead mitigation techniques, in *Proceedings of the 2016 International Symposium on Low Power Electronics and Design - ISLPED'16* (2016)

65. S. Ghosh, Spintronics and security: prospects, vulnerabilities, attack models, and preventions, *Proceedings of the IEEE* **104**(10), 1864–1893 (2016)

66. A. De, et al., Replacing eFlash with STTRAM in IoTs: security challenges and solutions. J. Hardware Syst. Secur. **1**(4), 328–339 (2017)

67. M.N.I. Khan, S. Ghosh, Information leakage attacks on emerging non-volatile memory and countermeasures, in *Proceedings of the International Symposium on Low Power Electronics and Design* (2018)

68. T.D. Happ, H.L. Lung, T. Nirschl, Current compliant sensing architecture for multilevel phase change memory, 2009. Patent No. US7515461B2, Filed January 5th, 2007, Issued April 7th. (2009)

69. C. Chavda, et al., Vulnerability analysis of on-chip access-control memory, in *9th USENIX Workshop on Hot Topics in Storage and File Systems (HotStorage), Santa Clara* (2017)

70. S. Chhabra, Y. Solihin, i-NVMM: a secure non-volatile main memory system with incremental encryption, in *Proceeding of the 38th Annual International Symposium on Computer Architecture - ISCA'11* (2011), pp. 177–188

71. M. Steil, 17 mistakes microsoft made in the xbox security system, in *22nd Chaos Communication Congress* (2005)

72. M.K. Qureshi, et al., Practical and secure PCM systems by online detection of malicious write streams, *2011 IEEE 17th International Symposium on High Performance Computer Architecture* (2011), pp. 478–489

73. M.M. Franceschini, et al., Adaptive wear leveling via monitoring the properties of memory reference stream. US Patent: US8356153B2

74. N.H. Seong, A reliable, secure phase-change memory as a main memory. Ph.D. Dissertation, Georgia Institute of Technology (2012)

75. N.H. Seong, et al., SAFER: stuck-at-fault error recovery for memories, in *2010 43rd Annual IEEE/ACM International Symposium on Microarchitecture* (2010), pp. 115–124

76. A. Seznec, A phase change memory as a secure main memory. IEEE Comput. Archit. Lett. 99(RapidPosts) (2010)

Chapter 15
Security of Analog, Mixed-Signal, and RF Devices

Debayan Das, Baibhab Chatterjee, and Shreyas Sen

15.1 Introduction: Security in the Analog Domain

Over the last decade, we have witnessed a steady growth in the domains of e-banking, smart cards, remote health monitoring, e-governance, and block-chain transactions (crypto-currencies). All these applications rely on the exchange of confidential data and hence data security and trust become extremely critical. To provide data security and trust between the communication parties, cryptographic protocols are developed which ensure the confidentiality and integrity of data. For secure data transfer, the *root of trust* is a small shared secret *key* which is used by the crypto algorithm. Now, with the secure foundation of the root of trust, the security properties need to disseminate throughout the entire chain of trust involving hardware, software, and application layers. The hardware chain of trust ranges from the transistor level to the architecture level, and forms the root of a physical design. In this chapter, we focus on hardware security in the transistor level, wherein the behavior of the device (both the analog properties and non-idealities of the transistor, even for a digital application) manifest themselves in the form of certain vulnerabilities as well as possible security mechanisms.

15.1.1 *Motivation*

The Internet of Things (IoT) has been expanding continuously and is forecasted to reach a $10 trillion economy by 2025 [35]. As these devices remain inter-connected

D. Das (✉) · B. Chatterjee · S. Sen
Purdue University, West Lafayette, IN, USA
e-mail: das60@purdue.edu; bchatte@purdue.edu; shreyas@purdue.edu

© The Author(s), under exclusive license to Springer Nature Switzerland AG 2021
M. Tehranipoor (ed.), *Emerging Topics in Hardware Security*,
https://doi.org/10.1007/978-3-030-64448-2_15

391

within a wireless network, one weak point of entry might spread catastrophically over large areas leaking confidential information to the attacker [46]. Recent distributed denial-of-service (DDoS) attack by the Dyn DNS company demonstrated taking control over millions of inter-connected webcams [33]. Also, attack on the commonly used Philips Hue light bulbs housing Atmel micro-controllers was demonstrated utilizing what is known as power side-channel analysis (SCA) [46]. These attacks only show a small subset of the wide range of vulnerabilities existing in many of the existing commercial resource-constrained IoT devices.

In addition to the security of the IoT devices, with the globalization of the IC manufacturing, trust has also become a major threat to the supply chain such as counterfeiting [22], IP theft, and hardware trojans [31, 48, 56, 60].

Hence, it becomes extremely important to consider the trust and security implications, and being pro-actively aware of the attacks and consequently develop sufficient protection during the design life-cycle of these embedded devices. All the attacks are fundamentally on the transistor level within the *root of trust* [3], and hence preventions are the most efficient when dealt at its source, that is, at the transistor level itself.

15.1.2 Analog Vulnerabilities

To provide the security and authenticity of data, most devices today employ cryptographic algorithms. Traditional crypto algorithms revolve around the concepts of one-wayness and trapdoor functions. One-wayness means that the function is easy to compute, but hard to invert. A trapdoor one-way function is easy to compute, but is also invertible only in presence of the secret "key." Classical cryptanalysis of these algorithms has shown great success, and has proven to be mathematically secure. However, these algorithms are implemented on a physical substrate, which leak critical information in the form of "side-channel leakage" that can be utilized by attackers to extract the secret key operating in the device. Side-channel attacks are examples of analog creating vulnerabilities.

On the other hand, attackers also utilize the inherent analog vulnerabilities in the form of counterfeit ICs [22], DoS attacks like jamming wireless signals [39].

15.1.3 Analog Preventions

Analog circuits can also be utilized to protect against the side-channel attacks and to provide PUF-based authentication using the inherent device characteristics. Since, analog circuits form the basis of every circuit design and hence the source of any attack, it is most efficient to provide preventions in the analog domain.

As shown in Fig. 15.1, mixed-signal (discrete time continuous signal) and continuous time are commonly referred to as analog, while the discrete time discrete

Fig. 15.1 Analog and digital classification: Both mixed-signal mode (discrete time continuous signal) and the continuous time are referred to as analog

Analog & Digital Modalities	
Continuous Time Discrete Signal	Continuous Time Continuous Signal
Discrete Time Discrete Signal	Discrete Time Continuous Signal

signals are considered digital. In the next section, we will extensively study the EM and power side-channel analysis, and see how analog circuits can be utilized to help prevent such attacks.

15.2 Electromagnetic and Power Side-Channels

Now let us study the different hardware SCA attacks that can be mounted on embedded devices to extract the secret key from the encryption engines. As discussed earlier, an encryption algorithm involves a trapdoor one-way function which can only be decrypted using the secret "key." Hence, the goal of an attacker is to recover the key using power/EM SCA. Physical attacks typically occur in 2 phases: the data (encryption traces) collection phase and the attack phase. During the trace collection phase, the attacker measures the power consumption or the EM radiation of the device under attack while it performs the encryption. Next, once the traces are collected, during the attack phase, the attacker performs correlational/differential EM/power analysis (CEMA/DEMA/CPA/DPA), and the correct key emerges out after multiple traces are analyzed, as shown in Fig. 15.2.

Also, the attacks can be classified as non-profiled and profiled attacks (Fig. 15.3a), which we will study in the following subsection.

In 1998, Kocher et al. demonstrated the first power analysis attack [32] in the form of non-profiled simple power analysis (SPA) and differential power analysis (DPA). Since then, several attack vectors have emerged making it an active research domain even today.

15.2.1 Non-profiled Attacks

Non-profiled SCA attack is a direct attack on a target device using the hamming weight (HW) or the hamming distance (HD) leakage model. It includes the conventional CEMA/DEMA/CPA/DPA attacks.

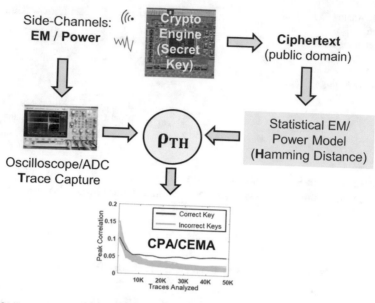

Fig. 15.2 Overview of power/EM side-channel analysis

Figure 15.3b shows the evolution of the EM/power SCA attacks. Following the inception of SPA/DPA, non-invasive EM attack was studied extensively in 2001 [44]. In 2002, Chari et al. developed the template-based profiling attacks [10]. Subsequently, correlational attack was developed by Brier et al. in 2004 [8]. In 2011, machine learning (ML) based SCA attacks were first introduced by Hospodar et al. [28].

EM SCA is very similar to power SCA and is a more practical attack as it does not require any modification of the target device and is non-invasive [44]. Power SCA requires insertion of a small series resistor to measure the current consumption of the target device while it performs encryptions. EM SCA only requires a H-field or E-field probe that can be kept on top of the IC to collect the traces. The main challenge involved in EM SCA attacks is finding the best leakage location on the chip.

Recently, the SCNIFFER framework has been proposed [14] which uses a gradient search algorithm to converge to one of the high leakage points on the chip by performing the test vector leakage assessment (TVLA) or measuring signal-to-noise ratio (SNR). As shown in Fig. 15.4a,b, SCNIFFER makes the EM scanning and attack in the loop thereby reducing the attack time drastically by $\sim 100\times$.

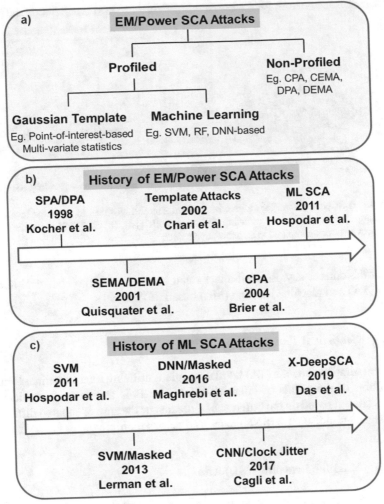

Fig. 15.3 (a) Classification of EM/power SCA attacks, (b) history of SCA, (c) evolution of ML SCA

15.2.2 Profiled Attacks

Profiling SCA attack comprises two phases—the profiling phase and the attack phase. During the profiling phase, an offline template is built using an identical device prior to the actual attack. The attack is then performed on a similar device with much fewer traces leading to a more powerful attack. The entire heavy-lifting is thus offloaded to the training phase which happens offline prior to the actual attack. During the attack phase, the unseen traces are fed to the trained model which then

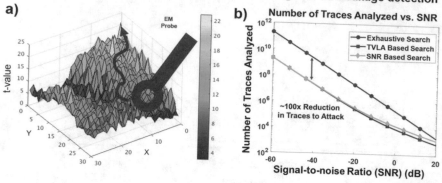

Fig. 15.4 (**a**) SCNIFFER utilizes a greedy gradient search algorithm to converge to one of the high leakage points on the chip by using TVLA or SNR [14], (**b**) SCNIFFER requires ~100× lower traces to attack compared to an exhaustive search

predicts the correct key. Profiled attacks can be classified into statistical template attacks (TA) and machine learning (ML) based SCA attacks.

15.2.2.1 Statistical Template Attacks (TA)

Statistical template attacks (TA) build a template utilizing a multi-variate Gaussian distribution of the points of interest (PoIs). The PoIs are the most critical time samples chosen based on the difference of means, or the sum of squared differences, or the signal-to-noise ratio (SNR) measured across multiple traces [10].

15.2.2.2 Machine Learning (ML) Attacks

The ML-based attacks use supervised techniques such as the support vector machine (SVM), random forest (RF), self-organizing map (SOM), and deep neural network (DNN). The evolution of the ML SCA attacks is shown in Fig. 15.3c.

In 2011, Hospodar et al. [28] for the first time used SVM to attack an unprotected AES implementation. In 2013, Lerman et al. [34] used SVMs and RF-based models to break the masked AES implementations of the DPA contest v4. Recently, DNNs have generated significant interest in the SCA community as it can even defeat SCA protected implementations. In 2016, Maghrebi et al. [36] showed the first deep learning based attacks on masked AES. Following this, in 2017, Cagli et al. demonstrated clock misalignment based countermeasures using convolutional neural (CNN) [9, 50]. Compared to the statistical template-based profiling attacks, deep learning frameworks are more automated and would aid an attacker mount successful attacks on crypto implementations without prior knowledge of the

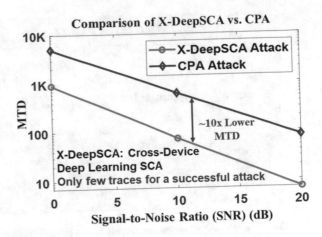

Fig. 15.5 X-DeepSCA utilizes multi-device training and a 256-class DNN to provide high attack accuracy across unseen devices, requiring as low as a single trace for a successful attack, thereby increasing the treat surface significantly

architecture. Moreover, DNNs are preferred over the statistical TA as they can handle large dimensionality of the data and do not require a precise PoI selection.

However, the main challenge involving the profiling attacks is the portability across different devices. A model trained with a trained device should work on an unseen device of the same architecture running the same algorithm. However, the inter-device variations are often significantly higher than the inter-class variations of one device. Hence, although training and testing a neural network on the same device gives a high accuracy, it does not necessarily guarantee a high accuracy for an unseen device.

Recently, in DAC 2019, X-DeepSCA demonstrated the first cross-device deep learning based side-channel attack on AES-128, showing the feasibility of even a single trace attack [16]. X-DeepSCA showed a $10\times$ improvement in the minimum traces to disclosure (MTD) even for low SNR scenarios compared to the traditional CPA attack, increasing the threat surface significantly (Fig. 15.5). Extensions of this work have resulted in even better models utilizing pre-processing techniques like principal component analysis (PCA) for dimensionality reduction and dynamic time warping (DTW) for trace alignment, scaling the ML SCA attack to a large range of devices [21].

15.2.3 State-of-the-Art Countermeasures

Countermeasures against EM/power SCA can be classified as logical, architectural, and physical (circuit-level) countermeasures (Fig. 15.6). Most of the logical and architectural countermeasures are design and algorithm-specific, while the circuit-level countermeasures are generic to any crypto algorithm and can be used as a wrapper around it. All of these countermeasures operate on the fundamental principle of decreasing the signal-to-noise ratio (SNR), and thus rely on the

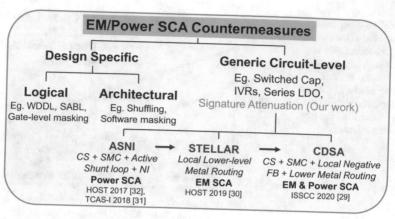

Fig. 15.6 Classification of countermeasures against power/EM SCA

combination of the two key techniques: (i) noise injection (NI) and (ii) critical correlated signature suppression.

15.2.4 Design-Specific Countermeasures

15.2.4.1 Logical Countermeasures

Logical countermeasures are mainly based on power balancing which include the wave dynamic differential logic (WDDL) [29], dual-rail pre-charge (DRP) circuits, sense amplifier based logic (SABL), and gate-level masking [50]. Dual-rail logic requires custom design of the logic gates to equalize the power consumption. In DRP cells, one of the outputs always switches its state (either the original output or its compliment), making the power consumption constant. SABL employs a dynamic and differential logic and requires the complete re-design of the standard cell library to ensure that all the four output transitions (0-0, 0-1, 1-0, 1-1) consume the same amount of power. WDDL appears to be the first protection technique validated in silicon and can be built using the single-rail standard library cells; however, it incurs a 3× area overhead, 4× power overhead, and a 4× performance degradation.

15.2.4.2 Architectural Countermeasures

Architectural countermeasures introduce amplitude or time distortions to obfuscate the power/EM trace. Time distortion is achieved by random insertion of dummy operations or by shuffling the operations. However, it does not provide high levels of protection (MTD) as the number of operations that can be shuffled are limited depending on the specific algorithm and its architecture. Also, clock skipping and dynamic voltage and frequency scaling (DVFS) based countermeasures have been shown to be defeated using advanced attacks [1]. Algorithmic masking techniques are commonly used [42], but it incurs $> 2\times$ area and power overheads.

Overall, the logical and architectural countermeasures explored till date, including the masking and hiding techniques, suffer from high area/power/throughput overheads and are specific to a crypto algorithm. Next, we will study the generic countermeasures that are applicable to any crypto algorithms.

15.2.5 Generic Countermeasures

15.2.5.1 Physical Circuit-Based Countermeasures

This class of countermeasures involves physical noise injection (NI) and supply isolation circuits. While NI has been used extensively in many countermeasures, NI alone suffers from large power and area overheads. Supply isolation techniques include switched capacitor current equalizer [57], integrated voltage regulator (IVR) [30], and series low-dropout (LDO) regulators [51]. Switched capacitor current equalizer based countermeasure is a novel technique and achieves high MTD, but suffers from multiple trade-offs leading to a $2\times$ performance degradation. IVRs using buck converters and series LDOs have been explored extensively; however, they suffer from large passives—inductors and on-chip capacitors. As we will discuss later, these on-chip MIM (metal–insulator–metal) capacitors can leak critical side-channel information through the higher-level metal layers in the form of EM leakage [2, 19]. Also, a series LDO-based implementation inherently leaks critical correlated information [18], as it instantaneously tracks the voltage fluctuations across the crypto core and regulates the current accordingly.

15.2.5.2 Needs

Most of the countermeasures discussed above suffer from high area, performance, and power overheads ($> 2\times$). Although the circuit-level techniques are generic, they treat the crypto engine as a black box and hence incur high overheads. Our goal is to develop a white-box understanding of the EM leakage (Fig. 15.7) from a crypto IC leading towards a low-overhead generic countermeasure. Additionally, we also want to design a synthesis-friendly countermeasure so that it can be integrated seamlessly into different technology nodes without much design effort.

In the sections that follow, we will present the current domain signature attenuation (CDSA) hardware along with low-level metal routing (inspired from the white-box analysis) to provide a low-overhead generic countermeasure, validated in 65nm CMOS technology against a parallel AES256 implementation [2]. Figure 15.8 shows the key techniques behind the proposed CDSA design. As seen from

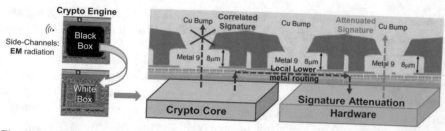

Fig. 15.7 White-box analysis of the crypto IC reveals that the high-level metal layers contribute more compared to the lower metals in terms of the EM leakage, leading to the STELLAR technique for EM SCA protection [19]. STELLAR (Signature aTtenuation Embedded CRYPTO with Low-Level metAl Routing) requires the crypto core to be routed in the lower metals and the signature attenuation hardware embeds the crypto core locally within the lower metal layers before it passes to the higher metal layers, which radiate significantly

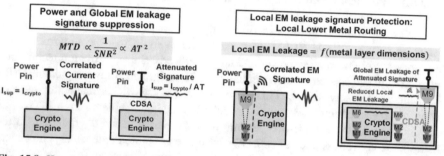

Fig. 15.8 Key techniques for the current domain signature attenuation (CDSA) hardware to provide both EM as well as power SCA protection

Fig. 15.8a, the MTD is proportional to the square of the attenuation factor (AT) providing resilience against both power as well as global EM SCA. Hence, our goal is to provide a very high signature attenuation with extremely low overheads. Figure 15.8b shows the key technique of local low-level metal routing to suppress the EM signature at its origin within the lower metal layers.

In the next subsection, we will study the white-box analysis of the EM leakage to develop a better understanding of the root-cause of the EM leakage.

15.2.6 White-Box Analysis of the EM Leakage

Most of the existing EM SCA attacks as well as countermeasures treat the crypto engine as a black box. However, to design a low-overhead countermeasure, we need to analyze and understand the root-cause of this EM leakage.

15.2.7 STELLAR: Effect of Metal Routing on EM Leakage

All crypto engines like AES256/SHA256/ECC consist of multiple digital gates. These transistors switch their state creating changing currents leading to the EM radiation, according to the Maxwell's equations. However, the main question then arises—what does this generated EM field depend on? Is it caused by the transistors itself?

Well, the EM fields depend on the metal layers carrying the current, and not the transistors. The transformation of the switching currents through the metal-interconnect stack creates the EM radiation which is then picked up by an external adversary, leading to EM SCA attacks. Higher metal layers are thicker (Fig. 15.7, metal-interconnect stack) and hence act as more efficient antennas at the operating frequency of the crypto cores, compared to the lower metal layers. Hence, the EM leakage from the top metal layers (M_9 and above for the Intel 32nm process [19]) has higher probability of detection using the commercially available EM probes. This is proven using 3D FEM system-level simulations of the Intel 32nm metal stack [19]. Hence, our goal is not to pass the correlated crypto current through the high-level metal layers. But, it needs to connect to the external power pin. So, we somehow need to restrict the correlated power signatures to the lower-level metal layers, such that the EM leakage is suppressed locally.

This quest led to the development of STELLAR (Signature aTtenuation Embedded CRYPTO with Low-Level metAl Routing) [19]. STELLAR proposes routing the crypto core within the lower-level metal layers and then embed it within a signature attenuation hardware (SAH) locally within the lower metal layers, such that the critical signature is significantly suppressed before it reaches the top-level metal layers which radiate significantly. This concept of signature suppression within the lower-level metal layers is shown in Fig. 15.7. The current from the crypto core (denoted by blue line) goes through the SAH, which embeds the crypto core locally within the lower metals and is then passed through the higher metal layers (denoted by green line) to connect to the external power pin.

Our work on STELLAR led to the first white-box analysis and developed a better understanding of the root-cause of the EM leakage. Now, combined with a signature attenuation hardware (SAH) with lower-level metal routing, we can develop a highly resilient countermeasure against both EM as well as power SCA attacks. The local routing is extremely critical to minimize long routing of the critical signals.

Looking into the future, we plan to develop further understanding of the genesis of the EM leakage so that we can kill it even closer to its source. Next, we will analyze the design of our signature attenuation hardware (SAH) and combine it with our STELLAR technique to prevent both EM and power SCA attacks.

15.2.8 *Signature Suppression*

In this section, we will study the details of the signature attenuation hardware (SAH). In 2017, we proposed the first concept of SAH design in the form of attenuated signature noise injection (ASNI) [17, 18] to prevent power side-channel analysis (SCA) attacks, generic for all cryptographic algorithms, without any performance overheads.

15.2.8.1 Development of the Signature Attenuation Hardware

The progression of the SAH is shown in Fig. 15.6. In ASNI, the key idea was to embed the crypto engine within a signature attenuation hardware (SAH) such that the correlated critical crypto signature is highly suppressed at the power supply node which an attacker can access, and then inject tiny amount of noise to protect against power SCA attacks. Next, *STELLAR* demonstrated the efficacy of local lower-level metal routing to prevent EM SCA attacks, as discussed in Section IV. Finally, we combine the concepts of signature attenuation from ASNI and the local lower metal routing from *STELLAR* leading to the current domain signature attenuation hardware (CDSA), which was demonstrated in a 65nm test-chip at the ISSCC 2020 [2].

15.2.8.2 ASNI

Let us now understand the design details of the ASNI circuit. ASNI combines a SAH along with a noise injection (NI) circuit (NI is not discussed here). The goal of developing a SAH is to have a constant supply current independent of the variations in the crypto current. The first thing that we can think of is a constant current source (CS). However, a constant CS cannot drive a variable current load (crypto engine). Hence, a load capacitor (C_{Load}) is required to account for the differences in the current, as shown in Fig. 15.9a. Now, as shown in Fig. 15.9b, a high bandwidth (BW) shunt LDO is used which bypasses any excess current through the bleed NMOS whenever the supply current (I_{CS}) is more than the crypto current (I_{Crypto}). A low-BW digital switched mode control (SMC) loop compensates for the process, voltage, and temperature (PVT) variations, and sets the I_{CS} to a quantization level closest to the average crypto current ($I_{Crypto_{avg}}$) by turning on or off required number of CS slices, such that $I_{CS} = I_{Crypto_{avg}} + \Delta$. The quantization error in the supply current Δ is bypassed through the shunt bleed. In steady state, once the top CS current is equal to the average crypto current, the SMC loop is disengaged and the attenuation is thus given by the load capacitance and the output resistance of the CS stage, $AT = \omega C_{Load} r_{ds}$. Now, as discussed previously in Fig. 15.8a, the MTD is proportional to AT^2, which means that a higher output resistance of the CS stage (r_{ds}) can reduce C_{Load}, lowering the area overhead for iso-attenuation (or

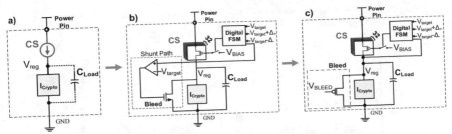

Fig. 15.9 CDSA Evolution: (**a**) Ideal Current Source (CS) driving the crypto core. (**b**) Shunt LDO-based circuit to provide power SCA protection [18], (**c**) Final CDSA architecture for the signature attenuation [2]

iso-MTD). Hence, a cascode CS stage with very high output impedance is chosen so that the load capacitance can be significantly reduced.

During steady state, the SMC loop is only engaged if the V_{reg} node voltage goes below $V_{target} - \Delta_-$ or is above $V_{target} + \Delta_+$, and remains disengaged as long as the voltage remains within the guard band. The low BW of the SMC loop ensures that the voltage fluctuations at the V_{reg} are not reflected instantaneously to the supply current, unlike series LDOs.

Finally, ASNI involves tiny amount of noise injection (NI) in the attenuated signature domain to further enhance the resilience against power SCA attacks [18].

15.2.8.3   CDSA

Current domain signature attenuation (CDSA) combines the signature attenuation hardware (SAH) from ASNI and the local lower metal routing from the *STELLAR* approach to develop the world's most secure SCA countermeasure with $< 1.5\times$ area and power overheads [2]. The main difference in the SAH design is the replacement of the active shunt LDO loop with a biased PMOS bleed, as shown in Fig. 15.9c. This reduces the power overhead while maintaining the same SCA security enhancement. The bleed PMOS provides the bypass path to drain the extra quantization error (Δ) in the CS current, and also provides an inherent local negative feedback (FB) allowing any average crypto current in between two quantized levels of the CS.

The cascode CS stage is designed such that the unit current per slice is higher than the key-dependent variation in $I_{Crypto_{avg}}$, so that the key-dependent information in the average crypto current is not transferred to the supply current and is leaked by the bleed path, providing information-theoretic security [2].

CDSA does not include noise injection and has been implemented in TSMC 65nm technology with local lower metal routing up to M_6. The parallel AES256 is encapsulated by the CDSA hardware providing both EM as well as power SCA immunity.

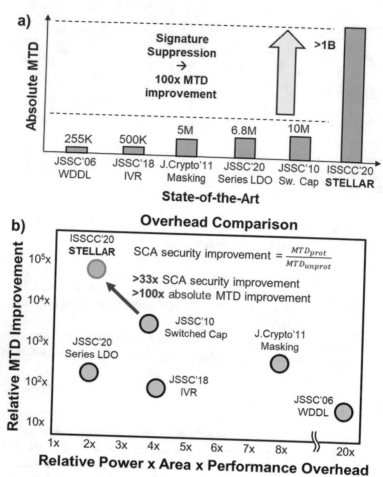

Fig. 15.10 Comparison of CDSA-AES with previously reported hardware security mechanisms

Measurements results of the CDSA-AES256 show an active signature attenuation of $> 350\times$. While the unprotected AES256 could be broken with only $8K$ and $12K$ traces respectively, for CPA and CEMA attacks, the protected CDSA-AES remains secure even after $1B$ encryptions, showing an MTD improvement of $100\times$ over the existing countermeasures [2] (Fig. 15.10, Table 15.1). The CPA and CEMA attacks were verified both in the time as well as frequency domain. Finally, to evaluate the effects of the metal layers on the EM leakage, fixed vs. random test vector leakage analysis (TVLA) was performed. With 200M total traces for the TVLA, the unprotected AES showed t-values of 1056 and 961 for power and EM TVLA, respectively, while the protected implementation with lower metal routing showed power and EM TVLA of 12 and 5.1, respectively (Table 15.2). CDSA-AES256 with high-level metal routing showed an EM TVLA of 8.9, which is much higher than

Table 15.1 CPA/CEMA attack summary

CPA/CEMA Attack Summary	Unprotected AES256	Protected CDSA-AES256
Power SCA	~8K	>1B
EM SCA	~12K	>1B

Table 15.2 Power/EM TVLA summary: Effect of metal layers

Power/EM TVLA Summary	Unprotected AES256	Protected CDSA (Higher metal routing)	Protected CDSA (Lower metal routing)
Power SCA	1056	12	12
EM SCA	961	8.9	5.1

the CDSA implementation with lower metal routing, proving for the first time the effects of metal routing on the EM SCA leakage using on-chip measurements.

The proposed CDSA has also been evaluated against the DNN-based profiling power SCA attacks [15]. While the DNN could be fully trained using only $< 5K$ power traces for the unprotected AES256, the protected CDSA-AES256 could not be trained even after $10M$ traces, demonstrating the efficacy of the proposed countermeasure against deep learning (DL) based SCA attacks. This is also the first countermeasure validated against the DL SCA attacks.

Overall, against the non-profiled attacks, the CDSA achieved $> 1B$ EM/power SCA MTD with $1.37\times$ area and $1.49\times$ power overhead. It is also a generic countermeasure and can be extended to any crypto algorithm providing both power and EM SCA protections without any performance overheads.

In this section, we have thoroughly studied the different EM/power SCA attacks including the countermeasures. In the next section, we will focus on PUF-based authentication mechanisms, and would primarily look into a new type of PUF named radio-frequency PUF (RF-PUF) which enables trust in radio-frequency data communications.

15.3 Physically Unclonable Functions (PUFs)

In the last decade, Physical Unclonable Functions (PUF) have emerged as a promising augmentation to key/token based cryptography, and certain techniques have even shown potential to even supersede basic key-based and software crypto-systems. PUFs leverage the inherent manufacturing process variations of the semiconductor technology to generate a unique and device-specific identity for a particular physical

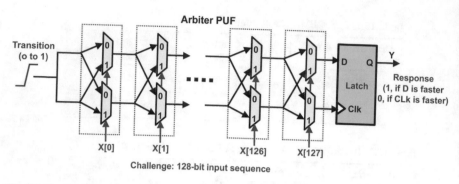

Fig. 15.11 128-bit Arbiter PUF [25]. Assuming same layout length, the circuit creates two delay paths for the data and clock of the latch, for each input X[0,1,...,127], and produces the 1-bit output Y based on which path is faster [11]

implementation [13, 25, 45]. The primary advantages of a PUF implementation over key-based systems lie in the fact that PUFs are usually simpler in terms of hardware and do not need to store the secret key in a battery-backed SRAM or in a non-volatile memory (NVM)/EEPROM to enable complex cryptographic algorithms. Traditional digital PUFs employ simple circuitry such as ring oscillators (RO) [54], Arbiters [40], SRAMs [26, 27, 53], and DRAMs [55] for efficient hardware implementation in a resource-constrained IoT system. Since PUFs rely on physical properties of the implementation (path delays for RO-PUF and Arbiter-PUF, charge distributions during power on for SRAM-PUF, etc.), any invasive tampering mechanism with the hardware usually changes the PUF's output, and hence the user becomes aware of any potential malicious activity.

As an example, an Arbiter PUF is shown in Fig. 15.11 [11], wherein a 128-bit input (called a "challenge" in terms of PUF terminology) is provided to the circuit using the 128 select inputs of the multiplexers. The final output (which is called the PUF response) is a 1-bit signal which is a function of the respective path delays in the data and clock paths due to the random manufacturing variations [25, 40]. By utilizing a large number of input challenges (and hence the high dimensionality of the input space), the Arbiter PUF can be utilized for device authentication. Even though it was shown later in [47] that the randomness of the output of the simple Arbiter PUF can be modeled by an attacker with reasonably low complexity, an improved design with XOR-ed outputs from multiple Arbiter PUFs demonstrated high tolerance against modeling attacks [47, 63].

15.3.1 The Genesis of PUF Properties

The genesis of the PUF properties can be traced back to the inherent manufacturing process variations of the devices and interconnects used in the system. In today's

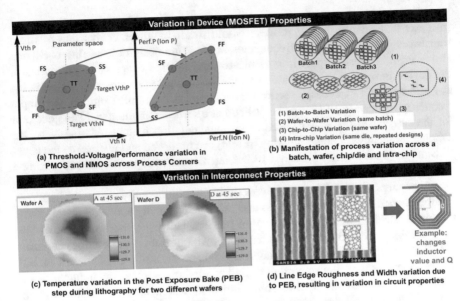

Fig. 15.12 Generation of PUF properties through variations in the semiconductor process technology and mapping to the device performance [43]. A lower threshold voltage results in faster performance and vice versa. (**b**) Manifestation of process variation across a batch, wafer, chip, and intra-chip [43]. (**c**) Temperature variation across two different wafers during PEB [52]. (**d**) Line-edge roughness and width variation due to PEB and its effects on circuit parameters [4]

complementary metal-oxide-semiconductor (CMOS) technologies, both P-type and N-Type MOSFET devices undergo process variations, as shown in Fig. 15.12a [43]. The process variation in the P and N-type devices can be described as a 2-dimensional (2-D) statistical space bounded by (1) slow NMOS and slow PMOS (SS), (2) slow NMOS and fast PMOS (SF), (3) fast NMOS and fast PMOS (FF), and (4) fast NMOS and slow PMOS (FS), where a fast device means that it has a lower threshold voltage (V_{th}) and a slow device means that it has a higher V_{th}. These 4 bounds are usually called process corners and the statistical average of this 2-D space is considered as the scenario with a typical NMOS and a typical PMOS (TT).

During the manufacturing process, the devices can lie anywhere in this 2-D space bounded by the 4 corners, which results in 4 different types of manufacturing variations in the same design as shown in Fig. 15.12b [43]. These variations observed in a standard process can be characterized into (1) batch-to-batch variation, (2) wafer-to-wafer variation, (3) chip-to-chip variation, and (4) intra-chip variation, and this characterization data is usually provided to the designers by the foundry prior to the actual design phase so that the designer can ensure that the functionality of the circuits remains within certain specifications across the 2-D space, subject to process variations. With a knowledge of these variations, it is quite intuitive to associate the device properties with the PUF properties. For example, the SS process would result in higher delays due to slower transistors, leading to lower frequencies in a RO-PUF, while the FF process would result in lower delays due to faster transistors, leading to higher frequencies in the same design of the ring oscillator. However, it remains almost impossible to replicate the exact same properties of the device through a second batch of manufacturing by an attacker, which is the key principle of a PUF-based security mechanism. As a simple example, the output frequency from an RO-PUF can be utilized as a seed in a randomized key generator for the case when the PUF is used as an augmentation feature. Alternatively, the average frequency (or some function of that) can be stored and utilized as the key.

Similar variations are observed in the interconnect properties as well during a manufacturing process. These variations primarily occur due to the steps involving post-exposure bake (PEB) or hard-bake during the lithography process. These procedures reduce the mechanical stress and promote thermally activated photo-resists to help in etching during the fabrication process. However, these procedures also cause temperature variations across the wafer, which results in line-edge roughness (LER) and width variations in the interconnects, as shown in Fig. 15.12c,d [4, 52]. These variations can alter important design parameters, such as the inductance and the quality factor of an on-chip spiral inductor, which creates variations in the center frequency and the phase noise in an oscillator design which uses such on-chip metal inductors. However, this also means that the center frequency and the phase noise of a radio-frequency (RF) transmitter, among other properties, can be utilized for identifying the transmitter, which is one of the primary concepts in RF fingerprinting.

15.3.2 Literature Survey: The Road from RF Fingerprinting to RF-PUF

RF fingerprinting [6, 7, 23, 38, 41, 49, 58, 59, 61] have been utilized extensively for RF device detection in a network, and uses time and frequency domain properties of individual transmitters as their signatures. However, certain features used for device identification need to be known a priori, and both time and frequency domain analysis have their own limitations in a practical scenario. For example, the time-domain analysis method requires detection of the start and end of the transients and needs fixed preambles before starting the data communication. On the other hand, frequency domain analysis requires high oversampling ratios which increases the power consumption in the receiver through the generation of a high-frequency clock.

Along with the physical layer, MAC layer and other upper layers in the communication protocol have also been used in the past for RF fingerprinting [62]. However, an attacker can easily spoof traditional device identifiers in upper layers like IMEI number, IP address, MAC address, etc. RF-DNA, which is an extension of RF fingerprinting, utilized the reflected/refracted electromagnetic (EM) waves based on the 3-dimensional positioning of unique scattering antennas [20]. This establishes the design of the specific antennas as the certificates of authenticity (COA). However, RF-DNA requires additional hardware and design effort in the form of careful implementation of the antennas for accurate device detection. This also increases the size of the transmitter device, thereby making it unsuitable for applications such as implantable wireless biosensors, brain-machine interfaces, etc. Our earlier work on RF-PUF [12, 13] aimed to solve this challenge by utilizing the intrinsic analog/RF properties of a transmitter (for example, LO frequency offset, I-Q offsets, non-linearities in the mixer and power amplifier) as a radio-frequency physical unclonable function (RF-PUF). A deep neural network based machine learning classifier at the receiver end detects the features of the PUF for accurate detection of the transmitting device. As compared to RF-DNA, RF-PUF does not require any additional analog/RF hardware for PUF implementation at the Transmitter as the features are selected such that feature generation and extraction is ingrained in the transceiver operation [12] (it is important to note that RF-DNA involved measurement of the reflected/refracted EM waves based on the 3D-positioning of scattering antennas which are custom-made for each RF unit). Additionally, error-correction and noise cancellation features are also inherent to the transceiver architecture, which increases the reliability without the need for any specific error-correction mechanism dedicated for the PUF operation [12]. Morin et al. [37] extended this concept and showed in their experimental work with 21 commercial transmitters, that up to 95% accuracy can be achieved. Hanna et al. have recently utilized power amplifier nonlinearity to fingerprint RF devices [24]. However, the model was tested mostly on simulation data and only 7 devices were utilized for experimental purposes.

15.3.3 RF-PUF: Concepts and Simulation

Traditional PUF designs such as Arbiter-PUF, RO-PUF, SRAM-PUF, or DRAM-PUF still require a moderate amount of additional hardware at the transmitter side for implementing the specific PUF design. With the proliferation of the resource-constrained asymmetric IoT devices, it is of paramount importance that the power consumption and size of the transmitter node are minimized for longer battery lifetime [11]. RF-PUF [12, 13] exploits the effects of inherent analog and RF process variations at the resource-constrained transmitter side by detecting them with an in situ machine learning hardware at the resource-rich receiver. This method embraces the already existing non-idealities at the Transmitter which are usually discarded in a traditional communication scenario, and hence do not require any additional hardware for PUF generation.

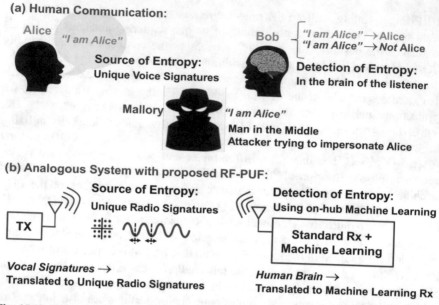

Fig. 15.13 Principle of RF-PUF [12]: (**a**) Authentication in Human Voice Communication: Bob (the receiver) can identify Alice (the transmitter) based on the unique voice signatures, and not based on the contents of what Alice speaks. Mallory (the impersonator) can also be identified (as not Alice), since his unique voice signatures would be different from Alice. (**b**) Analogous System that utilizes an RF-PUF framework for secure radio communication [11]

The principle of RF-PUF is inspired by the inherent authentication in human voice communication as shown in Fig. 15.13. As humans, we associate the inherent features of a person's voice (for example—phonation, pitch, loudness, and rate) with the identity of that person, irrespective of the contents of the person's speech. This is very different from today's digital radio communication, where the identity of the transmitter is included as a field in the packets that are transmitted and hence any transmitter can claim to be any other transmitter by changing the digital packet headers. In RF-PUF, however, the unique human voice is replaced by unique transmitter signatures in the analog and RF domain, and the human brain replaced by a neural network at the receiver. The holistic system-level view for RF-PUF implementation is shown in Fig. 15.14, while the number of unique Transmitters that can be identified with varying channel conditions and Rx signatures is shown in Fig. 15.15. It has been shown with simulation results that up to 8000 transmitting devices can be uniquely identified with 99% accuracy. Proof-of-concept hardware evaluations were also demonstrated in [12] with 10 emulated devices at various temperatures. Since this method does not require any additional hardware at the transmitter, the framework can be utilized as an extremely useful security feature for resource-constrained IoT devices for a small-to-medium scale smart system [11].

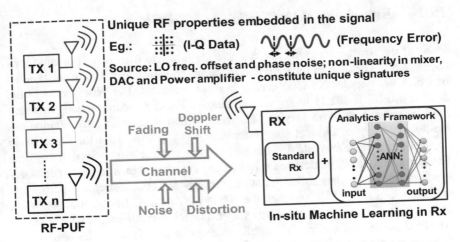

Fig. 15.14 Visualization of RF-PUF in an Asymmetric IoT network with multiple resource-constrained IoT devices as transmitters and one resource-rich receiver [12]

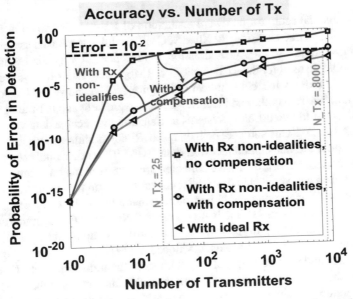

Fig. 15.15 Probability of False detection as a function of the total number of transmitters in the system, with and without receiver signature compensation [12]

15.3.4 Experimental Verification of RF-PUF

Many of the smart commodity devices nowadays utilize Bluetooth/Bluetooth low-energy/ANT/XBee/Wi-Fi for short and medium range wireless communication, all of which use popular 2.4 GHz carrier in most scenarios. For Wi-Fi, most of the devices use IEEE 802.11 b/g/n/ac/ax standards. These standards specify different modulation schemes with varying data rates, which is automatically decided on-the-fly with help of channel state information (CSI). Based on the variation in channel quality (for example, the available signal-to-noise ratio—SNR at any time during transmission), modulation schemes are switched which requires additional synchronization for data collection at the receiver end. For a simple experimental validation of the RF-PUF concept, the XBee S2C module has been chosen (that utilizes the IEEE 802.15.4 standard) which is designed for industrial use and is widely deployed on various IoT application scenarios. This can be used in the subsequent analysis without any loss of generality.

15.3.4.1 Physical Device Setup: Using 30 XBee Devices as Transmitters

The experimental setup for hardware evaluation of RF-PUF is shown in Fig. 15.16 [5]. For initial characterization, 30 Xbee S2C RF modules were used as transmitters. These transmitters use QPSK modulation (and hence all 4 quadrants in the I-Q domain could be used for device detection) at 1Mbps data rate which is pretty much a standard in today's IoT communication. Figure 15.16a shows the block diagram of the setup. The transmitter and receiver devices were kept about 1 meter apart and were connected to two different systems to transmit and receive data via a wireless channel. A HackRF One software defined radio (SDR) module was used as a sniffing device to spoof the transmitted/received data. Two different channel scenarios were considered: in scenario 1, data were collected directly from the transmitter while for scenario 2, data were collected from the receiver. Based on the specific scenario, the HackRF module was connected either to the transmitter or the receiver directly via an SMA cable. The HackRF One is set up using GNU radio that captures and saves the data continuously. Figure 15.16b shows the in-lab setup for the experiments.

From our initial set of features as introduced in [12] (carrier frequency offset and I-Q amplitude and phase offset in all 4 quadrants of the transmitted QPSK signal), it was found that the detection accuracy with these 30 transmitters at the same voltage and temperature is only about 70%. In the quest to better understand the features, a Principal Component Analysis (PCA) was performed to find dominant features, and it was found that the carrier frequency offset is indeed the most dominant feature. Next, different statistical properties of the frequency offset, namely the mean and the standard deviation, were analyzed for each transmitter. We noticed that the mean and standard deviation of the carrier frequency offset vary significantly from transmitter to transmitter, in general. Additionally, even if the mean frequency offset is similar

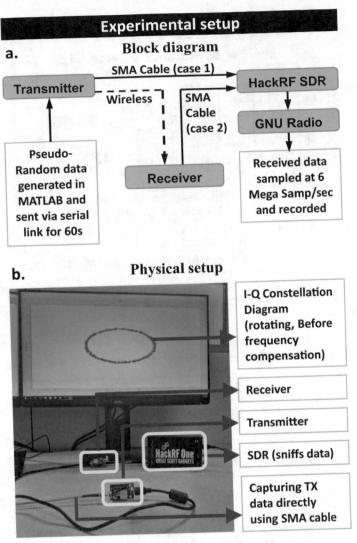

Fig. 15.16 (a) Conceptual Block diagram of the experimental setup shows data transmission and extraction process. (b) In-lab physical demonstration setup. The transmitter and the receiver were placed 1m apart and a HackRF SDR module was used to sniff data either from the Transmitter (case 1) or Receiver (case 2). The SDR records the collected data and shows live constellation [5]

for two different transmitters, the standard deviation is found to be different and vice versa. If these two statistical parameters can be combined to form a new feature, that can provide significant discrimination among transmitters and lead to much better accuracy [5]. The ratio of standard deviation and mean (which is known as COV—

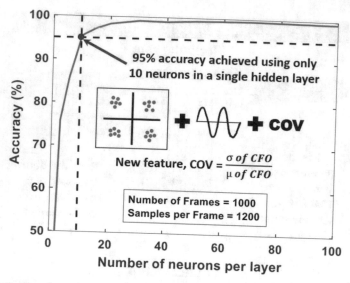

Fig. 15.17 Performance of the RF-PUF framework with COV on 30 Tx devices: Using a single hidden layer, 95% accuracy is achieved using only 10 neurons and > 99% accuracy is reached for > 50 neurons [5]

coefficient of variation) of the carrier frequency offset (CFO) was then introduced as a new feature.

$$COV = \frac{Standard\ deviation\ of\ CFO}{Mean\ of\ CFO}$$

15.3.4.2 Performance After Introducing COV as a Feature

After including COV as a feature, along with mean carrier frequency offset and I-Q amplitude and phase offsets as other features, our neural network was re-trained with the updated feature matrix and its performance was verified again. Figure 15.17 shows the performance of the network with the COV feature for 30 different transmitters. With only a single hidden layer, > 95% accuracy can be achieved using just 10 neurons, while the accuracy improves to a maximum of 99.6% by increasing the number of neurons. This analysis establishes RF-PUF as a strong contender for low-cost security enhancement (simply because it does not use any additional hardware at the transmitter side) for future IoT devices.

15.4 Conclusion

As the number of connected devices keep rising, the hardware vulnerabilities will become more prominent. Hence, it becomes extremely important to consider security during the design life-cycle of a product. In terms of EM/power SCA attacks, it is extremely important to design low-overhead scalable generic countermeasures that can be used to protect existing crypto implementations thereby maintaining legacy which is critical for an industry standpoint. Also, we need to understand the root-cause analysis of the EM SCA leakage with a more in-depth analysis to provide more efficient countermeasures against these attacks. At the same-time, low-cost PUF-based mechanisms augmented by techniques such as RF-PUF would lead to secure IoT systems that are tolerant to spoofing and man-in-the-middle attacks at the network level. It is extremely important to understand that most of the advanced physical-layer and network-layer attacks utilize the analog vulnerabilities of the devices, such as the SCA attacks as depicted in this chapter. Hence, the most efficient solutions to these attacks should also be designed from the analog properties of the system, and techniques such as STELLAR, CDSA, and RF-PUF are the stepping stones of solving this problem at its root.

References

1. M.L. Akkar, R. Bevan, P. Dischamp, D. Moyart, in *Advances in Cryptology — ASIACRYPT 2000*, ed. by T. Okamoto. Lecture Notes in Computer Science. (Springer, Berlin, Heidelberg, 2000), pp. 489–502. https://doi.org/10.1007/3-540-44448-3_38
2. D. Das et al., 27.3 EM and Power SCA-Resilient AES-256 in 65nm CMOS through >350× current-domain signature attenuation, in *2020 IEEE International Solid-State Circuits Conference - (ISSCC)* (2020), pp. 424–426. https://doi.org/10.1109/ISSCC19947.2020.9062997. ISSN: 2376-8606
3. M. Alioto, Trends in hardware security: From basics to ASICs. IEEE Solid-State Circuits Mag. **11**(3), 56–74 (2019). https://doi.org/10.1109/MSSC.2019.2923503 Conference Name: IEEE Solid-State Circuits Magazine
4. A. Asenov, *Statistical Nano CMOS Variability and Its Impact on SRAM* (Springer US, Boston, MA, 2010), pp. 17–49. https://doi.org/10.1007/978-1-4419-6606-3_3
5. M.F. Bari, B. Chatterjee, S. Sen, Hardware evaluation of RF-PUF on commodity devices. IEEE Internet Things J. (2020, under Review in)
6. T.J. Bihl, K.W. Bauer, M.A. Temple, Feature selection for RF fingerprinting with multiple discriminant analysis and using zigbee device emissions. IEEE Trans. Inf. Forensics Secur. **11**(8), 1862–1874 (2016)
7. K. Bonne Rasmussen, S. Capkun, Implications of radio fingerprinting on the security of sensor networks, in *2007 Third International Conference on Security and Privacy in Communications Networks and the Workshops - SecureComm 2007* (2007), pp. 331–340
8. E. Brier, C. Clavier, F. Olivier, Correlation power analysis with a leakage model, in *Cryptographic Hardware and Embedded Systems - CHES 2004*, ed. by M. Joye, J.J. Quisquater. Lecture Notes in Computer Science (Springer, Berlin, Heidelberg, 2004), pp. 16–29
9. E. Cagli, C. Dumas, E. Prouff, Convolutional Neural Networks with Data Augmentation against Jitter-Based Countermeasures – Profiling Attacks without Pre-Processing. Tech. Rep. 740 (2017). https://eprint.iacr.org/2017/740
10. S. Chari, J.R. Rao, P. Rohatgi, Template attacks, in *Cryptographic Hardware and Embedded Systems - CHES 2002*. Lecture Notes in Computer Science (Springer, Berlin, Heidelberg, 2002)
11. B. Chatterjee, N. Cao, A. Raychowdhury, S. Sen, Context-aware intelligence in resource-constrained IoT nodes: Opportunities and challenges. IEEE Des. Test **36**(2), 7–40 (2019)

12. B. Chatterjee, D. Das, S. Maity, S. Sen, RF-PUF: Enhancing IoT security through authentication of wireless nodes using in-situ machine learning. IEEE Internet Things J. (2018). https://doi.org/10.1109/JIOT.2018.2849324

13. B. Chatterjee, D. Das, S. Sen, RF-PUF: IoT security enhancement through authentication of wireless nodes using in-situ machine learning, in 2018 IEEE International Symposium on Hardware Oriented Security and Trust (HOST) (2018), pp. 205–208. https://doi.org/10.1109/HST.2018.8383916

14. J. Danial, D. Das, S. Ghosh, A. Raychowdhury, S. Sen, SCNIFFER: Low-cost, automated, efficient electromagnetic side-channel sniffing. arXiv:1908.09407 [cs] (2019). https://arxiv.org/abs/1908.09407

15. D. Das, J. Danial, A. Golder, S. Ghosh, A.R. Wdhury, S. Sen, Deep learning side-channel attack resilient AES-256 using current domain signature attenuation in 65nm CMOS, in 2020 IEEE Custom Integrated Circuits Conference (CICC) (2020), pp. 1–4. https://doi.org/10.1109/CICC48029.2020.9075889. ISSN: 2152-3630

16. D. Das, A. Golder, J. Danial, S. Ghosh, A. Raychowdhury, S. Sen, X-DeepSCA: Cross-device deep learning side channel attack, in 2019 56th ACM/IEEE Design Automation Conference (DAC) (2019), pp. 1–6. ISSN: 0738-100X

17. D. Das, S. Maity, S.B. Nasir, S. Ghosh, A. Raychowdhury, S. Sen, High efficiency power side-channel attack immunity using noise injection in attenuated signature domain, in 2017 IEEE International Symposium on Hardware Oriented Security and Trust (HOST) (2017), pp. 62–67. https://doi.org/10.1109/HST.2017.7951799

18. D. Das, S. Maity, S.B. Nasir, S. Ghosh, A. Raychowdhury, S. Sen, ASNI: Attenuated signature noise injection for low-overhead power side-channel attack immunity. IEEE Trans. Circuits Syst. I Regul. Pap. 65(10), 3300–3311 (2018). https://doi.org/10.1109/TCSI.2018.2819499

19. D. Das, M. Nath, B. Chatterjee, S. Ghosh, S. Sen, STELLAR: A generic EM side-channel attack protection through ground-up root-cause analysis, in 2019 IEEE International Symposium on Hardware Oriented Security and Trust (HOST) (2019), pp. 11–20. https://doi.org/10.1109/HST.2019.8740839

20. G. DeJean, D. Kirovski, RF-DNA: Radio-frequency certificates of authenticity, in Cryptographic Hardware and Embedded Systems - CHES 2007, ed. by P. Paillier, I. Verbauwhede (Springer, Berlin, Heidelberg, 2007), pp. 346–363

21. A. Golder, D. Das, J. Danial, S. Ghosh, S. Sen, A. Raychowdhury, Practical approaches toward deep-learning-based cross-device power side-channel attack. IEEE Trans. Very Large Scale Integr. (VLSI) Syst. 27(12), 2720–2733 (2019). https://doi.org/10.1109/TVLSI.2019.2926324

22. U. Guin, K. Huang, D. DiMase, J.M. Carulli, M. Tehranipoor, Y. Makris, Counterfeit integrated circuits: A rising threat in the global semiconductor supply chain. Proc. IEEE 102(8), 1207–1228 (2014). https://doi.org/10.1109/JPROC.2014.2332291. Conference Name: Proceedings of the IEEE

23. J. Hall, M. Barbeau, E. Kranakis, Detecting rogue devices in Bluetooth networks using radio frequency fingerprinting, in Communications and Computer Networks (2006)

24. S.S. Hanna, D. Cabric, Deep learning based transmitter identification using power amplifier nonlinearity, in 2019 International Conference on Computing, Networking and Communications (ICNC) (2019), pp. 674–680

25. C. Herder, M. Yu, F. Koushanfar, S. Devadas, Physical unclonable functions and applications: A tutorial. Proc. IEEE 102(8), 1126–1141 (2014). https://doi.org/10.1109/JPROC.2014.2320516

26. D.E. Holcomb, W.P. Burleson, K. Fu, Initial SRAM state as a fingerprint and source of true random numbers for RFID tags, in In Proceedings of the Conference on RFID Security (2007)

27. D.E. Holcomb, W.P. Burleson, K. Fu, Power-up SRAM state as an identifying fingerprint and source of true random numbers. IEEE Trans. Comput. 58(9), 1198–1210 (2009)

28. G. Hospodar, B. Gierlichs, E. De Mulder, I. Verbauwhede, J. Vandewalle, Machine learning in side-channel analysis: a first study. J. Cryptogr. Eng. 1(4), 293 (2011). https://doi.org/10.1007/s13389-011-0023-x

29. D.D. Hwang, K. Tiri, A. Hodjat, B.C. Lai, S. Yang, P. Schaumont, I. Verbauwhede, AES-based security coprocessor IC in 0.18um CMOS with resistance to differential power analysis side-channel attacks. IEEE J. Solid-State Circuits **41**(4), 781–792 (2006). https://doi.org/10.1109/JSSC.2006.870913

30. M. Kar, A. Singh, S.K. Mathew, A. Rajan, V. De, S. Mukhopadhyay, Reducing power side-channel information leakage of AES engines using fully integrated inductive voltage regulator. IEEE J. Solid-State Circuits **53**(8), 2399–2414 (2018). https://doi.org/10.1109/JSSC.2018.2822691

31. R. Karri, J. Rajendran, K. Rosenfeld, M. Tehranipoor, Trustworthy hardware: Identifying and classifying hardware Trojans. Computer **43**(10), 39–46 (2010). https://doi.org/10.1109/MC.2010.299. Conference Name: Computer

32. P. Kocher, J. Jaffe, B. Jun, Differential power analysis, in *Advances in Cryptology CRYPTO 99*, no. 1666, ed. by M. Wiener. Lecture Notes in Computer Science (Springer, Berlin, Heidelberg, 1999), pp. 388–397. https://doi.org/10.1007/3-540-48405-1_25. https://link.springer.com/chapter/10.1007/3-540-48405-1_25

33. B. Krebs, Hacked Cameras, DVRs Powered Today's Massive Internet Outage. https://krebsonsecurity.com/2016/10/hacked-cameras-dvrs-powered-todays-massive-internet-outage/. Library Catalog: https://krebsonsecurity.com

34. L. Lerman, G. Bontempi, O. Markowitch, A machine learning approach against a masked AES. J. Cryptogr. Eng. **5**(2), 123–139 (2015). https://doi.org/10.1007/s13389-014-0089-3

35. M. G. Institute, The internet of things: Mapping the value beyond the hype. Tech. rep., McKinsey Global Institute (2015)

36. H. Maghrebi, T. Portigliatti, E. Prouff, Breaking cryptographic implementations using deep learning techniques. Tech. Rep. 921 (2016). https://eprint.iacr.org/2016/921

37. C. Morin, L.S. Cardoso, J. Hoydis, J.M. Gorce, T. Vial, Transmitter classification with supervised deep learning, in *CrownCom* (2019)

38. N.T. Nguyen, G. Zheng, Z. Han, R. Zheng, Device fingerprinting to enhance wireless security using nonparametric Bayesian method, in *2011 Proceedings IEEE INFOCOM* (2011), pp. 1404–1412

39. C.P. O'Flynn, Message denial and alteration on IEEE 802.15.4 low-power radio networks, in *2011 4th IFIP International Conference on New Technologies, Mobility and Security* (2011), pp. 1–5. https://doi.org/10.1109/NTMS.2011.5720580. ISSN: 2157-4960

40. Z. Paral, S. Devadas, Reliable and efficient PUF-based key generation using pattern matching, in *2011 IEEE International Symposium on Hardware-Oriented Security and Trust* (2011), pp. 128–133

41. L. Peng, A. Hu, J. Zhang, Y. Jiang, J. Yu, Y. Yan, Design of a hybrid rf fingerprint extraction and device classification scheme. IEEE Internet Things J. **6**(1), 349–360 (2019)

42. A. Poschmann, A. Moradi, K. Khoo, C.W. Lim, H. Wang, S. Ling, Side-channel resistant crypto for less than 2,300 GE. J. Cryptol. **24**(2), 322–345 (2011). https://doi.org/10.1007/s00145-010-9086-6

43. K. Qian, Variability modeling and statistical parameter extraction for CMOS devices, Ph.D. Thesis, UC Berkeley (2015)

44. J.J. Quisquater, D. Samyde, ElectroMagnetic analysis (EMA): Measures and counter-measures for smart cards, in *Smart Card Programming and Security*, Lecture Notes in Computer Science (Springer, Berlin, Heidelberg, 2001), pp. 200–210. https://doi.org/10.1007/3-540-45418-7_17. https://link.springer.com/chapter/10.1007/3-540-45418-7_17

45. U. Rührmair, M. van Dijk, PUFs in security protocols: Attack models and security evaluations, in *2013 IEEE Symposium on Security and Privacy* (2013), pp. 286–300. https://doi.org/10.1109/SP.2013.27

46. E. Ronen, A. Shamir, A.O. Weingarten, C. O'Flynn, IoT goes nuclear: Creating a ZigBee chain reaction, in *2017 IEEE Symposium on Security and Privacy (SP)* (2017), pp. 195–212. https://doi.org/10.1109/SP.2017.14. ISSN: 2375-1207

47. U. Rührmair, F. Sehnke, J. Sölter, G. Dror, S. Devadas, J. Schmidhuber, Modeling attacks on physical unclonable functions, in *Proceedings of the 17th ACM Conference on Computer and*

Communications Security (ACM, New York, NY, USA, 2010), CCS '10, pp. 237–249. https://doi.org/10.1145/1866307.1866335

48. H. Salmani, M. Tehranipoor, J. Plusquellic, A novel technique for improving hardware Trojan detection and reducing Trojan activation time. IEEE Trans. Very Large Scale Integr. (VLSI) Syst. **20**(1), 112–125 (2012). https://doi.org/10.1109/TVLSI.2010.2093547. Conference Name: IEEE Transactions on Very Large Scale Integration (VLSI) Systems

49. P. Scanlon, I.O. Kennedy, Y. Liu, Feature extraction approaches to RF fingerprinting for device identification in femtocells. Bell Labs Tech. J. **15**(3), 141–151 (2010)

50. S. Sen, A. Raychowdhury, Electromagnetic and machine learning side-channel attacks and low-overhead generic countermeasures (2019). https://ches.iacr.org/2019/src/tutorials/ches2019tutorial_Sen.pdf

51. A. Singh, M. Kar, V.C.K. Chekuri, S.K. Mathew, A. Rajan, V. De, S. Mukhopadhyay, Enhanced power and electromagnetic SCA resistance of encryption engines via a security-aware integrated all-digital LDO. IEEE J. Solid-State Circuits **55**(2), 478–493 (2020). https://doi.org/10.1109/JSSC.2019.2945944

52. D.A. Steele, A. Coniglio, C. Tang, B. Singh, S. Nip, C.J. Spanos, Characterizing post-exposure bake processing for transient- and steady-state conditions, in the context of critical dimension control, in *Metrology, Inspection, and Process Control for Microlithography XVI*, vol. 4689, ed. by D.J.C. Herr. International Society for Optics and Photonics (SPIE, 2002), vol. 4689, pp. 517–530. https://doi.org/10.1117/12.473491

53. Y. Su, J. Holleman, B.P. Otis, A digital 1.6 pJ/bit chip identification circuit using process variations. IEEE J. Solid-State Circuits **43**(1), 69–77 (2008)

54. G.E. Suh, S. Devadas, Physical unclonable functions for device authentication and secret key generation, in *2007 44th ACM/IEEE Design Automation Conference* (2007), pp. 9–14

55. S. Sutar, A. Raha, D. Kulkarni, R. Shorey, J. Tew, V. Raghunathan, D-PUF: An intrinsically reconfigurable DRAM PUF for device authentication and random number generation. ACM Trans. Embed. Comput. Syst. **17**(1), 17:1–17:31 (2017). https://doi.org/10.1145/3105915

56. M. Tehranipoor, F. Koushanfar, A survey of hardware Trojan taxonomy and detection. IEEE Des. Test Comput. **27**(1), 10–25 (2010). https://doi.org/10.1109/MDT.2010.7. Conference Name: IEEE Design Test of Computers

57. C. Tokunaga, D. Blaauw, Securing encryption systems with a switched capacitor current equalizer. IEEE J. Solid-State Circuits **45**(1), 23–31 (2010). https://doi.org/10.1109/JSSC.2009.2034081

58. B. Vladimir, B. Suman, G. Marco, O. Sangho, Wireless device identification with radiometric signatures, in *MobiCom '08* (2008)

59. T.D. Vo-Huu, T.D. Vo-Huu, G. Noubir, Fingerprinting Wi-Fi devices using software defined radios. in *WiSec '16* (2016)

60. X. Wang, H. Salmani, M. Tehranipoor, J. Plusquellic, Hardware Trojan detection and isolation using current integration and localized current analysis, in *2008 IEEE International Symposium on Defect and Fault Tolerance of VLSI Systems* (2008), pp. 87–95. https://doi.org/10.1109/DFT.2008.61. ISSN: 2377-7966

61. F. Xie, H. Wen, Y. Li, S. Chen, L. Hu, Y. Chen, H. Song, Optimized coherent integration-based radio frequency fingerprinting in internet of things. IEEE Internet Things J. **5**(5), 3967–3977 (2018)

62. Q. Xu, R. Zheng, W. Saad, Z. Han, Device fingerprinting in wireless networks: Challenges and opportunities. IEEE Commun. Surv. Tutorials **18**(1), 94–104 (2016)

63. C. Zhou, K.K. Parhi, C.H. Kim, Secure and reliable XOR arbiter PUF design: An experimental study based on 1 Trillion challenge response pair measurements, in *Proceedings of the 54th Annual Design Automation Conference 2017* (ACM, New York, NY, USA, 2017), DAC '17, pp. 10:1–10:6. https://doi.org/10.1145/3061639.3062315

Chapter 16
Analog IP Protection and Evaluation

N. G. Jayasankaran, A. Sanabria-Borbón, E. Sánchez-Sinencio, J. Hu, and J. Rajendran

16.1 Introduction

The increasing cost of manufacturing integrated circuits (IC) has forced many companies to go fabless. With the outsourcing of IC fabrication in a globalized/distributed design flow, including multiple (potentially untrusted) entities, the semiconductor industry faces several challenging security threats. This fragility in the face of weak state-of-the-art intellectual property (IP) protection has resulted in hardware security vulnerabilities, such as IP piracy, overbuilding, reverse engineering, and hardware Trojans [1]. To address these issues at the hardware level [2], different design-for-trust (DfTr) techniques, such as IC metering, watermarking, IC camouflaging, split manufacturing, and logic locking [3–7] have been proposed to secure digital circuits. Though there are many DfTr techniques to secure digital circuits, there is a great dearth of techniques for analog and mixed-signal (AMS) IP protection. However, analog ICs are more prone to supply-chain attacks than digital ICs as they are easier to reverse engineer [8]. This high vulnerability is due to their low transistor count compared to their digital counterparts. To address the impact of process variations, they also have predefined layout patterns, e.g., common-centroid [9]. Analog ICs are not simple, although they have less number of transistors. Even with only hundreds of transistors, analog IC design requires highly experienced designers and a long time, as analog behaviors are quite complicated. Hence, it involves more capital in designing analog ICs [10]. Also, as explained in [8], analog ICs rank one in the top five counterfeited parts and cost several million dollars loss annually. Hence, it is necessary to develop a provable defense technique to secure analog-only and AMS circuits, and the existing defense techniques have

---
N. G. Jayasankaran · A. Sanabria-Borbón · E. Sánchez-Sinencio · J. Hu · J. Rajendran (✉)
Texas A&M University, College Station, TX, USA
e-mail: gjn@tamu.edu; adca.sanabria@tamu.edu; s-sanchez@tamu.edu; jianghu@tamu.edu; jv.rajendran@tamu.edu

to be validated. The first section of this chapter covers the analog IP protection techniques, and the following section describes the different evaluation techniques available to determine the resilience offered by these defenses.

16.2 Analog IP Protection

This section explains the existing digital IP protection techniques and the reasons why they cannot be leveraged to protect analog-only and mixed-signal circuits. The following section describes the different analog-only and AMS defense techniques proposed by the researchers.

16.2.1 Digital IP Protection

The DfTr techniques are categorized based on the different supply-chain attacks they thwart. The different techniques proposed by the researchers are: watermarking, IC camouflaging, split manufacturing, and logic locking [3–7]. While watermarking [11] is used for tracking the ownership of the IP, IC camouflaging [12], split manufacturing [13], and logic locking [3, 7] are used to secure the IP design from the attackers. In **watermarking**, the designer embeds his/her signature in the intellectual protocol (IP) design. This signature is later revealed by the designer to claim ownership of his/her IP. In **split manufacturing**, the netlist of the design is split into multiple sections. Each section is fabricated at different foundries. As none of the foundries gets access to the full design, it is not feasible for the attacker to pirate the IC. Likewise, **camouflaging** thwarts an attacker from reverse engineering an IC by introducing dummy contacts (or vias) into the layout. With the help of both real and dummy contacts, the standard cell is camouflaged, i.e., the number of functionality assumed by this cell is more than one. If an attacker cannot resolve the correct functionality of a camouflaged gate, he/she extracts an incorrect netlist. Unlike split manufacturing and camouflaging, which can protect the IC only from an untrusted foundry or end-user, **logic locking** [3, 7] can protect an IC when both the foundry and end-user are untrusted entities. Hence, logic locking is the preferred DfTr technique, as it protects from the attackers across the supply-chain. This technique inserts additional gates in the original circuit that are controlled by key inputs. Hence, only for the correct key is the output response of the unlocked circuit equal to the original response. Otherwise, for an incorrect key, it provides an incorrect output response. The key inputs are stored in an on-chip tamper-proof memory [14, 15]. These DfTr techniques cannot be used to secure analog circuits.

16.2.1.1   Applicability to Analog Circuits

The existing DfTr techniques are applicable to only digital circuits and cannot secure analog circuits. This is because,

1. These techniques are implemented in either gate-level [7, 12, 16–18] or RTL-level [19]. However, in the analog circuits, the design is in the transistor-level.
2. These techniques can be used only on digital circuits as the inputs/outputs to these circuits are Boolean variables. However, the variables associated with analog circuits are non-Boolean, such as gain, bandwidth, and center frequency.
3. Unlike digital circuits that give incorrect responses even if one bit of the key varies from the original, analog circuits can easily provide close to the desired performance for partially correct keys, if not designed carefully.

Hence, the researchers focus on developing new DfTr techniques that apply to analog-only and mixed-signal circuits. A simple approach to lock an analog circuit is to insert extra transistors controlled by key inputs. These key-transistors can be inserted at random locations in the circuit. When the key inputs are provided with the correct value, the analog circuit provides the desired response. However, such a simple approach cannot be used due to the following issues:

- As this includes a minimal number of key-transistors, the attacker can brute-force for the correct key by simulating the output response for each of the key combinations.
- Analog circuits have a smaller number of devices (only a few hundred). Hence it is relatively simple to reverse engineer than digital circuits, which have millions of transistors on a single chip.
- The attacker determines the key-transistors from the reverse-engineered netlist by tracking the key inputs. He/She can then remove these transistors to obtain the original circuit [20].

Therefore, stronger defense techniques are proposed, which are explained in the subsequent sections.

16.2.2   Threat Model

Depending on where the attacker is located in the supply-chain, he/she has access to various resources that help to attack the protected chip. Hence, based on where the attacker is and the resources he/she has access to, the threat models for the analog IP protection are classified as follows:

1. **Threat model for IP piracy.** The defense techniques [21–25] provide resilience against **IP piracy**. Hence, it considers both the foundry and end-user as untrusted entities. The attacker can modify the layout/mask of the victim's design based on his/her requirements.

2. **Threat model for overproduction.** [26] considers resilience against **overproduction**. This threat model is similar to the threat model for IP piracy. However, the attacker in the foundry can only overproduce the chip but does not have the necessary resources to modify the layout/mask.

3. **Threat model for reverse engineering.** [27] considers resilience against **reverse engineering.** This technique considers a trusted foundry and an untrusted end-user.

4. **Threat model for illegitimate access.** [28, 29] considers resilience against **illegitimate access.** The attacker has access to the locked chip.

16.2.2.1 Resources Available to the Attacker

The attacker has access to various resources depending on where he/she is located in the supply-chain. As stated in Table 16.1, the attacker in the untrusted foundry has access to the layout of the design provided by the designer, the process design kit (PDK) documentation, and the locking algorithm used as this information is public. He/She can overproduce the chip and sell the excess chips in the black market. Likewise, an end-user as an attacker has access to the reverse engineering tools to obtain the netlist of a locked chip [33]. It is relatively easier to reverse engineer analog circuits with several hundreds of transistors compared to SoCs with multi-million transistors. Also, compared to digital circuits, the analog circuits have a bigger transistor size [34] and predefined layout patterns, rendering them easier to reverse engineer. Similar to the attacker in the foundry, the untrusted end-user has access to the locking algorithm used. He/She also purchases a chip that has the

Table 16.1 Sources of information available to the attacker. Resistor (R), capacitor (C), width (W) and length (L) of the transistor, mobility (μ), oxide capacitance (C_{ox}), oxide thickness (t_{ox}), threshold voltage (V_{th}), bias current (I_B), bias voltage (V_B), input reference current (I_{ref}) and voltage (V_{ref}), transconductance (g_m), bandwidth (BW), and oscillation frequency (ω_{osc})

Source	Information acquired
Layout file from the foundry or the reverse engineered netlist using the oracle [30]	1. Sizes of passive components (R, C)
	2. Key size and transistor count
	3. W and L of the transistors
	4. Key connectivity to transistor switches or memristors
Technology library [31] (PDK documentation)	1. Values of passive components (R, C) and transistor details (μ, C_{ox}, t_{ox}, V_{th})
	2. Availability of different V_{th} transistors
Circuit specification [32]	1. Minimum and maximum values of I_B and V_B, which are the output of the bias circuit
	2. Values of I_{ref} and V_{ref}, which are the input to the bias circuit
	3. Minimum and maximum values of the resistance that can be programmed into the memristors
	4. Values of circuit parameters (BW, ω_{osc})

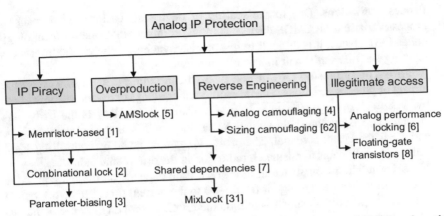

Fig. 16.1 The analog-only and mixed-signal IP protection techniques classified based on the supply-chain attack they thwart

correct key loaded. This chip serves as an oracle, where the attacker observes the output for a given input. The manufacturer provides the specification along with the purchased chip; thus, an untrusted end-user can access it. The defenses are classified based on the threat model they assume. Figure 16.1 shares the list of defenses to protect analog-only and AMS circuits. The following section discusses each of these techniques.

16.2.3 Memristor-Based Protection [23]

This technique proposes to secure the memory by leveraging the properties of emerging technology devices such as memristors and the impact of process variations. Each memory cell is designed using a sense amplifier whose working is controlled by the memristor-based voltage divider. The key inputs control the correct configuration of the memristors in the voltage divider. This secure sense amplifier functions as intended when the correct key in applied. Without the correct key, an attacker cannot identify whether the stored value in the memory is a logic 0 or 1.

16.2.3.1 Background

This work leverages (1) the process variation impact on the sense amplifiers, (2) the adaptive body biasing technique, and (3) the memristance property of the memristor to secure the memory.

1. **Process variations.** Due to the scaling of device dimensions, the impact of process variations on CMOS devices is more prominent in the newer technology nodes [35]. Hence, it is difficult to get precise matching between the transistors forming the differential pair in a sense amplifier.

2. **Adaptive body biasing [36].** It is a post-silicon tuning method to address the impact of process variations. The threshold voltage (V_{th}) of the transistor is given by, $V_{th} = V_{th0} + \gamma(\sqrt{|2\phi_F| - V_{BB}} - \sqrt{|2\phi_F|})$. Here, V_{BB} is the body bias voltage, V_{th0} is the threshold voltage when $V_{BB} = 0$, γ is the body coefficient, and ϕ_F is the Fermi potential. Therefore, V_{th} can be tuned by controlling V_{BB}. This V_{th} tuning helps in mismatch reduction in the differential pairs.

3. **Memristor.** The memristance at any given time depends on the integral value of the current/voltage through it from $-\infty$ to the current time. This work builds the security primitives using this unique memristance property offered by metal-oxide memristors.

16.2.3.2 Locking Architecture

Threat Model The attacker has physical access and write-access to the memory. However, he/she has read-access, only when the correct key is applied to the memory.

Differential Pair and Process Variations The differential pair illustrated in Fig. 16.2c forms the input stage of the sense amplifiers. The sizing of transistor dimensions for M_1 and M_2 should be identical to have zero output voltage when the differential inputs are equal ($V_n = V_p$). Layout methods such as common-centroid and interdigitization are followed to ensure M_1 and M_2 match as closely as possible. However, due to process variations, there exists a mismatch in their dimensions. This mismatch causes finite offset voltage at the output for equal input voltages. The body bias voltage (V_{BB}) of M_1 and M_2 are tuned to change the V_{th} of the respective

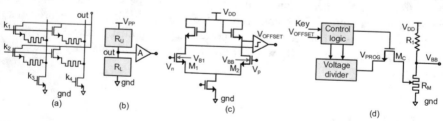

Fig. 16.2 Memristor-based obfuscation technique [23]. (**a**) The memristor crossbar architecture used in the voltage divider circuit. (**b**) The output of the memristor-based voltage divider is amplified by a factor of A. (**c**) Sense amplifier containing the differential pair M_1 and M_2. This amplifier generates the V_{OFFSET}. (**d**) For a non-zero V_{OFFSET}, the control logic turns on the transistor M_c. This passes the V_{PROG} generated by the memristor array to program the memristor R_M

transistors, as explained in Sect. 16.2.3.1. This tuning is required to cancel the output offset voltage for zero input differential voltage.

Mitigating Process Variations in the Differential Pairs The necessary V_{BB} is generated by the voltage divider formed using the resistor R_1 and the memristor R_M shown in Fig. 16.2d. This divider controls the body terminal of one of the transistors in the differential pair. This circuit can be an efficient analog security primitive, owing to the following:

- The increase in parasitic capacitance due to the memristor is relatively small. Hence, its impact on the analog circuit performance is minimal.
- The memristance of R_M ranges between $1.2M\Omega$ and $120k\Omega$. This memristance can be precisely tuned using the low-voltage programming pulses [37].
- The memristor will experience a breakdown and permanent shorting if the value of the programming voltage is more than the breakdown voltage (3.2V) [38]. This feature is leveraged by the defender to ensure that only the authorized user having access to the correct key can generate the precise programming voltages.

Memristor-Based Voltage Divider to Generate the Programming Voltage (V_{prog}s) This circuit produces the programming voltage (V_{prog}) using two memristor crossbars. Each crossbar is constructed using an array of memristors, indicated in Fig. 16.2a. The n-bit key $\mathbf{k} = (k_1, k_2, \cdots k_n)$ determines the connectivity among these memristors and the effective resistances of the upper (R_U) and lower (R_L) memristor arrays. Applying the correct key configures the crossbars' resistivity to provide the required V_{prog}. An incorrect key provides an undesired V_{prog}, which does not produce the correct body bias voltage V_{BB}. This voltage affects the sensitivity and reliability of the sense amplifiers and hence, the memory cells that are built using them.

16.2.3.3   Results

The memristor-based voltage divider and the body-bias voltage generator are required to tune V_{th} that helps in transistor matching in the presence of process variation. However, to ensure that the user suffers from the degraded performance of the sense amplifier for an incorrect key, the device dimension of one of the transistors in the differential pair is skewed to include an additional mismatch. For example, M_2 is made 50% larger than M_1. This difference in transistor sizes requires a $V_{BB} = 100\,mV$ to fix the mismatch. The body bias voltage varies between $-100\,mV$ to $+100\,mV$ when the memristance of R_M varies between $411764\,\Omega$ to $142857\,\Omega$ with $R_1 = 1M\,\Omega$. Therefore, the correct key is necessary to program the R_M with the required memristance, which controls the value of V_{BB}.

16.2.4 Combinational Lock [21]

The work demonstrates a satisfiability modulo theories (SMT)-based combinational locking. This technique locks the circuit that generates the bias current required for the proper operation of the analog circuit-under-protection. Here, a single transistor is obfuscated by replacing it with an array of transistors, as illustrated in Fig. 16.3a and b. Only for the correct key, the effective width of the obfuscated transistors is equal to the required width that provides the necessary bias current/voltage. Otherwise, for an incorrect key, it generates incorrect bias. Hence, the analog circuit's response does not meet the specifications.

16.2.4.1 Background

This technique controls the bias current produced by the configurable current mirror (CCM) via the secret key. As the circuit's performance depends on the bias current, the key controls the analog circuit performance.

1. **Current mirrors.** A current mirror circuit copies the current flowing into or out of an input terminal (I_{REF}) in the output terminal (I_B), as illustrated in Fig. 16.3a. The magnitude of I_B is the product of I_{REF} and the ratio of the aspect ratio of the transistors. This is given in Eq. (16.1). Here, the aspect ratio is the ratio of the width (W) and length (L) of a transistor. This current mirror is used to generate the bias current that is essential for the proper operation of the analog circuit.

$$I_B = \frac{\left(\frac{W}{L}\right)_{M_1}}{\left(\frac{W}{L}\right)_{M_0}} \times I_{REF} \tag{16.1}$$

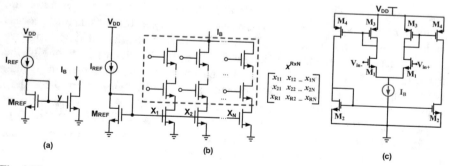

Fig. 16.3 (a) A simple current mirror. (b) In [21], the copying transistor is replaced by an array of transistors. (c) The CCM provides the bias current (I_B) required by the operational transconductance amplifier (OTA)

2. **Importance of precise bias currents.** The performance of the analog circuits relies on its precise biasing condition. Figure 16.3 shows the operational transconductance amplifier (OTA) biased using I_B. The transconductance of this OTA is,

$$g_m = \sqrt{2\mu C_{ox}\left(\frac{W}{L}\right)I_D}$$ (16.2)

where the drain current $I_D = I_B/2$. Hence, any variation in I_B leads to deviation in g_m that results in an error of the circuit's performance.

16.2.4.2 Defense Architecture

This technique, uses a CCM, as illustrated in Fig. 16.3b. The CCM consists of a $R \times N$ transistor array. An N-dimensional ratio vector represents the sizes $\alpha = (\alpha_1 \alpha_2 \cdots \alpha_N)$, where $\alpha_j > 0$ and $\alpha_j < 0$ represent the NMOS branch and PMOS branch, respectively. The output current can be written as

$$I_{out} = \sum_{j=1}^{N} \alpha_j \prod_{i=1}^{R} \phi(x_{ij}) \times I_{REF}$$ (16.3)

where $\phi(x_{ij})$ is the control signal at transistor of row i and column j that is expressed by,

$$\phi(x_{ij}) = \begin{cases} q_k, & \text{if } x_{ij} = k \neq 0; \\ \\ 1, & \text{else } x_{ij} = 0 \end{cases}$$ (16.4)

Circuit parameter, such as f_c, BW, and ω_{osc}, depends on the precise value of the bias current (I_B) generated by the CCM. The key input configures the effective width of the mirroring transistor in the CCM. The transistor sizes are modeled using the SMT formulations such that on a correct key, the CCM gives the desired I_B. Otherwise, I_B is outside the range $((1 - \Delta)I_B, (1 + \theta)I_B)$, where Δ and θ are lower and upper bounds, respectively. This technique ensures that each fabricated chip has a unique key by including a physical unclonable function (PUF) in the design. As shown in Fig. 16.4a, the unique chip key is XORed with the PUF output to generate the common key. This common key is the same for all the fabricated chips of the same design. This key controls the CCM and hence, the bias current value. A technique similar to the combinational lock is the parameter-biasing obfuscation [22]. This technique replaces the transistors in the voltage divider with a programmable transistor array, as illustrated in Fig. 16.4b.

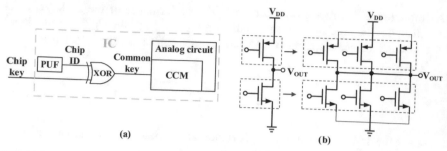

Fig. 16.4 (a) The common key is generated by XORing the unique chip key with the chip ID [21]. (b) Each transistor in the bias circuit is replaced by an array of transistors in [22]. The key input controls the effective width of the transistor array

16.2.4.3 Results

The combinational lock is implemented on (1) a fourth-order Gm-C band-pass filter, (2) quadrature oscillator, (3) LC oscillator, and (4) a triangular signal generator used in the class-D amplifier. The CCMs replace the current mirrors in each of these circuits. These CCMs are controlled by the common key whose size ranges from 16 to 74 bits. The results indicate that only for the correct key is the unlocked analog circuits' performance matches the circuit specifications. Otherwise, for an incorrect key, the performance deviates from the expected performance.

16.2.5 Analog Camouflaging [27]

This technique leverages the multi-threshold voltage (V_{th}) design for securing analog circuits from reverse engineering-based attacks. As the transistors' operating region depends on the V_{th}, analog circuits are sensitive to V_{th}. Adding to this sensitivity, the V_{th} of each transistor cannot be found using the RE process. Only brute-forcing helps in determining the V_{th} of the transistors. However, the brute-forcing effort increases exponentially with respect to the number of transistors in the design. Therefore, by judiciously including the multi-V_{th} transistors, the circuit specifications such as gain and bandwidth are protected.

16.2.5.1 Background

1. **Multi-V_{th} design.** The latest technology nodes support the multi-V_{th} design that helps in performance and leakage improvements. The low-V_{th} and high-V_{th} transistors are used for improving the performance and leakage, respectively.

2. **Camouflaging digital circuits.** A mix of real and dummy contacts are used to camouflage a standard cell (gate) [12]. Hence, standard cells of different functionalities share the same layout appearance such that it is difficult for an attacker to decide its functionality from the reverse-engineered netlist, thereby making the design resilient against reverse engineering attacks.

16.2.5.2 Defense Architecture

Threat Model This technique considers a trusted foundry and an untrusted end-user. As the foundry fabricates the transistors with different V_{th}, it must be a trusted entity. The end-user reverse-engineers a purchased chip using tools such as Chipworks [33] and Degate [39] to determine the functionality of the chip.

Working The performance of an analog circuit depends on (1) the bias current/-voltage inputs and (2) the correct operating regions of each transistor. The value of the bias and the operating region is controlled by the V_{th} of the transistors. Hence, in this technique, few of the nominal-V_{th} (NVT) transistors are replaced with resized low-/high-V_{th} (LVT/HVT) transistors. This resizing is necessary to ensure that the circuit specifications do not vary after changing the V_{th} type. The attacker can purchase a chip and reverse engineer it to obtain the transistor sizes and connections. He/She simulates this reverse-engineered netlist. However, all the transistors are assumed to be NVT transistors, as it is not possible to determine the V_{th} type from the netlist. This assumption leads to improper biasing voltage and deviation in the circuit specifications. Camouflaging the analog circuits is done using three steps given below:

1. **Design based on NVT transistors.** The analog circuit is designed with NVT transistors and simulated to determine the values of the circuit specifications.
2. **Securing the design.** Few of the NVT transistors are replaced with LVT/HVT transistors. The replaced transistors are resized accordingly to maintain the performance of the original design.
3. **Adversarial observation.** The secured design with resized transistors is simulated with the incorrect V_{th} assumptions. This shows degraded performance compared to the base design.

16.2.5.3 Results

This work is illustrated on an operational amplifier implemented using the 65nm predictive technology model [40]. The objective of this experiment is to (1) camouflage a sufficient number of transistors that make the RE, time-intensive, and (2) there is severe degradation in the circuit performance even if the attacker assumes one of the V_{th} type incorrectly.

16.2.6 Mixed-Signal Locking [25, 26]

The high sensitivity of analog circuit to process, voltage, and temperature (PVT) variations, increases the complexity of designing analog circuits in recent fabrication processes. Hence, analog circuits are often aided by digital circuits that perform automatic post-fabrication calibration. The works [41, 42] proposed a built-in self-tuning (BIST) architecture for on-chip compensation of PVT variations. In this approach, an optimization loop minimizes the effect of process variation by choosing the optimal tuning knob settings of the configurable analog block. The BIST platform applies a stimulus input to the analog circuit and extracts their performance characteristics. An analog-to-digital converter (ADC) and quadrature sampling are used in the self-testing procedure [42]. AMSlock protects AMS circuits by locking their digital section using SFLL technique [7]. Out of the different digital logic-locking techniques available, the SFLL technique provides provable security against SAT and removal attacks. Moreover, it gives freedom to the designer to choose the input patterns to protect. Here, only on applying the correct key, the logic-locked optimizer sets the tuning knobs such that the analog circuit meets the specifications. Otherwise, for an incorrect key, the response deviates from the desired value. This defense technique was demonstrated in the protection of three representative circuits: an active-RC band-pass filter (BPF), a low-dropout (LDO) voltage regulator, and a low-noise amplifier (LNA).

16.2.6.1 Background

This technique uses the stripped-functionality logic locking to lock the digital section of the AMS circuits. There are different variants in SFLL [7], such as SFLL-HD^0, SFLL-HD^h, SFLL-flex, and SFLL-fault. It protects only a certain number of input patterns called the protected input patterns (PIPs). The output of the original circuit, O1, is inverted only when the input pattern (IN) is a PIP. The functionality-stripped circuit comprises the original circuit, the inversion logic, and the logic which checks if the IN is a PIP. Depending on the variant of SFLL, the corruption injected by the inversion logic is restored, when

1. The Hamming distance (HD) between the external key (k) and the IN equals 0 in SFLL-HD^0.
2. The HD between K and the IN equals h in SFLL-HD^h, as indicated in Fig. 16.5.
3. The input equals one of the PIPs that are stored in a content addressable memory in SFLL-flex.

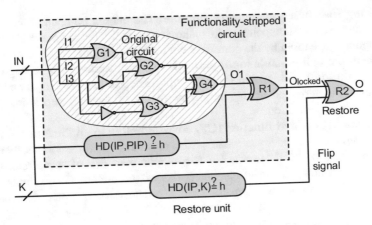

Fig. 16.5 Stripped-functionality logic locking [7]. Hamming distance (HD), key (k), input pattern (IN), protected input pattern (PIP)

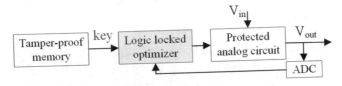

Fig. 16.6 Logic locking of the AMS circuit [26]

16.2.6.2 Locking Architecture

This technique protects the analog circuit by logic locking the optimizer (digital section) of the AMS circuit, as illustrated in Fig. 16.6. By performing *judicious* logic locking on the optimizer, only on applying the correct key, the optimization core configures the tuning knobs of the analog circuit such that the effect of process variations is nullified. In consequence, the analog circuit performs as desired and its performance meets the specifications. On applying an incorrect key, the digital circuit produces incorrect output, thereby setting the tuning knobs of the analog circuit to non-optimal values. This improper tuning deteriorates the performance of the analog circuit.

The locking architecture consists of the following steps:

1. **Choosing the tunable components using sensitivity analysis.** Making all the passive components tunable has huge impact on the area overhead. Hence, only a few of the components are made tunable. To ensure that the attacker suffers maximum degradation in the performance on applying an incorrect key, sensitivity analysis is performed to determine those components on which the output response of the analog circuit is highly dependent [43].

2. **Making the chosen component tunable.** The chosen component is replaced with an array of components to make it tunable. The optimal value of the component is chosen by the optimizer via the tuning knobs.

3. **Determining all possible input patterns to the optimizer.** The analog circuit-under-protection is simulated for each tuning knob setting. The input and output values are determined at the frequency points of interest, depending on the circuit-under-protection.

4. **Determining the cost function (CF) corresponding to all the input patterns.** The defender calculates the value of the cost function for each possible input pattern. This value determines if the output response of the circuit-under-protection follows the ideal characteristics of the circuit. For example, in an active-RC BPF its passive elements Rs and Cs determine its center frequency and bandwidth. Hence, the BPF's cost function (CF) is calculated by matching the gains $G_{f_1}, G_{f_2}, G_{f_3}$, and G_{f_4} at the frequencies f_1, f_2, f_3, and f_4, respectively.

$$CF = (G_{f_2} - (\sqrt{2} \times G_{f_1})) + (G_{f_2} - G_{f_3}) + (G_{f_3} - (\sqrt{2} \times G_{f_4}))$$
$$+ (G_{f_1} - G_{f_4}) \tag{16.5}$$

Here, f_1 and f_4 are the lower and upper cut-off frequencies, and f_2 and f_3 are two frequencies in the pass-band around the center frequency. The G_{f_2} should be equal to $\sqrt{2} \times G_{f_1}$ and the G_{f_3} should be equal to $\sqrt{2} \times G_{f_4}$ in the ideal transfer function. Likewise, G_{f_2} and G_{f_3} should be equal, and G_{f_1} and G_{f_4} should be equal in an ideal BP characteristic.

5. **Choosing the input patterns to protect.** Based on the minimum deviation in the output response the attacker has to encounter for an incorrect key, the designer protects all those input patterns that correspond to a deviation less than the chosen deviation. This deviation is quantified by the error in the cost function. If the error is close to zero, the output response is close to the ideal characteristics. Otherwise, it deviates from the ideal characteristics. In case of *HD-0*, since only one input pattern can be protected, it is obvious to select the input pattern of the optimizer that results in the minimum cost function. In case of *HD-h*, a designer can increase the number of input patterns protected by increasing value of h. However, this decreases the security level of an n-bit design by $2^{n-k} \cdot \binom{k}{h}$. Hence, one can increase the value of h only to an extent. The designer calculates the cost for each tuning knob setting. He/She will then choose those input patterns which corresponds to the least cost as protected input patterns (PIPs).

6. **Locking the optimizer.** Using the PIPs selected in the previous step, the optimizer is locked either using SFLL-HD^0, SFLL-HD^h, or SFLL-flex.

16.2.6.3   Setup and Results

The AMSlock technique is demonstrated on three different AMS circuits: BPF, LNA, and LDO [44–46]. This technique protects the frequency response of the BPF, particularly its center frequency (ω_o) and bandwidth (BW). The tuning knobs of this circuit are the passive elements Rs and Cs. The BPF's specifications are $\omega_o = 2 \cdot \pi \cdot 74$ Mrad/s and $BW = 13$ MHz. The input to the optimizer has 220 bits, which include the BPF frequency response data from ADC and the weights used in the cost calculation. In the LNA test case, the protected performance metrics are the gain $S_{21} > 20$ dB, and the input matching $S_{11} < -20$ dB, at a resonance frequency $f_R = 6$ GHz. Its response is represented by a 154-bit word given to the optimizer. Similarly, the protected LDO's specifications are $PSR \leq -50$ dB and a phase margin larger than $45°$ with optimizer input size equal to 234 bits. The optimizer implements a simulated annealing algorithm.

The response of the circuits for a correct and an incorrect key are compared for the protected performance metrics of the analog circuits-under-protection. Figure 16.7a shows the difference in the frequency response of the BPF. In this case,

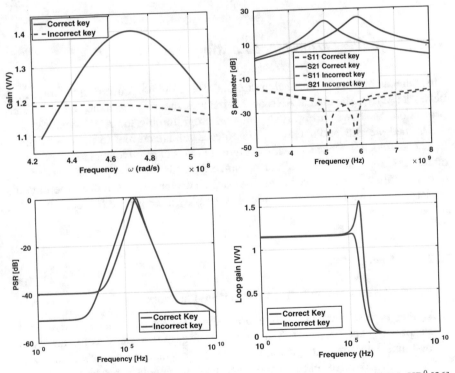

Fig. 16.7 Behavior of the analog circuits for correct and incorrect keys on using SFLL-HD^0 [26]. The key size used for BPF, LNA, and LDO are 87, 81, and 109, respectively. (**a**) Frequency response of BPF, (**b**) S-parameters of LNA, (**c**) Power supply rejection (PSR) of LDO, and (**d**) Loop gain of LDO

Fig. 16.8 The bias to the analog circuit is controlled by key inputs to the neural network [29]

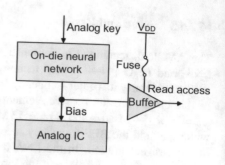

the correct key allows the optimizer tune the circuit to the target ω_o and BW, while an incorrect key forces the optimizer to tune to a lower frequency and also reduces the Q and gain values. Figure 16.7b compares the difference in the S-parameters of the LNA targeting $f_R = 6$ GHz. One can observe an error of 1 GHz in f_R and an error on S_{11} and S_{21} of at least 15 dB. Finally, the deviation on the LDO performance for the two cases was evaluated. Figure 16.7c shows a degradation close to 10 dB in the PSR. Figure 16.7d shows a large peaking on the loop gain for the incorrect key, which indicates a low phase margin and potential instability.

16.2.7 Analog Performance Locking [29]

This defense obfuscates the analog circuits' operating point using an analog neural network (ANN). The trained ANN acts as the lock securing the analog circuit-under-protection, as shown in Fig. 16.8. Only on programming the ANN synapses with the trained weights and supplying the correct voltage/current to the ANN's input layer can unlock it. Hence, provide the required bias current/voltage to the protected circuit. As the analog circuits' performance depends on these precise bias values, only the authorized user having access to the correct key can unlock the circuit's performance. This is the first technique to employ analog keys for locking analog IC performances.

16.2.7.1 Background

This work creates locks with the help of a neural network.

1. **Analog neural network.** This neural network consists of an $n \times n$ array of synapses (S) and neurons (N), as shown in Fig. 16.9. Here, n is the number of rows and columns in the neural network. Each synapse implements an analog multiplier. Likewise, each of the neurons implements a nonlinear activation function, e.g., tanh. The network is trained so that for a given set of input voltages, it determines the weights of each synapse to generate the required

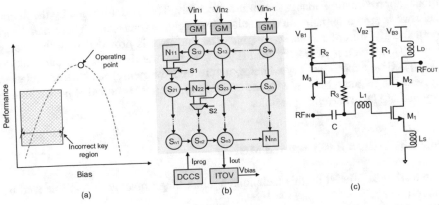

Fig. 16.9 (**a**) Only for a specific bias value, the performance of the analog circuit is maximum [29]. The analog neural network is trained such that the bias generated for an incorrect key leads to degraded performance. (**b**) The analog neural network along with the input differential transconductance (GM), a digitally controlled current source (DCCS), and current to voltage converter (ITOV). The core of the neural network consists of the neurons (N) and synapses (S). (**c**) The low-noise amplifier (LNA) requires three bias voltages V_{B1}, V_{B2}, and V_{B3} for its working. The unlocked ANN feeds these voltages

output. It is also possible to train the neural network to provide the same bias voltages for different inputs.

2. **Impact of biasing on analog circuit performance.** The bias inputs to the analog circuits are required for the proper functioning of these circuits. Circuit specifications, such as bandwidth and center frequency, depend on the current circuits' bias current. Any change in these bias values impacts the performance of the circuits.

16.2.7.2 Locking Architecture

Threat Model The attacker is an unauthorized user having illegitimate access to an analog IC. This analog IC can be a recycled IC, or an IC used in the military ended up in the wrong hands.

Circuit Operation In this technique, the precise bias input required by the analog circuit is not provided directly; instead, it is generated by the ANN, illustrated in Fig. 16.9b. This ANN behaves like a lock, and the input voltage (V_{in}) is the key. The ANN is trained in such a way that only applying the correct key (V_{in}) provides the required bias value necessary for the proper operation of the analog circuit-under-protection. Hence, a correct pair of lock and key is essential for the operation of the circuit. If the value of the key changes, it will produce a bias that does provide optimal performance, as shown in Fig. 16.9a. Adding to this requirement, the ANN should be modeled to leak minimal information that can be leveraged by model approximation attacks for an incorrect key. The model learned by the neural network is stored on-chip via the analog floating-gate transistors. These transistors permanently store the synapse weights. The end-user is provided with the analog key (V_{IN}), which is necessary for the ANN to provide the necessary bias voltage (V_{bias}). Given the input key's analog nature and the neural network modeled to provide the required bias only for the correct key, each IC is bound to have a unique key making key sharing impossible.

16.2.7.3 Attack Resilience and Metrics

The resilience offered by this technique against illegitimate access is achieved by the following:

- Owing to the analog nature of the input key, the key search space is infinite. This helps in prohibiting brute-force attacks.
- The neural network can be trained such that only for a correct key, the desired bias is generated. Hence, even if the value of the input key changes marginally, the desired bias is not generated, leading to performance degradation.
- The neural network in each chip is trained for a different correct key. Consequently, key sharing is not a viable attack.
- The bias voltages can be observed only while the neural network is getting trained. After training the ANN, the observability of the bias is removed by blowing a fuse.

The metric used for measuring the attack difficulty is,

$$\text{metric} = \frac{N_p}{N_s} \tag{16.6}$$

Here, N_s is the search space, where the analog key draws values from, and N_p is the number of options in N_s that provides the bias which brings the circuit to its desired or close to the desired operating point.

16.2.7.4 Results

The proof-of-concept of this defense is shown on a low-noise amplifier (LNA). It is fabricated using the 130 nm Global Foundries RF CMOS process. The analog circuit-under-protection consists of a cascade LNA with inductive source degeneration, as indicated in Fig. 16.9c. The transistors M_1, M_2, and M_3 work as a common-source, common-gate, and bias transistor, respectively. Each of these transistors requires a unique bias voltage. These voltages (V_{B1}, V_{B2}, and V_{B3}) are generated by the ANN that is modeled to provide the precise bias value when correct analog input voltages (key) are provided. The impact of these bias voltages on the LNA performance is determined by measuring the S-parameters (S_{11}, S_{12}, S_{21}, and S_{22}). Only when the bias voltages equal (1.35, 2.55, 2.25)V, the S-parameters equal $(-8, 11.2, -31.4, -7.5)$ dB corresponding to the desired circuit performance. For any other combination of bias voltages, at least one S-parameter fails its specifications.

16.2.8 Shared Dependencies [24]

This defense's resilience is improved by locking both the analog and digital sections of the AMS circuit. The analog section is locked using the parameter-biasing obfuscation [22], and the digital section is locked using the random logic locking (RLL) [3] and stripped-functionality logic locking [7]. This technique also includes functional dependency between the analog and digital domains. This dependency increases the number of iterations required by the SAT attack [47] 3× to determine the correct key.

16.2.8.1 Background

1. **Parameter-biasing obfuscation technique [22].** In this technique, the transistors' effective width in the bias circuit is controlled by the key inputs. An array of transistor switches replaces each transistor, and each of these switches is connected to a unique key input. Only the authorized user having the correct key sets the required transistor width that provides the required bias current/voltage.
2. **Random logic locking (RLL) [3].** This technique locks a design by inserting XOR/XNOR gates at random locations in a netlist. Each of these gates has one key input and one design input. Only on applying the correct key, the desired output is generated. Otherwise, an incorrect output is generated.
3. **Stripped-functionality logic locking (SFLL) [7].** It protects only a certain number of input patterns called the PIPs. The output of the original circuit is corrupted for an incorrect key. Only when the key is equal to the PIP, the corrupted output is restored.

16.2.8.2 Defense Architecture

Threat Model This technique considers resilience against IP piracy. Hence, it assumes that the foundry and the end-users are untrusted entities. The attacker has access to the layout/mask of the circuit in the foundry. He/She as an untrusted end-user can reverse engineer the chip using services such as Chipworks [33] and Degate [39]. He/She also possesses an oracle (chip with the correct key loaded) used to obtain the output for a given input. These input–output values are used for pruning the incorrect keys.

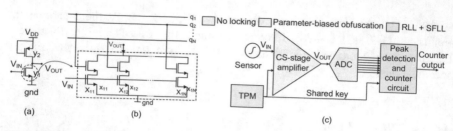

Fig. 16.10 (a) Common-source (CS) amplifier. (b) The CS amplifier is locked by replacing the transistor y_1 by an array of transistors [22]. (c) In shared dependencies lock [24], the CS-stage amplifier is locked using parameter-biasing obfuscation technique [22]. The digital circuit is locked using random logic locking (RLL) and stripped-functionality logic locking (SFLL) [7]. Tamper-proof memory (TPM) and 7-bit analog-to-digital converter (ADC)

Example Analog and Mixed-Signal (AMS) Circuit The sensor, along with the data processing circuit, is used to illustrate this locking methodology. The analog front end consists of a common-source amplifier shown in Fig. 16.10a. This amplifier amplifies the weak signals from the sensor. These analog signals are then converted to digital using a 7-bit analog-to-digital converter (ADC). These signals are sent to the digital section that consists of peak detection and counter circuit. As the input to the peak detection circuit is from the ADC, the output is set to one when the ADC output is maximum and resets to zero when the ADC output falls below the maximum value. Finally, the counter counts the number of times the output of the peak detector goes high. This setup is used in many applications, including heart rate monitoring devices, mass spectrometers, X-ray machines, and image processing applications.

Locking the Analog Section The CS amplifier in the analog front end is locked using parameter-biasing obfuscation technique [22]. The transistor y_1 in CS amplifier is replaced by N transistors in parallel, $\{x_{11}, x_{12}, \cdots, x_{1N}\}$. Likewise, y_2 is replaced by $\{x_{21}, x_{22}, \cdots, x_{2N}\}$, as depicted in Fig. 16.10b. The effective aspect ratio of the obfuscated transistors y_1 and y_2 are y_{1_v} and y_{2_v}, respectively. Here,

$$\tilde{y1}_v = \left(\frac{W}{L}\right)_{y_1} = \sum_{i=1}^{N} q_k \times x_{1i_v}$$

$$\tilde{y2}_v = \left(\frac{W}{L}\right)_{y_2} = \sum_{i=N+1}^{2 \times N} q_l \times x_{2i_v}$$

(16.7)

where $\mathbf{q}$ is the key of size $2 \times N$. Here, $k \in \{1, 2, \cdots, N\}, \forall i$ and $l \in \{N + 1, N + 2, \cdots, 2 \times N\}, \forall i$. x_{1i_v} and x_{2i_v} are the aspect ratios of transistors x_{1i} and x_{2i}, respectively. The effective width of y_1 and y_2 is controlled by $\mathbf{q}$. Hence, the amplifier's gain is given by $G_{CS} = -\frac{1}{1+\eta}\sqrt{\frac{\tilde{y1}_v}{\tilde{y2}_v}}$, controlled by the key input. Here, η is the back-gate transconductance.

Locking the Digital Section The peak detection circuit and the counter are locked using SFLL-HD^0 [7] and RLL [3], respectively, as shown in Fig. 16.10c. The peak detection output is set to one when the input corresponds to the maximum ADC value. Hence, this value is considered as a protected input pattern. In SFLL-HD^0, the key is equivalent to the protected input pattern.

16.2.8.3 Results

As explained in Sect. 16.2.8.2, the CS-stage amplifier, the peak detection, and counter circuits are locked using the parameter-biasing obfuscation [22], SFLL-HD^0, and RLL [3], respectively. **Adding key dependencies between analog and digital circuits.** To introduce a dependency between the analog and digital sections of the circuit-under-protection, the defender includes 2-input XOR key gates. Here, one input is driven by key connected to the analog circuit and other from the key connected to the digital circuit. The output of these key gates controls the SFLL and RLL locked digital circuits. This ensures that the attacker cannot break the analog and digital sections in isolation.

16.2.9 *Range-Controlled Floating-Gate Transistors [28]*

As analog circuits are sensitive to process variations, many pre- and post-fabrication techniques are proposed to address this issue. The pre-fabrication technique includes layout designs such as common-centroid and interdigitization [9]. Likewise, the impact of process variations is addressed post-fabrication using techniques such as tuning the component parameters [42] and analog floating-gate transistors (AFGTs) [48]. They fine-tune the performance of analog ICs post-fabrication, thereby enabling high yield even in the presence of process variations. This technique leverages AGFTs to prevent unauthorized use of analog ICs. It consists of a two-step unlock-&-calibrate process. A sequence of input voltages within the specified range, called the waypoints program the AFTGs. These waypoints unlock the ability to program the AFGTs over the entire range. Therefore, in the second step, the typical AFGT-based post-silicon calibration process can be applied to adjust the IC's performance within its specifications.

16.2.9.1 Background

Analog Floating-Gate Transistors and Process Variations These are standard MOS transistors with an electrically isolated gate called the floating gate (f_g). This gate is formed using a control gate capacitor (C_{CG}) and a tunneling capacitor (C_{Tun}). The charges trapped in this gate are, hence, stored permanently. The threshold voltage (V_{th}) of this transistor depends on this charge stored in the floating

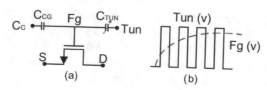

Fig. 16.11 (**a**) Analog floating-gate transistor and (**b**) the programming voltage pulses (*Tun*) required to change the value of charge stored in the floating gate (F_g) [28]

gate. By adding or removing charges in this gate, the V_{th} of the transistor can be varied accordingly. Owing to the programmability of AFGTs, they can be used as tunable components for calibrating an IC to its desired operating point [48]. The voltage pulses are applied to the *Tun* terminal, as shown in Fig. 16.11b to program the AFTGs in Fig. 16.11a. The AFTG-based calibration is illustrated on the operational transconductor amplifier (OTA). This OTA consists of a differential pair (M_1 and M_2) and three current mirrors. This current mirror reflects the current of the input pair to the output. Post-fabrication, the transistors M_1 and M_2 are not perfectly matched, and hence, the threshold voltage can vary slightly between them. This variation results in higher drain-to-source current through M_1 compared to M_2. The V_{th} of the differential pair in the OTA is calibrated using the AFGT pair. The calibration continues until the output offset voltage is zero when the input differential voltages IN_1 and IN_2 are equal and opposite.

16.2.9.2   Defense Architecture

Threat Model   The attacker is an unauthorized user having illegitimate access to an analog IC. This analog IC can be a recycled IC, or an IC used in the military ended up in the wrong hands.

Locking Architecture   This technique introduces a mechanism to control the programming range of the AFTGs, as shown in Fig. 16.12a. It includes n AFGTs and m waypoints that determine the programming range. The key checker controls the *Inhibit* signal that unlocks the full-range programming of the AFGT when the correct sequence of waypoints are fed to it.

Inhibit Circuit   It consists of a Zener diode, a resistor, and an NMOS switch connected in series, as illustrated in Fig. 16.12b. The Zener diodes work as voltage regulators when reverse-biased and fed with reverse bias voltage equal to the reverse-breakdown voltage. In this stage, the diode has low impedance and acts as a sink. Hence, the current through the load that is in parallel to the diode remains constant. Thus, the voltage across the *Tun* terminal and ground is also constant. The resistor limits the current through the diode and the load. If the *Inhibit* signal is low (high), this circuit's output is equal to the programming pulse input (zero). Hence, the *Inhibit* signal controls the amount of charge trapped in the floating gate.

Key Checker This circuit consists of a window comparator, a finite state machine (FSM), and two multiplexers. The FSM generates the *Inhibit* signal that is set to high initially. Hence, the AFGTs will not be programmed. The FSM sequentially compares each of the floating-gate voltages of the AFTGs with the corresponding input voltages, i.e., waypoints. If the AFGTs are programmed such that their floating-gate voltages match the waypoints, the FSM produces a low *Inhibit* signal that allows the programming of the AFGTs in their entire range. Otherwise, the FSM will have the *Inhibit* signal asserted if the compared voltages are not equal.

16.2.9.3 Results

This defense was illustrated on the operational transconductance amplifier, the appropriate range controller, and the key checker. This circuit is fabricated using the GlobalFoundries' 130 nm CMOS process. The impact of input offset is simulated by driving 15 mV to one of the differential inputs. When all the waypoints are given correctly, the output offset is zero even for non-zero input voltage. However, when the attacker programs even one incorrect waypoint, the *Inhibit* signal remains high even when the following waypoints are programmed correctly.

16.2.10 AMS Security via Sizing Camouflaging [49]

Analog layout design uses layout techniques such as common-centroid and inter-digitization to address the impact of process variations in sub-micron technology nodes [9]. Therefore, the non-minimum size transistors are designed using many minimum-sized transistors connected in parallel. These smaller transistors share the drain/source regions and are known as gate fingers. Likewise, the resistors and capacitors are laid out using unit resistors and capacitors in a serpentine fashion. This defense technique uses fake contacts in these layout designs to add seemingly connected yet in reality, inactive and electrically disabled gate fingers, unit resistors, and unit capacitors. Therefore, if an attacker reverse-engineers the chip, he/she cannot determine the exact size of these components, thereby protecting the design.

16.2.10.1 Background

1. **Analog layout techniques.** There are different layout techniques proposed to address the impact of process variations in analog circuits. This impact is prominent in sub-micron technology nodes. Hence, techniques such as common-centroid and interdigitization are established to mitigate this issue [9].
2. **Camouflaging digital circuits.** A mix of real and dummy contacts are used to camouflage a standard cell (gate) [12]. Hence, standard cells of different functionalities share the same layout appearance such that it is difficult for an

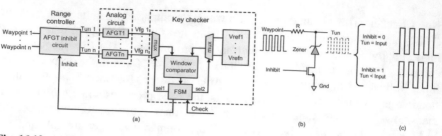

Fig. 16.12 (**a**) Floating-gate transistor-based analog IP protection [28]. (**b**) *Inhibit* signal generator circuit (**c**) The value of tuning pulses based on the *Inhibit* signal value

attacker to decide its functionality from the reverse-engineered netlist, thereby making the design resilient against reverse engineering attacks.

16.2.10.2 Defense Architecture

This work proposes two distinguished design flows, namely camouflaging of an existing design and involving camouflaging already from the design phase, as illustrated in Fig. 16.13.

- **Camouflaging an existing design (modifying an existing layout).** The defender chooses a few of the transistors in random from the original design and resizes it. This resizing is iterated until the modified design fails in one or more circuit specifications. The defender converts this resized netlist by replacing (1) the unmodified components with standard cells and (2) the modified components with camouflaged cells. Hence, the obfuscated netlist has a mixture of standard and camouflaged cells. The defender performs post-layout simulation on the obfuscated netlist to ensure that the design performs as per the desired specifications. Otherwise, the resizing process is iterated until the circuit meets the specifications.
- **Camouflaging during the design phase (designing a new layout based on the camouflaged library).** Here, the defender already knows the existence of the camouflaged library. Hence, as the first step, the defender designs the circuit at the schematic-level. The schematics is simulated to ensure it meets the design specifications. As the second step, using the camouflaged library, the layout-level design is drawn. The defender then performs the resizing operation on the components chosen at random, as shown in Fig. 16.13.

The goals considered while designing the camouflaged design are:

1. The obfuscated design has the maximum performance penalty with respect to the original design in both the design flows.
2. The obfuscation overhead is minimized.
3. The obfuscation design effort is minimized as well.

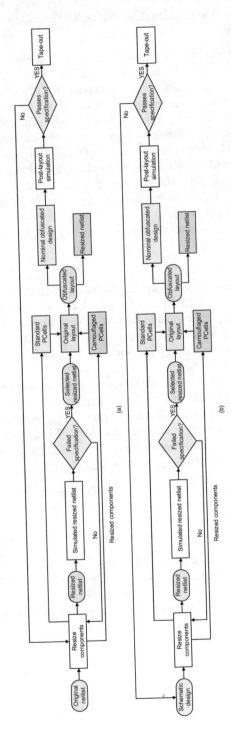

Fig. 16.13 The two variants of camouflaging proposed in [49]. (**a**) The design flow for camouflaging an existing design. (**b**) The design flow for camouflaging the circuit at the design phase

The design metrics considered by the defender while generating the camouflaged design is as follows:

- **Number of components to resize and degree of resizing.** As the attacker considers all the components to be camouflaged in a camouflaged netlist, the hardness of reverse engineering does not depend on the number of resized components. Hence, even resizing only a small fraction of the components is sufficient for achieving resiliency against reverse engineering-based attacks. By camouflaging the minimum number of transistors, the defender achieves (1) the minimum obfuscation area overhead, (2) the minimum camouflaged layout-induced parasitics, and (3) the minimum floor planning and routing changes.
- **Degree of performance degradation.** The defender must introduce a performance penalty so that the circuit has a significant error in its output response, thereby making the chip unusable.
- **Selection of components to resize.** This selection is made based on the following factors:

  1. The components can be selected based on sensitivity analysis that ranks them according to the impact it has on the circuit specifications. However, the attacker can use the same analysis to trace back the resized components. Hence, this work suggests choosing the components to camouflage randomly.
  2. The components that are connected to sensitive or high-frequency nodes are avoided. It helps in minimizing the impact of camouflaged layout-induced parasitics on the performance.
  3. The transistor obfuscation is prioritized to resistor/capacitor obfuscation. This is because adding extra inactive fingers to transistors results in a lower area overhead than adding extra inactive unit capacitors and resistors.
  4. A complex system can have multiple sub-blocks. Obfuscating the components in each block impacts the overhead. Hence, it is sufficient to obfuscate a few sub-blocks and ensure that they provide all-true contact design with degraded performance.

16.2.10.3 Results

This technique is demonstrated on Miller op-amp and an $\Sigma \Delta$ ADC. The simulation experiments were performed on an Intel(R) Xeon E5-2640 @ 2.5 GHz with 128 GB of RAM. The results indicate that camouflaging during the design flow did not increase design iterations, and the nominal obfuscated design met the target specifications.

16.2.11 *Techniques for Key Provisioning*

In analog locking techniques [21–23, 25, 26, 28, 29], the locked circuit's output response meets the specifications only when the correct key is applied. This key is either stored in a tamper-proof memory [7] or generated by using a key provisioning unit [3, 21]. The correct key value is the same for all the same design instances and hence, called common key (CK). Therefore, if an attacker manages to find the CK, he/she can unlock all the instances of the same design [47, 50]. The key provisioning unit enables each chip instance to have a unique key, a.k.a. user key (UK), to mitigate this issue. Hence, this block takes in the UK, and generates the CK, which is equal for all the instances of the locked analog circuit. The properties of a key provisioning unit are:

- unique UK for each chip instance.
- probability of the attacker recovering CK from UK must be negligible.
- low area overhead.

16.2.11.1 Background

Ending piracy of integrated circuits (EPIC) [3] uses a physically unclonable function (PUF) or a true random number generator (TRNG), and RSA encryption to remotely activate a locked chip. This approach follows four steps:

Step 1 The designer locks the circuit with a CK and embeds his/her public master key (MK-Pub) in the circuit. CK is the secret key known only to the designer. The locked design is also embedded with a PUF/TRNG and an RSA module. The locked design is sent to the untrusted foundry, where the chip is manufactured and tested. The testing process does not require to load the key into the chip [51].

Step 2 The manufactured chip on the first power-up generates the public and the private random chip keys RCK-Pub and RCK-Pri, respectively, using the PUF/TRNG. The foundry sends the RCK-Pub to the designer.

Step 3 The designer encrypts the CK with RCK-Pub. This can be decrypted only with the RCK-Pri generated inside the locked chip by a PUF/TRNG. For the purpose of authentication, the encrypted CK is signed using the MK-Pri to generate the UK.

Step 4 The UK is sent to the foundry to activate the locked chip. The RSA module inside the locked chip authenticates the UK with MK-Pub and then decrypts it using RCK-Pri to obtain CK, thereby activating the locked chip.

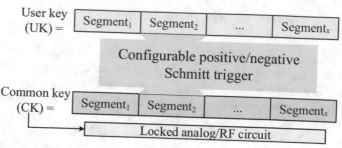

Fig. 16.14 The key provisioning unit generates the common key using the user key, which is unique to that chip instance [52]. Both keys are divided into segments that reuse the same circuitry resulting in an area overhead that is independent of the key size

16.2.11.2 Key Provisioning Architecture [52]

This technique is built on a Schmitt trigger (ST) circuit that has configurable hysteresis. The ST circuit is a comparator with low and high thresholds that define the hysteresis region. Its output is a function of the input voltages, the threshold values, and the previous output. The conventional ST operation is enhanced with dynamic hysteresis and inversion of the thresholds to create a CK with the desired security properties. As shown in Fig. 16.14, the UK is divided into segments; each of them contains a particular configuration of the threshold values and a series of input values. This operation enables circuitry reuse, thereby keeping the area constant and independent of the key size. The CK is the concatenation of the outputs of all the segments. This circuit block receives an analog input and generates a digital output. However, digital-to-analog converters enable both analog and digital representations of the UK and CK. A small portion of the ST configuration is stored in on-chip fuses, written before the chip's distribution, to increase security. This technique supports two different implementations. One implementation uses digital circuits such as inverters, while the other uses analog circuits such as amplifiers. This technique is tested on three analog/RF circuits to demonstrate its area scalability and security effectiveness. The Hamming distance characterizes the uniqueness of the UK, and the entropy estimates the effective key size of the key generated by this technique.

16.2.11.3 Results

This technique generates the CK for different UKs and is robust to PVT variations. It is compatible with existing analog locking techniques that have either digital or analog keys. Finally, this technique has the least area overhead compared to the previously reported techniques [3, 21]. This technique requires a UK that is at least 5.75X the size of the CK. Hence, the key provisioning increases the security level without increasing the area of the analog locks.

16.3   Evaluation of Analog IP Protection Techniques

Having discussed the different analog-only and AMS defense techniques, it is necessary to evaluate the resilience offered by these techniques against the supply-chain attacks. As explained in the following section, there are different evaluation schemes to study the resilience of digital logic-locking techniques. However, they cannot be leveraged to evaluate the analog locking schemes.

16.3.1   Attacks Against Digital IP Protection

Attacks such as SAT [47], removal [53], bypass [6], and SMT [54] evaluate the resilience offered by the digital logic-locking techniques against the supply-chain attacks.

1. **Brute-forcing** is the simplest attack that breaks the locked digital circuit. The attacker accesses the locked netlist either from the foundry (layout) or as an end-user (reverse-engineers the chip). He/She then simulates this netlist for all possible key combinations. The output response for each key combination is compared with the oracle's output response for the same input. The key corresponding to the same output responses from the oracle and the unlocked chip is the correct key. If the key size is n, then the total number of combinations is 2^n. Hence, by ensuring a minimum key size of 80, the defense will be resilient against brute-force attacks.

2. **SAT attack** is a Boolean satisfiability-based attack on combinational logic-locked circuits [47]. It uses a miter circuit that consists of two copies of the locked netlist. These netlists are fed with the same inputs but different keys. The outputs of the locked netlists differ for the inputs called distinguishing input patterns (DIP). The oracle, i.e., the functional chip loaded with the secret (correct) key, is queried for the output corresponding to this DIP. The keys for which the netlist does not give the desired output are pruned from the search space. This process is iterated until all the incorrect keys are pruned-off, leaving only the correct key.

3. **SMT attack** is a superset of SAT attack [54]. While the SAT attack can handle only Boolean variables, this can handle non-Boolean variables, such as logic delay. This property enables them to break delay logic locking [55] that cannot be broken by the SAT attack.

4. **Removal attack** determines the protection logic and removes them to recover the original circuit [53].

5. **Bypass attack** finds the DIPs that generate incorrect output for an incorrect key. The attacker adds a bypass circuitry around the protection block to restore the corrupted output for those DIPs.

6. **SFLL-hd–Unlocked [56] and FALL attack [57]** break SFLL-HD^h[7]. These attacks determine the PIPs by structurally analyzing the locked netlist. Using these PIPs, it determines the key using SAT formulations.

These techniques cannot be used to evaluate the analog locking techniques due to the following reasons:

1. The output of the logic-locked digital circuits is a Boolean variable, for a given input-key combination. However, the analog circuits' output is a non-Boolean variable such as bias current, bias voltage, and transconductance. As the SAT attack can handle only Boolean variables, it cannot break analog logic locking.
2. The analog-only locking techniques [21–23] lock the bias circuit. Launching removal attack on them removes the bias, which is required for the circuit to be functional.
3. The attacker performing the bypass attack has to replace the locked bias circuit with the current/voltage source that generates the required bias. This attack is feasible only if the attacker can get the precise value of the bias inputs and can pirate the design. If he/she does not have the resources to pirate the design or is not aware of the precise value of the bias, this attack is not feasible.
4. The SMT formulation in [54] breaks the delay logic locking [55] and shares an algorithm to speed up the SAT attack. The attack constraints of delay logic locking are different from the constraints required to model the analog locks. Hence, new SMT formulations are required to break analog locks.
5. The SFLL-HD^0 and SFLL-HD^h used in [24–26] are resilient against all the attacks mentioned above, except [56] and [57]. In [56] and [57], for a given input size, key size, and Hamming distance (HD), to determine the key, all possible PIPs are considered. However, the attacker does not have access to all possible PIPs. Hence, new SAT formulations are required that returns the correct key.

As the existing evaluation schemes cannot be used for breaking the analog-only and AMS locks, researchers have developed a new evaluation technique that uses SMT/SAT formulations [58] and genetic algorithm (GA) [59]. The resilience of all these techniques against this attack is explained in the following section.

16.3.2 Security Analysis of Analog IP Protection Techniques

16.3.2.1 Resilience of Memristor-Based Protection Against Supply-Chain Attacks

- **Brute-force.** The key size must be ≥ 80 bits to be resilient against brute-force attack. A memristor crossbar size of 7×7 and above will require a minimum key size of 98 bits and above. This key size ensures resilience against the brute-force attack.
- **Removal attack.** Launching this attack removes the memristor array that is controlled by the key inputs. This attack removes the body bias voltage (V_{BB}), which is essential for the offset cancellation in the differential pairs of the sense amplifiers. Hence, this technique is resilient against the removal attack.

- **Bypass attack.** To launch this attack, the crossbar should be replaced by a voltage source that generates V_{PROG}. The attacker having access to the memristor's datasheet knows the V_{PROG} value and hence, replaces the crossbar with the voltage source that generates V_{PROG} making the attack successful.
- **Attack using SAT formulations (SAT attack, SFLL-HD–Unlocked, and FALL attack).** Boolean satisfiability algorithms handle only Boolean variables. Hence, this technique is resilient against all these attacks as the crossbar's output corresponds to non-Boolean value (voltage).
- **IP piracy.** This technique is not resilient against IP piracy and is discussed in Sect. 16.3.3.

16.3.2.2 Resilience of Combinational Lock [21] Against Supply-Chain Attacks

As explained in Sect. 16.2.4, parameter-biasing obfuscation [22] is similar to combinational lock. Hence, the following is applicable to both techniques.

- **Brute-force.** A key of size ≥ 80 bits and above that controls the configurable current mirrors ensures resilience against brute-force attacks.
- **Removal attack.** Launching this attack removes the configurable current mirror (CCM) that generates the bias current. This bias current is necessary for the proper operation of the analog circuit-under-protection. Hence, removing the CCM makes the circuit non-functional.
- **Bypass attack.** To launch this attack, the attacker must know the bias current value required for the analog circuit to perform as per the specifications. In most cases, the attacker cannot determine this precise value of the bias current making this attack unsuccessful.
- **Attack using SAT formulations (SAT attack, SFLL-HD–Unlocked, and FALL attack).** This technique is resilient against all these attacks as the analog circuits' input and output variables are non-Boolean variables, such as transconductance, bandwidth, and gain.
- **IP piracy.** The SMT-based [58] and GA-based attacks break this technique, and hence, it is not resilient against IP piracy. The attack formulations for SMT-based attack are discussed in Sect. 16.3.3.

16.3.2.3 Resilience of Analog Camouflaging [27] Against Supply-Chain Attacks

- **Brute-force.** The attacker has to simulate the circuit for all combinations of V_{th} for each transistor in the circuit. If the number of V_{th} variants is three, and if there are n transistors in the circuit, the total number of possible combinations the attacker evaluates is 3^n. Hence, this technique is preferred for larger analog designs such as class D amplifiers with more number of transistors. n should be chosen such that $3^n \geq 2^{80}$, i.e., $n \geq 51$. Hence, the minimum number of transistors required to be camouflaged is 51 to have resilience against brute-force attack.

- **Reverse engineering attack.** The V_{th} of the transistor cannot be determined by RE techniques such as delayering and imaging the ICs. The end-user can purchase a chip, depackage it, and take multiple high-resolution images of each layer of the chip. However, the V_{th} type of the transistor cannot be determined from the images. Hence, this technique is resilient to reverse engineering attacks.
- **Micro-probing attack.** The V_{th} of the transistor cannot be determined by RE techniques such as delayering and imaging the ICs. However, dopant profiling techniques such as spreading resistance profiling can measure the channel doping concentration. These techniques are more resource-intensive and hence, not suitable for small ad hoc attackers.
- **Removal and bypass attacks.** As the entire design is camouflaged, there is no separate protection logic that can be removed or bypassed to make this attack successful. Hence, these attacks are not applicable to camouflaged analog circuits.
- **Attack using SAT formulations (SAT attack, SFLL-HD–Unlocked, and FALL attack).** This technique is resilient against all these attacks as the analog circuits' input and output variables are non-Boolean variables, such as transconductance, bandwidth, and gain.
- **IP piracy.** The camouflaged analog circuit can be transformed into a logic-locked netlist using [60]. The SMT-based attack [58] can then break the transformed netlist. The attack formulations are discussed in Sect. 16.3.3.

16.3.2.4 Resilience of Analog Performance Locking [29] Against Supply-Chain Attacks

- **Brute-force.** As the input to the ANN is analog in nature, this allows for a very large input keyspace. Adding to this, the analog floating-gate transistors (FGTs) are used to store the synapse weights. These FGTs have 10-bit precision that corresponds to 2^{10} unique weight values. As the number of synapses is more than 10 in a 30×20 ANN, this technique is resilient against brute-force attack.
- **Removal attack.** Launching this attack removes the ANN that provides the necessary bias voltages required for the LNA's proper operation. Also, the threat model of this technique considers a trusted foundry, and hence, the removal attack is not possible as it requires layout-level modification.
- **Bypass attack.** As this attack requires layout-level modification, it cannot break this technique as the threat model considers trusted foundry. Even if the foundry is an untrusted entity, the attacker must know the precise value of the bias voltages. Only then he/she could replace the ANN with the voltage source generating the desired bias value. However, it is not always possible to determine the precise bias values hence, rendering this attack unsuccessful.

- **Attack using SAT formulations (SAT attack, SFLL-HD–Unlocked, and FALL attack).** This technique is resilient against the SAT-based attacks as the ANN handles non-Boolean variables, such as bias voltages and currents.
- **IP piracy.** The ANN can be represented in SMT formulations, making it vulnerable to the SMT-based attack [58]. The attack formulations and its results are discussed in Sect. 16.3.3.

16.3.2.5   Resilience of Mixed-Signal Protections [25, 26] Against Supply-Chain Attacks

As explained in Sect. 16.2.6, MixLock [25] is similar to AMSlock. Hence, the following is applicable to both techniques.

- **Brute-force attack.** The optimizer is locked using SFLL-HD^0 and SFLL-HD^h with a minimum key size of 80 bits, making it resilient against brute-force attack.
- **Removal attack.** If the locked optimizer is removed, the tuning knobs are set to the default value. The probability of this value being equal to the desired value to address process variations is negligible. Hence, launching the removal attack makes the analog circuit-under-protection non-functional.
- **SAT attack.** The resiliency against SAT attack offered by SFLL-HD^0 and SFLL-HD^h are k and $k - log_2\binom{k}{h}$, respectively [7]. The security level s achieved for the BPF is the maximum when $h = 0$ or $h = 220$ and the minimum when $h = 110$. To ensure that the locked circuit is SAT attack resilient, h and k are chosen such that the security level is greater than 80. Hence, the allowable h values can be $0 \leq h \leq 37$ or $183 \leq h \leq 220$, and the corresponding number of input patterns which can be protected are $1 <$ # of patterns protected $< 1.37 \times 10^{42}$. Also, the security increases with the increase in key size, whereas the number of input patterns protected reduces with the increase in key size. A key size, $k \geq 80$, ensures resilience against the SAT attack. The time required for the attack increases exponentially with the input size. For the input size of 14, the attack takes close to 1.5 h to identify the key. This trend indicates that the AMSlock technique is secure against the SAT attack. Similarly, it is also secure against AppSAT [61], as it protects only a linear number of input patterns.
- **Bypass attack.** This attack finds the PIPs that give an incorrect output for an incorrect key. The attacker adds a bypass circuitry around the protection block to restore the output for those PIPs. However, the bypass attack cannot compute all the PIPs from the circuit protected using SFLL. This is because, in SFLL, a PIP produces the same incorrect output for most of the incorrect key values. Also, the output corresponding to the PIP may be restored correctly, even for an incorrect key. Hence, the bypass attack does not consider the corresponding input pattern as PIP. Therefore, the construction of the bypass circuitry using the incomplete set of PIPs will be erroneous. Thus, it renders the attack unsuccessful.

- **IP piracy and overproduction.** This technique is not resilient against IP piracy and overproduction. The new SAT formulations required to break this technique are explained in Sect. 16.3.3.7.

16.3.2.6 Resilience of Shared Dependencies [24] Against Supply-Chain Attacks

Though this technique locks both the analog and digital sections of the AMS circuits, the attacker can conquer each section individually, making the defense not resilient against IP piracy.

- **Brute-force.** A key size of 80 bits and above will ensure that this technique is resilient against the brute-force attack.
- **Removal and bypass attacks.** Rather than including a separate protection logic, this technique locks directly the analog and digital sections of the circuit. Hence, the removal attack does not apply to this defense.
- **SAT-based attacks.** The part of the digital section locked using RLL [3] is broken by SAT attack [47]. The section of the digital circuit that is locked using SFLL-HD^0 is unlocked by determining the PIPs, as explained in Sect. 16.3.3.8.
- **IP piracy.** The locked analog circuit is vulnerable to the SMT-based attack. Its attack formulations and the corresponding results are discussed in Sect. 16.3.3.

16.3.2.7 Resilience of Analog Floating-Gate Transistor Technique [28] Against Supply-Chain Attacks

- **Brute-force.** This technique has a sequence of "secret" analog values, called waypoints. These waypoints (a.k.a. keys) are programmed into the AFGTs and are analog values. Hence, this technique is resilient against the brute-force attack as the keyspace is analog.
- **Removal and bypass attacks.** This design is secured based on the charge stored in the floating-gate transistors. It does not include a separate protection logic. Hence, it is not possible to launch removal or bypass attacks on this technique.
- **Attack using SAT formulations (SAT attack, SFLL-HD–Unlocked, and FALL attack).** This technique is resilient against the SAT-based attacks as the analog circuit handles non-Boolean variables, such as bias voltages and currents.
- **Illegitimate access.** Only the attacker who has access to the waypoints can unlock the circuit's performance. As these waypoints are analog in nature, the probability of finding the precise points is negligible, making it resilient against illegitimate access.
- **IP piracy.** This technique is not resilient against piracy-based attacks, as explained in this work [28]. The attacker can program a new set of waypoints of his/her choice to make the circuit functional.

16.3.2.8 Resilience of Mixed-Signal Camouflaging [49] Against Supply-Chain Attacks

- **Brute-force.** This technique ensures resilience against the brute-force attack if the possible number of camouflaged contacts is more than 2^{80}.
- **Reverse engineering attack.** As it is not possible to determine the contacts that connect the gate fingers, unit capacitors, and unit resistors, this technique is resilient against reverse engineering-based attacks.
- **Removal and bypass attacks.** As the entire design is camouflaged, there is no separate protection logic that can be removed or bypassed to make this attack successful. Hence, these attacks do not apply to camouflaged analog circuits.
- **Attack using SAT formulations (SAT attack, SFLL-HD–Unlocked, and FALL attack).** This technique is resilient against all these attacks as the input and output variables of the analog circuits are non-Boolean variables, such as transconductance, bandwidth, and gain.
- **SMT-based attack.** If the attacker knows the equations linking the transistor dimensions, resistors, and capacitor values with the circuit specifications, this technique can be broken using [58] as indicated in [49].

The resilience offered by each technique is summarized in Table 16.2. Except for the recent techniques [28, 49], the remaining techniques are broken by either SAT-or SMT-based attacks.

16.3.3 Breaking Analog-Only and Mixed-Signal Locks Using Boolean Satisfiability (SAT) and Satisfiability Modulo Theories (SMT)

16.3.3.1 Attack Methodology on Analog-Only Locks

The analog locking techniques obfuscate the circuit components' effective value, such as the transistor's width, resistance, and capacitance. These components are used in the bias circuit, as the precise bias current (I_B) or voltage (V_B) is required for the proper operation of the analog circuits. These components are called **obfuscated components** as they are made configurable, and their effective value is hidden from the attacker. The key input determines this value that sets the precise biasing condition. The following attack aims to find the key that provides the required bias to make the analog circuit functional using the SMT formulations. The generalized attack methodology shown in Fig. 16.15 to break the existing analog-only locks is discussed below:

1. **Identifying the obfuscated circuit components and their dependency on the key.** From the locked netlist, the attacker can determine the obfuscated components, such as transistors, resistors (Rs), and capacitors (Cs) [30], by tracing the wire connections from the key input. A component y in the original

Table 16.2 The attack success on various analog locks. ○ denotes locked netlist. ● denotes locked netlist. ⊥ denotes the oracle access. ⊥ denotes that the resource availability aiding overproduction but does not aid piracy. ∅ denotes reverse-engineered locked netlist. * denotes availability of digitally controlled current source. √ denotes that the defense is broken. ≈ denotes the attack reduce key search space

Type	Defense	Circuit locked	Claimed resilience	TM (defense)	TM (attack)	Brute-force	Removal	Bypass	RE-based	SAT-based	SFLL-HD –Unlocked	FALL	SMT-based
Analog-only locks	Combinational lock [21]	Analog	IP piracy	○●	○●	×	×	×	NA	×	×	×	√
	Parameter-biasing obfuscation [22]		IP piracy	○●	○●⊥	×	×	×	NA	×	×	×	√
	Parameter-biasing obfuscation [22]		IP piracy	○●	○●⊥	×	×	×	NA	×	×	×	√
	Memristor-based protection [23]		IP piracy	○●	○●⊥	×	×	×	NA	×	×	×	√
	Analog camouflaging [27]		Reverse engineering	∅●	∅●	×	×	×	×	×	×	×	√
	Analog neural network [29]		Illegitimate access	○●	○●*	×	×	×	NA	×	×	×	≈
AMS locks	AMS lock [26]	Digital	Overproduction	○●⊥	○●⊥	×	×	×	NA	√	×	×	√
	MixLock [25]		IP piracy	○●	○●⊥	×	×	×	NA	√	×	×	×
	Shared dependencies [24]	Analog and digital	IP piracy	○●	○●⊥	×	×	×	NA	√	×	×	×

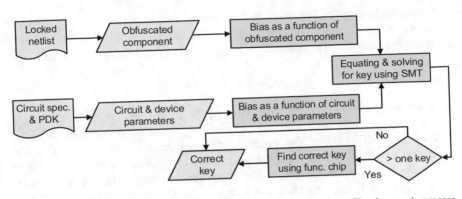

Fig. 16.15 SMT-based attack methodology [58]. Circuit specification (Circuit spec.), process development kit (PDK), and functional chip (func. chip)

design is replaced by a set of n obfuscation components $\mathbf{x} = \{x_1, x_2, \ldots, x_n\}$, which are controlled by an m-bit key vector $\mathbf{q} = (q_1, q_2, \ldots, q_m)$. We denote the values of y and $x_i, i \in \{1, 2, \ldots, n\}$, by y_v and $x_{i_v}, i \in \{1, 2, \ldots, n\}$, respectively. Then, the effective value of the obfuscated components $\mathbf{x}$ is,

$$\tilde{y}_v = \phi(\mathbf{x_v}, \mathbf{q}) \tag{16.8}$$

Here, $\mathbf{x_v} = (x_{1_v}, x_{2_v}, \ldots, x_{n_v})$, and the function ϕ depends on how the obfuscation circuit is constructed. For the correct key $\mathbf{q}^*$, $\tilde{y}_v - y_v$, i.e., the effective value of the obfuscated component is equal to that of the original one.

2. **Finding the equation that shows the dependence of the obfuscated component's value $\tilde{y}_v$ on the bias z_{ob_comp}.** The bias z_{ob_comp} (e.g., I_B or V_B) is a function ψ of the value of the obfuscated component $\tilde{y}_v$.

$$z_{ob\_comp} = \psi(\tilde{y}_v) \tag{16.9}$$

Substituting Eq. (16.8) in (16.9) gives the dependency of z_{ob_comp} on the key $\mathbf{q}$, $z_{ob\_comp} = \psi(\phi(\mathbf{x_v}, \mathbf{q}))$.

3. **Derive bias z_{spec} from the circuit parameter p of the protected analog IC.** The attacker can obtain the circuit parameters, such as g_m, BW, and ω_{osc}, from the circuit specification. He/she then analyzes the target circuit and extracts its characteristic equations. Solving these equations yields the bias point I_B or V_B,

$$z_{spec} = \theta(p) \tag{16.10}$$

Here, θ is the function to compute the circuit parameter p [62–64]. If the attacker knows the bias point, he/she can redesign the entire analog circuit. However, it is sometimes possible to determine only the bias range and not the precise z_{spec}. This is because there may not be a direct equation linking z_{spec} and p. Instead, equations linking the minimum and maximum values of the bias with different circuit parameters are available in the specification [65]. Here, we calculate a range for bias using the equations $z_{spec_{min}} = \theta_1(p_1)$ and $z_{spec_{max}} = \theta_2(p_2)$.

$$z_{spec_{min}} \leq z_{spec} \leq z_{spec_{max}} \tag{16.11}$$

where θ_1 and θ_2 are the functions to compute p_1 and p_2, respectively. $z_{spec_{min}}$ and $z_{spec_{max}}$ are the minimum and maximum values of the bias, respectively.

4. **Putting it all together.** The I_B or V_B obtained from circuit specification and the I_B or V_B estimated from the obfuscated components should be equal or approximately equal for the analog circuit to be functional. Hence, the bias z_{ob_comp} equals z_{spec}, or alternatively, z_{ob_comp} satisfies the inequality specified by z_{spec}. Solving the Eqs. (16.8), (16.9), and (16.11) using the SMT solver [66] computes the correct key $\mathbf{q}^*$. This key sets the effective value of the obfuscated component equal to the value of the original component, i.e., $\tilde{y}_v = y_v$.

16.3.3.2 Attack Methodology on Digital Locks Used in Mixed-Signal Circuits

SFLL [7] explained in Sect. 16.2.6.1 has been extensively used in locking the digital section of the AMS circuits in [24–26, 67]. As this technique protects only a handful of input patterns, the probability of finding them is negligible[7]. However, when this technique is used to lock the digital section of the AMS circuits, the PIPs can be found by analyzing the analog–digital interface signals. The key is then determined using SAT formulations, which are discussed next.

1. **Finding PIPs.** In [24–26], the analog section of the AMS circuit drives the input to the locked digital section. By simulating the analog section of the netlist, the attacker can determine the INs that drive the digital section. The inputs from the analog circuit are the only INs to the locked digital circuit. Hence, the PIPs should be the subset of these INs.
2. **SAT formulation to determine the correct key.** The attack formulations are unique based on the locking technique used [24, 26]. They are explained in the respective attack Sects. 16.3.3.7 and 16.3.3.8. The SAT formulations are solved using a SAT solver to determine the correct key.

The attacks on analog-only locks [21–23, 27, 29] are first demonstrated, followed by attack demonstration on mixed-signal locks [24–26, 67].

16.3.3.3 Resilience of Memristor-Based Protection Against SMT-Based Attack

The security of this technique lies in the pre-programmed memristor crossbar. This technique can be compromised if the attacker finds: (1) the pre-programmed resistance's value in each crossbar, and (2) the connectivity among the memristors, controlled by the key inputs. The SMT formulations for this technique are given below.

1. **Obfuscated component.** The effective resistance of the obfuscated upper and lower memristor crossbars are $\tilde{y1}_v$ and $\tilde{y2}_v$, respectively.

$$\tilde{y1}_v = R_U = \left(\sum_{i=1}^{U} \frac{k_q}{x_{i_v}} \right)^{-1} \tag{16.12}$$

where $q \in \{1, 2, \cdots, n\}, \forall i$.

$$\tilde{y2}_v = R_L = \left(\sum_{i=1}^{L} \frac{k_q}{x_{i_v}} \right)^{-1} \tag{16.13}$$

where $q \in \{1, 2, \cdots, n\}, \forall i$. Here, $\mathbf{k} = (k_1, k_2, \ldots, k_n)$ is an n-bit key. x_{i_v} is the pre-programmed memresistance value of memristor i.

2. **Equations linking the bias z_{ob_comp} with obfuscated resistivities $\tilde{y1}_v$ and $\tilde{y2}_v$ of the crossbars.** The voltage divider generates the necessary programming voltage, V_{PROG} for the memristor M. The equation linking the bias V_{PROG} with $\tilde{y1}_v$ and $\tilde{y2}_v$ is $z_{ob\_comp} = V_{PROG} = \psi(\tilde{y1}_v, \tilde{y2}_v)$.

$$
\begin{aligned}
\psi(\tilde{y1}_v, \tilde{y2}_v) &= \frac{\tilde{y2}_v}{\tilde{y1}_v + \tilde{y2}_v} \times AV_{PP} \\
&= \frac{\left(\sum_{i=1}^{L} \frac{k_q}{x_{i_v}} \right)^{-1}}{\left(\sum_{i=1}^{U} \frac{k_q}{x_{i_v}} \right)^{-1} + \left(\sum_{i=1}^{L} \frac{k_q}{x_{i_v}} \right)^{-1}} \times AV_{PP}
\end{aligned}
\tag{16.14}
$$

3. **Equation linking the bias z_{spec} with circuit parameter p.** The equations connecting V_{PROG} with V_{BB} are,

$$V_{BB} = \frac{R_M}{R_1 + R_M} \times V_{DD} \tag{16.15}$$

$$R_M = \left(\sqrt{\frac{\gamma}{\varphi}} \right) \times \rho \times V_{PROG} \frac{2\pi}{\omega_{PT}}$$

Combining the equations gives $z_{spec} = V_{PROG} = \theta(p) = \theta(V_{BB})$. γ is a constant depending on device parameters such as carrier mobility and device thickness, φ is the flux, ρ is the duty cycle, and ω_{PT} is the frequency of programming pulse. The values of V_B, V_{DD}, φ, γ, ρ, ω_{PT}, and R_1 are available in the circuit specification. The resistivity range with which the memristors can be pre-programmed is (R_{min}, R_{max}), where R_{min} and R_{max} are the minimum and maximum resistivity of the memristor M. The attacker can obtain the value of R_{min} and R_{max} from the circuit specification of the memristor.

4. **Putting it all together.** Solving these equations gives the correct key and each memristor's resistance with which it has to be pre-programmed. There can be more than one correct key that gives the same V_{BB} due to the memristor array configuration. Hence, the SMT solver is called only once to determine one correct key and one set of memristors' resistance values.

16.3.3.4 Resilience of Combinational Lock Against SMT-Based Attack

This technique obfuscates the transistor's width to secure the analog circuit-under-protection. Hence, the attacker must determine the key that sets the transistor's effective width to produce the desired bias current value. The SMT formulations required for this attack are:

1. **Obfuscated component.** y_v is the ratio of $\left(\frac{W}{L}\right)$ of transistor y with respect to $\left(\frac{W}{L}\right)$ of M_{REF} in Fig. 16.3b, where $\left(\frac{W}{L}\right)$ is the aspect ratio of the transistor. y is replaced by n NMOS transistor switches. The gate terminals of these switches are controlled by the key **q**. x_{i_v} is the ratio of the aspect ratio of transistor x_i with respect to the aspect ratio of M_{REF}, where $i \in \{1, 2, \cdots, n\}$. Hence, the obfuscated size ratio is,

$$\tilde{y}_v = \sum_{i=1}^{n} x_{i_v} q_k \tag{16.16}$$

where $q_k \in \{0, 1\}$ and $k \in \{1, 2, \cdots, n\}$.

2. **Equations linking the bias z_{ob_comp} with obfuscated widths $\tilde{y_{1_v}}$ and $\tilde{y_{2_v}}$ of the transistors.** In the combinational lock [21], the bias circuit is designed to be a current mirror. If the bias circuit is a current mirror as in [21], I_B of CCM is,

$$\psi(\tilde{y}_v) = I_B = \tilde{y}_v \times I_{REF} \tag{16.17}$$

where I_{REF} is the reference current obtained from the circuit specification.

3. **Equation linking the bias z_{spec} with circuit parameter p.** The g_m of the OTA shown in Fig. 16.3c is $\frac{g_{m1}g_{m4}}{g_{m3}}$. Here, g_{mi} is the transconductance of transistor Mi. If x_v is the ratio of the aspect ratio of transistor $M4$ to the aspect ratio of $M3$, then $g_m = x_v \times g_{m1}$. The attacker finds I_B using,

$$\theta(g_m) = I_B = \frac{g_m^2}{\mu C_{ox} \frac{W}{L}} \qquad (16.18)$$

4. **Putting it all together.** Solving the Eqs. (16.16)–(16.18) gives the correct key q^*. This key sets the required g_m in the OTA.

The above attack methodology can also break the parameter-biasing obfuscation technique [22]. The locking technique proposed in parameter-biasing obfuscation [22] is similar to [21]. In [22], the width of the transistor in the bias circuit is obfuscated. It is achieved by replacing the transistor with multiple transistors connected in parallel, as illustrated in Fig. 16.4b.

16.3.3.5 Resilience of Camouflaged Analog IPs Against SMT-Based Attack

In this technique, few of the NVT transistors are replaced by LVT and HVT transistors [27]. The functionality is maintained by resizing the replaced transistors. Thus, each camouflaged transistor can be modeled as,

$$\left(\frac{W}{L}\right)_i = x_i \times \left(\frac{W}{L}\right)_{NVT} \qquad (16.19)$$

Here, $i \in \{NVT, LVT, HVT\}$. The attacker can transform individual camouflaged transistors to logic-locked transistors [60].

$$\left(\frac{W}{L}\right)_{camouflaged} = \begin{cases} \left(\frac{W}{L}\right)_{NVT} & \text{if } q_1 = 1, \\ \left(\frac{W}{L}\right)_{LVT} & \text{if } q_2 = 1, \\ \left(\frac{W}{L}\right)_{HVT} & \text{if } q_3 = 1. \end{cases} \qquad (16.20)$$

Here, $q_1, q_2, q_3 \in \{0, 1\}$ are the key-bits controlling transistors of type NVT, LVT, and HVT, respectively. Also, $q_1 + q_2 + q_3 = 1$, as each camouflaged transistor can be only one of the three types. The attacker performs the SMT-based attack on the transformed logic-locked circuit as follows:

1. **Obfuscated component.** The transistors in the design are the obfuscated components, as their V_{th} type is unknown to the attacker. From Eqs. (16.19) and (16.20), the aspect ratio of the camouflaged transistor $\tilde{y} = \phi(x, q)$, is

$$\tilde{y} = \frac{W}{L}_{camouflaged} = \left(\sum_{\forall i} x_i q_i\right) \times \left(\frac{W}{L}\right)_{NVT} \qquad (16.21)$$

Here, $i \in \{NVT, LVT, HVT\}$ and $q_i \in \{0, 1\}$ is the key-bit controlling transistor of type i.

2. **Linking $\tilde{y}$ to the circuit specification of the locked analog IC.** Unlike techniques that obfuscate the bias circuits, analog camouflaging obfuscates the transistors in the design that affect the circuit specification. Considering the fourth-order Gm-C BPF, the transconductance as a function of the obfuscated width $\tilde{y}$ is,

$$g_m = \sqrt{2\mu C_{ox} \tilde{y} I_D}$$

(16.22)

where $I_D = I_B/2$. Solving Eqs. (16.19)–(16.22) gives the required key for the precise operation of the BPF. From this key, the attacker determines the V_{th} of the transistor.

16.3.3.6 Attack on Analog Performance Locking [29]

The attack mathematically models the ANN using the SMT formulations. The synapse outputs the product of the inputs along with the weight associated with it, as shown in Fig. 16.9b. S_{ij} is the synapse output, where i and j are the row and column number of the synapse considered. Sw_{ij} is the weight associated with the synapse S_{ij}. $IN_{1_{ij}}$ and $IN_{2_{ij}}$ are the two inputs to S_{ij}. Then, S_{ij} is modeled as, $S_{ij} = Sw_{ij} \times IN_{1_{ij}} \times IN_{2_{ij}}, i \neq j$ where, $i, j \in (1, \cdots, n)$. The output of the neurons (N_{ij}) which forms the diagonal elements of the ANN matrix depends on the select signal s_i, given by

$$N_{ij} = \begin{cases} \tanh(S_{i(j+1)}) & \text{if } s_i = 1 \wedge i = j = 1 \\ \tanh(S_{i(j-1)}) & \text{if } s_i = 1 \wedge i = j = n \\ \tanh(S_{i(j-1)} \times S_{i(j+1)}) & \text{if } s_i = 1 \wedge \text{otherwise} \\ S_{i(j+1)} & \text{if } s_i = 0 \end{cases}$$

The inputs to the synapses in the input layer ($i = 1$) are given by $IN_{1_{1j}} = IN_{i-1}$ and $IN_{2_{ij}} = S_{1(j+1)}$ and for other layers, the inputs are given by the following equations:

$$IN_{1_{ij}} = \begin{cases} S_{(i-1)j}, \text{if } j \neq i - 1 \\ N_{(i-1)j}, \text{if } j = i - 1 \end{cases} \qquad IN_{2_{ij}} = \begin{cases} 1, \text{if } j = 1 \vee j = n \\ S_{i(j-1)}, \text{if } j < i \\ S_{i(j+1)}, \text{if } j > i \end{cases}$$

(16.23) (16.24)

The SMT formulations for the ANN are fed to the SMT solver along with the required V_B range. This bias range is essential for the proper operation of the OTA. The solver returns the input voltages to the ANN, weights associated with each synapse, the type of neuron, and the value of V_B. The attacker can procure an off-the-shelf digitally controlled current source to program the new weights into the synapses. The input voltages returned by the attack are fed to the input layer synapses. The ANN thus produces the required I_B or V_B, thereby rendering the attack successful. More than one correct configuration gives the same V_B due to the neural network topology [29]. Hence, the SMT solver is called once to determine one correct configuration that provides the necessary bias.

16.3.3.7 Attack on Mixed-Signal Locks [25, 26]

Attack Methodology This defense thwarts an attacker from overproducing the chip but cannot thwart him/her from modifying the layout of the design and pirate (IP piracy). Hence, the attacker can assume access only to the layout but cannot modify the same. This design has 1024 tuning knob settings corresponding to 1024 unique resistor settings. These tuning knobs are internal to the chip and are not available as top-level ports. Adding to this, the optimal settings of the tuning knobs vary chip to chip due to the process variations. Therefore, the attacker cannot simulate the analog circuit-under-protection for all the tuning knob settings to determine the correct settings as it changes chip to chip. Only the optimizer can control the tuning knob settings. As the attacker cannot modify the tuning knob settings directly, it is necessary to determine the correct key to unlock the locked optimizer.

1. **Obfuscated component.** To reduce the impact of PVT variations and mismatch, passive components, such as R and C, are often implemented as banks of elements to enable calibration. The correct value of the passive components is chosen by the locked optimizer and cannot be computed by analyzing the netlist. Hence, we identify this component as the obfuscated component.
2. **Equation linking the obfuscated component and the key inputs to the optimizer.** In the BPF circuit, the resistors R_1 and R_2 are the obfuscated components. The correct value of these resistors is chosen by two 5-bit tuning knobs controlled by the locked optimizer. Each of the 1024 tuning knob settings corresponds to a unique resistor value. The output response of the analog circuit can be determined from the transfer function given by $H(s) = \frac{s/(R_1 C)}{s^2 + s/(R_1 C) + 1/(R_2^2 C)}$. Here, C is the fixed capacitor. The attacker can simulate the output response of the circuit for unique resistor settings, via transistor-level simulations. As there are only 1024 unique tuning knob settings, the analog circuit can have 1024 unique output responses. The output response is digitized using the analog-to-digital converter. These digitized output responses are the input patterns (INs) that are fed to the locked optimizer. The optimizer chooses the tuning knobs based on these inputs. These INs are required to determine the SAT formulations for the attack.
3. **Breaking SFLL-HD^h [7].** The SFLL-HD^h technique can have more than one correct key for a PIP when $h > 0$ [7]. If the attacker finds one key that ensures the correct output for all 1024 PIPs, this key can unlock the overproduced chip. **Finding PIPs.** The attacker can determine the PIPs in the 1024 INs with the help of oracle. The entire AMS chip loaded with the correct key constitutes the oracle.

Only the input and output ports of the analog circuit-under-protection in this chip are available to the attacker. Hence, the attacker has to simulate the analog circuit for different tuning knob settings to determine the input patterns to the locked optimizer. The signal generator gives the required input to the oracle [41], and he/she can observe the oracle's response on the output port. If the locked optimizer gives an incorrect output for an IN, then it is a PIP. Otherwise, it is not a PIP. Hence, if there are p PIPs out of 1024, then the remaining n patterns $(1024 - p)$ are unprotected IPs.

SAT formulations. Along with the locked netlist (N_{locked}) and the HD used by the defender, the following are the other constraints added to the SAT formulations:

(a) The output response (O) of the analog circuit corresponding to each PIP is found using the oracle. The corresponding constraint is given by $(PIP_1 \Rightarrow O_1) \wedge (PIP_2 \Rightarrow O_2) \wedge \cdots \wedge (PIP_p \Rightarrow O_p)$.

(b) HD between each PIP and the key (K) must be equal to h, $\sum_{i=1}^{p} \wedge$ $(HD(PIP_i, K) = h)$.

(c) HD between other n INs and K should not be equal to h, $\sum_{i=1}^{n} \wedge$ $(HD(IN_i, K) \neq h$.

$$\sum_{i=1}^{n} \wedge (HD(IP_i, K) \neq h \qquad (16.25)$$

Combining all the above constraints gives,

$$N_{locked} \wedge PIP_p \wedge O_p \wedge (PIP_1 \Rightarrow O_1)$$
$$\wedge (PIP_2 \Rightarrow O_2) \cdots \wedge (PIP_p \Rightarrow O_p)$$
$$\wedge \sum_{i=1}^{p} \wedge (HD(PIP_i, K) = h) \qquad (16.26)$$
$$\wedge \sum_{i=1}^{n} \wedge (HD(IP_i, K) \neq h)$$

Equation (16.26) helps in determining the correct key to unlock the locked optimizer.

Similarity between AMSlock [26] and MixLock [25] Both these techniques use SFLL-HD0/h [7] to lock the digital section of the AMS circuit. The digital optimizer in AMSlock and the digital decimation filter in the MixLock are locked using SFLL-HD^0 or SFLL-HD^h [7]. These locked circuits receive inputs from the ADC in AMSlock and from the $\Delta\Sigma$ ADC in MixLock. In [26], the ADC is fed by the analog circuits, such as BPF, LC oscillator, or triangular waveform generator. Whereas, in [25], the $\Delta\Sigma$ ADC is fed with the audio input, which has to be modulated. The defender chooses the PIPs by analyzing the inputs from the ADC or the $\Delta\Sigma$ ADC. Therefore, the attacker determines the PIPs by simulating the analog circuit and ADC in [26] and analyzing the audio signals sent to the $\Delta\Sigma$ ADC in [25]. He/She then uses SAT formulations to determine the correct key. The following section explains our attack on AMSlock [26]. The only difference between these two techniques is the circuit over which the defense is implemented. However, the underlying defense algorithm remains the same. Therefore, this attack can break MixLock too.

16.3.3.8 Attack on Shared Dependencies [24]

This technique improves the resiliency against IP piracy and overproduction by locking the analog and digital parts of an AMS circuit. Hence, the attacker has to find the correct key required to unlock both the analog and digital sections of the AMS circuit. This technique's threat model assumes an untrusted foundry and an untrusted end-user. Hence, he/she can access the layout, the oracle, and the circuit specification. The attacker targets the analog and digital locks separately.

Breaking the Digital Lock The output of the ADC is the input to the peak detection circuit locked using SFLL-HD^0. This circuit sets the peak detection signal to one when the ADC output is maximum and resets to zero when the ADC output falls below the maximum value. The PIP corresponds to the maximum ADC value. In SFLL-HD^0, as the key is equal to the PIP, the attacker can unlock the locked peak detection circuit. He/She could verify the correctness of the key found, using the SAT formulations in Sect. 16.3.3.7. The SAT attack [47] can unlock the counter circuit. Thus, the attacker has determined the key to unlock the digital part of the AMS circuit without unlocking the locked analog circuit.

Breaking the Analog Lock The following SMT formulations help to determine the key to unlock the analog circuit.

1. **Obfuscated component.** The effective aspect ratio of the obfuscated transistors y_1 and y_2 are y_{1_v} and y_{2_v}, respectively. Here,

$$\tilde{y1}_v = \left(\frac{W}{L}\right)_{y_1} = \sum_{i=1}^{N} q_k \times x_{1i_v} \tag{16.27}$$

where $k \in \{1, 2, \cdots, N\}, \forall i$.

$$\tilde{y2}_v = \left(\frac{W}{L}\right)_{y_2} = \sum_{i=N+1}^{2 \times N} q_k \times x_{2i_v} \tag{16.28}$$

where $k \in \{N + 1, N + 2, \cdots, 2 \times N\}, \forall i. \, x_{1i_v}$ and x_{2i_v} are the aspect ratios of transistors x_{1i} and x_{2i}, respectively. The attacker knows the value of N from the layout.

2. **Equations linking the gain z_{ob_comp} with obfuscated aspect ratios y_{1_v} and y_{2_v} of the transistors.** The CS amplifier generates the necessary analog input to the ADC. The equation linking the amplifier's gain (G_{CS}) with y_{1_v} and y_{2_v} is,

$$z_{ob\_comp} = G_{CS} = -\frac{1}{1 + \eta} \sqrt{\frac{\tilde{y_{1_v}}}{\tilde{y_{2_v}}}} \tag{16.29}$$

Here, η is the back-gate transconductance available in PDK.

3. **Equation linking the gain z_{spec} with circuit parameter.** The gain of the CS amplifier is in the specification.

4. **Putting it all together.** Solving the above equations gives the correct key and hence, the effective aspect ratios of the obfuscated transistors.

Using the above formulations, the attacker can find the correct key to unlock the AMS circuit.

16.3.3.9 Summary

The attack equations based on SMT and SAT formulations are shared in this Section. As the floating-gate transistor-based technique [28] and AMS camouflaging [49] are recently proposed techniques, their resilience against supply-chain attacks is not analyzed by the SMT-based attack [58].

16.3.4 Breaking Combinational Locks [21] and Parameter-Biasing Obfuscation [22] Using Genetic Algorithm (GA) [59]

This technique breaks the analog circuits locked using combinational locks [21] and parameter-biasing obfuscation [22] techniques using genetic algorithm (GA). It either finds the obfuscation key or the obfuscated parameters in the locked analog circuits. This technique requires only the locked netlist and an unlocked chip (oracle). However, it does not require the circuit specification as in SMT-based attack.

16.3.4.1   Attack Methodology

This technique takes a locked netlist and an oracle output as inputs. Based on the locking architecture, the GA will either provide various combinations of key bits or various values of obfuscated circuit parameters to obtain an output that matches the oracle's. The following steps explain the GA attack:

1. The initial candidate solutions are created by generating N random chromosomes. Here, N is chosen to be small, which enables the algorithm to converge faster. The chromosome is encoded with real number/binary if the algorithm is used for finding the correct circuit parameter/key.
2. The objective function, i.e., the fitness function is defined. This function estimates how close is a given design solution (chromosome) to the required solution. Factors like frequency and transient response of outputs are used to generate the fitness function.
3. The chromosomes corresponding to the least value of the fitness function is chosen for the next steps of evolution while the others are discarded.
4. The selected chromosomes are used to create new chromosomes or offspring using the crossover operator.
5. The new chromosomes are mutated to determine how many bits in the offspring will be flipped from 0 to 1 and vice versa.
6. The steps 2 to 5 are iterated until the stopping criterion is met. Each iteration is called a run, and at the end of each run, there is usually at least one chromosome with the smallest fitness value. The number of iterations is one of the stopping criteria for halting the optimization process. Depending on the progress made, these parameters can be changed, including the mutation rate, the selection criteria, the crossover point, etc., for better optimization.

16.3.4.2   Results

The attack is validated on five analog circuits: operational transconductance amplifier, fourth-order Gm-C band-pass filter, phase-locked loop, triangular waveform generator used in class-D amplifiers, and superheterodyne receivers. The circuits are designed and obfuscated with a k-bit key where $k = 16, 32, 40$, and 64 using combinational lock [21] and parameter-biasing obfuscation [22]. For all the cases, the attack converges in a few minutes. Hence, this technique successfully extracts the value of the obfuscated parameter and the correct key from the locked circuits by only using the locked netlist and an oracle.

16.4 Limitations and Future Works

16.4.1 Limitations

There have been several analog-only and mixed-signal locks proposed in the last few years. However, each of them has its limitation that is explained in the following section:

1. As given in Table 16.2, almost all the existing techniques are not resilient against IP piracy. Hence, it is necessary to develop new techniques that are resilient against piracy-based attacks.
2. The current locking techniques consider only the security metrics and do not consider the overhead in power, performance, and area.
3. The protection techniques against IP piracy and overproduction share only simulation or MATLAB-based results. It is suggested to share on-board results for future techniques makes it concrete.
4. The existing SMT-based evaluation [58] shows the resilience offered by each analog-only and mixed-signal protection technique. However, it is required to develop security metrics to quantify the resilience offered by each of these techniques.

16.4.2 Future Works

The future works in analog and mixed-signal IP protection must address the current limitations.

1. The researchers should develop techniques that are resilient against IP piracy. It ensures resilience against both the untrusted foundry and untrusted end-users.
2. New security metrics should be introduced to evaluate the resilience offered by the defense techniques.
3. The new locking techniques must consider minimum overhead in power, area, and performance.

Acknowledgments This work is funded by the National Science Foundation C(CF-1815583, CNS-1618824, CNS-1828840, STARSS-1618797, and SATC CAREER-1822848), Semiconductor Research Corporation (2016-T3S-2688 and 2016-T3S-2689), and Intel. The authors thank Qualcomm and Synopsys for their support in this work.

References

1. M. Rostami, F. Koushanfar, R. Karri, A primer on hardware security: models, methods, and metrics. Proc. IEEE **102**(8), 1283–1295 (2014)

2. T.S. Perry, Why hardware engineers have to think like cybercriminals, and why engineers are easy to fool (2017), Accessed 21 August 2020 [Online]. Available https://bit.ly/2RfoBkS

3. J.A. Roy, F. Koushanfar, I.L. Markov, Ending piracy of integrated circuits. IEEE Comput. **43**(10), 30–38 (2010)

4. J. Rajendran, Y. Pino, O. Sinanoglu, R. Karri, Security analysis of logic obfuscation, in *IEEE/ACM Design Automation Conference* (2012), pp. 83–89

5. Y. Xie, A. Srivastava, Mitigating SAT attack on logic locking *International Conference on Cryptographic Hardware and Embedded Systems* (2016), pp. 127–146

6. X. Xu, B. Shakya, M.M. Tehranipoor, D. Forte, Novel bypass attack and BDD-based tradeoff analysis against all known logic locking attacks, in *Cryptographic Hardware and Embedded Systems* (2017), pp. 189–210

7. M. Yasin, A. Sengupta, M.T. Nabeel, M. Ashraf, J. Rajendran, O. Sinanoglu, Provably-secure logic locking: from theory to practice, in *ACM SIGSAC Conference on Computer & Communications Security* (2017), pp. 1601–1618

8. IHS Technology Press Release, Top 5 most counterfeited parts represent a $169 billion potential challenge for global semiconductor industry (2012) [Online]. Available http://technology.ihs.com/405654/top-

9. A. Hastings, *The Art of Analog Layout* (Pearson, London, 2005)

10. R.A. Rutenbar, Design automation for analog: the next generation of tool challenges, in *IEEE/ACM International Conference on Computer-Aided Design* (2006), pp. 458–460

11. Y.M. Alkabani, F. Koushanfar, Active hardware metering for intellectual property protection and security, in *USENIX Security Symposium* (2007)

12. J. Rajendran, M. Sam, O. Sinanoglu, R. Karri, Security analysis of integrated circuit cam-ouflaging, in *ACM SIGSAC Conference on Computer & Communications Security* (2013), pp. 709–720

13. S. Garg, J. Rajendran, *Split Manufacturing* (Springer International Publishing, Berlin, 2017)

14. P. Tuyls, G. Schrijen, B. Škorić, J. van Geloven, N. Verhaegh, R. Wolters, Read-proof hardware from protective coatings, in *Cryptographic Hardware and Embedded Systems* (2006), pp. 369–383

15. M. Integrated, DeepCover security manager for low-voltage operation with 1KB secure memory and programmable tamper hierarchy (2010), https://www.maximintegrated.com/en/products/embedded-security/security-managers/DS3660.html. Accessed 21 August 2020

16. A. Kahng, S. Mantik, I. Markov, M. Potkonjak, P. Tucker, H. Wang, G. Wolfe, Robust IP watermarking methodologies for physical design, in *IEEE/ACM Design Automation Conference* (1998), pp. 782–787

17. G. Wolfe, J.L. Wong, M. Potkonjak, Watermarking graph partitioning solutions, *IEEE/ACM Design Automation Conference* (2001), pp. 486–489

18. C.J. Alpert, A.B. Kahng, Recent directions in netlist partitioning: a survey. Integr. VLSI J. **19**(1–2), 1–81 (1995)

19. J. Lach, W.H. Mangione-Smith, M. Potkonjak, FPGA fingerprinting techniques for protecting intellectual property, in *IEEE Custom Integrated Circuits Conference* (1998), pp. 299–302

20. M. Yasin, B. Mazumdar, O. Sinanoglu, J. Rajendran, Security analysis of anti-SAT, in *IEEE Asia and South Pacific Design Automation Conference* (2017), pp. 342–347

21. J. Wang, C. Shi, A. Sanabria-Borbon, E. Sanchez-Sinencio, J. Hu, Thwarting analog IC piracy via combinational locking, in *IEEE International Test Conference* (2017), pp. 1–10

22. V.V. Rao, I. Savidis, Parameter biasing obfuscation for analog IP protection, in *IEEE Latin American Test Symposium* (2017), pp. 1–6

23. D.H.K. Hoe, J. Rajendran, R. Karri, Towards secure analog designs: a secure sense amplifier using memristors, in *IEEE Computer Society Annual Symposium on VLSI* (2014), pp. 516–521

24. K. Juretus, V. Venugopal Rao, I. Savidis, Securing analog mixed-signal integrated circuits through shared dependencies, in *ACM Great Lakes Symposium on VLSI* (2019), pp. 483–488

25. J. Leonhard, M. Yasin, S. Turk, M.T. Nabeel, M.-M. Louërat, R. Chotin-Avot, H. Aboushad, O. Sinanoglu, H.-G. Stratigopoulos, MixLock: securing mixed-signal circuits via logic locking, in *IEEE/ACM Design Automation and Test in Europe* (2019)

26. N.G. Jayasankaran, A.S. Borbon, E. Sanchez-Sinencio, J. Hu, J. Rajendran, Towards provably-secure analog and mixed-signal locking against overproduction, in *IEEE/ACM International Conference on Computer-Aided Design* (2018), pp. 7:1–7:8
27. A. Ash-Saki, S. Ghosh, How multi-threshold designs can protect analog IPs, in *IEEE International Conference on Computer Design* (2018), pp. 464–471
28. S.G. Rao Nimmalapudi, G. Volanis, Y. Lu, A. Antonopoulos, A. Marshall, Y. Makris, Range-controlled floating-gate transistors: a unified solution for unlocking and calibrating analog ICs, in *IEEE/ACM Design Automation and Test in Europe* (2020), pp. 286–289
29. G. Volanis, Y. Lu, S.G.R. Nimmalapudi, A. Antonopoulos, A. Marshall, Y. Makris, Analog performance locking through neural network-based biasing, in *IEEE VLSI Test Symposium* (2019), pp. 1–6
30. R. Torrance, D. James, The state-of-the-art in semiconductor reverse engineering, in *IEEE/ACM Design Automation Conference* (2011), pp. 333–338
31. C.S. Chang, C.P. Chao, J.G.J. Chern, J.Y.C. Sun, Advanced CMOS technology portfolio for RF IC applications. IEEE Trans. Electron Devices 52(7), 1324–1334 (2005)
32. Texas Instruments, Universal active filter. https://www.ti.com/lit/ds/symlink/uaf42.pdf (2010). Accessed 25 April 2020
33. Chipworks, Reverse engineering software. http://www.chipworks.com/en/technical-competitive-analysis/resources/reerse-engineering-software (2016)
34. T. Iizuka, *CMOS Technology Scaling and Its Implications* (Cambridge University Press, Cambridge, 2015)
35. C. Toumazou, G. Moschytz, B. Gilbert, *Trade-Offs in Analog Circuit Design* (Kluwer Academic Publishers, Dordrecht, 2002)
36. J.W. Tschanz, J.T. Kao, S.G. Narendra, R. Nair, D.A. Antoniadis, A.P. Chandrakasan, V. De, "Adaptive body bias for reducing impacts of die-to-die and within-die parameter variations on microprocessor frequency and leakage. IEEE J. Solid State Circ. 37(11), 1396–1402 (2002)
37. S. Shin, K. Kim, S. Kang, Memristor applications for programmable analog ICs. IEEE Trans. Nanotechnol. 10(2), 266–274 (2011)
38. B. Long, J. Ordosgoitti, R. Jha, C. Melkonian, Understanding the charge transport mechanism in VRS and BRS states of transition metal oxide nanoelectronic memristor devices. IEEE Trans. Electron Devices 58(11), 3912–3919 (2011)
39. Degate, http://www.degate.org/documentation/
40. Berkeley Predictive Technology Model (PTM), http://ptm.asu.edu/
41. S. Lee, C. Shi, J. Wang, A. Sanabria, H. Osman, J. Hu, E. Sánchez-Sinencio, A built-in self-test and in situ analog circuit optimization platform. IEEE Trans. Circuits Syst. Regul. Pap. PP(99), 1–14 (2018)
42. J. Wang, C. Shi, E. Sanchez-Sinencio, J. Hu, Built-in self optimization for variation resilience of analog filters, in *IEEE Computer Society Annual Symposium on VLSI* (2015), pp. 656–661
43. I. Guerra-Gómez, E. Tlelo-Cuautle, L.G. De La Fraga, Richardson extrapolation-based sensitivity analysis in the multi-objective optimization of analog circuits. Appl. Math. Comput. 222, 167–176 (2013)
44. B. Razavi, *RF Microelectronics* (Pearson Education, London, 2011)
45. M.V.V. Rolf Schaumann, Haiqiao Xiao, *Design of Analog Filters* (Oxford University Press, Oxford, 2009)
46. T. Instruments, Understanding low drop out (LDO) regulators, https://bit.ly/37bvvyE (2006). Accessed 24 Jan 2020
47. P. Subramanyan, S. Ray, S. Malik, Evaluating the security of logic encryption algorithms, in *IEEE International Symposium on Hardware Oriented Security and Trust* (2015), pp. 137–143
48. V. Srinivasan, G.J. Serrano, J. Gray, P. Hasler, A precision CMOS amplifier using floating-gate transistors for offset cancellation. IEEE J. Solid State Circuits 42(2), 280–291 (2007)
49. J. Leonhard, A. Sayed, M. Louërat, H. Aboushady, H. Stratigopoulos, Analog and mixed-signal IC security via sizing camouflaging, in *IEEE Transactions on Computer-Aided Design of Integrated Circuits and Systems* (2020), pp. 1–1

50. N.G. Jayasankaran, A.S. Borbon, A. Abuellil, E. Sanchez-Sinencio, J. Hu, J. Rajendran, Breaking analog locking techniques via satisfiability Modulo theories, in *IEEE International Test Conference* (2019)

51. M. Yasin, S.M. Saeed, J. Rajendran, O. Sinanoglu, Activation of logic encrypted chips: pre-test or post-test?, in *IEEE/ACM Design, Automation Test in Europe* (2016), pp. 139–144

52. A. Sanabria-Borbon, N.G. Jayasankaran, S. Lee, E. Sanchez-Sinencio, J. Hu, J. Rajendran, Schmitt trigger-based key provisioning for locking analog/RF integrated circuits, in *IEEE International Test Conference* (2020).

53. M. Yasin, B. Mazumdar, O. Sinanoglu, J. Rajendran, Removal attacks on logic locking and camouflaging techniques, in *IEEE Transactions on Emerging Topics in Computing* (2017), pp. 1–1

54. K. Azar, H. Kamali, H. Homayoun, A. Sasan, SMT attack: next generation attack on obfuscated circuits with capabilities and performance beyond the SAT attacks. IACR Trans. Cryptographic Hardw. Embed. Syst. **2019**(1), 97–122 (2018)

55. Y. Xie, A. Srivastava, Delay locking: security enhancement of logic locking against IC counterfeiting and overproduction, in *IEEE/ACM Design Automation Conference* (2017), pp. 1–6

56. F. Yang, M. Tang, O. Sinanoglu, Stripped functionality logic locking with hamming distance-based restore unit (SFLL-hd) – unlocked. IEEE Trans. Inf. Forensics Secur. **14**(10), 2778–2786 (2019)

57. D. Sirone, P. Subramanyan, Functional analysis attacks on logic locking, in *IEEE/ACM Design Automation and Test in Europe* (2019), pp. 936–939

58. N.G. Jayasankaran, A. Sanabria-Borbón, A. Abuellil, E. Sánchez-Sinencio, J. Hu, J. Rajendran, Breaking analog locking techniques, in *IEEE Transactions on Very Large Scale Integration Systems* (2020), pp. 1–14

59. R.Y. Acharya, S. Chowdhury, F. Ganji, D. Forte, Attack of the genes: finding keys and parameters of locked analog ICs using genetic algorithm (2020). arXiv:2003.13904

60. M. Yasin, O. Sinanoglu, Transforming between logic locking and IC camouflaging, in *IEEE International Design Test Symposium* (2015), pp. 1–4

61. K. Shamsi, M. Li, T. Meade, Z. Zhao, D.Z. Pan, Y. Jin, AppSAT: approximately deobfuscating integrated circuits, in *IEEE International Symposium on Hardware Oriented Security and Trust* (2017), pp. 95–100

62. R. Sotner, J. Jerabek, N. Herencsar, K. Vrba, T. Dostal, Features of multi-loop structures with OTAs and adjustable current amplifier for second-order multiphase/quadrature oscillators. Int. J. Electron. Commun. **69**(5), 814–822 (2015)

63. Z. Zahir, G. Banerjee, A multi-tap inductor based 2.0-4.1 GHz wideband LC-oscillator, in *IEEE Asia Pacific Conference on Circuits and Systems* (2016), pp. 330–333

64. R. Senani, D.R. Bhaskar, V.K. Singh, R.K. Sharma, *Sinusoidal Oscillators and Waveform Generators using Modern Electronic Circuit Building Blocks* (Springer International Publishing, Berlin, 2015)

65. Texas Instruments, AWR1243 Single-Chip 77-GHz and 79-GHz FMCW Transceiver, http://www.ti.com/lit/ds/symlink/awr1243.pdf (2012), Accessed 6 Jan 2020

66. J.P. Lang, iSAT3 (2017). https://projects.informatik.uni-freiburg.de/projects/isat3/wiki/ISAT3_004. Accessed 6 Jan 2020

67. J. Leonhard, M.-M. Louërat, H. Aboushady, O. Sinanoglu, H.-G. Stratigopoulos, Mixed-signal hardware security using MixLock: demonstration in an audio application, in *IEEE International Conference on Synthesis, Modeling, Analysis and Simulation Methods and Applications to Circuit Design* (2019)

Chapter 17
Application of Optical Techniques to Hardware Assurance

Leonidas Lavdas, M. Tanjidur Rahman, and Navid Asadizanjani

17.1 Introduction

Over the past few years, it has become commonplace for ICs to be produced with over a dozen metal interconnect layers. Aside from mainly connecting circuit components, the increasingly complex interconnect also provides a side benefit of additional protection against an adversary optically invading the die from chip frontside. The downside is that from the FA point of view, previous silicon debugging and diagnosis (SDD) techniques from the frontside have thus become ineffective and expensive in interacting with the substrate layer.

To accommodate this downfall in debugging, new FA techniques have been invented that take advantage of the silicon backside. Different optical FA techniques, which include methods like PEA [37, 43], LFI [39, 42], Electro-Optical Probing (EOP) [17, 25], Electro-Optical Frequency Mapping (EOFM) [17, 25], Optical Beam Induced Resistance Change (OBIRCH) [36], etc., can directly interact with the active device layer in the substrate without having to navigate through the vast interconnect layers. The FA techniques are based on the principle of silicon transparency to near-infrared (NIR) photons. The techniques also arise at a time when compact design requirements, advanced integration, and heterogeneous packaging are inevitable. In flip-chip packaging, the die is inserted into the package upside-down, with the highest metal layers being closer to the bottom of the package. Therefore, removing the heatsink placed over the flip-chip is enough to expose the silicon backside. Furthermore, with surface polishing, the FA methods can interface with the active substrate regions.

L. Lavdas · M. T. Rahman · N. Asadizanjani (✉)
Department of ECE, University of Florida, Gainesville, FL, USA
e-mail: leonidas.lavdas@ufl.edu; mir.rahman@ufl.edu; nasadi@ece.ufl.edu

© The Author(s), under exclusive license to Springer Nature Switzerland AG 2021
M. Tehranipoor (ed.), *Emerging Topics in Hardware Security*,
https://doi.org/10.1007/978-3-030-64448-2_17

Unfortunately, this leads back to the original problem securitywise, where the adversaries can take advantage of methods to optically probe real-time circuit activity with greater ease. This can have devastating consequences, as confidential assets previously shown to be secure against other physical attacks are now susceptible to optical attacks [8, 21, 29, 38, 42, 47, 50].

This chapter starts with a general background of optical FA techniques. Next we discuss innovative security threats through the application of optical methods. Logic locking and state space obfuscation attacks are explained in detail. Finally, a case study on hardware assurance through the same FA methods is explored, pertaining to the prevention of sequential hardware Trojans.

17.2 Taxonomy and Overview of Optical Techniques

Optical inspection and attacks are based on the fact that silicon is transparent to near-infrared (NIR) photons. Information about circuit activities is gained through the use of photons emitted from a transistor, photons modulated by the activity of a transistor, or perturbations in circuit activity from photon stimulation [5, 6, 53]. As device complexity increases and feature size decreases, these non-invasive to semi-invasive attacks have gained a comparative advantage over invasive attacks, like electrical probing [28]. Figure 17.1 depicts different classes of optical inspection/attacks.

17.2.1 Photon Emission Analysis (PEA)

PEA is a passive optical inspection/attack approach. The principle of PEA comes from the nature of transistors. When logic gates are in the process of switching values, the internal transistors are undergoing a change in their region of operation. Briefly, a given transistor will pass through the saturation region. During this time, the charge carriers—electrons and holes—within the transistor gain an increase in

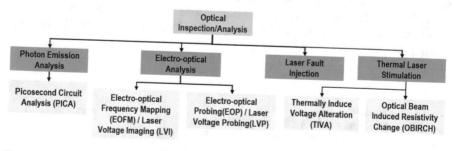

Fig. 17.1 Taxonomy of optical attacks/inspection techniques [27]

kinetic energy. This energy is then released near the channel pinch-off region as hot-carrier luminescence (see Fig. 17.2a), i.e., the emission of photons, which is picked up by a Si-CCD or InGaAs detector [26, 37]. With the help of additional computer processing, a 2-D image is constructed that maps the switching activity of a Region of Interest (RoI). Combining this with the fact that NMOS switching yields noticeably more energy than PMOS switching due to the higher mobility of the charge carrier, an attacker can utilize the switching map to understand logical changes and signal propagation in the given area [27, 37, 46]. While PEA reveals the time-integrated spatial activities of the circuit, picosecond circuit analysis (PICA) can extract the time-resolved spatial activity information of the chip. A microchannel plate imaging photomultiplier detector is used for PICA imaging [22, 38]. The chip is stimulated electrically with a tester repeatedly to collect the transistor switching information in time-domain in PICA analysis.

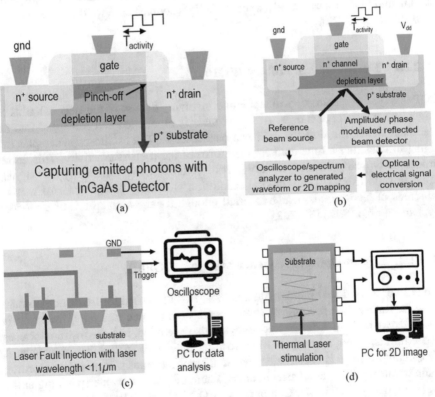

Fig. 17.2 (a) Setup used for Photon Emission Analysis. During N-type MOSFET switching, emitted photons are captured using an InGaAS detector and a 2D mapping of the device is generated, (b) optically probing a transistor operating at T_{active} frequency. The reflected modulated laser signal is converted into an electrical signal and fed to an oscilloscope/spectrum analyzer for EOFM/EOP analysis, (c) Laser Fault Injection setup using photo-current stimulation and oscilloscope, (d) setup for thermal laser stimulation to observe OBIRCH or Seebeck effect [27]

17.2.2 Electro-Optical Technique

Electro-optical technique, or optical probing, is an active method used to determine the state of transistors or logic gates. Rather than passively measuring energy like in PEA, it instead utilizes a laser stimulus that travels through the silicon backside. Once passed through the substrate, the laser is reflected at the next interface in the transistor it is currently pointed at, e.g., active region or interconnects. The reflected laser undergoes an amplitude and phase modulation due to the switching activities of the electric field and free carrier density [17, 18, 60]. The modulated optical signal is then converted with an avalanche photodiode into an electrical signal [17, 25]. The electrical signal is fed to an oscilloscope or spectrum analyzer depending on the modality of the electro-optical measurement—EOP or EOFM (see Fig. 17.2b). Other approaches for optical probing are Laser Voltage Probing (LVP) and Laser Voltage Imaging (LVI). The LVP and LVI methods are equivalent to EOP and EOFM, respectively, except the light sources used for the latter ones are incoherent [6, 27, 28].

17.2.2.1 Electro-Optical Probing (EOP)

EOP is a type of electro-optical technique involving the use of the laser stimulus to target one transistor at a time. After the laser is modulated and converted into an electrical signal, the signal is fed to an oscilloscope. The oscilloscope samples the signal to construct a time-domain waveform of the transistor activity. This entire process is typically done repeatedly in a loop to overcome a naturally low signal-to-noise ratio (SNR). The temporal resolution is usually on the order of tens or hundreds of picoseconds, which is small enough for the high speeds of a modern System-on-chip (SoC) [3, 17, 25].

17.2.2.2 Electro-Optical Frequency Mapping (EOFM)

While EOP is used for time-domain analysis, EOFM is used for frequency-domain analysis. Rather than focusing the laser beam on only one transistor, the beam is swept across an RoI. A target frequency is also specified, and the instrument uses a spectrum analyzer, a bandpass filter, to filter the signal around that frequency. A 2D computer image is constructed that maps out the transistor activity according to the frequency filter, emphasizing areas where the transistors are switching at the target frequency [3, 25, 27]. Compared to EOP, EOFM offers faster measurement with higher SNR [17].

17.2.3 Laser Fault Injection (LFI)

Unlike the observational role of the aforementioned techniques, LFI introduces the ability to modify the logical state of a target region. LFI relies on the manipulation of an infrared laser with higher energy than silicon's bandgap energy, typically 1.1 eV. This property allows for the formation of electron–hole pairs inside the silicon. This phenomenon is commonly referred to as Photoelectric Laser Stimulation (PLS) (see Fig. 17.2c). When the laser is directed toward the source or drain of a transistor in a CMOS circuit, the PLS effect causes a change in the CMOS' logical state, introducing a fault [39, 42]. The overall success of LFI is not guaranteed because it depends on variables such as exposure time and power. However, there are powerful ramifications when properly executed, like in previous works where faults were injected in embedded microcontrollers and FPGAs [27, 47, 53].

17.2.4 Thermal Laser Stimulation (TLS)

TLS is an optical technique that, unlike LFI, utilizes a laser with lower energy than silicon's bandgap energy. Thus, instead of stimulating a PLS effect, a thermal effect is stimulated in the form of localized heating. This heating influences the test device's electrical parameters, e.g., the voltages and currents at select pins, which are then monitored, analyzed by a computer, and plotted (see Fig. 17.2d). TLS is commonly used in OBIRCH and thermally induced voltage alteration (TIVA) [36] and is capable of successfully exposing secret keys and other assets in the IC [21].

17.3 Essential Steps for Chip Backside Attacks

There is a series of crucial steps an attacker performs to successfully execute an optical backside attack.

Primarily, the adversary must ensure that there is a non-obstructive path from the instrument's objective lens to the target transistors. This would not be the case if the frontside of the die were used, due to the interconnect layers and additional shield structures used for protection [6]. The backside may have a laser-engraved device marking, which can reduce the optical quality, but this can be removed by polishing. This follows for other passive obstructions [27].

Additionally, a successful optical backside attack involves asset localization. This typically is dependent on whether the adversary has access to the GDSII layout for the chip. Ideally, the layout is available, and subsequently, full-blown reverse engineering is possible. Otherwise, the toggling of various combinational logic and sequential elements is critical to revealing the assets through PEA, EOFM, EOP, and OBIRCH [5, 27, 30].

Finally, the main portion of the attack is performed, in which one of the aforementioned optical analysis methods is chosen to extract circuit information (refer to Sects. 17.2 and 17.4 for attack details) [27].

17.4 Security Threats Through Optical Techniques

In recent years, optical attacks have been successful in breaking into different security-critical applications like cryptography modules [38, 39, 50], secure boot-up [54], cache or non-volatile memory [21, 41, 43], physically unclonable functions (PUFs) [46–48], neural network chips [19], registers, and logic gates [20, 29]. In response, new protective countermeasures have been developed.

Logic locking [11, 14, 23, 34, 51] is a method that has become a renewed area of focus from both the protection and attack perspectives. The method consists of inserting additional logic elements into a circuit in an attempt to obscure its gate-level functionality and prevent IP piracy, counterfeiting, and overproduction. Logic locking has two main implementations: combinational and sequential. Recently, it has been proven to be vulnerable to optical probing attacks [29, 30].

17.4.1 Combinational Logic Locking

17.4.1.1 Background and Implementation

In combinational logic locking, new logic gates called *key gates* are inserted in between the preexisting gates in combinational blocks [34] (see Fig. 17.3). Each of the obfuscating key gates involves a new control signal that, when combined with all the other key-gate control signals, forms a general *key* for "unlocking" the circuit's original functionality.

The key is not available during the IC fabrication process. It is programmed into an NVM (e.g., Flash, EEPROM, e-fuse/one-time programmable memory) inside the chip before releasing to market. Consequently, the correct functionality is hidden from an untrusted foundry during manufacturing. Key-gate insertion into the IP needs to consider the design implementation and functionality of the chip to maximize security. Therefore, several key-gate insertion algorithms, based on the XOR/XNOR gates [31, 33, 34], multiplexers [33], and lookup tables [4], have been proposed. However, XOR/XNOR based logic locking offers higher output corruptibility at lower area, power, and delay overhead [56].

Overall, the end product of a logic locked circuit is an entanglement of the original gates and the obfuscation gates, along with signal paths to intermediary key registers and then the tamper-evident memory where the key is stored [7, 9, 24, 58].

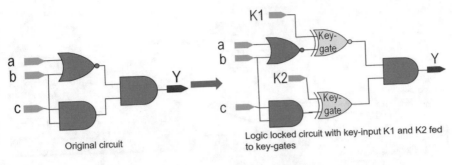

Fig. 17.3 Simplified example of combinational logic locking [30]

17.4.1.2   Attack Model

Before detailing possible attack methods, we first establish an attack model. An untrusted foundry serves as a realistic adversary against logic locking. Foundries typically have access to FA equipment that conducts the previously discussed optical methods, along with unlocked chips and their GDSII netlists [29]. Unfortunately, due to time-to-market and cost constraints, part of the IC design and fabrication process is often outsourced to other foundries in the world. Many of the outsourced foundries are not considered reliable entities in the IC development process and pose a serious risk to protected IP in terms of reverse engineering, overselling, counterfeiting, etc. [12]. Another aspect of the attack model is motivation. Complete reverse engineering of the IP is a rewarding goal for an attacker, as the IP functionality could be stolen for ransom and/or counterfeit production. In such cases, an SoC integrator or a competitor can also be a potential attacker. This threat model is true for both combinational and sequential logic obfuscation.

17.4.1.3   Security Assessment Against Optical Probing

Optical technique factors in to logic locking when attempting to extract the key data. Previously, vulnerability analysis of logic obfuscation has focused on algorithmic attacks, such as key sensitization [32], Boolean satisfiability (SAT) [45], Signal Probability SKey attacks (SPS) [59], and bypass attacks [57]. However, optical attacks leave a wide attack surface available for an adversary to break into key-based IP protection schemes such as logic locking. The security of the tamper-evident memory may be a valid assumption, but the propagation of the key from memory to the gate-driving signals has been shown to be compromised [29].

Physical access to an unlocked chip is required to extract the unlocking key from the IC. This is straightforward both for an untrusted foundry that produces the chips, and most other adversaries since unlocked chips can be purchased from the open market or other untrusted entities in the supply chain. The second requirement is access to the chip's backside, which will be the interface for the optical attack.

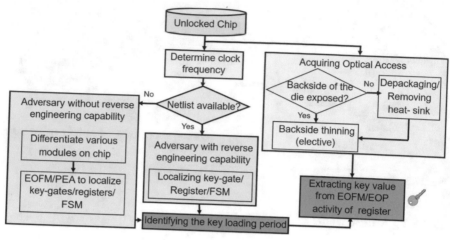

Fig. 17.4 Simplified illustration of key extraction methodology from logic obfuscated circuitry using optical contactless probing [29]

Again, this is easier to obtain if the package is flip-chip. If it is not flip-chip, the adversary can still polish away the packaging material and acid-etch the bulk silicon if necessary.

Once proper physical access is acquired, the attacker will move on to the next part of the attack: localizing the key gates, and the key registers that propagate the key from memory to the key gates. In the untrusted foundry attack model, the adversary has access to the GDSII layout file and thus is able to fully reverse engineer the location of the key gates and key registers. For an attacker without the capability of performing full-blown reverse engineering (like an end user), partial reverse engineering is still possible to find the key registers. To do this, the attacker first performs a visual analysis, using reflected light images from a 1.3 μm laser probe to differentiate target logic areas from memory and cache blocks. Then, EOFM is used to measure activity during secure boot-up of the chip. The key registers always load the same key values independent of other input signals, so their EOFM boot-up activity will differentiate them from other data registers [10, 29, 30]. Figure 17.4 shows the necessary steps for extracting the unlocking key using optical probing.

Finally, the attacker performs another EOFM analysis of the newly localized key registers, while the chip is stimulated with the reset loop in Fig. 17.5a. With this waveform being inputted to the chip's reset pin, the chip is forced to reset repeatedly with a reset frequency that is the new EOFM target frequency. The propagated key bit values of logical "1" will stand out as active nodes in the EOFM map, since the flip flops are toggling at the reset frequency. Meanwhile, the logical "0" key bits appear as inactive nodes since they are not changing from the reset value of "0" (see Fig. 17.5b). With the advantage of such knowledge, the attacker is capable of extracting all bit values in the secret key, which can now be used to complete the reverse engineering of the functional IP, or even just be used for ransom. Similar to

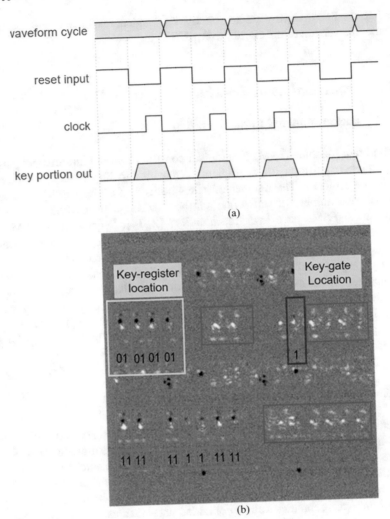

Fig. 17.5 (a) The chip is reset at a certain frequency in a loop. The EOFM signal is collected at the reset frequency. (b) The EOFM activity at clock frequency and reset frequency, where black and white dots represent clock and logic element activity, respectively. The value stored in each register is mentioned at the bottom of the corresponding register. The registers in the yellow box represent the key-register location, while the registers in the red box represent the registers responsible for the reset signal [29]

EOFM analysis, EOP can also be used to extract each bit of the key. However, if EOP were used for the last step instead of EOFM, separate measurements would be necessary for every key bit [20, 29, 50].

Another recent work [15] has demonstrated a different methodology in extracting key data from combinational logic obfuscation. Laser Fault Injection is used to inject faults into the lines carrying the key-bit signals. Logical response is then compared to a fault-free version of the circuit, using ATPG test tools.

17.4.2 Sequential Logic Locking

17.4.2.1 Background and Implementation

A Finite State Machine (FSM) is a logical construct representing distinct "states" and associated transition logic, implemented with flip flop based state-bits and combinational blocks. The number of states utilized in a given FSM is often far below the total number of states that can be encoded with the FSM's state bits, meaning most FSMs have low "reachability" [2, 14]. Sequential or FSM logic locking takes advantage of the commonly low reachability by inserting additional states and state transitions into the state space. The additions do not add useful functional purpose but exist to obscure the original design. This methodology, sometimes referred to as state space obfuscation, complicates malicious attempts at IP piracy or tampering of the original circuit, while providing control over when and how the FSM can be unlocked into its original functional mode [11]. Three types of state space obfuscation will be considered: HARPOON [7], Interlocking [9], and Entangled [24] (see Fig. 17.6).

17.4.2.2 HARPOON Obfuscation

Obfuscation Methodology HARPOON [7] aims to obfuscate and authenticate designs by modifying transition logic and internal logic structures. In HARPOON, the state space is divided into two regions, the "pre-initialization" space and the "post-initialization" space (see Fig. 17.6a). The "post-initialization" space simply contains the states of the original, unobfuscated FSM. The other region, the "pre-initialization" space, contains a cluster of added states that lead up to the first state of post-initialization. A specific order of states must be followed in pre-initialization to get to post-initialization. If any incorrect transition is taken, the state machine will be redirected over the next few cycles toward the very beginning of the pre-initialization region.

The correct path traversal is accomplished with the application of a key input sequence. The portions of the key are sequentially loaded as part of the primary inputs to the transition logic.

Security Assessment Against Optical Attack The HARPOON-locked circuit is considered broken once the attacker localizes the pre-initialization region and obtains the key input sequence used in pre-initialization. Once the adversary has access to the IP netlist, then he/she focuses on locating the registers that implement pre-initialization. Such registers are identified using gate-level reverse engineering

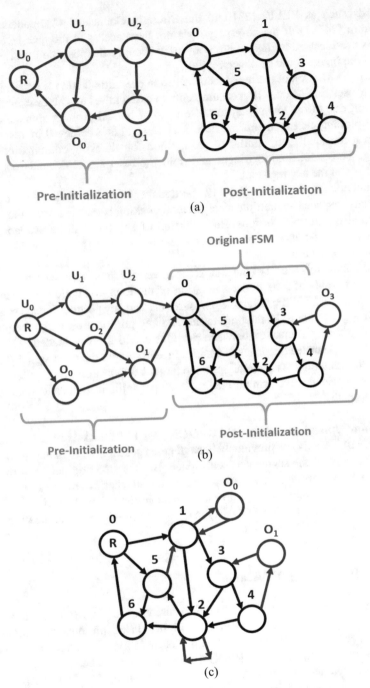

Fig. 17.6 (a) An example HARPOON-type state space, (b) an example Interlocking-type state space, (c) an example Entangled-type state space

methods such as RELIC [23] and the Strongly Connected Components (SCC) algorithm [24, 51]. If the adversary does not have access to the gate-level netlist, PEA is used instead during secure boot-up to identify the possible locations, or an optical die image could be observed [30].

Once the location of the registers involved in pre-initialization is determined, the attacker uses EOFM or EOP measurements to extract the key values from the chip. However, due to the low signal-to-noise ratio of optical probing signals, multiple optical probing measurements are necessary [29]. This is achieved by resetting the chip in a repeated loop, similar to the method for attacking combinational logic locking. The exact time to trigger and hold the reset signal depends on the key availability at the key registers.

To understand the key availability, the attacker must acknowledge that the pre-initialization space has multiple states in its unlocking path. Therefore, the key will be fetched in multiple portions. Each portion of key bits will be available at the state-element input once the previous pre-initialization state has been unlocked with the previous key portion.

Recognizing this aspect of the attack, the adversary must keep the circuit enabled for longer periods of time in each iteration of the loop, to "build up" to each key portion (see Fig. 17.7). For example, consider the 2nd key portion. If each waveform cycle simply lets the FSM fetch up to the 2nd key portion repeatedly, then there is no distinguishable frequency between portions 1 and 2 being loaded. Thus, each waveform cycle instead consists of multiple resets, where the reset signal is held at "0" for different lengths to iterate up to the target key portion. Overall, there will be a separate repeated waveform utilized for each pre-initialization state in the correct unlocking path.

Additional Optical Attack on HARPOON An LFI attack is also important to consider. HARPOON has previously been deemed susceptible to fault injection that takes advantage of the segregated state space. Because the original FSM region is clearly distinct from the new startup region, an attacker is able to inject a laser fault into the state register to jump into a state in the original FSM [24]. This is another disadvantage of HARPOON-type obfuscation—the pre-initialization region is distinct enough to be bypassed.

17.4.2.3 Interlocking Obfuscation

Obfuscation Methodology Interlocking obfuscation [9] is a type of state space obfuscation developed after HARPOON. Figure 17.6b presents an example of the state space implemented in Interlocking obfuscation. It contains two separate state space regions, similar to HARPOON's pre-initialization and post-initialization. The main difference in Interlocking is that additional obfuscation states are also inserted in the post-initialization region with the original FSM. Furthermore, in the startup region, every path eventually leads to entering the original FSM's reset

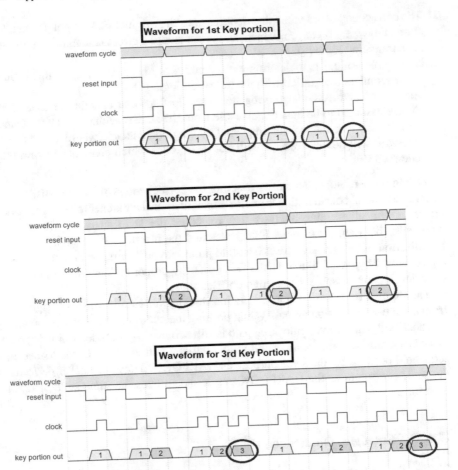

Fig. 17.7 The reset signal going into the chip is forced high and low with different input waveforms. For the first key portion, simply alternating the reset signal symmetrically is sufficient, for the same reasons as the combinational attack. Subsequent key portions, however, have to be "built up" to. Notice how the frequency of each target key portion being loaded matches the frequency of each waveform cycle

state. However, only one path unlocks the original FSM's correct functionality, by disabling the extra obfuscation states in post-initialization.

Security Assessment Against Optical Probing An attacker may initially consider targeting a new feature in Interlocking obfuscation: codeword generation. As the startup states are traversed, a "codeword" is dynamically developed to potentially deobfuscate the original FSM [9]. While there is a predefined "correct" value for the codeword to unlock the FSM, there are three factors that prevent an attacker from discovering the correct codeword:

1. The correct codeword value that deobfuscates the original FSM is not physically stored anywhere on the chip. Instead, the codeword's effect is integrated into the state transition logic [9].
2. No comparator exists to determine if the codeword is correct. Again, this is due to the seamless integration with the transition logic [9].
3. There is no defined codeword register. The flip flops that contain the codeword bits are assumed to be indistinguishable from the state machine elements. Thus, even though we are considering an Oracle-guided attack where the correct codeword will be developed, it will not be obvious which elements contain the codeword [9].

Due to these reasons, optically extracting the correct codeword is unrealistic.

Another initial consideration is an LFI attack in a similar manner to HARPOONs. However, the algorithmic analysis that recreates the state space would fail to distinguish between the original FSM states and the inserted obfuscations in post-initialization. If the attacker injects a fault to jump into post-initialization, the startup region is still bypassed like in HARPOON, but the proper codeword will not have been developed to disable the inserted obfuscations.

The solution to successfully attacking Interlocking obfuscation instead originates from the pre-initialization unlock. While probing the codeword has been eliminated as a possibility, the correct unlocking path is still achieved by a predetermined series of primary inputs. Namely, a general unlocking key still exists, and it is loaded in sequential portions for the same purpose as HARPOON. Thus, the hacker performs the same procedure as the HARPOON attack and utilizes the same style of input waveform.

17.4.2.4 Entangled Obfuscation

Obfuscation Methodology Lastly, entangled obfuscation [24] is a recent yet conceptually simpler form of obfuscation. Unlike the aforementioned obfuscation types, the Entangled form contains only one region in the state space, consisting of the original FSM interwoven with additional obfuscation states (see Fig. 17.6c). This more direct form of obfuscation is an effective countermeasure to purely algorithmic attacks.

Security Assessment Against Optical Probing The HARPOON and Interlocking waveform is reduced to a simpler form for Entangled, to accommodate for the lack of a pre-initialization path. We make the assumption that a generic unlocking key still exists for Entangled since there are additional obfuscation states that must be unlocked upon reset [24]. A design that fetches the key in portions would be more complicated than simply unlocking everything upon reset, so we also assume there is a one-time key loading for Entangled. Therefore, the attack waveform does not need different iterations of reset cycles for various key portions. Instead, a simple on-and-off reset pattern suffices. While the application is different than the combinational logic encryption, it is effectively the same input waveform that is utilized to extract the key data, along with the same general procedure.

If the assumption of a one-time key loading is incorrect, the attacker instead utilizes the same waveform methodology as HARPOON and Interlocking, with the exception being that there are multiple paths that unlock different portions of the FSM.

17.5 Potential Countermeasures Against Optical Attacks

Optical attacks are hindered by protecting the backside of the chip at all levels. In terms of packaging, opaque active layers can be added on to the backside as extra protection. Since they are still removable, it is best to include an additional monitoring scheme at the device level [1]. Photosensors are a sufficient first step to detect optical probing but can be circumvented by thermal lasers. To identify unauthorized thermal stimulation, circuits like Ring Oscillators (ROs) with a Phase Locked Loop (PLL) monitor [13, 49] are useful at detecting and responding to temperature and current variations. An alternative protection scheme uses pyramid-shaped nanostructures in sensible locations to scatter the reflected laser beam and scramble its measurements [40] (see Fig. 17.8).

Introducing randomization to the reset state of the key registers is another way to dismantle an optical probing attack. This promising countermeasure is based on the use of a True Random Number Generator (TRNG) and additional control circuitry. Similarly, the clock frequency during chip boot-up can also be randomized. The disadvantage is the additional area and delay overhead, along with the fact that an untrusted foundry is capable of localizing and removing the TRNG through reverse engineering and circuit edit [28].

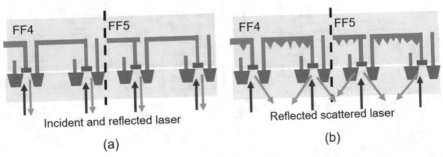

Fig. 17.8 Incident laser beam is scrambled by nanopyramids to disturb the signal for optical probing [40]

17.6 Hardware Assurance Through Optical Techniques

While the previous discussion demonstrated how optical techniques are a security threat, they are also a means of hardware assurance. We now shift the view point from an attacker trying to deobfuscate a system to an IC design entity concerned about hardware breaches. A recent paper [44] will be the highlight of this section, since it is an example of new consequential optical research involving the detection of sequential hardware Trojans.

Hardware Trojans are a type of malicious element inserted into an existing circuit, often used to steal protected information or disable the circuit's functionality after a period of time [44]. Hardware Trojans are also characterized by their relatively rare operation. In order to remain as inconspicuous as possible, they are designed to be activated by low-occurrence triggers, like sensing a specific temperature signal [52].

17.6.1 Hardware Trojans

Hardware Trojans are also categorized as combinational or sequential Trojans, as shown in Fig. 17.9. Unsurprisingly, this correlates to whether they are implemented in combinational or sequential circuitry [52]. SPARTA-COTS [44] focuses on

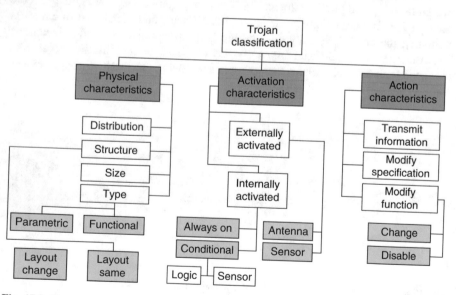

Fig. 17.9 Taxonomy showing various characteristics and classifications of hardware Trojans [52]

sequential Trojans, introducing an optical probing method for detecting Trojan insertion into a preexisting design.

17.6.2 SPARTA-COTS

Previous research before SPARTA-COTS yielded mediocre methods for detecting and deterring sequential Trojans [16, 35, 55]. For example, design methodologies were proposed where various modules are designed separately and test points are added to circuit nets. This has several disadvantages in terms of cost, design-constraint, and even confidentiality risks. Furthermore, many detection methods after fabrication are destructive to the chips being studied, another large drawback in both cost and time [44]. SPARTA-COTS introduces a straightforward, non-destructive approach to sequential Trojan detection. Without having full knowledge of a given commercial-off-the-shelf (COTS) chip's functionality, optical probing is used to uncover the malware [44].

SPARTA-COTS relies on the scan chain, a common design tool used to verify digital IC components. The scan chain consists of a long array of flip flops that rapidly propagate an automatically generated test pattern. The flip flops are connected to the logic gates that form the designer's IP. Obviously, the designer has no motive to insert a Trojan into their own design. Furthermore, the malicious entity who inserts the Trojan later on in development wants it to remain hidden. Thus, the adversary generally does not connect the Trojan's components to the scan chain, as the scan chain allows for visibility of the circuit and tests for logical flaws [44].

This distinction between authentic components connected to the scan chain, and malicious components absent from the scan chain, is the key to surfacing the sequential Trojans. Recall how EOFM analyzes repeated toggling of logic elements and maps a 2D image based on a target frequency. The scan chain elements are constantly changing values as the current test pattern is propagated. By manipulating the test pattern to create an observable toggle frequency, an EOFM scan will emphasize the scan chain.

To begin a SPARTA-COTS analysis, the IC in question is powered and given a clock signal. Considering the clock frequency, EOFM maps out all of the sequential elements in the circuit, including the malicious ones. Next, the clock signal is kept the same, but a scan pattern that alternates 101010... is inputted into the scan chain. This creates an observable frequency in the scan chain elements of half the clock signal's frequency. EOFM maps out this new frequency, and then the two images are compared [44].

In reality, an IC may contain a few additional benign components that are only present in the original image, such as clock buffers. However, after multiple image processing steps and basic analysis, the image comparison reveals which sequential elements are present outside the scan chain and are likely part of a sequential Trojan (see Fig. 17.10) [44].

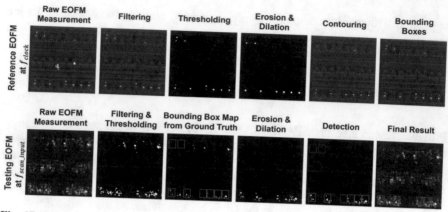

Fig. 17.10 Hardware Trojan detection using the EOFM measurements [44]

References

1. E. Amini, A. Beyreuther, N. Herfurth, A. Steigert, B. Szyszka, C. Boit, Assessment of a chip backside protection. J. Hardw. Syst. Secur. **2**(4), 345–352 (2018)
2. S. Amir, B. Shakya, X. Xu, Y. Jin, S. Bhunia, M. Tehranipoor, D. Forte, Development and evaluation of hardware obfuscation benchmarks. J. Hardw. Syst. Secur. **2**, 1–20 (2018)
3. N. Asadi, M. Tehranipoor, Optical inspection and attacks, in *Physical Assurance for Electrical Devices and Systems* (Springer, New York, 2020), pp. 53–72
4. A. Baumgarten, A. Tyagi, J. Zambreno, Preventing IC piracy using reconfigurable logic barriers. IEEE Des. Test Comput. **27**(1), 66–75 (2010)
5. C. Boit, C. Helfmeier, U. Kerst, Security risks posed by modern IC debug and diagnosis tools, in *2013 Workshop on Fault Diagnosis and Tolerance in Cryptography* (IEEE, New York, 2013), pp. 3–11
6. C. Boit, S. Tajik, P. Scholz, E. Amini, A. Beyreuther, H. Lohrke, J.P. Seifert, From IC debug to hardware security risk: the power of backside access and optical interaction, in *2016 IEEE 23rd International Symposium on the Physical and Failure Analysis of Integrated Circuits (IPFA)*. IEEE, New York (2016), pp. 365–369
7. Chakraborty, R.S., Bhunia, S.: Harpoon: an obfuscation-based soc design methodology for hardware protection. IEEE Trans. Comput.-Aided Des. Integr. Circ. Syst. **28**(10), 1493–1502 (2009)
8. F. Courbon, P. Loubet-Moundi, J.J. Fournier, A. Tria, Increasing the efficiency of laser fault injections using fast gate level reverse engineering, in *2014 IEEE International Symposium on Hardware-Oriented Security and Trust (HOST)* (IEEE, New York, 2014), pp. 60–63
9. A.R. Desai, M.S. Hsiao, C. Wang, L. Nazhandali, S. Hall, Interlocking obfuscation for anti-tamper hardware, in *Proceedings of the Eighth Annual Cyber Security and Information Intelligence Research Workshop* (2013), pp. 1–4
10. S. Engels, M. Hoffmann, C. Paar, The end of logic locking? A critical view on the security of logic locking. IACR Cryptol. ePrint Arch. **2019**, 796 (2019)
11. D. Forte, S. Bhunia, M.M. Tehranipoor, Hardware Protection Through Obfuscation (Springer, New York, 2017)
12. U. Guin, K. Huang, D. DiMase, J.M. Carulli, M. Tehranipoor, Y. Makris, Counterfeit integrated circuits: a rising threat in the global semiconductor supply chain. Proc. IEEE **102**(8), 1207–1228 (2014)

13. W. He, J. Breier, S. Bhasin, N. Miura, M. Nagata, Ring oscillator under laser: potential of PLL-based countermeasure against laser fault injection, in *2016 Workshop on Fault Diagnosis and Tolerance in Cryptography (FDTC)* (IEEE, New York, 2016), pp. 102–113

14. T. Hoque, R.S. Chakraborty, S. Bhunia, Hardware obfuscation and logic locking: a tutorial introduction. IEEE Des. Test **37**(3), 59–77 (2020)

15. A. Jain, T. Rahman, U. Guin, ATPG-guided fault injection attacks on logic locking (preprint, 2020). arXiv:2007.10512

16. Y. Jin, Y. Makris, Hardware Trojan detection using path delay fingerprint, in *2008 IEEE International Workshop on Hardware-Oriented Security and Trust* (IEEE, New York, 2008), pp. 51–57

17. U. Kindereit, Fundamentals and future applications of laser voltage probing, in *2014 IEEE International Reliability Physics Symposium* (IEEE, New York, 2014), pp. 3F–1

18. U. Kindereit, G. Woods, J. Tian, U. Kerst, R. Leihkauf, C. Boit, Quantitative investigation of laser beam modulation in electrically active devices as used in laser voltage probing. IEEE Trans. Dev. Mater. Reliab. **7**(1), 19–30 (2007)

19. Y. Liu, L. Wei, B. Luo, Q. Xu, Fault injection attack on deep neural network, in *2017 IEEE/ACM International Conference on Computer-Aided Design (ICCAD)* (IEEE, New York, 2017), pp. 131–138

20. H. Lohrke, S. Tajik, C. Boit, J.P. Seifert, No place to hide: contactless probing of secret data on FPGAs, in *International Conference on Cryptographic Hardware and Embedded Systems* (Springer, New York, 2016), pp. 147–167

21. H. Lohrke, S. Tajik, T. Krachenfels, C. Boit, J.P. Seifert, Key extraction using thermal laser stimulation, in *IACR Transactions on Cryptographic Hardware and Embedded Systems* (2018), pp. 573–595

22. M. Mc Manus, J. Kash, S. Steen, S. Polonsky, J. Tsang, D. Knebel, W. Huott, PICA: backside failure analysis of CMOS circuits using picosecond imaging circuit analysis. Microelectron. Reliab. **40**(8–10), 1353–1358 (2000)

23. T. Meade, Y. Jin, M. Tehranipoor, S. Zhang, Gate-level netlist reverse engineering for Trojan detection and hardware security, in *The IEEE International Symposium on Circuits and Systems (ISCAS)* (2016), pp. 1334–1337

24. T. Meade, Z. Zhao, S. Zhang, D. Pan, Y. Jin, Revisit sequential logic obfuscation: attacks and defenses, in *2017 IEEE International Symposium on Circuits and Systems (ISCAS)* (IEEE, New York, 2017), pp. 1–4

25. P. Perdu, G. Bascoul, S. Chef, G. Celi, K. Sanchez, Optical probing (EOFM/TRI): a large set of complementary applications for ultimate VLSI, in *Proceedings of the 20th IEEE International Symposium on the Physical and Failure Analysis of Integrated Circuits (IPFA)* (IEEE, New York, 2013), pp. 119–126

26. J.C. Phang, D. Chan, S. Tan, W. Len, K. Yim, L. Koh, C. Chua, L. Balk, A review of near infrared photon emission microscopy and spectroscopy, in *Proceedings of the 12th International Symposium on the Physical and Failure Analysis of Integrated Circuits, 2005, IPFA 2005* (IEEE, New York, 2005), pp. 275–281

27. M.T. Rahman, N. Asadizanjani, Backside security assessment of modern SoCs, in *2019 20th International Workshop on Microprocessor/SoC Test, Security and Verification (MTV)* (IEEE, New York, 2019), pp. 18–24

28. M.T. Rahman, Q. Shi, S. Tajik, H. Shen, D.L. Woodard, M. Tehranipoor, N. Asadizanjani, Physical inspection & attacks: new frontier in hardware security, in *2018 IEEE 3rd International Verification and Security Workshop (IVSW)* (IEEE, New York, 2018), pp. 93–102

29. M.T. Rahman, S. Tajik, M.S. Rahman, M. Tehranipoor, N. Asadizanjani, The key is left under the mat: on the inappropriate security assumption of logic locking schemes. IACR Cryptol. ePrint Arch. **2019**, 719 (2019)

30. M.T. Rahman, M.S. Rahman, H. Wang, S. Tajik, W. Khalil, F. Farahmandi, D. Forte, N. Asadizanjani, M. Tehranipoor, Defense-in-depth: a recipe for logic locking to prevail. Integration **72**, 39–57 (2020)

31. J. Rajendran, Y. Pino, O. Sinanoglu, R. Karri, Logic encryption: a fault analysis perspective, in *2012 Design, Automation & Test in Europe Conference & Exhibition (DATE)* (IEEE, New York, 2012), pp. 953–958

32. J. Rajendran, Y. Pino, O. Sinanoglu, R. Karri, Security analysis of logic obfuscation, in *Proceedings of the 49th Annual Design Automation Conference* (ACM, New York, 2012), pp. 83–89

33. J. Rajendran, H. Zhang, C. Zhang, G.S. Rose, Y. Pino, O. Sinanoglu, R. Karri, Fault analysis-based logic encryption. IEEE Trans. Comput. **64**(2), 410–424 (2013)

34. J.A. Roy, F. Koushanfar, I.L. Markov, EPIC: Ending piracy of integrated circuits, in *Proceedings of the Conference on Design, Automation and Test in Europe* (2008), pp. 1069–1074

35. H. Salmani, M. Tehranipoor, J. Plusquellic, A novel technique for improving hardware Trojan detection and reducing Trojan activation time. IEEE Trans. Very Large Scale Integr. Syst. **20**(1), 112–125 (2011)

36. K. Sanchez, R. Desplats, F. Beaudoin, P. Perdu, S. Dudit, M. Vallet, D. Lewis, Dynamic thermal laser stimulation theory and applications, in *2006 IEEE International Reliability Physics Symposium Proceedings* (IEEE, New York, 2006), pp. 574–584

37. A. Schlösser, D. Nedospasov, J. Krämer, S. Orlic, J.P. Seifert, Simple photonic emission analysis of AES, in *Cryptographic Hardware and Embedded Systems – CHES 2012*ed. by E. Prouff, P. Schaumont (Springer, Berlin, Heidelberg, 2012), pp. 41–57

38. A. Schlösser, D. Nedospasov, J. Krämer, S. Orlic, J.P. Seifert, Simple photonic emission analysis of AES, in *International Workshop on Cryptographic Hardware and Embedded Systems* (Springer, New York, 2012), pp. 41–57

39. J.M. Schmidt, M. Hutter, T. Plos, Optical fault attacks on AES: a threat in violet, in *2009 Workshop on Fault Diagnosis and Tolerance in Cryptography (FDTC)* (IEEE, New York, 2009), pp. 13–22

40. H. Shen, N. Asadizanjani, M. Tehranipoor, D. Forte, Nanopyramid: an optical scrambler against backside probing attacks, in *ISTFA 2018: Proceedings from the 44th International Symposium for Testing and Failure Analysis* (ASM International, Cleveland, 2018), p. 280

41. S. Skorobogatov, Optical fault masking attacks. In: 2010 Workshop on Fault Diagnosis and Tolerance in Cryptography (IEEE, New York, 2010), pp. 23–29

42. S.P. Skorobogatov, R.J. Anderson, Optical fault induction attacks, in *International Workshop on Cryptographic Hardware and Embedded Systems* (Springer, New York, 2002), pp. 2–12

43. F. Stellari, P. Song, M. Villalobos, J. Sylvestri, Revealing SRAM memory content using spontaneous photon emission, in *2016 IEEE 34th VLSI Test Symposium (VTS)* (IEEE, New York, 2016), pp. 1–6

44. T.G.F.T. Stern Mehta, SPARTA-COTS: a laser probing approach for sequential Trojan detection in cots integrated circuits, in *2020 PAINE Conference* (2020)

45. P. Subramanyan, S. Ray, S. Malik, Evaluating the security of logic encryption algorithms, in *2015 IEEE International Symposium on Hardware Oriented Security and Trust (HOST)* (IEEE, New York, 2015), pp. 137–143

46. S. Tajik, D. Nedospasov, C. Helfmeier, J.P. Seifert, C. Boit, Emission analysis of hardware implementations, in *2014 17th Euromicro Conference on Digital System Design* (IEEE, New York, 2014), pp. 528–534

47. S. Tajik, H. Lohrke, F. Ganji, J.P. Seifert, C. Boit, Laser fault attack on physically unclonable functions, in *2015 Workshop on Fault Diagnosis and Tolerance in Cryptography (FDTC)* (IEEE, New York, 2015), pp. 85–96

48. S.Tajik, E. Dietz, S. Frohmann, H. Dittrich, D. Nedospasov, C. Helfmeier, J.P. Seifert, C. Boit, H.W. Hübers, Photonic side-channel analysis of arbiter PUFs. J. Cryptol. **30**(2), 550–571 (2017)

49. S. Tajik, J. Fietkau, H. Lohrke, J.P. Seifert, C. Boit, PUFMon: security monitoring of FPGAs using physically unclonable functions, in *2017 IEEE 23rd International Symposium on On-Line Testing and Robust System Design (IOLTS)* (IEEE, New York, 2017), pp. 186–191

50. S. Tajik, H. Lohrke, J.P. Seifert, C. Boit, On the power of optical contactless probing: attacking bitstream encryption of FPGAS, in *Proceedings of the 2017 ACM SIGSAC Conference on Computer and Communications Security* (ACM, New York, 2017), pp. 1661–1674

51. R. Tarjan, Depth-first search and linear graph algorithms. SIAM J. Comput. **1**(2), 146–160 (1972)

52. M. Tehranipoor, F. Koushanfar, A survey of hardware Trojan taxonomy and detection. IEEE Des. Test Comput. **27**(1), 10–25 (2010)

53. N. Vashistha, M.T. Rahman, O.P. Paradis, N. Asadizanjani, Is backside the new backdoor in modern SoCs? Invited paper, in *2019 IEEE International Test Conference (ITC)* (2019), pp. 1–10

54. A. Vasselle, H. Thiebeauld, Q. Maouhoub, A. Morisset, S. Ermeneux, Laser-induced fault injection on smartphone bypassing the secure boot. IEEE Trans. Comput. **69**(10), 1449–1459 (2018)

55. K. Xiao, M. Tehranipoor, BISA: Built-in self-authentication for preventing hardware Trojan insertion. In: 2013 IEEE International Symposium on Hardware-Oriented Security and Trust (HOST) (IEEE, New York, 2013), pp. 45–50

56. Y. Xie, A. Srivastava, Anti-SAT: mitigating SAT attack on logic locking. IEEE Trans. Comput.-Aided Des. Integr. Circ. Syst. **38**(2), 199–207 (2018)

57. X. Xu, B. Shakya, M.M. Tehranipoor, D. Forte, Novel bypass attack and BDD-based tradeoff analysis against all known logic locking attacks, in *International Conference on Cryptographic Hardware and Embedded Systems* (Springer, New York, 2017), pp. 189–210

58. M. Yasin, J.J. Rajendran, O. Sinanoglu, R. Karri, On improving the security of logic locking. IEEE Trans. Comput.-Aided Des. Integr. Circ. Syst. **35**(9), 1411–1424 (2015)

59. M. Yasin, B. Mazumdar, O. Sinanoglu, J. Rajendran, Removal attacks on logic locking and camouflaging techniques. IEEE Trans. Emerg. Top. Comput. **14**(8) (2017)

60. H. Zhang, P. Tian, X. Qian, W. Wang, Electro optical probing/frequency mapping (EOP/E-OFM) application in failure isolation of advanced analogue devices, in *2017 IEEE 24th International Symposium on the Physical and Failure Analysis of Integrated Circuits (IPFA)* (IEEE, New York, 2017), pp. 1–5

Chapter 18
Computer Vision for Hardware Security

Hangwei Lu, Daniel E. Capecci, Pallabi Ghosh, Domenic Forte,
and Damon L. Woodard

18.1 Introduction to Computer Vision

Computer vision is defined as a field of artificial intelligence that focuses on training computers to interpret and understand the visual world. Images used for computer vision applications can include those captured in unstructured settings such as those acquired using a digital optical camera or by non-optical microscopy, e.g., X-ray, electron, etc. These systems typically incorporate image analysis to extract high-level information and utilize machine learning algorithms to gain an understanding of the digital image content. Recent advances in computer vision have only increased its use across many application domains, including the electronics industry for security and/or failure analysis purposes. Traditional inspection of electronics relies heavily on subject matter experts (SMEs), and due to the high precision requirements, these results can be prone-to-error. Computer vision aided automatic inspection systems can potentially address this issue by providing a fatigue-free inspection with predictable failure rates. Before introducing and reviewing the computer vision methods used in such systems, the fundamental basics of digital images will be briefly introduced. This section will also discuss the motivation for applying computer vision to hardware security research and summarize a typical computer vision pipeline used for three hardware security applications.

H. Lu · D. E. Capecci · P. Ghosh · D. Forte · D. L. Woodard (✉)
University of Florida, Gainesville, FL, USA
e-mail: qslvhw@ufl.edu; dcapecci@ufl.edu; pallabighosh@ufl.edu; dforte@ece.ufl.edu; dwoodard@ece.ufl.edu

18.1.1 Digital Image Basics

As mentioned above, the goal of computer vision is to interpret or gain an *understanding* of digital image content. Here we review the basic definition of digital images. A digital image is a discrete two-dimensional (2D) array of values, or namely, a matrix, where each element of the matrix is called a pixel, and each pixel represents the intensity of sensor information (light or otherwise) captured at its location. Such a 2D matrix can be a grayscale image, binary image, or single color-channel. An 8-bit grayscale image is a monochrome that contains pixel values scalar between 0 and 255, where a larger value represents a higher response at that sensor location; 0 and 255 indicating black and white, respectively. Similarly, the pixels of binary image are logical values stored using one bit per pixel, where 0 for "off" and 1 for "on." A color image has three values associated with each pixel where each value contains the amount of light captured in red, green, and blue (RGB) color channels. More specifically, a single RGB image consists of three 2D matrices. Examples of these three images are shown in Fig. 18.1. Besides, there are many other multi-channel image configurations, such as Hue–Saturation–Value (HSV), Hue–Saturation–Intensity (HSI), YCbCr, etc. It is often the case that a specific channel configuration is preferred based on a specific application or use-case, such as the use of YCbCr for television applications.

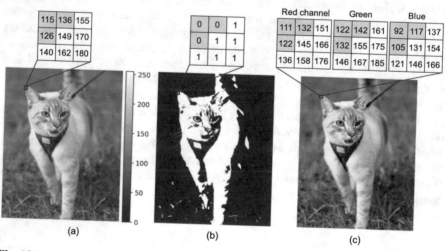

Fig. 18.1 Digital image examples with their area 2D matrices (red squares). Note, gray blocks in all matrices are the pixel "off" in the binary image, and white pixels are "on." The color image (c) has three color channels red, green, and blue (RGB) where each pixel value represents the amount of that channel present in the rendered image. (**a**) Grayscale image. (**b**) Binary image. (**c**) RGB image

18.1.2   Computer Vision-Based Hardware Security Motivation

Outsourcing helps minimize manufacturing time and cost, and it increases the flexibility and adaptability of the hardware supply chain to meet customer demands in the ever-expanding electronics industry. However, outsourcing the supply chain can also raise security concerns due to the possible use of counterfeit electronics and malicious Trojan implantation (e.g., Trojan circuits or malicious/hidden components) by untrusted elements of the supply chain. Researchers have been devoting efforts to develop a reliable inspection system for decades to reduce security breaches, which can result from these threats. The methods can be broadly categorized into two categories: electrical testing and physical inspection. Electrical testing usually compares the performance parameters of a working electronic component to the design specification; when the parameter variations are within the specified tolerance (such as an amount of process variation), this component would pass the inspection and be considered as authentic. Nevertheless, this type of method could miss modifications that do not alter the hardware function. Even when electrical testing methods are carefully designed to achieve accurate performance for high-end applications, the required test time, cost, and equipment make them impractical to meet industry demands. Also, these methods may not be suitable for the short development cycle for new design technologies.

Physical inspection compares the hardware under inspection from the aspects of exterior appearance, interior structure, and material composition to the original design. Since most hardware vulnerabilities involve changes of at least one of these aspects, a thorough physical inspection could enable the detection of counterfeits and/or hardware Trojans that may or may not alter the hardware function. Full reverse engineering is such a practice. Unfortunately, traditional physical inspection is labor-intensive, and the time, cost, and decision consistency are entirely dependent on the SMEs. However, these can be addressed by developing automatic inspection systems. Such a system involves using computer vision algorithms and could achieve high performance when coupled with Machine Learning (ML). For example, object detection algorithms are much faster than a human to point out a specific area of inspection; and a fine-tuned decision model helps generate inspection results with a predictable error rate, which reduces the inconsistency and subjective errors attributable to SMEs. In addition, computer vision-based hardware method would be more scalable compared to existing manual inspection approaches. Computer vision-based inspection methods have been proposed for a number of hardware security applications, including counterfeit integrated circuit (IC) detection, hardware Trojan detection in integrated circuits (IC), and printed circuit board (PCB) assurance. Each of these applications is achieved by following the computer vision pipeline.

18.1.3 Computer Vision Pipeline for Hardware Security Applications

Computer vision tasks typically follow a general pipeline similar to the one shown in Fig. 18.2a. The steps are image acquisition, preprocessing, feature extraction and selection, and the use of an application-specific decision algorithm. The methodologies selected for each step highly depend on the specific problem. Therefore, before providing details of the aforementioned applications, the computer vision pipeline steps are introduced from the perspective of the hardware security research literature.

18.1.3.1 Image Acquisition

As the first step in the computer vision pipeline, image acquisition involves the appropriate imaging modality and imaging environment for the specific application. This decision requires considering many aspects of the hardware under inspection, including its physical size, the area/structure of interest, and its material properties. The following image modalities are commonly used in existing automatic inspection systems.

- **Optical Microscope**—The most commonly used technique for visual inspection. Due to the high magnification, scratches, alignment, ghost markings, and device damage can be detected [6]. The range of magnification ranges from $2\times$ to $200\times$, but higher magnifications can be reached with specialized lenses. Images are typically RGB images based on the electromagnetic radiation (EMR) of the visible light. These types of image modalities are popular in inspection systems for their low cost, fast image speed, and the preservation of RGB color characteristics. Also, optical cameras can reconstruct a depth map from two focused images of the same region with different sensor heights. On the other hand, optical microscopy can only be used to examine the exterior appearance of an object, thus resulting in destructive sample preparation if the internal structure

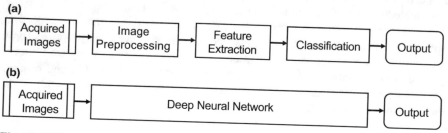

Fig. 18.2 Computer vision pipeline. (**a**) General computer vision pipeline. (**b**) Deep learning involved pipeline

inspection is required. Also, images obtained from these image modalities are affected by illumination variations (lighting source location and intensity).

- **Infrared (IR) Imaging**—Infrared imaging can be used in multiple ways to detect multiple types of defects [11, 32]. One of the well-known uses of infrared imaging is Fourier transform infrared spectroscopy (FTIR) [32]. This technique is used for material analysis where the chemical composition of the components is examined. This technique is mainly useful to detect cloned ICs. Infrared imaging or SQUID microscopy is another non-invasive technique used to monitor the active parts of an IC with questionable components on it [11]. This type of technique is particularly useful in detecting tampered ICs. The other type of infrared imaging technique available in the state of the art is Infrared Thermography [11]. Anything with temperature above absolute zero emits heat in the form of IR radiation. Higher is the temperature, more IR radiation is emitted. Infrared Thermography helps in visualizing this energy emission. Hence this type of energy is also very useful in detecting tampered counterfeit electronics. Then there is the exterior package inspection using laser IR imaging which can be used to detect recycled and remarked counterfeit electronics [29]. Unlike optical images, IR images have very low noise.
- **Scanning Electron Microscope (SEM)**—Scanning electron microscopy is another very powerful tool which is used for counterfeit IC detection. The main advantage of this type of imaging is that it can take image at nanometer resolution and form 3D images from that. This type of imaging scans the surface with a focused beam of electrons. It has large depth field which makes it possible to keep surface features, at radically different heights, in focus simultaneously. This technology is very useful in determining scratches and some other range of surface defects like package material and texture. One such work is described in [57] where the authors have used SEM images to find texture inconsistency by doing stereo-photogrammetry.
- **X-Ray Imaging**—X-ray Imaging is another non-destructive imaging technique which mainly investigates the internal defects that are commonly found in counterfeit components. Both 2D X-ray imaging, called radiography, and 3D imaging, called Computational Tomography, are used in various techniques to identify counterfeit traits. Some important contributions in this domain are the work published in [3, 44], in which the authors used X-ray images of internal structures of ICs, such as bond wires and die, to detect defects and thus to imply counterfeit ICs. It can also be used in a similar way as the NIR/IR imagery for material analysis. Drawbacks of this imaging modality are the high cost of X-ray devices, and the damage X-rays can cause to semiconductors [3]. By incorporating Energy Disruptive Spectroscopy (EDS) from X-ray imagery, the 3D SEM images can be used to analyze the external surface properties. The drawbacks of this modality are the high cost of the SEM imaging device and the challenge of determining the optimal imaging parameter settings. Once the raw image is acquired, it may not be suitable for use for computer vision applications, therefore, requiring preprocessing.

18.1.3.2 Preprocessing

The collected images typically require image preprocessing to suppress unwanted distortions and enhance desired features so that the resulting image is better suited for the given application. The preprocessing techniques can be as simple as image resizing or other operations, such as contrast enhancement, color or geometrical transformation, denoising, etc. Typically, more than one technique is required during preprocessing. In short, an appropriate selection of preprocessing techniques is dependent on the acquired images and the specific computer vision aided hardware security application. The overall goal of image processing is to enhance the image such that feature extraction can be performed reliably.

18.1.3.3 Feature Extraction

Feature extraction is one of the most essential steps in many computer vision tasks, as it can determine the performance of the overall system. This step aims to obtain representative characteristics of an image object to ensure the accuracy, efficiency, generalizability, and interpretability of the computer vision model. Ideally, a useful feature representation is invariant to scale, lighting, position, orientation, and even occlusion conditions. Traditionally, this feature representation is obtained by feature engineering, often performed manually for a given image. The extraction of features is highly dependent on the application task and can be broadly classified as color, shape, and texture features. For example, the color histogram has been used for PCB assurance [68], Fourier shape descriptors have been applied for detecting hardware Trojans [64], and Local Binary Pattern (LBP) was adopted for texture analysis in counterfeit IC detection [25].

A more advanced method of feature extraction is achieved by deep learning, where the end-to-end neural network automatically identifies the most discriminatory features through the training process and directly outputs the results, as shown in Fig. 18.2b. The learned features are typically more representative than the engineered features; however, this self-learning process requires training on a very large labeled dataset to guarantee the generalizability of the learned features and the testing accuracy on new (previously unseen) data. For example, the ImageNet dataset for natural scene object classification consists of more than 14 million samples [22], which supports the development of several well-known neural network architectures. Once salient features have been extracted, they can be used for classification.

18.1.3.4 Classification

The final stage of the computer vision pipeline before output involves the classification of features extracted from the image to extract high-level information. Since features and not the entire image is used during classification, this results in much lower computational cost. Classification is a well-studied task in computer vision, and as a result, several classification algorithms have been developed over the years. Two of the most popular approaches for classification are the k-nearest neighbor classifier (KNN) and support vector machines (SVM). Recently, due to the availability of large image datasets, deep learning-based approaches for classification have been applied with great success to many computer vision applications. The most common approach of the deep learning-based methods involves the use of convolutional neural networks (CNNs). Deep learning-based approaches incorporate the feature extraction and classification into a single stage within the computer vision pipeline, as shown in Fig. 18.2b. However, due to the requirement of large scale labeled image datasets, deep learning-based classification approaches have seen limited application in hardware security-related applications thus far.

In the following sections, examples of the application of computer vision to hardware security are provided. Specifically, the applications of integrated circuit counterfeit detection, hardware Trojan detection, and printed circuit board hardware assurance are discussed. This chapter concludes with a discussion of the research challenges of computer vision-based hardware security and future research opportunities.

18.2   Integrated Circuit Counterfeit Detection

A counterfeit part has been defined by the Bureau of Industry and Security's Office of technology evaluation in 2010 as: (1) an unauthorized copy; (2) does not conform to original chip manufacturer design, model, and/or performance standards; (3) not produced by the OCM or is produced by unauthorized contractors; (4) an off-specification, defective, or used OCM product sold as "new" or working; or (5) possesses incorrect or false markings and/or documentation [17]. Based on these, [33] categorized counterfeit ICs into seven types, including recycled, remarked, overproduced, cloned, defective, forged documentation, and tampered where the recycled and remarked ICs are the most common types found in the market. Figure 18.3 presents a few visual examples of such counterfeit ICs [54]. Counterfeit ICs decrease the longevity of the electronics, bring a huge negative repercussion on the revenue of semiconductor companies, and create life-threatening risks in safety-critical systems, such as aircraft, telecommunication networks, smart grids, etc. According to [12], the reported counterfeit parts increased by a factor of 4 from 2009 to 2012. In 2012, [5] reported that counterfeit ICs could bring $169 billion potential economic risks for the electronics industry annually. Therefore, it is critical to have accurate procedures to detect counterfeit ICs.

As discussed earlier, computer vision induced physical inspection can success-fully address many limitations of electrical testing and SME-required physical inspection. Specifically, for counterfeit IC detection, computer vision can provide a faster, lower cost, automated, and more consistent method of physical inspection than SMEs. Following the general computer vision pipeline given in Fig. 18.2, the steps on counterfeit detection are summarized as follows:

- **Image Acquisition**—Images are acquired using optical microscopy, infrared (IR) imaging infrastructure, and X-ray machines with non-destructive sample preparation.
- **Image Preprocessing**—Color transformation (e.g., RGB to grayscale) and thresholding are typical preprocessing steps for further texture and shape feature extraction. Denoising methods could also be applied as needed.
- **Feature Extraction**—Textural feature extraction is one of the widely used feature extraction technique present in the state of the art. Other features, including color and shape, were also analyzed for this application.
- **Classification**—Counterfeit and golden ICs are classified using various machine learning, deep learning, and classical computer vision techniques.

In this section, a review of computer vision and ML techniques that have been used for feature extraction and classification of counterfeit IC in the literature is introduced; most notably, image texture analysis, color and shape feature representations, and finally, supervised and unsupervised classification methods for detection.

Counterfeit IC detection often requires more than cursory glances at a device in order to detect impostor devices. Sophisticated adversarial techniques make discerning clones and/or counterfeit devices incredibly challenging to even the most discerning SMEs. Because of this, image acquisition techniques are centered around modalities that produce imagery that is not easily captured by the naked eye, such as optical microscopy.

Though the preprocessing of the above modalities leverages many of the same techniques described in the sections below, the need for image alignment and stitching is unique to the modalities in this section. The multiple instances of high magnification or volumetric image data need to be accurately aligned and stitched. These tasks are often non-trivial and require complex algorithms and parameter tuning. Many commercial systems include these algorithms as part of their imaging solutions.

18.2.1 Image Texture Analysis

Among state of the art, texture features have been the most widely used type of feature to identify counterfeit ICs. Texture is an image property that describes the spatial arrangement of image intensities. As shown in Fig. 18.3d, the surface of counterfeit IC appears to be rougher than the authenticated IC; this associates with the arrangement of pixels in different values. Since the package surface of a counterfeit IC typically undergoes sanding, re-coating, or remarking, texture differences are often present between authentic ICs and the highly prevalent remarked/recycled counterfeit ICs. Similarly, the pins/leads of recycled ICs contain additional materials (solder, paint, etc.) or are often worn, leading to different textures and orientations.

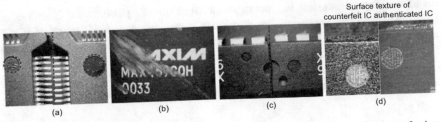

Fig. 18.3 Examples of counterfeit ICs. (**a**) Country of origin mismatch. (**b**) Marking imperfection. (**c**) Indent mismatch. (**d**) Texture mismatch

These differences can be identified in images captured by optical cameras, IR cameras, or scanning electronic microscopes (SEMs). Specifically, there are three texture representations from recent works that have been successful. They are discussed in more detail within the subsections below.

$$L5 \text{ (level)} = [\ 1\ 4\ 6\ 5\ 1\ ]$$
$$E5 \text{ (edge)} = [\ -1\ -2\ 0\ 2\ 1\ ]$$
$$S5 \text{ (spot)} = [\ -1\ 0\ 2\ 0\ -1\ ]$$
$$R5 \text{ (ripple)} = [\ 1\ -4\ 6\ -4\ 1\ ]$$

18.2.2   Surface Texture Parameters and Image Similarity

Five surface texture parameters and image similarity measurement were first used in [58] to quantify the surface properties of the authentic and counterfeit ICs from 3D images. The texture parameters are defined in [56] as the following for 3D SEM imaging surface analysis:

- **Average roughness** or arithmetic mean of the absolute height of the surface area
- **RMS roughness** or root-mean-square of the surface area height
- **Peak** or maximum height value of the surface area
- **Valley** or the minimum height value of the surface area
- **Peakedness** or kurtosis of the surface area

In [58], the parameter variations between the authentic ICs were observed as much smaller than the counterfeit ICs, and parameter differences between the counterfeit and the authentic ICs were also presented, where the first two roughness parameters are less discriminatory compared to the last three. These observations are consistent with the fact that sanding and grinding affects the textural properties of any surface.

Image similarity measurement was used in [58] and [27] to compare the intensity differences between surface images of the counterfeit and authentic ICs. The area-auto-correlation is applied in [58], and it was observed to have similar results on identifying counterfeit ICs using the surface texture parameters. Ghosh et al. [27] calculated the Sorensen-dice similarity score between depth maps of ICs to detect bent pins using optical microscopy, where a depth map is generated by two focused images obtained with different lens heights. The existence of defects is then determined by a threshold value, i.e., the defective pins exist when the similarity score is larger than the threshold, which indicates the detection of a counterfeit IC.

18.2.3 Law's Texture Energy Measurement (LTEM)

LTEM is an energy map that describes multiple types of image texture features. Each feature is defined by a five-dimensional vector, including level (L5), edge (E5), spot (S5), and ripple (R5), which detects the texture feature corresponding to their names. These five vectors are defined below [38]:

Their pair-combinations, which are the product of two vectors (e.g., L5E5), can form sixteen 5×5 masks. These masks are correlated with the image, and some symmetric pairs are combined, resulting in a total of nine masks. This produces a full image with nine-dimensional feature vectors at each pixel, i.e. nine energy maps in terms of full image. LTEM represents sufficient texture features and has been successfully applied in [25] and [26], where the authentic ICs presented similar energy that were distinguished from the counterfeit ICs. LTEM is one of the very few texture feature extraction methods, where each of the feature can be directly related to the defects commonly found on counterfeit IC package surface. For example, the feature edge can be used for quantifying scratches present on the IC surface.

18.2.4 Local Binary Pattern (LBP) and Local Directional Pattern (LDP)

LBP and LDP extract texture features within a pre-defined window (local image area) using binary code, and they are derived with similar processes. Figure 18.4 presents the texture detection principle of LBP. For each image pixel (center), the eight neighboring pixels that have values larger than the center are denoted as "1," otherwise, denoted as "0." This results in an LBP encoded texture image, where its pixel distribution can distinguish it from different LBP images. For example, [25, 26] used the LBP histogram to analyze the differences between authentic and counterfeit ICs, which can be visualized in Fig. 18.5 (left: counterfeit; right: authentic). In [44], the dimensional reduced LBP feature vectors were normalized and input to train a machine learning classifier to recognize such differences of LBP codes.

LDP originates from LBP but is more robust to image noise by applying eight directional masks (Kirsch masks) to describe the edge significance in corresponding

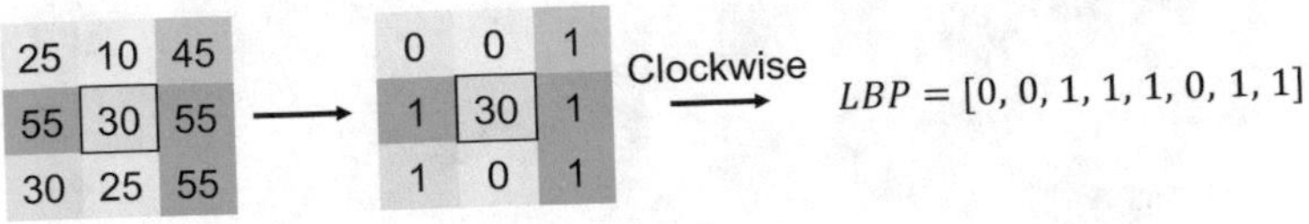

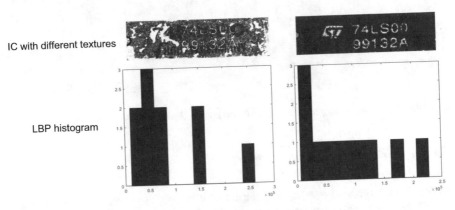

Fig. 18.4 LBP feature extraction

Fig. 18.5 Example of LBP histogram on different textured IC images

directions. For implementation details, interested readers are referred to [34]. In [29], another variant of LDP, Local Gabor Directional Pattern (LGDiP) is extracted. In this technique, a LDP operator is applied on Gabor magnitude images. This technique combines the appearance information over a broader range of scales from Gabor filters with the local fine details from the LDP operator. Hence, it is useful in counterfeit IC detection as it can detect a wide range of textural defects. The differences between LGDiP feature vectors are then analyzed by intra-class and inter-class similarity scores to identify and locate IC defects.

18.2.5 Image Entropy

Entropy analyzes the randomness of the image pixels to represent the texture characteristics, and the local entropy map has been used for detecting corroded pins in counterfeit ICs. This energy map is obtained by applying image filtering using a 9×9 mask on the side view of an optical image. Then, pixels of energy map that have similar values are grouped by clustering method. Since the images of non-corroded pins should have uniform texture, the number of groups larger than "1" indicates the presence of corrode pins, as shown in Fig. 18.6.

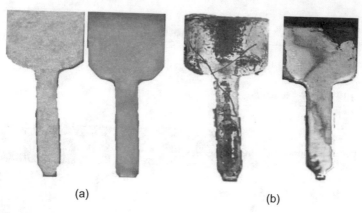

(a)   (b)

Fig. 18.6 Examples of non-corroded pins (**a**) and corroded pins (**b**) after clustering

18.2.6 Shape and Color Properties

Researchers have also investigated the capability of shape and color features for counterfeit IC detection, which also aims to find the separation between counterfeit and authenticated ICs. In [26], the active contour is applied to determine the shape properties, the center position and the radius, of an IC's indents (indent differences are shown in Fig. 18.3c). Ghosh et al. [28] used RGB color to distinguish ICs, where the principle component analysis (PCA) and linear discriminative analysis (LDA) are used to reduce the dimension of color vectors. These features are also demonstrated as accurate for distinguishing counterfeit ICs.

18.2.7 Classification

The detection methods as well as the features used in corresponding works are summarized in Table 18.1. As it is shown, the supervised methods, including thresholding, SVM, KNN, CNN, and an autoencoder-induced neural network, were used for recognizing counterfeit ICs and detecting defective pins. Note, as mentioned previously, the thresholding is typically used in the similarity score-based methods. This threshold is usually determined according to the experiments on a training set. Typically, the threshold is selected when the false alarm rate equals the false positive rate; however, this criteria does not always hold when a system is required to be highly secure. Meanwhile, from [44] and [28], neural networks showed higher accuracy than other supervised methods; however, it is worth noting that the implemented network in [44] only consists of four layers and was only trained on eight authenticated and eight counterfeit ICs, while the CNN in [28] consists of five layers and trained on more than 100 samples. Since the

Table 18.1 Summary of automated counterfeit IC detection. ANN—artificial neural network. AE—autoencoder

Defects	Images	Works	Data size[a]	Features	Detection	Accuracy[b]
Surface	Optical microscopy (depth map)	[25]	3 ICs (1)	LTEM, LBP	Threshold	100%[c]
		[58]	10 ICs (5)	Texture param. image sim.	Threshold	100%[c]
	Optical microscopy (2D image)	[26]	14 ICs (6)	LTEM, LBP, shape	Threshold	100%
		[4]	6 ICs	Learned	ANN	100%[c]
	IR camera (2D image)	[29]	25 ICs (15)	LGDiP	Threshold	100%
	X-ray (2D image)	[44]	32 ICs (18)	LBP	SVM, AE	100% (AE)
Pins	Optical microscopy depth map (bent)	[27]	90 depth 6 side view (2)	Image sim. entropy	Threshold	100%[c]
	Optical microscopy 2D image (corroded)	[28]	163 depth (72) 144 side view (92)	LTEM, color	SVM, KNN, CNN, K-means	100% 95.7% (CNN)

[a] The number in brackets is the number of defect samples within the dataset

[b] The highest reported accuracy

[c] The exact accuracy rate was not reported, but the work presented images or states that their methods performed well on all test cases

experimental data size is relatively small compared to experiments on developing a few well-known networks (e.g., AlexNet on ImageNet dataset), it is unknown if such a network has good generalizability on different ICs; neither is it known if it can maintain a similar performance when testing on a larger dataset. Besides, since most works achieved 100% accuracy on counterfeit detection using the reviewed texture features, it is also worth to investigate if developing a deeper network or exploring more discriminatory features helps improve performance. The promising performances reported for these two networks and these unsolved questions present prospective research potential when a large dataset is available. The unsupervised method, K-means clustering, has also been applied to separate corroded and non-corroded pins. This simulates the real-world scenario where the ground-truth label is not available for training.

18.3 Hardware Trojan Detection

IC design details must be given to the outsourced IC fabricator, who has access to the transistor-level designs from the intellectual property (IP) holder [60, 66]. With raw design materials in hand, an adversarial entity can fabricate a hardware Trojan (also called IC Trojans in this section) into the ICs and result in irreparable security breaches, such as confidential data leakage, system reliability compromise, or system control by an adversary [2, 37, 55]. Therefore, it is critical to apply inspection practices on the fabricated ICs before their deployment.

Hardware Trojans are malicious changes that exist in ICs, and can be as simple as an extra insertion, deletion, or modification/replacement of IC cells (gates) or circuit connection that leads to appearance changes on IC layers, e.g., Fig. 18.7, and some of them can alter IC's function or/and side-channel signals. Electrical testing for detecting hardware Trojans can be categorized into either side-channel signal analysis or Trojan activation methods [63], which are used to supervise IC design parameters, such as circuits timing or power consumption. These methods have been demonstrated to be efficient in detecting chip-level or architecture-level hardware Trojans. However, two limitations can impact the reliability of these methods. First,

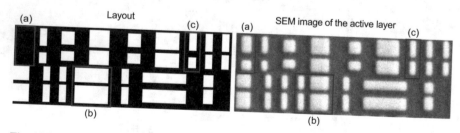

Fig. 18.7 Examples of synthesized hardware Trojans in the active layer. (**a**) Cell insertion, (**b**) cell deletion and replacement, (**c**) cell replacement

as discussed earlier, is the limitation of detecting hardware Trojans that do not alter functionality. Some hardware Trojan activation methods may not be triggered in testing. The small signal changes of a hardware Trojan can be small enough to stay within process variation tolerances [64].

Secondly, today's circuits consist of millions of paths; it is impractical to apply these methods on all paths to prevent such hardware Trojan scenarios [63].

To improve the inspection reliability and reduce the cost, some researchers have adopted ideas from electrical testing to inspect if a functioning IC has an abnormal activation. Although these methods could miss small hardware Trojans, they significantly reduce the time cost by avoiding the delayering process. As mentioned earlier, hardware Trojans consists of altered cells or circuit connections; another inspection practice is to check only the active layer or metal layer. This type of method reduces delayering cost compared to full reverse engineering and has been demonstrated as accurate and robust to detect various hardware Trojans. The state-of-the-art hardware Trojan detection systems using these two types of method follow the general computer vision pipeline, as summarized below:

- **Image Acquisition**—Since the reference (golden) IC rarely exists in practice, many hardware Trojan detection systems were developed using synthesized datasets. The tool and standard library used for obtaining these images will be reviewed. Other researchers have collected SEM and NIR images from real ICs to detect Trojans at the transistor-level and architecture-level.
- **Image Preprocessing**—Many different preprocessing techniques have been applied in previous work involving the use of SEM images. Examples of techniques for image denoising include the convolution of SEM images with 2D Fourier, Gaussian, bilateral, and median filters. Other preprocessing methods employed include image rotation correction, image registration, and histogram equalization.
- **Feature Extraction**—Shape features have mostly been used for detecting transistor-level hardware Trojans, while thermal map properties have been used for detecting architecture-level hardware Trojans.
- **Classification**—Image matching has been used for the detection of both transistor and architecture hardware Trojans. Methods for the detection of transistor-level hardware Trojans have involved the use of k-nearest neighbor and support vector machine classifiers.

In the following sections, an overview of computer vision-based hardware Trojan detection methods is provided.

18.3.1 Obtaining Reference Data

The reference data is the ground-truth for hardware Trojan detection and considered as a golden IC image. The images of ICs under authentication (ICUA) or synthesized

Table 18.2 Summary of reference data used in existing works

Benchmark	Works	Image area	Technology	Synthesizing tool	Added noise
Golden IC	[14, 15]	Active layer	180 nm	–	–
	[65]	Active layer	130 nm	–	–
	[16]	Backside of IC	65 nm	–	–
Fabricated IC	[64]	Active layer	130 nm	–	–
	[59]	Active layer	90 nm	–	–
ISCAS89	[7, 8]	Metal layer	Synopsys 90 nm	RC	Random
ICS99	[7]	Metal layer	Synopsys 90 nm	RC	Random
TrustHub[a]	[71]	Backside of IC	Synopsys 32 nm	DC, ICC, Primetime-PX, HotSpot	Gaussian
	[75]	Backside of IC	Nangate 45 nm	RC, FDTD response	Gaussian
Opencores[b]	[49]	Backside of IC	Nangate 45 nm	DC, RC, Primetime-PX, HotSpot	Gaussian

[a] Specific benchmark is no longer available
[b] Four benchmarks from Opencores were used in [49], including 128-bit AES cipher, 32-bit microprocessor without interlocked pipeline stages (MIPS) Processor, Reed–Solomon (RS) decoder, and joint photographic experts group (JPEG)

images are used for comparison purposes. A summary of reference data used in hardware Trojan detection research efforts is presented in Table 18.2.

The golden IC is one that comes from a trusted fabricator or manually authenticated by the outsourced fabricator before the inspection. Since the golden IC needs to be processed the same way as the ICUA for image acquisition, all obtained images have the same image properties, which allows a direct matching between the golden IC image and ICUA images to find the potential hardware Trojan areas [14, 16, 65]. Using the SEM images of a golden IC, [64] established a small, private dataset that can be used for recognizing hardware Trojan and Trojan-free cells. They segmented individual cell images and manually annotated them to different groups by cell's appearance.

Researchers have proposed an on-chip golden cell design that combines electrical testing and computer vision [59]. This method involves filling the circuit's empty spaces with extra cells that are copied from the originally designed cells. After receiving the ICUA, electrical testing is first applied to authenticate these inserted cells. Next, these cells can be located from IC's SEM images by applying computer vision algorithms that perform object localization. The images of these cells can serve as the golden references for further inspection. Since the reference obtained by this method is from the ICUA, the system can be re-trained or fine-tuned with new technologies. As mentioned previously, synthesized images of the golden image and ICUA can be used for comparison purposes, thereby avoiding the need for expensive imaging equipment. Once suitable images have been obtained, they can be used to determine if difference exists between the trusted design and what is produced.

State-of-the-art SEM and NIR imaging modalities still result in somewhat noisy image data, so filtering and denoising are important initial steps in preprocessing image results. Further, rotation correction and alignment is usually performed to ensure accuracy when localizing and segmenting logic cells. SEM and NIR imaging produces single-channel grayscale images, which are then thresholded to separate foreground (logic cells) and background. The binarized images are then used for the following steps as direct input, or as masks to define the pixels of interest in further computations.

18.3.2   Thermal Map Analysis

Features of a functioning IC are related to IC's thermal properties. [49] applied **2D PCA** on the IC's thermal map to identify abnormal activities on the hardware Trojan ICs. 2D PCA is a linear dimensionality reduction method for image samples, and its reduced dimensions are ordered by the significance of retained image formation. The most significant dimension of ICs can be considered as the discriminatory features for classifying hardware Trojan and Trojan-free ICs.

Researchers have explored the relationship between **time and temperature** changes. In their work, the functioning hardware Trojan ICs are assumed to spend a shorter time to reach a temperature threshold; the time versus temperature changes was analyzed [71]. This method has low cost and easy to implement; however, it could be affected by environmental changes. Another work of using IC's thermal properties is presented in [75]. This method does not require the IC to be in the status of working; instead, the method detects metal layer modification on NIR reflection images that are obtained under a NIR lighting source. Since the metal layer is highly reflective of the NIR light, the area containing tampers will present different reflections compared to the authentic ICs. As with the previously described approach, it could also be limited by the detection environment. Also, as an image of a complete IC can be quite large, the comparison of two images for hardware Trojan detection can be computationally expensive, limiting the approach's potential for scalability.

18.3.3   Shape Representations

A hardware Trojans changes the appearance of IC layers, where some transistor-level hardware Trojans appear as small modifications within a specific area. In this scenario, global image features, such as the texture features used in counterfeit IC detection, is not capable of detecting these changes. Existing research commonly extracts shape features to compare the layer structures between ICUA and reference

Binary cell image   2D FD shape approximation

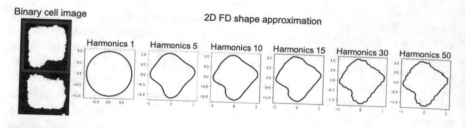

Fig. 18.8 Shape approximation using FD descriptor for the cell region in the red box. Increasing coefficients/harmonics leads to more shape details

samples. In [64] and [59], the authors applied **2D Fourier shape descriptor** (2D-FD) to detect cell modification/replacement hardware Trojans. The 2D-FD transforms the coordinates of shape edge points to approximate closed shapes in N-well and P-well of a binarized cell image. The result of 2D-FD approximation is decided by the number of coefficients/harmonics included in the feature vector, where higher dimensional coefficients represent more shape detail, but also include more unwanted information, such as unclear edges caused by image noise. This approximation can be visualized in Fig. 18.8. In these two works, the authors concatenated two 40-dimensional features (10 harmonics FD) that represent N and P wells to form a shape feature vector for one cell. The 2D-FD is invariant to PV, image noise, and image shifting/rotation and has been demonstrated accurate and computationally efficient for hardware Trojan detection. Shape features were also used for detecting metal layer hardware Trojans. Bao et al. [7, 8] **engineered geometrical features** to represent the area and centroid differences on the image intersection of cell layout and cells under inspection. These result in 5D features vectors that help reduce the computational cost of comparison. Meanwhile, these shape features are fairly robust to illumination variations, which extend their applications in inspection systems under different lighting conditions.

The various classification methods and features that have been explored for hardware Trojan detection are summarized in Table 18.3. The application of a one-class SVM to detect outliers as hardware Trojans, in which the hardware Trojan sample has a feature vector dissimilar to most Trojan-free samples, has been explored in multiple previous efforts [7, 8, 59, 64]. One research effort evaluated the hardware Trojan detection performance using one-class SVMs with linear and nonlinear kernels. Meanwhile, the KNN classifier was applied and obtained similar performances as using SVM [59]. In [49], the Euclidean distance between feature matrices was calculated, where the threshold is determined when FPR is less than 1%. This work also involved the use of the density-based clustering of applications with noise (DBSCAN) algorithm to classify hardware Trojan and hardware Trojan-free ICs into separate groups. Deep learning has not yet been attempted for this application due to the current lack of large image databases of hardware Trojan images.

Table 18.3 Summary of feature extraction and Trojan detection methods

Image area	Works	Data size	Features	Detection
Active layer	[14]	1 golden, 1 Trojan IC	Image overlay	Pixel differences
	[65]	1 golden, 1 Trojan IC	SSIM	Score map
	[64]	350 cell im.	2D FD	SVM
	[59]	1232 cell im.	2D FD	SVM, KNN
Metal layer	[7]	57~122,559 cell im.[a]	Geometrical feature	SVM
	[8]	57~122,559 cell im.[a]	Geometrical feature	SVM, k-means
Backside of IC	[49]	4 IC im. containing[a] 8661~269,970 cells	2D PCA	Threshold, DBSCAN
	[71]	3 IC im.[a]	Time vs temperature	Threshold
	[75]	6 IC im.[a]	2D correlation	Threshold
	[16]	3 synthesized,1 IC im.	Image overlay	Pixel differences

[a] Synthesized images

18.4 Printed Circuit Board Hardware Assurance

PCB and printed circuit board assembly industries are expected to grow to around $68 billion by 2020, reaching $84.6 billion by the end of 2026 [1, 51]. Such PCB vulnerabilities include but are not limited to hardware Trojans, piracy, counterfeiting, and In/Out-field alteration (peripheral and test/debug exploitation) [46]. Exploiting vulnerabilities not only provides adversaries with illegal advantages, but also impacts the system's integrity, results in device functional failure, or leaves hidden back-doors for future stealth attacks. Facilities often attempt to achieve 100% assurance of all products using SMEs to monitor and inspect the over 50 process steps required for PCB fabrication. This high standard of assurance is required for any facility which hopes to stay competitive within their respective market.

PCB inspection can be generally classified into electrical testing and physical inspection[1] Though there are many cases where electrical inspection can validate design specifications, it usually requires a PCB-specific test setup, which is costly and time-consuming to develop. Even so, electrical inspection has limitations that can validate an unqualified product, i.e., when defects or tampering do not alter PCB function at the tested locations. The physical inspection can partially address these limitations by comparing PCB samples with bills of materials, design files, or a golden PCB sample when available. Existing state-of-the-art computer vision aided inspection methods have demonstrated high reliability and efficiency in defect detection through trace and via detection on the bare PCB board, text and logo markings, component placement checks, and component classification and

---

[1]The interested reader can find a more thorough treatment of this topic in [47] and [48].

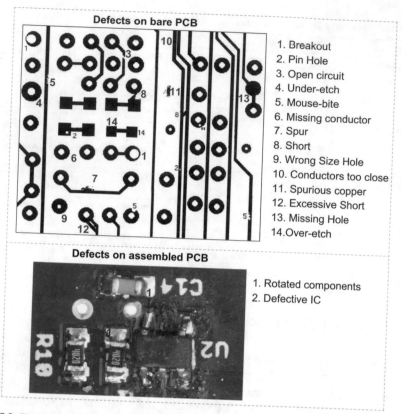

Fig. 18.9 Examples of bare PCB board and assembled PCB board containing defects

authentication (Fig. 18.9 shows a few PCB defects under detection). PCB image analysis follows the four steps of the general computer vision pipeline as follows:

- **Image Acquisition**—Based on individual application demands, the most appropriate imaging modality and imaging environment is chosen. Considerations such as material properties (e.g., reflectivity, specularity), environmental lighting conditions can greatly impact the quality of results in future pipeline steps.
- **Image Preprocessing**—Raw imaging measurements are processed to further enhance information that supports the application. This can include filtering, thresholding, or other transformations which isolates the necessary information for feature extraction or, in some cases, classification.
- **Feature Extraction**—Using the results from preprocessing, higher-order *feature representations* are derived from the image information which are meant to be representative of the underlying question or problem statement in the application.
- **Classification**—using either higher-order features or preprocessed image information, algorithms are used to provide a final "decision," or classification result.

Table 18.4 Summary of imaging modalities. A thorough description of these modalities is given in [6]

Category		Modality
Surface	Reflective surface	Digital optical microscope patterned light optical cameras (e.g., DSLR) white light interferometry
	Penetrative surface imaging	Tetrahertz imaging scanning acoustic imaging
Volumetric		X-ray imaging neutron imaging

For example, whether a given PCB should be classified as coming from a specific source factory, or whether component placement is within tolerance.

18.4.1   Modality Selection

Choosing an imaging modality and environment defines the information constraints which the following algorithms can utilize, and thereby affecting the performance of future computer vision pipeline steps. PCB imaging techniques can be broadly grouped as surface or volumetric imaging. Each modality has different acquisition speeds and other environmental requirements and ramifications to image output. A more thorough treatment of PCB imaging modalities presented in [6], and common modalities are shown in Table 18.4.

The typical PCB imaging process includes a controlled lighting environment and a mounted optical camera for image acquisition. A PCB under inspection can either be placed manually or can be moved under the camera via an automated system such as a conveyor belt. The lighting environment is controlled using ring lights, or other diffuse lighting techniques to create a uniform, consistent illumination across all images. Or in other cases, where high-resolution images are required, optical microscopy is used to take close up images of the PCB board. Then the images are stitched together to create a very large, high-resolution images.

18.4.2   Filtering and Enhancement

Preprocessing steps include filtering, denoising, and image enhancement. The techniques used here are intended to isolate salient information and minimize non-salient information in an image for the given task.

Low-pass filters, such as median and Gaussian filters, help to remove high-frequency information and improve the robustness of later computer vision pipeline steps to small variations in pixel values. Denoising techniques help to recover original image information after being corrupted by transmission or sensor noise.

However, almost without question, preprocessing techniques leverage some sort of morphological operation to enhance object detection tasks.

Morphological operations are a collection of nonlinear set operations that operate on binarized images (pixel values of 0 or 1) and manipulate the structure of these pixels within the image [30]. The primitive operations of image morphology are *erosion* and *dilation*. More complex operations such as *opening*, *closing*, and *thinning* can be created by using combinations of erosion and dilation. In the case of bare PCB visual analysis, morphological operations have been used extensively to improve binarized PCB trace images [45, 47], and they are usually the first attempt used towards identifying background pixels for background subtraction.

Segmentation (discussed in more detail below) in its simplest/most common form is the separation of foreground and background. *Background subtraction* attempts to remove non-salient parts of the image [50], e.g., the PCB board. Such methods used in hardware inspection include image averaging, image differencing, and eigen background subtraction [50]. However, they require controlled inspection environments during imaging to be predictably effective since they are sensitive to environmental condition changes (e.g., illuminations).

Generally, these approaches do not often find the salient object within the image. They can include non-relevant pixels within a region, but their accuracy for a specific PCB inspection task usually can be improved by adjusting parameters of the methods experimentally.

The presence of many small traces, vias, and components causes a diverse set of variations in focus, color/illumination, occlusion, which adds unique complexities to PCB image analysis. To address these, some approaches have extract higher-order features in an attempt to robustly represent image characteristics and improve the reliability of inspection systems. The key property of these features, which aids in reliably improvement is *invariance*. Image features are derived and selected if the feature is invariant to one if not more image properties, i.e., scale, position, pose, etc. Researchers have explored various types of property-invariant features, and some notable examples are presented in the following sections.

18.4.3 Color Features

Similar to human vision, color is one type of rich information used for identifying objects within images. For example, metal materials, such as component solder joints and vias, usually have specular reflections. In contrast, the component packages, although their color not entirely associated with component type, can still present differences that indicate their broad categories or manufacturer. Applied to PCB image analysis, [69, 70] uses the red channel of an RGB image to identify solder joints to then localize resistors, while the red and green channels are compared to distinguish between resistors and capacitors. Similarly, [74] identified solder joints by looking for specular highlights in the LUV color space, and also used Gaussian mixture modeling based on color distributions to detect PCB components. However, color information does not generalize well because sensor color response is sensitive to environmental lighting conditions. Due to this, PCB imaging approaches have attempted leveraged material reflective properties, such as specular reflections. For example, [9, 36, 72, 74] utilized the reflection from solder joints to localize components, and the positions of solder joints can be used to determine the existence of component placement defects.

18.4.4 Boundary Properties

Along with edge-based segmentation, methods such as the Hough line/circle detection algorithms can provide shape information of elements in PCB images. Shape information tends to be invariant to scale, i.e., both a small circle and a large circle stay circles. So, when the object of interest has predictable geometric shape properties, these methods can be used to locate objects within an image more reliably. In one such case, [10] combined Hough circle detection and active contouring to detect the circular vias on PCBs, which are designed to connect internal and external layers of a multi-layer PCB. While in [21], the component contours were extracted to obtain the component rotation angles and the mass centers, while Fourier shape descriptors were obtained based on the contours to distinguish components with or without text. Other properties, such as component height derived from 3D reconstructed PCB images has been used for classifying five types of components [31].

Some hybrid systems also combine boundary and color features to improve inspection capability. Specifically, the boundary features are used to describe component placement, and the color features are used for identifying component types. For example, [73] applied a Canny edge detector and obtained a histogram of the component's contour. They also incorporated with HSI color histogram to achieve component classification, as shown in Fig. 18.10.

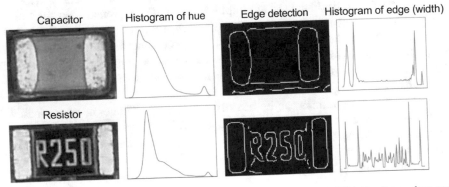

Fig. 18.10 Examples of color and edge features (histogram) used in [73] (the figure does not include all features used in their work)

Table 18.5 Summary of invariant descriptor relative performance from [43]

Method	Time	Scale	Rotation	Blur	Illumination	Affine translation
SIFT	Common	Best	Best	Best	Common	Good
PCA-SIFT	Good	Common	Good	Common	Good	Good
SURF	Best	Good	common	Good	Best	Good

18.4.5 Keypoint Extraction

Another widely used property-invariant feature is the aptly named scale-invariant feature transform (SIFT). As the name implies, this feature is scale-invariant and uses a set of extracted keypoints at different scale-levels that describe elements in an image. It has been used for PCB image alignment and text detection based component authentication [24, 62]. Another popular descriptor is the speeded-up robust features (SURF), which is invariant to both scale and rotation. SURF also produces a set of keypoints as the feature descriptors but is based on Hessian approximations, which is described as the main driver behind its compute speed. Along with a variant of the SIFT descriptor called PCA-SIFT, [43] summarized the performance characteristics between the descriptors in terms of their robustness to each variation, which is shown in Table 18.5. As can be seen in Table 18.5, no single feature descriptor is best in all cases. Descriptors are often combined with other features or as well as other descriptors to improve efficacy.

18.4.6 Image Segmentation

Almost invariably, image segmentation of some kind is performed during prepro-cessing. In essence, image segmentation attempts to group related pixels. For the case of PCB image segmentation, the goal is to separate critical elements under inspection from other image regions (thresholding higher intensity pixels from lower intensity pixels can be thought of as simple foreground/background segmentation). Images are grouped into sub-regions, where each group of pixels satisfies some condition(s) of similarity or discontinuity [30]. Depending on the application and algorithms being used, segmentation can be performed as a preprocessing step for feature extraction, or the resulting partitions themselves can be used as features for classification.

Segmentation algorithms can be broadly categorized as edge- or region-based. Region-based methods separate image pixels based on neighboring pixels or fea-tures and include methods such as watershed [40], region splitting and merging [30], graph-based methods [35], mean shift filtering [23]. Edge-based methods (a.k.a. active contouring), such as snake contouring, seek the boundary paths based on the minimization of an energy objective function. Computer vision solutions tend to leverage more than one of these approaches or combine them at different points in preprocessing or feature extraction.

After segmentation, algorithms often use morphological operators to refine segmentation results and to interpret connected regions in the image.

18.4.7 Template Matching

Image template matching compares images of the PCB under inspection to the images of a golden PCB. One of the earliest and simplest techniques used in PCB image comparison is an *XOR image operator* [39], which applies the "XOR" operation on two binary images pixel-wise, and any turned-on pixel in the output correspond to defects. This method is still heavily used with morphological operations for inspecting bare PCB boards, such as in [20] and [45], due to its speed and simplicity.

In template matching, a correlation function, most commonly normalized cross-correlation, is used to compare an image region interest with entries in an image database of labeled components. Essentially, the correlation function measures how "similar" the input images are to one another. A significant drawback of template matching is computationally costly, and optimization methods for component defect detection have also been explored [18, 67]. Also, template matching has a substantial storage footprint, because a template must exist for every potential valid variation of the object under inspection. Template matching methods usually extract higher-order invariant features, which are then stored into templates for future comparisons, to inspect PCB component placement. This often both reduces the memory requirements for the templates as well as the computational costs associated with the correlation computation.

Features used in this capacity include boundary/edge features [18], frequency and phase features [13], and geometric or structural information [19, 61]. Leveraging feature representations often improves template matching robustness to variance in illumination and image contrast. It also reduces the search space of potential templates for computing the NCC metric, i.e., reducing computational requirements.

However, there is a strong assumption made when using template matching. Correlation computations only find similarities between a sample image and template images and assume similarity thresholds for validation are reliable. When considering the automation of hardware security, this assumption is likely to be too strong to rely solely on template matching detection approaches.

Table 18.6 Image data used in deep learning-based PCB assurance

Task	Works	Classes	Data size[a]	Data accessibility	Performance[b]
Classification	[41]	14	7659/4822	Private	90.8% Acc
Detection	[53]	1	417/66 PCBs (5000 ICs)[c]	Public	89.84% IoU
Authentication	[53]	1	8000/286 IC pairs[c]	Private	92.31% Acc
Classification	[42]	2	55,856/13,964	Public	97.76% Acc

[a] Training/testing

[b] IoU (intersection over union) is the evaluation metrics used for object detection (the higher the better). Acc is classification accuracy. The table presents the best reported performance.

[c] Web images

18.4.8 Artificial Neural Networks

The process of feature extraction can be complex and nuanced. As computational resources have become more readily available, deep learning has grown in popularity for image processing applications because feature extraction is implicitly done by the model and algorithms automatically. Artificial neural networks (ANNs) are, more or less, function approximators. They "learn" (adjust weights/parameters within the model) to map an input to some desired output. CNNs, include several well-known architectures such as AlexNet, ResNet, and Inception-v3, and have shown promising results when trained for component classification/authentication [41, 42]. [53] used a network combining VGG16 and AlexNet backbones (called a Siamese network, because they share internal parameters) along with a "You-Only-Look-Once" network [52] for IC detection, localization, and authentication. The authors observe that the traditional forms of these networks do not perform nearly as well on object detection tasks that include many small, highly dense objects such as in the case of PCB component detection, and make adjustments to optimize for the case of PCBs.

Yet, as is the case with ANNs in general, the models need to train on large datasets to improve performance accuracy (examples shown in Table 18.6). Also, this training process on a large dataset is time-consuming; a high-performance machine is required to guarantee the method's efficiency. Another key drawback of deep learning is the problem of explainability, i.e., even when it works; we cannot necessarily explain *why* it works, nor can we guarantee the generalizability of a trained model to new image data.

Though ANNs are a powerful and promising approach, this demonstrates the need for more research in the application of ANNs for PCB-specific applications, especially in the case of hardware assurance.

18.5   Challenges and Future Research Directions

Although the use of computer vision to address various problems in hardware security shows great potential, several challenges must be overcome before widespread adoption is a reality. Computer vision-based challenges can be categorized into two groups: general and domain-specific challenges.

General Challenges A significant challenge to computer vision-based hardware security is the lack of large ground-truth datasets. Such datasets are required for the application of sophisticated deep learning approaches, which have been demonstrated state of the art in other computer vision applications. Currently, there are no suitable SEM, optical, or optical microscopy datasets of computer hardware components publicly available to researchers to leverage recent advances in deep learning. Another general challenge is the dynamic nature of the hardware security problem. Adversaries are continually seeking new means of compromising hardware security requiring researchers and practitioners to discover more innovative solutions. Further complicating the situation, hardware security problems suffer from both the general challenges mentioned above, and domain-specific nuances. The evolving and diverse nature of hardware security makes a static, general solution infeasible.

Domain-Specific Challenges Of the hardware security problems discussed, integrated circuit hardware Trojan detection is most affected by the presence of noise in images. Scanned electron microscopy images are very noisy unless the appropriate values for magnification, dwelling time, and resolution are chosen, as depicted in Fig. 18.11. One could maximize SEM quality by choosing the largest value for all parameters. However, this would result in a dramatic increase in the time required to obtain images, thereby reducing the utility of the computer vision-based hardware security solution.

Today's electronics are comprised of printed circuit boards containing many more components as compared to those of the past. A computer vision-based hardware security solution must be able to accurately localize each of the components. However, due to the cluttered nature of the images, as depicted in Fig. 18.12, this is extremely difficult. The presence of clutter not only affects feature extraction but most certainly will cause a degradation in classification performance.

One possible approach to the problem of dealing with the clutter of the printed circuit board image is to acquire high magnification images of separate sections of the PCB and then to stitch them together into a large image. Some objects in the smaller images will appear more distanced compared to if the PCB was captured within a single image. However, improvement comes at a cost, as shown in Fig. 18.13. Specifically, preprocessing artifacts due to stitching the smaller images include misaligned stitching, parallax errors, and non-uniform illumination. Unless these preprocessing artifacts are reduced, the performance of later computer vision pipeline steps will suffer. As a result of the numerous challenges of developing a

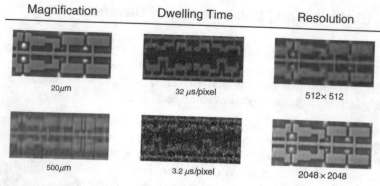

Fig. 18.11 Choice of scanned electron microscope parameter settings has a significant effect on the quality of the image

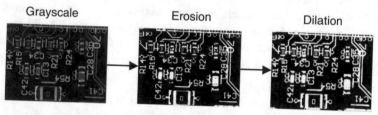

Fig. 18.12 PCB image converted from grayscale to binary image, and morphological operations applied for segmentation

computer vision-based hardware security solution, there are areas where researchers can make impactful contributions to the hardware security community.

Firstly, researchers should consider the use of multi-modal imagery to address hardware security challenges. No single modality is without shortcomings. Therefore by utilizing multi-modal input, researchers can address the inherent limitations of solutions based on unimodal imaging. Secondly, there is a need for large, multimodal, ground-truthed, publicly available image datasets. This would not only allow for leveraging deep learning advances but also serve as a means of evaluating and comparing developed computer vision-based hardware security solutions. Thirdly, suitable image datasets would allow for the application of deep learning methods earlier in the computer vision pipeline to address issues related to preprocessing (e.g., noisy data), thereby improving the performance of later pipeline steps. Lastly, because computer vision-based hardware security solutions require the extraction of high-level information from an image to make a decision, it may be the case that solely the available image data is insufficient for decision making. Therefore, research should explore a means of incorporating domain information into solution development. This incorporation would require a multi-disciplinary approach involving hardware design, imaging, computer vision, and machine learning experts.

Fig. 18.13 (**a**) Parallax causing pins to face different directions. (**b**) Non-uniform illumination between the top and bottom part of IC. (**c**) Misaligned stitching causing repetition of components

References

1. 360 Market Updates: Global PCB PCBA Market. https://www.360marketupdates.com/global-pcb-pcba-market-14845751 (2020). Accessed 8 June 2020
2. S. Adee, The hunt for the kill switch. IEEE Spectrum **45**(5), 34–39 (2008)
3. M. Alam, H. Shen, N. Asadizanjani, M. Tehranipoor, D. Forte, Impact of X-ray tomography on the reliability of integrated circuits. IEEE Trans. Device Mater. Reliab. **17**(1), 59–68 (2017)
4. N. Asadizanjani, M. Tehranipoor, D. Forte, Counterfeit electronics detection using image processing and machine learning. J. Phys. Conf. Ser. **787**, 012023 (2017)
5. S.I. Association, Others, Winning the battle against counterfeit semiconductor products. Tech. rep. (2013)
6. M. Azhagan, D. Mehta, H. Lu, S. Agrawal, M. Tehranipoor, D.L. Woodard, N. Asadizanjani, P. Chawla, A review on automatic bill of material generation and visual inspection on PCBs, in *ISTFA 2019: Proceedings of the 45th International Symposium for Testing and Failure Analysis* (2019), p. 256. books.google.com
7. C. Bao, D. Forte, A. Srivastava, On application of one-class SVM to reverse engineering-based hardware Trojan detection, in *Fifteenth International Symposium on Quality Electronic Design* (2014), pp. 47–54
8. C. Bao, D. Forte, A. Srivastava, On reverse engineering-based hardware Trojan detection. IEEE Trans. Comput. Aided Des. Integr. Circ. Syst. **35**(1), 49–57 (2015)
9. C. Benedek, Detection of soldering defects in Printed Circuit Boards with Hierarchical Marked Point Processes. Pattern Recognit. Lett. **32**(13), 1535–1543 (2011). https://doi.org/10.1016/j.patrec.2011.06.006
10. U.J. Botero, R. Wilson, H. Lu, M.T. Rahman, M.A. Mallaiyan, F. Ganji, N. Asadizanjani, M.M. Tehranipoor, D.L. Woodard, D. Forte, Hardware Trust and Assurance through Reverse Engineering: A Survey and Outlook from Image Analysis and Machine Learning Perspectives (2020). arXiv [eess.IV]
11. G. Caswell, Counterfeit detection strategies: when to do it/how to do it, in International Symposium on Microelectronics, vol. 2010 (2010), pp. 000227–000233
12. J. Cassell, Reports of Counterfeit Parts Quadruple Since 2009, Challenging US Defense Industry and National Security. https://technology.informa.com/389481/reports-of-counterfeit-parts-quadruple-since-2009-challenging-us-defense-industry-and-national-security. Accessed 21 July 2020

13. H.J. Cho, T.H. Park, Wavelet transform based image template matching for automatic component inspection. Int. J. Adv. Manuf. Technol. **50**(9–12), 1033–1039 (2010)

14. F. Courbon, P. Loubet-Moundi, J.J.A. Fournier, A. Tria, A high efficiency hardware trojan detection technique based on fast SEM imaging, in *2015 Design, Automation & Test in Europe Conference & Exhibition (DATE)* (2015), pp. 788–793

15. F. Courbon, P. Loubet-Moundi, J.J.A. Fournier, A. Tria, SEMBA: a SEM based acquisition technique for fast invasive Hardware Trojan detection, in *2015 European Conference on Circuit Theory and Design (ECCTD)* (2015), pp. 1–4

16. M. Cozzi, J.M. Galliere, P. Maurine, Thermal scans for detecting hardware Trojans, in *International Workshop on Constructive Side-Channel Analysis and Secure Design* (2018), pp. 117–132

17. M. Crawford, Defense Industrial Base Assessment: Counterfeit Electronics Prepared by U.S. Department of Commerce Bureau of Industry and Security Office of Technology Evaluation (2010). http://bis.doc.gov/defenseindustrialbaseprograms/

18. A.J. Crispin, V. Rankov, Automated inspection of PCB components using a genetic algorithm template-matching approach. Int. J. Adv. Manuf. Technol. **35**(3), 293–300 (2007). https://doi.org/10.1007/s00170-006-0730-0

19. A.M. Darwish, A.K. Jain, A rule based approach for visual pattern inspection. IEEE Trans. Pattern Anal. Mach. Intell. **10**(1), 56–68 (1988). https://doi.org/10.1109/34.3867

20. N. Dave, V. Tambade, B. Pandhare, S. Saurav, PCB defect detection using image processing and embedded system. Int. Res. J. Eng. Technol. **3**(5), 1897–1901 (2016)

21. A.R. de Mello, M.R. Stemmer, Inspecting surface mounted devices using k nearest neighbor and Multilayer Perceptron, in *2015 IEEE 24th International Symposium on Industrial Electronics (ISIE)* (2015), pp. 950–955. https://doi.org/10.1109/ISIE.2015.7281599

22. J. Deng, W. Dong, R. Socher, L.J. Li, K. Li, L. Fei-Fei, ImageNet: a large-scale hierarchical image database, in *CVPR09* (2009)

23. N. Dhanachandra, K. Manglem, Y.J. Chanu, Image segmentation using K-means clustering algorithm and subtractive clustering algorithm. Proc. Comput. Sci. **54**, 764–771 (2015). https://doi.org/10.1016/j.procs.2015.06.090

24. C. Fonseka, J. Jayasinghe, Feature extraction and template matching algorithm classification for PCB fiducial verification. J. Achiev. Mater. Manuf. Eng. **86**(1), 14–32 (2018)

25. P. Ghosh, R.S. Chakraborty, Counterfeit IC detection by image texture analysis, in *2017 Euromicro Conference on Digital System Design (DSD)*, pp. 283–286 (2017)

26. P. Ghosh, R.S. Chakraborty, Recycled and remarked counterfeit integrated circuit detection by image-processing-based package texture and indent analysis. IEEE Trans. Ind. Inf. **15**(4), 1966–1974 (2018)

27. P. Ghosh, D. Forte, D.L. Woodard, R.S. Chakraborty, Automated detection of pin defects on counterfeit microelectronics, in *ISTFA 2018: Proceedings from the 44th International Symposium for Testing and Failure Analysis* (2018), p. 57

28. P. Ghosh, A. Bhattacharya, D. Forte, R.S. Chakraborty, Automated defective pin detection for recycled microelectronics identification. J. Hardw. Syst. Secur. **3**(3), 250–260 (2019)

29. P. Ghosh, U. Botero, F. Ganji, D. Woodard, R.S. Chakraborty, D. Forte, Automated detection and localization of counterfeit chip defects by texture analysis in infrared (IR) domain, in *IEEE International Conference on Physical Assurance and Inspection of Electronics (PAINE)* (2020)

30. R.C. Gonzalez, R.E. Woods, *Digital Image Processing*, 4th edn. (Pearson, London, 2017)

31. E. Guerra, J.R. Villalobos, A three-dimensional automated visual inspection system for SMT assembly. Comput. Ind. Eng. **40**(1–2), 175–190 (2001)

32. U. Guin, D. DiMase, M. Tehranipoor, A comprehensive framework for counterfeit defect coverage analysis and detection assessment. J. Electron. Test. **30**(1), 25–40 (2014)

33. U. Guin, K. Huang, D. DiMase, J.M. Carulli, M. Tehranipoor, Y. Makris, Counterfeit integrated circuits: a rising threat in the global semiconductor supply chain. Proc. IEEE **102**(8), 1207–1228 (2014)

34. T. Jabid, M.H. Kabir, O. Chae, Local directional pattern (LDP) for face recognition, in *2010 Digest of Technical Papers International Conference on Consumer Electronics (ICCE)* (2010), pp. 329–330
35. M.J. Jianbo Shi, Normalized cuts and image segmentation. IEEE Trans. Pattern Anal. Mach. Intell. **22**(8), 888–905 (2000). https://doi.org/10.1109/34.868688
36. B.C. Jiang, C.C. Wang, Y.N. Hsu, Machine vision and background remover-based approach for PCB solder joints inspection. Int. J. Prod. Res. **45**(2), 451–464 (2007). https://doi.org/10.1080/00207540600607184
37. R. Johnson, The Navy Bought Fake Chinese Microchips That Could Have Disarmed U.S. Missiles. Business Insider (2011)
38. K.I. Laws, Rapid texture identification, in *Image Processing for Missile Guidance*, vol. 0238 (1980), pp. 376–381
39. D.T. Lee, A computerized automatic inspection system for complex printed thick film patterns, in *Applications of Electronic Imaging Systems*, vol. 0143 (International Society for Optics and Photonics, Bellingham, WA, 1978), pp. 172–177. https://doi.org/10.1117/12.956563
40. W. Li, B. Esders, M. Breier, SMD segmentation for automated PCB recycling, in *2013 11th IEEE International Conference on Industrial Informatics (INDIN)*. ieeexplore.ieee.org (2013), pp. 65–70. https://doi.org/10.1109/INDIN.2013.6622859
41. D.U. Lim, Y.G. Kim, T.H. Park, SMD classification for automated optical inspection machine using convolution neural network, in *2019 Third IEEE International Conference on Robotic Computing (IRC)* (2019), pp. 395–398
42. H. Lu, D. Mehta, O.D. Paradis, N. Asadizanjani, M. Tehranipoor, D.L. Woodard, FICS-PCB: a multi-modal image dataset for automated printed circuit board visual inspection (2020)
43. O.G. Luo Juan, A comparison of SIFT, PCA-SIFT and SURF. Int. J. Image Process **3**(4), 143–152 (2009)
44. K. Mahmood, P.L. Carmona, S. Shahbazmohamadi, F. Pla, B. Javidi, Real-time automated counterfeit integrated circuit detection using X-ray microscopy. Appl. Opt. **54**(13), D25–D32 (2015)
45. P.S. Malge, R.S. Nadaf, PCB defect detection, classification and localization using mathematical morphology and image processing tools. Int. J. Comput. Appl. Technol. **87**(9), 40–45 (2014)
46. D. Mehta, H. Lu, O.D. Paradis, M. Azhagan, T. Rahman, P. Chawla, D.L. Woodard, M. Tehranipoor, N. Asadizanjani, The big hack explained: detection and prevention of PCB supply chain implants. ACM J. Emerg. Technol. Comput. Syst. https://doi.org/10.1145/3401980
47. M. Moganti, F. Ercal, C.H. Dagli, S. Tsunekawa, Automatic PCB inspection algorithms: a survey. Comput. Vis. Image Underst. **63**(2), 287–313 (1996). https://doi.org/10.1006/cviu.1996.0020
48. T.S. Newman, A.K. Jain, A survey of automated visual inspection. Comput. Vis. Image Underst. **61**(2), 231–262 (1995). https://doi.org/10.1006/cviu.1995.1017
49. A.N. Nowroz, K. Hu, F. Koushanfar, S. Reda, Novel techniques for high-sensitivity hardware Trojan detection using thermal and power maps. IEEE Trans. Comput. Aided Des. Integr. Circ. Syst. **33**(12), 1792–1805 (2014)
50. M. Piccardi, Background subtraction techniques: a review, in *2004 IEEE International Conference on Systems, Man and Cybernetics (IEEE Cat. No.04CH37583)*, vol. 4. ieeexplore.ieee.org (2004), pp. 3099–3104. https://doi.org/10.1109/ICSMC.2004.1400815
51. B.A. Procurement, Printed Circuit Board (PCB) Forecast Report for 2020. Tech. rep. (2019)
52. J. Li, J. Gu, Z. Huang, J. Wen, Application research of improved YOLO V3 algorithm in PCB electronic component detection. NATO Adv. Sci. Inst. Ser. E Appl. Sci. **9**(18), 3750 (2019) https://doi.org/10.3390/app9183750
53. M. Reza, Z. Chen, D. Crandall, Deep neural network based detection and verification of microelectronic images. J. Hardw. Syst. Secur. **4**, 44–54 (2020)
54. H. Rob, Counterfeit electronic component detection. https://www.aeri.com/counterfeit-electronic-component-detection/. Accessed 21 July 2020

55. K. Rosenfeld, R. Karri, Attacks and defenses for JTAG. IEEE Des. Test Comput. **27**(1), 36–47 (2010)
56. S. Shahbazmohamadi, E.H. Jordan, Optimizing an SEM-based 3D surface imaging technique for recording bond coat surface geometry in thermal barrier coatings. Meas. Sci. Technol. **23**(12), 125601 (2012)
57. S. Shahbazmohamadi, D. Forte, M. Tehranipoor, Advanced physical inspection methods for counterfeit detection, in *Proceedings of International Symposium for Testing and Failure Analysis (ISFTA)* (2014), pp. 55–64
58. S. Shahbazmohamadi, D. Forte, M. Tehranipoor, Advanced physical inspection methods for counterfeit IC detection, in *ISTFA 2014: Conference Proceedings from the 40th International Symposium for Testing and Failure Analysis* (2014), p. 55
59. Q. Shi, N. Vashistha, H. Lu, H. Shen, B. Tehranipoor, D.L. Woodard, N. Asadizanjani, Golden gates: a new hybrid approach for rapid hardware trojan detection using testing and imaging, in *2019 IEEE International Symposium on Hardware Oriented Security and Trust (HOST)* (2019), pp. 61–71
60. Y. Shi, C.W. Ting, B.H. Gwee, Y. Ren, A highly efficient method for extracting FSMs from flattened gate-level netlist, in *Proceedings of 2010 IEEE International Symposium on Circuits and Systems* (2010), pp. 2610–2613
61. Y.N. Sun, C.T. Tsai, A new model-based approach for industrial visual inspection. Pattern Recognit. **25**(11), 1327–1336 (1992). https://doi.org/10.1016/0031-3203(92)90145-9
62. C. Szymanski, M.R. Stemmer, Automated PCB inspection in small series production based on SIFT algorithm, in *2015 IEEE 24th International Symposium on Industrial Electronics (ISIE)*. ieeexplore.ieee.org (2015), pp. 594–599. https://doi.org/10.1109/ISIE.2015.7281535
63. M. Tehranipoor, F. Koushanfar, A survey of hardware Trojan taxonomy and detection. IEEE Des. Test Comput. **27**(1), 10–25 (2010)
64. N. Vashistha, H. Lu, Q. Shi, M.T. Rahman, H. Shen, D.L. Woodard, N. Asadizanjani, M. Tehranipoor, Trojan scanner: detecting hardware trojans with rapid SEM imaging combined with image processing and machine learning, in *ISTFA 2018: Proceedings from the 44th International Symposium for Testing and Failure Analysis* (2018), p. 256
65. N. Vashistha, M.T. Rahman, H. Shen, D.L. Woodard, N. Asadizanjani, M. Tehranipoor, Detecting hardware Trojans inserted by untrusted foundry using physical inspection and advanced image processing. J. Hardw. Syst. Secur. **2**(4), 333–344 (2018)
66. T.J. Wagner, Hierarchical layout verification. IEEE Des. Test Comput. **2**(1), 31–37 (1985)
67. D.Z. Wang, C.H. Wu, A. Ip, C.Y. Chan, D.W. Wang, Fast multi-template matching using a particle swarm optimization algorithm for PCB inspection, in *Applications of Evolutionary Computing* (Springer, Berlin, Heidelberg, 2008), pp. 365–370. https://doi.org/10.1007/978-3-540-78761-7_39
68. F. Wu, X. Zhang, Feature-extraction-based inspection algorithm for IC solder joints. IEEE Trans. Compon. Packaging Manuf. Technol. **1**(5), 689–694 (2011). https://doi.org/10.1109/TCPMT.2011.2118208
69. H.H. Wu, X.M. Zhang, S.L. Hong, A visual inspection system for surface mounted components based on color features. in *2009 International Conference on Information and Automation* (2009), pp. 571–576
70. H. Wu, G. Feng, H. Li, X. Zeng, Automated visual inspection of surface mounted chip components, in *2010 IEEE International Conference on Mechatronics and Automation* (2010), pp. 1789–1794
71. L. Yang, X. Li, H. Li, Hardware Trojan detection method based on time feature of chip temperature, in *2020 10th Annual Computing and Communication Workshop and Conference (CCWC)* (2020), pp. 1029–1032
72. C.H. Yeh, T.C. Shen, F.C. Wu, A case study: passive component inspection using a 1D wavelet transform. Int. J. Adv. Manuf. Technol. **22**(11), 899–910 (2003). https://doi.org/10.1007/s00170-003-1608-z

73. S. Youn, Y. Lee, T. Park, Automatic classification of SMD packages using neural network, in *2014 IEEE/SICE International Symposium on System Integration* (2014), pp. 790–795
74. Z. Zeng, Z.M. Li, Z. Zheng, Extracting PCB components based on color distribution of highlight areas. Comput. Sci. Inf. Syst. **7**(1), 13–30 (2010)
75. B. Zhou, A. Aksoylar, K. Vigil, R. Adato, J. Tan, B. Goldberg, M.S. Ünlü, A. Joshi, Hardware Trojan detection using backside optical imaging. IEEE Trans. Comput. Aided Des. Integr. Circ. Syst. (2020). https://doi.org/10.1109/TCAD.2020.2991680

Chapter 19
Asynchronous Circuits and Their Applications in Hardware Security

Eslam Yahya Tawfik and Waleed Khalil

19.1 Introduction

19.1.1 The Rise of Hardware Security and Assurance

Cyberattacks are the fastest growing crime, and they are increasing in size, attack space, and cost. It is estimated that malicious cyber activity cost the US economy between $57 billion and $109 billion in 2016 [1]. Hardware assets (processors, crypto accelerators, transceivers, data converters, etc.) are the backbone of the cyber space, and they must be secure and must assure to enable cyber security mechanisms at the firmware/software layers. There are different definitions for hardware security and assurance. Within this chapter, we define *Hardware Security* as "design techniques and practices enabling the hardware to protect security assets from unlawful-access and side-channel leakage." In contrast, *Hardware Assurance* is defined as "techniques and practices to disable/detect any malicious insertion, design alteration, and IP-infringement that can be applied to the hardware throughout the supply chain."

Computing hardware had evolved significantly in the last few decades (Fig. 19.1a). In 1950s, computing platforms started as huge mainframes which are captivated in secure buildings and maintained by "trusted" and specialized teams. During this era, user interaction with computation resources was done through Operating System (OS) terminals. As a result, all access controls and security measures were implemented into the OS and physical security of the hardware itself was not thought of. In 1970s and 1980s, microcomputers and PCs were introduced

E. Y. Tawfik (✉) · W. Khalil
Electrical and Computer Engineering Department, The Ohio State University,
Columbus, OH, USA
e-mail: tawfik.10@osu.edu

527

Fig. 19.1 Hardware Security and Assurance: (**a**) Exposed mobile computing introduced a variety of physical attack threats; (**b**) Fabrication as a service, fabless model, and IP reuse introduced different threats to the supply chain

to the market, and even with the Internet invention, the only gate to maliciously interact with the computing resources was software applications, OS, and networks. All security defenses were built "only" on the software level due to the limited access to the physical hardware. Starting with the second millennium, the game was completely changed by introducing different forms of mobile computing, such as mobile handheld, laptops, and smart/wearable devices. In the new computing model, we take our computation resources to the public, exposing them to malicious physical access through near-field eavesdropping and invasive/noninvasive physical attacks. This situation drives a paradigm shift in security to protect the physical hardware itself against different potential attacks such as side-channel analysis, fault injection, and EMI eavesdropping (Fig. 19.1).

Microelectronics industry started with a fully vertical model, in which fabrication firms such as Intel and IBM owned the entire supply chain including silicon fabrication, packaging, system design, and testing. In 1987, Taiwan Semiconductor Manufacturing Company (TSMC) was introduced as "silicon fabrication as a service" company; its main business is to fabricate designs of other companies (Fig. 19.1b). This business model lowered the industry barrier and encouraged "Fabless" companies to enter the market. To shorten the time to market, design houses adopted the "IP reuse" concept, in which they integrated third-party IPs into different products. Fabrication as service, fabless model, and IP reuse introduced a variety of potential threats to the entire supply chain; examples of these threats are Hardware Trojan insertion, IP theft, and counterfeited ICs. Microelectronics industry ecosystem (including design methods, EDA tools, and fabrication supply-chain) evolved throughout the years while being agnostic to hardware security and assurance. The whole community is working very hard to tackle this challenging situation on all different levels.

19.1.2 Side-Channel Attacks

Most of the current microelectronics ICs are designed and fabricated while having different vulnerabilities that can be exploited by highly skilled attacker to

Fig. 19.2 Common types of side-channel attacks: fault injection, timing attacks, power analysis, EMI analysis, acoustic attacks

extract confidential information. The main reason behind these vulnerabilities is the unintentional emissions which are not typically modeled/measured as part of traditional verification and validation processes. Side-channel attacks are security breaches exploiting these emanations to extract sensitive information out of the physical implementation of the hardware [2–4]. Side-channel attacks do not break the theoretical strength of the cryptographic algorithm; they eavesdrop leaked information using physical inspection of the hardware. Commonly used methods for side-channel attack, shown in Fig. 19.2, are fault injection, timing attacks, power analysis, electromagnetic analysis, and acoustic attacks [5–7].

The objective of the attacker in these methods is to extract some security assets (encryption keys, program instructions, design IPs) out of the hardware through one of the above phenomena. For example, by monitoring the power consumption of the hardware, the attacker can correlate different power traces with different instructions (multiplication, addition, etc.). Many countermeasures are introduced in the literature to hide hardware emanations or avoid correlation between these emanations and the data being processed by the hardware [8–10].

Security of the hardware can be significantly increased by reducing power consumption and EMI emissions so that attackers cannot use them to extract useful information. It is paramount to protect designs with the minimum size, performance, and power consumption overheads. Throughout the chapter, we will discuss the application of asynchronous circuits as a countermeasure for different hardware security threats.

19.2 Asynchronous Circuits

Data flow in asynchronous circuits is based on local handshaking between registers, which distributes the switching activity over time producing a nicely shaped power and EMI distribution. This section compares synchronous and asynchronous circuit architectures, explaining the potential benefits for using asynchronous circuits.

19.2.1 *Data Synchronization: Clocked Versus Clockless*

19.2.1.1 Synchronous Circuit: Drawbacks of Global Timing

Figure 19.3 depicts a pipeline that is implemented in synchronous and asynchronous styles. In synchronous design (Fig. 19.3a) circuit functionality is implemented by combinational function blocks; synchronous registers are sampling the output of these blocks at clock edges which determine the sampling time of the registers. Clock period is fixed so that all function blocks complete their operations and have their outputs stable at the active clock edge. Presence of clock edges at each register guarantees the satisfaction of two important digital design assumptions. First, no register is permitted to change its output data until the logic in front uses current data and produces stable output (no data overwriting). Second, no register is permitted to capture input data until the delays of previous logic is satisfied (capturing valid data only). This operation concept implies a global timing assumption which is applied to the whole circuit. Delay uncertainty is drastically increased in advanced CMOS nodes. In this case, global clock timing assumption reduces performance due to the

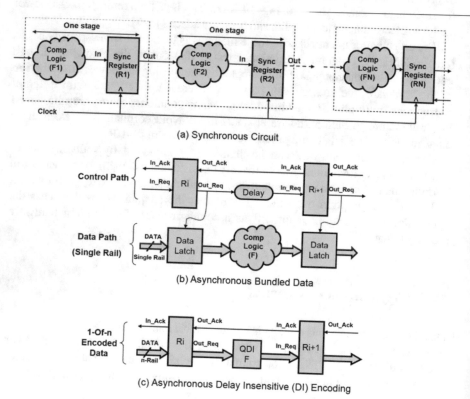

(a) Synchronous Circuit

(b) Asynchronous Bundled Data

(c) Asynchronous Delay Insensitive (DI) Encoding

Fig. 19.3 Synchronous circuits versus asynchronous circuits: (**a**) Synchronous circuits (basic view) (**b**) Asynchronous bundled data (**c**) Asynchronous delay insensitive (DI) encoding

increased pessimism added to compensate this uncertainty. In addition to this, clock trees in considerably large SoCs are power hungry; they are getting clock transitions regardless of whether the circuit is computing or not. Therefore, omitting the clock tree is a way to significantly reduce power consumption. Active edge of the clock signal represents a synchronization event on which all circuit activities start and supposedly finish before the next edge. This synchronization creates current bursts with each clock edge. These bursts are used in side-channel analysis (power and EMI) to infer knowledge about the circuit activities and the data being processed.

19.2.1.2 Asynchronous Circuit Handshaking

Asynchronous circuits are attracting the interest of both research and industry due to their low power and intrinsic security. Asynchronous circuits (also known as self-timed and clockless) are a class of circuits that contains no global synchronization signal (equivalent to the clock in synchronous circuits). In contrast to synchronous circuits, asynchronous circuit registers (Fig. 19.3b) have a local handshaking protocol which synchronizes the register with preceding and following registers. By using this local handshaking protocol, registers acknowledge the inputs coming from preceding stage (using *InAck* signal). On the output side, they use handshaking signals (*OutReq* signal) to request the following stage to process new data. This localization of the circuit synchronization avoids problems caused by global timing assumption presented by the clock in synchronous circuit, such as clock distribution and clock skew and jitter. The following section elaborates on fundamentals of asynchronous design to help understanding the operation of the handshaking protocols.

19.2.2 Basics of Asynchronous Circuit Design

19.2.2.1 Bundled Data Circuits

Asynchronous circuits can be classified based on their architecture, timing assumptions, or their handshaking protocol [11, 12]. The two main architecture classes are depicted in Fig. 19.3. In bundled data circuits (Fig. 19.3b), there are two paths: *datapath* and *control path*. The datapath is a single rail implementation (similar to synchronous circuits) and functions are implemented using normal combinational logic. The control path contains *request* and *acknowledgment* signals for implementing the handshaking protocol. These two signals replace the function of the clock signal in synchronous designs. *Request* signal triggers the register to capture a new data, guaranteeing that the register inputs are stable, and logic delays are satisfied (capturing valid data only). *Acknowledgment* signal guarantees that data sent by previous register is computed and properly captured (no data overwriting). To maintain correct behavior, matching delays must be inserted in the request signal paths to compensate the propagation delays of the function blocks in the datapath. The concept behind bundled data is to replace the global clock timing assumption with various timing assumptions placed in the circuit. Indeed, this enhances the circuit performance as it reduces the pessimism and avoids the distribution of a global signal with all of buffering, jitter, and skew issues.

19.2.2.2 Delay Insensitive (DI) Encoding Circuits

The second asynchronous circuit architecture is the delay insensitive (DI) encoding style (Fig. 19.3c). The concept behind this architecture is to avoid any timing assumptions (no global or local timing assumption), which is known as *delay insensitivity*. Since there is no predetermined assumption on when functional blocks will finish their computation, the challenge in DI-encoded circuits is to detect the presence of a valid data at the registers' inputs, which is known as *completion detection*. To achieve this, datapath uses some coding in which jumping from one code word to another indicates the presence of new valid data. In this case, the design of combinational logic has to be hazard-free (no intermediate results). The most common implementation of DI-encoded circuits is the dual-rail asynchronous logic. In dual-rail logic, each data bit is encoded by two wires, D0 and D1. As shown in Fig. 19.4, when $D0 = 1$ and $D1 = 0$, the dual-rail output is equal to logic-0. However, when $D0 = 0$ and $D1 = 1$ the dual-rail output is equal to logic-1. To distinguish between two consecutive identical data (cascaded 1s or cascaded 0s), dual-rail protocol returns both wires (D0, D1) to zeros, which is known as *return-to-zero* phase (RTZ) or *spacer*. This communication protocol is the most common in DI-encoded circuits and known as "dual-rail four-phase-protocol" [3].

19.2.2.3 The Muller-Gate and Handshaking Protocols

Muller-gate (also known as C-element) is the basic component used to build asynchronous circuits. The output of the Muller-gate switches high when all inputs are high, and switches low when all inputs are low; otherwise, the Mueller-gate output holds its state (the truth table is shown in Fig. 19.5). Muller-gates are used as data synchronization element that guarantees the output of the logic will not switch until its input is in a predetermined state. Designers can build very complex

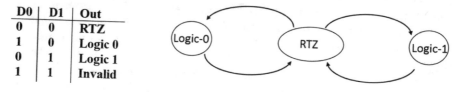

D0	D1	Out
0	0	RTZ
1	0	Logic 0
0	1	Logic 1
1	1	Invalid

Fig. 19.4 Dual-rail encoding (truth-table and state-machine)

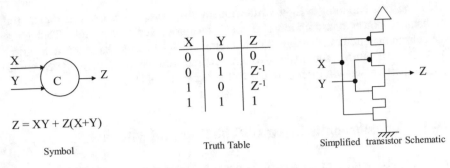

$$Z = XY + Z(X+Y)$$

Symbol   Truth Table   Simplified transistor Schematic

Fig. 19.5 The Muller-gate (C-element)

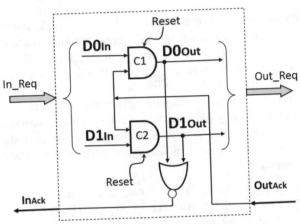

Fig. 19.6 Dual-rail asynchronous register

functions which are hazard-free using Muller-gates. A simple asynchronous register is shown in Fig. 19.6 [13, 14], input and output data of the register are dual-rail. The operation of the register is explained as follows:

- At reset, all Muller-gates are zeros, indicating an RTZ code on the register output. That applies to all the system registers.
- Register outputs: $D0 = 0$, $D1 = 0$, and the NOR gate output $= 1$ which is the acknowledgment signal sent to the previous stage ($In_{ACK} = 1$).
- Register inputs: $D0 = 0$, $D1 = 0$, and the acknowledgment signal coming from the next register$=1$ ($Out_{ACK} = 1$).
- This initial state makes all the system registers ready to capture new data.
- When the main input inserts new data, comes to the input register as dual-rail data (Fig. 19.4), and one of the register Muller-gates will switch to one depending weather the input is zero ($D0 = 1$, $D1 = 0$), or one ($D0 = 0$, $D1 = 1$).

The number of transitions needed to switch from RTZ to logic-0 is equal to the number of transitions needed to switch from RTZ to Logic-1 and vice versa.

This constant hamming weight while transmitting/computing data has a potential to balance power consumption and mask the processed data values. That gives asynchronous circuits advantages in reducing side-channel leakage through power and EMI signatures and making them secured compared to their synchronous counterparts as we will discuss in the following sections [15, 16].

19.3 Asynchronous Circuits: The Plus and Minus

19.3.1 Advantages of Asynchronous Circuits

Digital circuits are mostly synchronous, and their operation is based on two important assumptions: discrete voltage (all signals are binary) and discrete timing (computation results are valid "only" at specific points in time). Asynchronous circuits hold the same two assumptions, except that their notion of "discreet time" is different. Synchronous circuits have a global event (the clock signal) synchronizing the moment of capturing the valid data across the whole design. In contrast, asynchronous circuits locally generate this synchronization event using equivalent delays in bundled data or using completion detection of encoded data in dual rail circuits. The removal of the global timing signal introduces the following intrinsic advantages to asynchronous circuits:

- Propagation delays are localized to the individual combinational blocks, which gives asynchronous circuit performance advantages. Generally, they can compute within the average delay of their functional blocks, compared to the worst delay in case of synchronous circuits. A commercial example is a 72-port 10G asynchronous Ethernet switch chip designed by Davies et al. [17]. In their work they used a class of DI-encoded circuits known as *quasi-delay insensitive* (QDI) to implement 1.2 billion transistor chip with 90% of the transistors in QDI circuits. The product shows how asynchronous circuits can be used for pushing the performance envelope; the company got acquired by Intel.
- Switching activities occur randomly based on the individual function block delays. This behavior significantly enhances the EMI spectral properties compared to synchronous counterparts. In synchronous circuits, switching activities are centered on the clock frequency and its harmonics. Fan et al. exploited this advantage to attenuate spectral peaks by 12 db in frequency-modulated continuous wave (FMCW) RADAR chip [18].
- Removal of the clock signal and its distribution tree saves a lot of power and enhances EMI properties. In addition, it reliefs design issues related to the clock, such as skew and jitter. As an example, Sheikh and Manohar designed asynchronous IEEE 754 double-precision floating-point adder that showed 56.7% reduction in energy for comparable throughput [19].

The above advantages are common between different asynchronous circuits' styles. Dual-rail circuits have more advantages due to their "delay insensitivity":

- Robustness towards process variations and operating condition fluctuations (voltage, temperature). The fact that dual-rail circuits have no predetermined timing assumption makes them naturally respond to PVT variations. They trade performance versus their PVT parameters in a *correct by construction* fashion. The work done in [20] by Hollosi et al. demonstrated a reliable operation of QDI ALU under a temperature range of 2 K (−271 °C) to 297 K (23 °C), as well as over wide supply voltage variations.
- The above fact makes the use of voltage scaling a natural trend in dual-rail circuits to enable them working on extreme low supply voltage to save even more power. Bailey et al. [21] combined multi-threshold CMOS (MTCMOS) with QDI asynchronous logic to solve major problems of synchronous MTCMOS circuits.

19.3.2 Asynchronous Circuits Challenges and Drawbacks

Adopting asynchronous circuits implies some challenges; some of them are due to the concept of asynchronous logic itself, and most of them are due to the availability of design and test frameworks. As mentioned above, one of the main advantages of asynchronous circuits is their low power consumption. Achieving considerable power reduction in asynchronous circuits is not straightforward particularly in DI-encoded circuits. In bundled data circuits (Fig. 19.3b), the overhead of the control path (asynchronous registers and equivalent delays) represents an extra hardware that is going to consume power during the operation. In general, this power overhead is not equivalent to the saved power due to the removal of the clock signal and its distribution tree. Bundled data generally are more power efficient compared to their synchronous counterparts.

In DI-encoded circuits, such as dual-rail, registers have to use complex trees of completion detection circuits. Let us consider the dual-rail logic circuit in Fig. 19.7, per each output data we must use an OR gate to detect the presence of logic-zero (10) or logic-one (01). The outputs of the whole group of OR gates have to be collected by a Muller-gate to ensure that all the output data are valid before producing the request signal to the next register. Complexity of this completion detection tree grows linearly with the number of data bits in the datapath. In cryptographic accelerators, which are one of the most attractive applications of asynchronous logic, datapath is always very wide, as an example, 128-AES datapath contains 128-bits for the plain text and 128-bits for the key inputs. The asynchronous research community introduced different methods to efficiently implement completion detection circuits [22, 23].

Another challenge in asynchronous circuit design is their performance analysis and optimization. Performance analysis in synchronous circuits is performed by finding the longest latency path between any two registers; this determines the

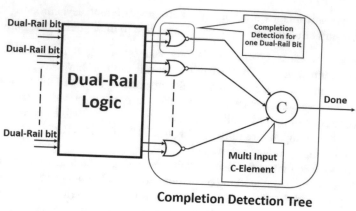

Fig. 19.7 Completion detection circuit in dual-rail

period of the clock. The global clock partitions the circuit into islands of combinational circuits that can be analyzed individually. In contrast, asynchronous circuit is much more complex, since each stage has its own individual delay that can change based on the data value in processing. Asynchronous handshaking makes the timing in one block dependent on the timing of its neighbors, which again are dependent on the timing of their neighbors. This problem is tackled in many articles such as the work done by Yahya et al. [24], in which they developed a statistical static timing analysis framework that can analyze complex asynchronous circuits.

Even with the massive theoretical work in the literature addressing asynchronous circuit design automation (synthesis, performance analysis, verification, etc.), fully automated design/verification flow for asynchronous circuits is still inaccessible. There are some success stories such as the design flow provided by Tiempo [25, 26]. Their commercial tools are used to build different secure microcontrollers and contactless chips which showed low power and robustness towards side-channel attacks.

19.4 Applications of Asynchronous Circuits in Hardware Security and Assurance

Asynchrony features such as reduced power, enhanced EMI spectral, and robustness towards PVT variations are exploited by many designers to implement hardware security countermeasures into circuits using asynchronous architectures [27–31]. The following sections elaborate on applications of asynchronous circuits in hardware security and assurance.

19.4.1 Security Evaluation of Asynchronous Circuits

Before elaborating on applications and use cases in hardware security, this section conducts a security evaluation of asynchronous circuit. The best strategy is to compare asynchronous designs with their synchronous counterparts in terms of their efficiency to countermeasure security threats.

Fournier et al. designed Springbank test chip for this purpose [32]. Springbank contains five different implementations of 16-bit microcontroller processors. The five implementations are using the same architecture where the five versions are:

- Synchronous
- Bundled data
- 1-of-4 DI-encoded circuit
- Dual-rail circuit (1-of-2 DI-encoded)
- A secure variant of the dual-rail

The fact that all five versions are on the same chip makes the comparison as fair as possible with identical silicon and fabrication variations. In their design, the secured dual-rail design was approximately twice the area of the synchronous base design. Power analysis of the dual-rail processor revealed some imbalance in power consumption due to physical design optimizations. Tools optimize to minimize wirelengths which sometimes imbalance dual-rail gates allowing some data-dependent side-channel leakage. Comparison between the synchronous base design and the secure dual-rail design showed a reduction in information leakage by 22 dB.

Kulikowski et al. conducted a deeper analysis of asynchronous circuit security [33]. They prove by simulation that the RTZ phase in DI-encoded protocol has two fundamental effects:

- Information about the state of each gate can be leaked, since the gate has transitions with every data (either one, zero, consecutive ones, consecutive zeros).
- The complexity of the system transitions is reduced from n^2 to n, where n is the number of possible inputs.

To evaluate the effect of RTZ phase on deferential power analysis (DPA), they conducted the attack on a DES Sbox. The implementation of the Sbox is done in three different versions:

- Synchronous logic.
- Synchronous dual-rail logic. The function of the RTZ phase is to distinguish between two consecutive similar code words. The idea behind this style is to enable the circuit to work with and without the RTZ phase.
- Asynchronous dual-rail logic.

The work used synchronous implementation as the baseline to compare the other implementations. It is good to note that this experiment is conducted over the Sbox "only," which is a pure combinational logic. Simulation results showed that synchronous Sbox revealed its secret key after 5000 traces. Synchronous dual-rail suffered from the same problem highlighted earlier in [32], their physical design is not perfect and some power imbalance is still observed. Even with this imbalance, the same number of traces (5000 traces) was not enough to guess the correct key. Simulation shows, even with the straightforward imbalanced dual-rail implementation, performing successful DPA attack will require larger number of power traces. Simulating the asynchronous dual-rail circuit shows a significant effect of the RTZ phase, and the correct key can be guessed after only 500 traces. The presence of the RTZ phase after each data reduced the possible system input states from about half a million to only 1024 possible states, full proof is provided in [33]. To solve this problem, a balanced dual-rail circuit is designed and tested. Same simulation setup is repeated on the balanced dual-rail circuit, and results clearly showed that with even using all the possible 1024 input combinations of the Sbox, correct key cannot be extracted using all the available power traces. The balanced implementation produced data-independent traces, and the RTZ phase reduced the attack space, reducing the opportunity of the attacker to extract information out of the hardware.

Evaluation of asynchronous circuit security showed that they provide high protection against side-channel attacks. However, this result is a subject to how balanced the circuit is. There are various design techniques in the literature which are developed to balance asynchronous logic to guarantee their efficacy as side-channel countermeasure, examples of this type of work can be found in [34–37].

19.4.2 Differential Power Analysis (DPA)-Resistant Asynchronous Circuits

Since its introduction in 1998 by Kocher et al. [38], power analysis became an important metric in evaluating the security level of any hardware implementation. Power analysis is simple, noninvasive, and does need sophisticated setup to retrieve sensitive data. DPA countermeasures generally fall under one of three categories:

- Reducing leaked information SNR by injecting random noise, which prevents the attacker from extracting any useful information using the power traces [39, 40].
- Obscuring the crypto-core signature at the chip main power rails. In this case, even the implementation is not balanced, yet the attacker will not be able to observe any meaningful information on the power rails [41, 42].
- Balancing the power consumption to avoid any information leakage through the power traces [33, 34].

One of the advantages of dual-rail logic is the balanced hamming weight due to the 1-of-2 coding (this is generally true for any 1-of-n circuits). The observed imbalance is due to details in the transistor sizing and physical design. Considering its other security advantages, such as robustness against EMI and fault injection, balanced dual-rail circuits provides very attractive countermeasure for DPA with justified balancing overhead. However, balancing dual-rail logic needs detailed approach that can be time consuming and error prone if done manually. As a result, there are different researchers who developed models and techniques for automating this process. As an example, Bouesse et al. developed a model and proposed a flow to design dual-rail circuits what are DPA resistant [34]. Their model shows strong dependency between power consumption and the load capacitance of the individual gates in the design. Authors applied DPA analysis to their model; results showed the necessity to guide placement and routing to balance the individual nodes capacitance of the design. They proposed a hierarchical approach place and route the individual functions. The goal is to avoid fragmentation of connected gates, which avoids long wires and high capacitance nodes. This hierarchical physical design is applied to an AES core and compared with a flatten AES design. As expected, the hierarchical design showed an area overhead of 20% compared to the flatten design. On the other hand, the proposed design showed one order of magnitude enhancement in the proposed imbalance criteria.

In the same direction and to develop more understanding of the effects of physical design on DPA resistance, Guilley et al. designed 13 variants of AES SubByte implementations with various back-end constraints [28]. The comparison used unprotected Sbox, Wave Dynamic Differential Logic (WDDL) [43], and Secured Library (SecLib) [44]. The work presented two methods for evaluating security features of different designs:

- *Static evaluation*: This tries to find net asymmetries and any static unbalances, which are considered as a metric for potential leak. This is typical to the criteria presented in [34], discussed in the above paragraph.
- *Dynamic evaluation*: statistical information is collected through simulation and/or measurement of all possible input combinations of the Sbox. This is similar to the criteria used in [33], which is discussed in Sect. 19.4.1.

The study showed clearly that dual-rail logic is more immune towards DPA attacks. Another interesting conclusion was that the effect of relative arrival times of inputs is several orders of magnitude more important than the dispersion of routing balancing. In addition to the above, the study showed that the results obtained by "static evaluation" are unreliable when compared to results of the "dynamic evaluation."

As discussed in the introduction, asynchronous circuits generally are self-timed. The temporal progress of the data into the asynchronous datapath is subject to the local delays of the individual functional blocks. That distributes data processing in time, reducing the potential of having current peaks as observed in synchronous counterparts at the active clock edges. Building on this property, Bouesse et al. proposed a design technique to improve asynchronous circuit resistance to DPA. Their approach based on introducing temporal variations to the acknowledgment signal, which in return, injects some randomization to the temporal properties of the power traces [45]. This technique can be classified under the noise injection DPA countermeasures, which are trying to reduce the leakage SNR.

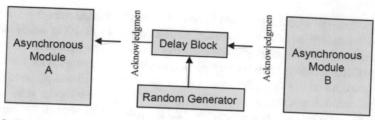

Fig. 19.8 The proposed randomly time-shifted acknowledgment circuit [45]

As shown in Fig. 19.3b, c, the acknowledgment signal is sent to the previous stage to guarantee that the current data is stored and previous stage can change the data value; acknowledgment signal is responsible of avoiding any data overwriting. The work in [45] proposed the circuit shown in Fig. 19.8. The concept of operation is to insert a controlled delay element in the acknowledgment signal path. This delay element is reconfigurable by using a random number generator. The proposed technique is applied to a DES encryption core. Simulation results showed that, after running all the possible input combinations of the DES Sbox, DPA attacks were not able to extract the correct key out of the hardened design.

19.4.3 Fault-Injection-Resistant Asynchronous Circuits

Injecting faults to hardware can be done with contact by using active probes, or without contact by using heavy ion radiation, also known as single event upset (SEU) [46, 47]. The effect of these injected faults may cause delay variation, single/multiple bit flips, and even permanent damage to the target transistors. Asynchronous circuits, particularly DI-encode circuits, are immune to delay variation. However, they can be very susceptible to bit flips. As explained in Fig. 19.4, dual-rail encoding must strictly follow the coding state diagram (that applies to any DI-encoding and the control path of bundled-data as well). Single bit flip can cause incorrect computation or the circuit to stall and enter a deadlock state [48]. Jang and Martin [47] studied this effect and proposed different asynchronous buffer designs which are SEU-tolerant.

Monnet et al. presented in [49] three fault models and studied the behavior of asynchronous circuit under these fault models. Their first fault model is "delay faults," against which asynchronous circuits are inherently immune. The second model is the "transient faults," which is defined as injected bit flips into the combinational part of the asynchronous circuit. The third model is "memory bit flips," in which the fault is directly injected to one (or more) of the circuit Muller-gates, or injected to the combinational part; however, it got propagated to the Muller-gates and memorized. Following the above classification of fault models, the authors developed a tool that can quantify the fault sensitivity of each gate in the design. By using the results of this tool, they hardened a DES design and showed 250% of fault tolerance with 7.7% area and 18% speed overhead.

In another work presented by Bastos et al. in [50], the robustness of asynchronous circuits towards injected faults, particularly long-duration faults, is proven. The presented work proposed to use *signal of end operation* (SOE) which is a primary output signal that indicates the end of operation of the circuit. Examples of SOE signals are done signals in synchronous designs, and output request signals in asynchronous circuits. Based on this, the work presented a different classification of the failures:

- Failure filtered naturally (FTN): This is an injected fault that is masked by the circuit without provoking soft errors.
- Failure detectable naturally (FDN): In this case, the failure provokes a soft error that causes the SOE signal to not appear at the primary output within the expected time. In this case, the failure can be detected without any additional hardware.
- Failure non-detectable naturally (FNN): In this case, the SOE still can appear at the output even though there is a soft error injected during the operation. This class needs additional hardware to detect the error.

As previously discussed, DI-encoded handshaking, in most cases, will show a deadlock in case of having faults injected to the circuits which cause bit-flips. The work suggested using this built-in response to error as an indication of a fault-injection. This work provides detailed analysis to prove the higher ability of asynchronous circuits to naturally detect errors compared to synchronous counterparts. Which demonstrates the efficiency and natural immunity of asynchronous circuit against fault injection.

Results of this case study are shown in Fig. 19.9, which illustrates the behavior of the two DES versions (synchronous and asynchronous), when transient faults are

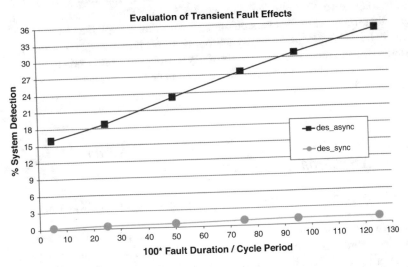

Fig. 19.9 DES crypto-processor case study: system's ability to detect faults in function of the transient-fault duration. The asynchronous DES detection rate is significantly higher than the synchronous version. "Reproduced with permission" Bastos et al. [50]

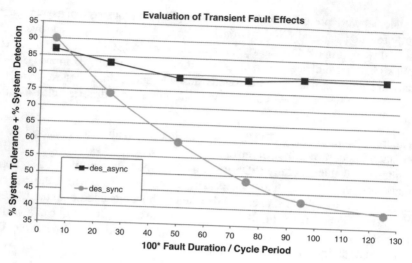

Fig. 19.10 Asynchronous DES demonstrated stronger intrinsic ability to tolerate/mitigate LDT faults. "Reproduced with permission" Bastos et al. [50]

injected. As shown in the graph, the asynchronous DES can naturally detect 15–30% of the injected faults compared to 1–2% in the synchronous version. Fig. 19.10 represents the ability of the two implementations to tolerate and detect faults in one curve. The synchronous DES demonstrates a downward trend with the increase of the fault duration. In contrast, the asynchronous DES shows almost a constant trend (around 80% of the injected faults). The curve shows the ability of synchronous DES to tolerate very narrow faults (around 5% of the clock period); this is the only region that the synchronous version outperforms; this is explained by the sensitivity of the handshaking protocol in asynchronous to bit flips. However, when the fault duration starts to increase, the benefits of asynchronous circuit starts to kick-in. This case study showed how asynchronous circuit can be effective towards long-duration transient faults.

19.4.4 Security-Primitives and Hardware Assurance Using Asynchronous Circuits

19.4.4.1 Asynchronous True Random Number Generator (TRNG)

Hardware security-primitives are essential components for building security and assurance frameworks into microelectronics circuits. Random number generators are one of the basic building blocks needed in hardware security applications, such as cryptography, and identity management. In addition to their use in cryptography,

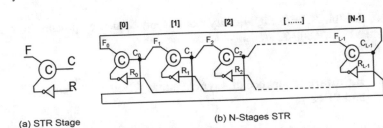

(a) STR Stage   (b) N-Stages STR

Fig. 19.11 The architecture of self-timed rings (STR). "Reproduced with permission" Yahya et al. [51]

they are frequently used to introduce randomness to the chip operation as a security countermeasure. One example is discussed earlier in Sect. 19.4.2, Fig. 19.8 [45].

Asynchronous circuits have very interesting self-timing features that can be used to construct highly efficient true random number generators (TRNGs). Yahya et al. studied in detail the behavior of self-timed rings (STR) [51, 52].

Fig. 19.11a shows the structure of a ring stage. It consists of a Muller-gate and an inverter. In each stage, the input which is connected to the previous stage is marked F (Forward) and the input which is connected to the following stage is marked R (Reverse); C denotes the output of the stage. Using this basic STR stage, one can construct N-stage ring as depicted in Fig. 19.11b. The operation of the ring is based on the concept of *Tokens* and *Bubbles*. Stage$_i$ contains a token if its output C_i is not equal to the output C_{i+1} of Stage$_{i+1}$. In contrast, Stage$_i$ contains a bubble if its output C_i is equal to the output C_{i+1} of Stage$_{i+1}$:

$$C_i \neq C_{i+1} \iff \text{Stage}_i \leftarrow \text{Token}$$
$$C_i = C_{i+1} \iff \text{Stage}_i \leftarrow \text{Bubble}$$

Denoting the number of tokens and bubbles as NT and NB, respectively, NT must be an even number to guarantee the oscillation of the ring. Each stage of the ring contains either a token or a bubble, $NT + NB = N$, where N is the number of ring stages. Suppose that there is a token in Stage$_i$, this token will move to Stage$_{i+1}$, if and only if Stage$_{i+1}$ contains a bubble. The behavior of STR ring stage has an analog phenomenon known as *Charlie Effect* and *Drafting Effect*. Due to these two effects, STR produces two different modes of oscillation: "evenly spaced" and "burst" modes. In the evenly spaced mode, the events inside the ring are equally spaced in time. In the burst mode, the events are spaced in time in a non-homogenous way. A detailed model is presented to capture these analog effects in digital simulations [51].

By using STR interesting features [52–54], Cherkaoui et al. [53] presented a design of true random number generator (TRNG). The presented design harvests the events jitter in STR to generate a random number at high data-rate. The presented architecture is implemented on FPGAs from Altera and Xilinx. The implementation provides 16 Mbits/s, with high-quality random bit sequences passing both FIPS 140-1 and NIST SP 800-22 statistical tests.

19.4.4.2 Hardware Assurance in Asynchronous Circuits: Trojans and PUFs

Complexity of the microelectronics industry creates a global supply chain which is very challenging to tightly control. Hardware Trojans, reverse engineering, and overproduction are examples of the threats that can be introduced within the supply chain. These threats apply to analog, digital, synchronous, and asynchronous hardware. Inaba et al. showed that a malicious skilled designer with enough knowledge about asynchronous circuit can insert a hardware Trojan without a significant overhead. In their proposed work, they also presented several Trojan detection methods which are already applied to synchronous circuits, and they showed their potential in detecting Trojans in asynchronous circuits as well [55].

Hardware Trojan is a malicious hardware with a payload and a trigger, which is not part of the design. Trojans are inserted with harmful intension and can be classified based on the adversary's targeted hardware metric:

- Functionality: Trojan changes the hardware function.
- Performance: Trojan degrades the target-hardware performance.
- Information leakage: stealthy Trojan that aims to leak sensitive information.
- Denial of service: Trojan that causes the devise to function.

The focus of the proposed work in [55] was on the last two categories of hardware Trojans. Exploiting the sensitivity of asynchronous circuit to the handshaking protocol, a MOUSETRAP [56] based pipeline is designed, and a Trojan is inserted to the acknowledgment path. The contaminated MOUSETRAP pipeline is then used within and asynchronous AES crypto-processor. The area overhead of the inserted Trojan is significantly small (<0.1%) compared to the total AES area. Same experiment is repeated with an asynchronous network on chip router (ANOC). The Trojan this time targeted to leak information. The overhead of the router Trojan is 1.8% of the total router area. The author after this applied some deep learning algorithms to detect the inserted Trojan. Both support vector machine (SVM) and neural network (NN) are applied to the test case. Results showed true positive ratio (TPR) between 41% and 100% and true negative ratio (TNR) between 55% and 65% for SVM classifier. The NN classifier results are TPR between 24% and 94%, and TNR between 78% and 80%.

Physical unclonable functions (PUFs) are another important hardware primitive that has many applications in hardware security and assurance. PUFs harvest the process-specific characteristics which are assumed to be *"unique"* in each die even from the same silicon wafer and *"reliable"* over a wide range of operating conditions and throughout the chip lifetime. PUF's purpose is to construct a fingerprint of the silicon die. Chowdhury et al. proposed a design of weak asynchronous RESet (ARES) PUF [57]. In this PUF design, null conventional logic (NCL) is used as the entropy source. The key idea is to use the startup conditions of NCL threshold gates as the source of entropy. The transistor sizing of the NCL gates are done very carefully to reach a perfect balance, from which process variability can create unbiased yet random bit. The proposed work developed a design methodology that is based on linear delay matching and genetic algorithms. The developed method is applied to NCL PUF, and results are obtained by HSPICE and silicon fabrication. Preliminary silicon results of fabricated 30 samples using the proposed PUF sizing method showed 34% improvement in uniqueness compared to the standard sizing.

19.4.5 Asynchronous Crypto-Processors

Encryption is one of the basic functions in any modern connected system. Software implementation on general purpose processors are not as efficient or secure as dedicated hardware accelerators [58]. Crypto-processors are commonly used to accelerate the encryption process, and their implementation has to be secured towards side-channel attacks. Implementation of secure crypto-processor has been always one of the most attractive areas for asynchronous hardware designers. This subsection covers some of these designs.

Yu et al. presented a dual-rail implementation of the AES algorithm that is DPA resistant [59, 61]. They are implemented using the two main asynchronous architectures; bundled-data and dual-rail circuits. Different methods are applied to protect the design against power and timing attacks. Results showed the importance of balanced physical design to fully prevent data leakage through power traces.

Bouesse et al. designed another AES core that is dual-rail implementation of the AES-128, AES-192, and AES-256 [60]. The design was fabricated using 130 nm CMOS technology from STMicroelectronics. At nominal voltage (1.2 volts), the fabricated chip consumes 10 nJ and encrypting 141 Mbits/s for the AES-128 operation. They demonstrated the efficiency of using dynamic voltage scaling (DVS), where the fabricated chip could work correctly at 0.4 volt.

Shang et al. proposed a highly balanced security latches in their pipeline structure to avoid data leakage in their AES implementation [61]. The work used Petri-Nets high-level description that is converted into David Cells (consists of memory element and control hardware). A balanced latch design was used to build the AES core pipeline with maximum power balance. Results showed that the proposed asynchronous AES core is 33% performance and 28% power consumption better than the synchronous counterpart.

Focusing on Wireless Sensor Networks (WSN), Otero et al. [62] proposed a hardware-software codesign of an AES crypto-processor. In their design, the AES core is implemented using dual-rail logic combined with a low power microcontroller. The proposed design offers 30× increase in performance, however, it reduces the node lifetime. To address this challenge, power gating techniques are applied, and the results showed reduction lifetime differences between pure software and hardware/software implementation to only 66%.

19.5 Case Study I: Low Power Asynchronous AES Core

In this subsection, we present application-specific integrated circuit (ASIC) design for an asynchronous AES crypto-processor. The design is implemented in dual-rail logic, and fabricated using UMC 130 nm CMOS process. Two versions of the AES core were fabricated on the same chip, one is asynchronous, and the other is synchronous. Results demonstrate how asynchronous circuits can be used to reduce power consumption and enhance DPA resistant [64, 65].

The design is an AES-128 encryption/decryption core in counter mode of operation (CTR) [66]. The first challenge designers face when it comes to crypto-processors is the chip level planning. The required number of input/output pins is enormous and usually unaffordable due to silicon area overhead and/or package and PCB complications. A common solution is to feed the processor with the input data serially and then use deserializer/serializer (SerDes) at the input/output IC interface. In our design, we wanted to be able to feed the data in real time to enable different security evaluation, including timing attack. We do not want to end up with a fast core and slow input/output interface. As a result, the number of pins and pipeline depth must be selected carefully so that the core determines the maximum throughput not the SerDes block.

The top-level interface between the SerDes and the core is shown in Fig. 19.12. Deserializer block has two 16-bit dual-rail ports for data and key. It constructs the full plain text and key inputs and feed them to the processor. In the same way, the serializer takes the encrypted data output and serialize it on the main chip IOs.

The core implements four main functions according to the AES algorithm (AddRound, SubByte, ShiftRow, and MixColumn) as shown in Fig. 19.13. Core design is based on one full round and then looping over this round to consume

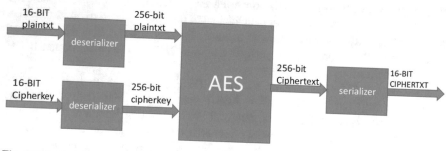

Fig. 19.12 The AES core interface with deserializer/serializer. "Reproduced with permission" Elmeligy et al. [65]

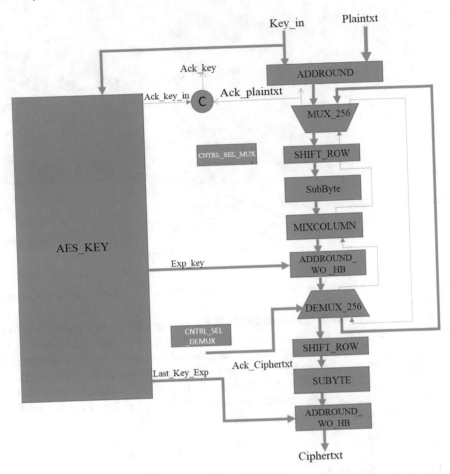

Fig. 19.13 The architecture of the AES core. "Reproduced with permission" Elmeligy et al. [65]

the minimum possible power. The core takes plaintext and cipher key as an input and produces ciphertext as an output after performing the above four functions 10 times (10 rounds); as specified in the standard, the last round does not contain the MixColumn function. Key expansion is done parallel to the encryption activity to achieve good throughput. The use of CTR mode enables the same core to perform encryption and decryption.

The presented architecture is modeled using Hardware Description Language (HDL) and synthesized targeting UMC 130 nm Faraday standard cells. Fig. 19.14 shows the micrograph of the chip including the AES Core surrounded by the deserializer/serializer hardware.

Table 19.1 shows the area and speed of the fabricated chip. The area is 0.64 mm^2 including *I/O*. The ciphering time (the time needed between starting the ciphering process until getting the cipher text) is measured at the typical voltage (@ 1.2 V)

Fig. 19.14 Micrograph of the chip including AES core, deserializer/serializer hardware. "Reproduced with permission" Elmeligy et al. [65]

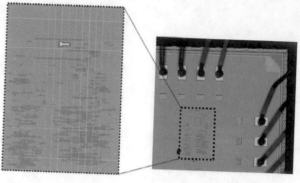

Table 19.1 The fabricated AES area, power, speed figures

Key length	Area	Power	Ciphering time	Technology
128	0.64 mm²	5.47 mw	300 ns	130 nm

and is equal to 300 ns. The fabricated core can encrypt 426 Mb/s under the typical conditions.

The fabricated core is tested under different supply voltage and guaranteed to work correctly while the supply voltage is reduced to 0.7v. By using dynamic voltage scaling (DVS), we could reduce the power consumption by 15% compared to operating the core on the typical voltage. This chip is one of the lowest ASIC implementations of the AES in the literature. It shows how asynchronous circuits can reduce the power of the crypto-processor, which implies lower current peaks with more immunity towards power analysis attacks.

19.6 Case Study II: LightSec (Lightweight Secure Enclave for Resource-Constrained Devices)

Resource-constrained devices are used in many applications such as smart grids, distributed control systems, and IoT. In many cases, these devices are connected and exchanging information in a decentralized network with an ad hoc architecture. This scenario introduces authentication issues of the nodes which dynamically connect to the network. The loose network architecture and the massive number of connected nodes require rigorous encryption and authentication primitives to be integrated into these devices. On the contrary, most of the current devices are deployed with little or no security to reduce their cost and/or to cope with their highly-constrained resources (particularly battery life). NIST initiated the lightweight cryptography project and announced, on April 2019, 56 algorithms as Round-1 candidates [67]. After analyzing the security of the proposed algorithms, NIST announced a shortlist that contains 32 algorithms in September 2019. Another independent effort was conducted in "CAESAR: Competition for Authenticated Encryption: Security,

Fig. 19.15 The secure enclave top-level diagram

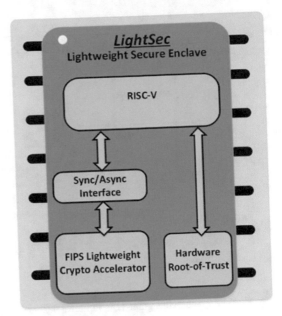

Applicability, and Robustness," in which six algorithms were announced as the final portfolio in February 2019 [68]. Out of the CAESAR six final algorithms, only "ASCON" is selected as a candidate in the NIST candidate list. Authors of the NIST candidates provided very basic hardware performance analysis of their algorithms. Aside from these studies, to the best of our knowledge, there is no independent evaluation of the hardware implementation of these algorithms. In this work, we are addressing this need by providing the community with independent, fair, silicon-proven results on the performance of some of these algorithms.

Secure enclaves are piece of hardware in which all sensitive encryption and authentication functions are performed. The purpose of a secure enclave is to isolate cryptographic functions from the general SoC fabric to provide HW security architects with a bounded problem to deal with. In this research, we propose a design of a lightweight secure enclave, shown in Fig. 19.15, which comprises an asynchronous lightweight crypto accelerator that is compliant with the coming FIPS for resource-constrained devices, and a physical unclonable function (PUF), which will act as an immutable root-of-trust (RoT). Both are integrated with an optimized RISC-V processor. Using asynchronous circuits for implementing the light-weight crypto (LWC) accelerator is giving the design different advantage in terms of power consumption, and immunity towards side-channel attacks as discussed earlier in the chapter.

One of the candidate algorithms is Grain-128AEAD. The algorithm consists of two main blocks, Linear Feedback Shift Register (LFSR), and a Non-linear Feedback Shift Register (NFSR). The design is similar to Grain-128a, with some

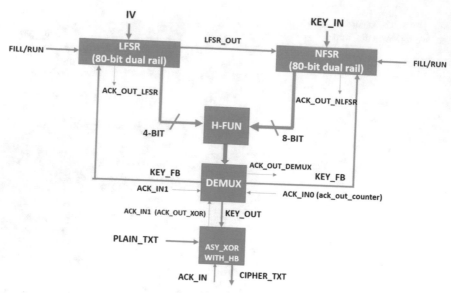

Fig. 19.16 Asynchronous grain-128a top-level design

modifications to support AEAD. Figure 19.16 shows the basic design of the Grain-128a algorithm using dual-rail channel interface.

The main challenges to efficiently implement this design are:

- Optimizing the acknowledgment tree, particularly in the LFSR, and NFSR
- Balance register design and layout to avoid any information leakage through power traces
- Balance the design and layout of the output stage (Demux, and XOR)

The design is currently under development. We are applying different mitigations to address any potential security flaw inside our asynchronous design. Functional simulation showed correct operation of the algorithm, and we are currently working on the synthesis phase targeting 22 nm CMOS process from Global Foundries.

19.7 Summary

In this chapter, we provided a quick tutorial on asynchronous circuit design, highlighting their advantages and drawbacks. In addition to this tutorial, the chapter presents a literature survey on the evolution of using asynchronous circuits in hardware security applications. Throughout the discussion, different security flaw that can happen during the design and their proposed solutions are highlighted and discussed. Finally, the chapter introduced two case studies, one for a low power asynchronous AES core and the other for an asynchronous lightweight secure enclave for resource-constrained devices.

Acknowledgments This work was partially supported by Semiconductor Research Corporation (SRC) and Battelle Memorial Institute.

References

1. The Cost of Malicious Cyber Activity to the U.S. Economy, The Council of Economic Advisers, February 2018
2. Y. Zhou, D. Feng, Side-channel attacks: Ten years after its publication and the impacts on cryptographic module security testing. IACR Cryptol. ePrint Arch. **2005**, 388 (2005)
3. M. Tehranipoor, C. Wang (eds.), *Introduction to Hardware Security and Trust* (Springer, 2011)
4. R. Focardi, R. Gorrieri (eds.), *Foundations of Security Analysis and Design: Tutorial Lectures* (Springer, 2003)
5. T. Popp, S. Mangard, E. Oswald, Power analysis attacks and countermeasures. IEEE Des. Test Comput. **24**(6), 535–543 (2007)
6. P. Kocher, J. Jaffe, B. Jun, Differential power analysis. Proceedings of the 19th annual international cryptology conference on Advances in Cryptology, in *CRYPTO '99*, (2017), pp. 388–397
7. S. Mangard, E. Oswald, T. Popp, *Power Analysis Attacks: Revealing the Secrets of Smart Cards* (Springer, New York, 2010)
8. W. Shan, F. Xingyuan, X. Zhipeng, A secure reconfigurable crypto IC with countermeasures against SPA, DPA, and EMA. IEEE Trans. Comp. Aided Des. Integr. Circ. Syst. **34**(7), 1201–1205 (2015)
9. C. Giraud, R.S.A. An, Implementation resistant to fault attacks and to simple power analysis. IEEE Trans. Comput. **55**(9), 1116–1120 (2006)
10. G. Ratanpal, R. Williams, T. Blalock, An on-chip signal suppression countermeasure to power analysis attacks. IEEE Trans. Dependable Secure Comput. **1**(3), 179–189 (2004)
11. J. Sparsø, *Principles of Asynchronous Circuit Design: A Systems Perspective* (Kluwer, Boston, 2001)
12. J. Sparsø, *Introduction to Asynchronous Circuit Design* (DTU Compute, Technical University of Denmark, 2020)
13. E. Yahya, M. Renaudin, QDI latches characteristics and asynchronous linear-pipeline performance analysis, in *Integrated Circuit and System Design. Power and Timing Modeling, Optimization and Simulation. PATMOS 2006. Lecture Notes in Computer Science*, ed. by J. Vounckx, N. Azemard, P. Maurine, vol. 4148, (Springer, Berlin/Heidelberg, 2006)
14. H. Zakaria, E. Yahya, L. Fesquet, Self adaption in SoCs, in *Autonomic Networking-on-Chip: Bio-inspired Specification, Development, and Verification*, ed. by P. Cong-Vinh, (CRC Press, Boca Raton, 2012)
15. E. Yahya, L. Fesquet, Asynchronous design: A promising paradigm for electronic circuits and systems, in *2009 16th IEEE International Conference on Electronics, Circuits and Systems – (ICECS 2009)*, (2009)
16. C.H. Van Berkel, M.B. Josephs, S.M. Nowick, Applications of asynchronous circuits. Proc. IEEE **87**(2), 223–233 (1999). https://doi.org/10.1109/5.740016
17. M. Davies, A. Lines, J. Dama, A. Gravel, R. Southworth, G. Dimou, P. Beerel, A 72-port 10G Ethernet switch/router using quasi-delay-insensitive asynchronous design, in *2014 20th IEEE International Symposium on Asynchronous Circuits and Systems, 12 May 2014*, (IEEE), pp. 103–104
18. X. Fan, O. Schrape, M. Marinkovic, P. Dähnert, M. Krstic, E. Grass, GALS design for spectral peak attenuation of switching current, in *2013 IEEE 19th International Symposium on Asynchronous Circuits and Systems, 19 May 2013*, (IEEE), pp. 83–90
19. Sheikh BR, Manohar R. An operand-optimized asynchronous IEEE 754 double-precision floating-point adder. In 2010 IEEE Symposium on Asynchronous Circuits and Systems, 3 May 2010 (pp. 151-162). IEEE

20. B. Hollosi, M. Barlow, G. Fu, C. Lee, J. Di, S.C. Smith, H.A. Mantooth, M. Schupbach, Delay-insensitive asynchronous ALU for cryogenic temperature environments, in *2008 51st Midwest Symposium on Circuits and Systems, 10 August 2008*, (IEEE), pp. 322–325

21. A. Bailey, A. Al Zahrani, G. Fu, J. Di, S. Smith, Multi-threshold asynchronous circuit design for ultra-low power. J. Low Power Electron. **4**(3), 337–348 (2008)

22. R. Zhou, K.S. Chong, B.H. Gwee, J.S. Chang, A low overhead quasi-delay-insensitive (QDI) asynchronous data path synthesis based on microcell-interleaving genetic algorithm (MIGA). IEEE Trans. Comput. Aided Des. Integr. Circ. Syst. **33**(7), 989–1002 (2014)

23. W.G. Ho, K.S. Chong, B.H. Gwee, J.S. Chang, M.F. Yee, A power-efficient integrated input/output completion detection circuit for asynchronous-logic quasi-delay-insensitive pre-charged half-buffer, in *2011 International Symposium on Integrated Circuits, 12 December 2011*, (IEEE), pp. 376–379

24. E. Yahya, L. Fesquet, Y. Ismail, M. Renaudin, Statistical static timing analysis of conditional asynchronous circuits using model-based simulation, in *2013 IEEE 19th International Symposium on Asynchronous Circuits and Systems, 19 May 2013*, (IEEE), pp. 67–74

25. A. Yakovlev, P. Vivet, M. Renaudin, Advances in asynchronous logic: From principles to GALS & NoC, recent industry applications, and commercial CAD tools, in *2013 Design, Automation & Test in Europe Conference & Exhibition (DATE), 18 March 2013*, (IEEE), pp. 1715–1724

26. M. Renaudin, A. Fonkoua, Tiempo asynchronous circuits system verilog modeling language, in *2012 IEEE 18th International Symposium on Asynchronous Circuits and Systems, 7 May 2012*, (IEEE), pp. 105–112

27. N.E.C. Akkaya, B. Erbagci, R. Carley, K. Mai, A DPA-resistant self-timed three-phase dual-rail pre-charge logic family. IEEE International Symposium on Hardware Oriented Security and Trust (HOST), 2015, pp. 112–117

28. S. Guilley, L. Sauvage, F. Flament, V.-N. Vong, P. Hoogvorst, R. Pacalet, Evaluation of power constant dual-rail logics countermeasures against DPA with design time security metrics. IEEE Trans. Comput. **59**(9), 1250–1263 (2010)

29. N.-H. Zhu, Y.-J. Zhou, H.-M. Liu, Employing symmetric dual-rail logic to thwart LPA attack. IEEE Embed. Syst. Lett. **5**(4), 61–64 (2013)

30. Y. Monnet, M. Renaudin, R. Leveugle, Designing resistant circuits against malicious faults injection using asynchronous logic. IEEE Trans. Comput. **55**(9), 1104–1115 (2006)

31. Q. Ou, F. Luo, S. Li, L. Chen, Circuit level defences against fault attacks in pipelined NCL circuits. IEEE Trans. Very Large Scale Integr. (VLSI) Syst. **23**(9), 1903–1913 (2015)

32. J.J. Fournier, S. Moore, H. Li, R. Mullins, G. Taylor, Security evaluation of asynchronous circuits, in *International Workshop on Cryptographic Hardware and Embedded Systems, 8 September 2003*, (Heidelberg, Springer/Berlin), pp. 137–151

33. K.J. Kulikowski, M. Su, A. Smirnov, A. Taubin, M.G. Karpovsky, D. MacDonald, Delay insensitive encoding and power analysis: a balancing act [cryptographic hardware protection], in *11th IEEE International Symposium on Asynchronous Circuits and Systems, 14 March 2005*, (IEEE), pp. 116–125

34. G.F. Bouesse, M. Renaudin, S. Dumont, F. Germain, DPA on quasi delay insensitive asynchronous circuits: formalization and improvement. Des. Autom. Test Eur. **1**, 424–429 (2005)

35. D. Sokolov, J. Murphy, A. Bystrov, A. Yakovlev, Improving the security of dual-rail circuits, in *International Workshop on Cryptographic Hardware and Embedded Systems, 11 August 2004*, (Heidelberg, Springer/Berlin), pp. 282–297

36. N. Liu, K.S. Chong, W.G. Ho, B.H. Gwee, J.S. Chang, Low normalized energy derivation asynchronous circuit synthesis flow through fork-join slack matching for cryptographic applications, in *2016 Design, Automation & Test in Europe Conference & Exhibition, 14 March 2016*, (IEEE), pp. 850–853

37. J. Wu, Y. Shi, M. Choi, Measurement and evaluation of power analysis attacks on asynchronous S-box. IEEE Trans. Instrum. Measur. **61**(10), 2765–2775 (2012)

38. P. Kocher, J. Jaffe, B. Jun, Differential power analysis, in *Annual International Cryptology Conference, 15 August 1999*, (Springer, Berlin/Heidelberg), pp. 388–397

39. D. Das, S. Maity, S.B. Nasir, S. Ghosh, A. Raychowdhury, S. Sen, High efficiency power side-channel attack immunity using noise injection in attenuated signature domain, in *2017 IEEE International Symposium on Hardware Oriented Security and Trust (HOST), 1 May 2017*, (IEEE), pp. 62–67

40. D. Das, S. Maity, S.B. Nasir, S. Ghosh, A. Raychowdhury, S. Sen, ASNI: Attenuated signature noise injection for low-overhead power side-channel attack immunity. IEEE Trans. Circ. Syst. I: Regul. Pap. **65**(10), 3300–3311 (2018)

41. A. Singh, M. Kar, S. Mathew, A. Rajan, V. De, S. Mukhopadhyay, 25.3 A 128b AES engine with higher resistance to power and electromagnetic side-channel attacks enabled by a security-aware integrated all-digital low-dropout regulator, in *2019 IEEE International Solid-State Circuits Conference-(ISSCC), 17 February 2019*, (IEEE), pp. 404–406

42. M. Kar, A. Singh, S.K. Mathew, A. Rajan, V. De, S. Mukhopadhyay, Reducing power side-channel information leakage of AES engines using fully integrated inductive voltage regulator. IEEE J. Solid-State Circ. **53**(8), 2399–2414 (2018)

43. K. Tiri, I. Verbauwhede, A logic level design methodology for a secure DPA resistant ASIC or FPGA implementation, in *Proceedings Design, Automation and Test in Europe Conference and Exhibition, 16 February 2004*, vol. 1, (IEEE), pp. 246–251

44. S. Guilley, F. Flament, R. Pacalet, P. Hoogvorst, Y. Mathieu, Security Evaluation of a Secured Quasi-Delay Insensitive Library. DCIS, full text in HAL. http://hal.archives-ouvertes.fr/hal-00283405/en. 2008

45. F. Bouesse, M. Renaudin, G. Sicard, Improving DPA resistance of quasi delay insensitive circuits using randomly time-shifted acknowledgment signals, in *Vlsi-Soc: From Systems to Silicon, 2007*, (Springer, Boston), pp. 11–24

46. M.C. Hsueh, T.K. Tsai, R.K. Iyer, Fault injection techniques and tools. Computer **30**(4), 75–82 (1997)

47. W. Jang, A.J. Martin, Seu-tolerant qdi circuits [quasi delay-insensitive asynchronous circuits], in *11th IEEE International Symposium on Asynchronous Circuits and Systems, 14 March 2005*, (IEEE), pp. 156–165

48. E. Yahya, H. Zakaria, Y. Ismail, Deadlock detection in conditional asynchronous circuits under mismatched branch selection, in *2015 IEEE International Conference on Electronics, Circuits, and Systems (ICECS), 6 December 2015*, (IEEE), pp. 596–600

49. Y. Monnet, M. Renaudin, R. Leveugle, Designing resistant circuits against malicious faults injection using asynchronous logic. IEEE Trans. Comput. **55**(9), 1104–1115 (2006)

50. R.P. Bastos, G. Sicard, F. Kastensmidt, M. Renaudin, R. Reis, Asynchronous circuits as alternative for mitigation of long-duration transient faults in deep-submicron technologies. Microelectron. Reliab. **50**(9-11), 1241–1246 (2010)

51. E. Yahya, O. Elissati, H. Zakaria, L. Fesquet, M. Renaudin, Programmable/stoppable oscillator based on self-timed rings, in *2009 15th IEEE Symposium on Asynchronous Circuits and Systems, 17 May 2009*, (IEEE), pp. 3–12

52. O. Elissati, S. Rieubon, E. Yahya, L. Fesquet, Self-timed rings: a promising solution for generating high-speed high-resolution low-phase noise clocks, in *IFIP/IEEE International Conference on Very Large Scale Integration-System on a Chip, 27 September 2010*, (Springer, Berlin/Heidelberg), pp. 22–42

53. A. Cherkaoui, V. Fischer, A. Aubert, L. Fesquet, A self-timed ring based true random number generator, in *2013 IEEE 19th International Symposium on Asynchronous Circuits and Systems, 19 May 2013*, (IEEE), pp. 99–106

54. Y. Zhang, J. Jiang, Q. Wang, N. Guan, A self-timed ring based true random number generator on FPGA, in *2018 14th IEEE International Conference on Solid-State and Integrated Circuit Technology (ICSICT), October 2018*, (IEEE), pp. 1–3

55. K. Inaba, T. Yoneda, T. Kanamoto, A. Kurokawa, M. Imai, Hardware Trojan insertion and detection in asynchronous circuits, in *2019 25th IEEE International Symposium on Asynchronous Circuits and Systems (ASYNC), 12 May 2019*, (IEEE), pp. 134–143

56. M. Singh, S.M. Nowick, MOUSETRAP: high-speed transition-signaling asynchronous pipelines. IEEE Trans. Very Large Scale Integr. (VLSI) Syst. **15**(6), 684–698 (2007)

57. S. Chowdhury, R. Acharya, W. Boullion, A. Felder, M. Howard, J. Di, D. Forte, A weak asynchronous RESet (ARES) PUF using start-up characteristics of null conventional logic gates, in *IEEE International Test Conference (ITC)*, (IEEE, 2020)

58. E. Yahya, Y. Ismail, Chapter 9: Hardware security: side-channel attacks and hardware Trojans (Security, 2018), in *Information Security: Foundations, Technologies and Applications'*, pp. 191–214. https://doi.org/10.1049/PBSE001E_ch9. IET Digital Library. https://digital-library.theiet.org/content/books/10.1049/pbse001e_ch9

59. Yu A, Brée DS. A clock-less implementation of the AES resists to power and timing attacks. In International Conference on Information Technology: Coding and Computing, 2004. Proceedings. ITCC 2004, 5 April 2004 (2, pp. 525-532). IEEE

60. G.F. Bouesse, M. Renaudin, A. Witon, F. Germain, A clock-less low-voltage AES crypto-processor, in *Proceedings of the 31st European Solid-State Circuits Conference, 2005. ESSCIRC 2005, 12 September 2005*, (IEEE), pp. 403–406

61. Z. Liu, Y. Zeng, X. Zou, Y. Han, Y. Chen, A high-security and low-power AES S-box full-custom design for wireless sensor network, in *2007 International Conference on Wireless Communications, Networking and Mobile Computing, 21 September 2007*, (IEEE), pp. 2499–2502

62. D. Shang, F. Burns, A. Bystrov, A. Koelmans, D. Sokolov, A. Yakovlev, High-security asynchronous circuit implementation of AES. IEEE Proc. Comput. Digit. Tech. **153**(2), 71–77 (2006)

63. C.T. Otero, J. Tse, R. Manohar, AES hardware-software co-design in WSN, in *2015 21st IEEE International Symposium on Asynchronous Circuits and Systems, 4 May 2015*, (IEEE), pp. 85–92

64. S. Agwa, E. Yahya, Y. Ismail, Power efficient AES core for IoT constrained devices imple-mented in 130nm CMOS, in *2017 IEEE International Symposium on Circuits and Systems (ISCAS)*, (Baltimore, 2017), pp. 1–4. https://doi.org/10.1109/ISCAS.2017.8050361

65. N. Elmeligy, M. Amin, E. Yahya, Y. Ismail, 130 nm low power asynchronous AES core. IEEE International Symposium on Circuits & Systems (ISCAS). 2017

66. F. Charot, E. Yahya, C. Wagner, Efficient modular-pipelined AES implementation in counter mode on ALTERA FPGA, in *International Conference on Field Programmable Logic and Applications, 1 September 2003*, (Springer, Berlin/Heidelberg), pp. 282–291

67. NIST Lightweight Cryptography Project. https://csrc.nist.gov/Projects/Lightweight-Cryptography. Retrieved 31 August 2020

68. CAESAR: Competition for Authenticated Encryption: Security, Applicability, and Robustness. https://competitions.cr.yp.to/caesar.html. Retrieved 31 August 2020

Chapter 20
Microfluidic Device Security

**Mohammed Shayan, Tung-Che Liang, Ramesh Karri,
and Krishnendu Chakrabarty**

20.1   Introduction

Functional diversification is expected to drive the growth of hardware computing beyond the end of Moore's law. Medical application is expected to be a system driver of such diversification [9, 20]. Microfluidic technologies enable miniaturization of laboratory-based biochemical protocols. A microfluidic biochip or lab-on-a-chip (LoC) performs biochemical reactions by consuming nano-/pico-liter volume of reagents [64]. These platforms use less volume of samples and reagents and provide quicker results than the traditional lab. Also, these platforms enable automation that reduces the reliance on high-skilled personnel. Digital microfluidic biochip (DMFB) and continuous flow-based microfluidic biochip (CFMB) are examples of such biochip platforms. CFMBs manipulate fluid flow through a network of micro-channel by actuating pressure-driven micro-valves [41]. DMFB offers a programmable fluidic platform in which discrete fluid droplets can be manipulated through electrical actuations [61].

Biochips have made a significant impact on health care by redefining point-of-care diagnostics [12], drug research, and development [26], as well as personalized medicine [59]. They make diagnostics affordable and accessible compared to traditional bio-labs. For example, the immunoassay platform shown in Fig. 20.1a,b performs low-cost detection of measles and rubella viruses using a single drop

M. Shayan (✉) · R. Karri
Tandon School of Engineering, New York University, Brooklyn, NY, USA
e-mail: mos283@nyu.edu; rkarri@nyu.edu

T.-C. Liang · K. Chakrabarty
Duke University, Durham, NC, USA
e-mail: tung.che.liang@duke.edu; krish@ecd.duke.edu

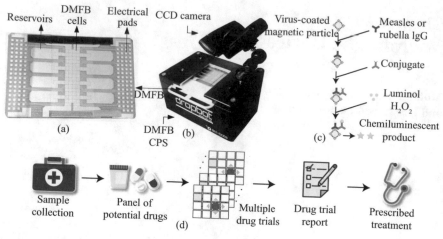

Fig. 20.1 An open-source DMFB system: (**a**) DMF biochip and (**b**) DropBot platform (source: https://sci-bots.com/). (**c**) Rubella/measles immunoassay description [40]. (**d**) A personalized drug development process for cancer patients (A high throughput screening system to identify actionable treatments for cancer patients, 2016; SEngine—Precision medicine, 2019)

of blood (Fig. 20.1c). This was deployed in refugee camps where many basic life necessities were inaccessible (Wheeler group at Kakuma refugee camp in Kenya, 2016) [16]. Further, biochips enable diagnostics that were not possible in traditional bio-labs. For example, microfluidics enables the development of personalized medicine needed for cancer patients by running thousands of parallel tests on patient's bio-sample, as shown in Fig. 20.1d. Using traditional lab methods, it is not possible to exactly determine which of the many clinical trials that are on offer is appropriate for a particular patient (A high throughput screening system to identify actionable treatments for cancer patients, 2016; SEngine—Precision medicine, 2019).

The global biochip market is projected to reach to $12.3 billion by 2025 from $5.7 billion in 2018 (Zion market research, 2019). This is corroborated by the sales (Shipment of 3,000,000th test, 2019), investment (10x genomics funding, 2019), and acquisitions (Illumina-press release, 2018) reported by microfluidic companies. Baebies' SEEKER, a DMFB-based immunoassay platform, received FDA approval in 2016 (FDA advisors back approval of Baebies' SEEKER analyzer for newborns, 2016). Since then, Baebies has shipped three million tests and raised $13 million in funding (Shipment of 3,000,000th test, 2019). SEEKER provides a high throughput quantitative measurement of deadly diseases from the dried blood spot of newborns. $10\times$ Genomics uses a combination of microfluidics, and bioinformatics for single-cell analysis. Since its founding in 2012, it has received $243 million in funding until 2018 (10x genomics funding, 2019).

20.1.1 Biochip Security: Motivation

As biochips penetrate the market, security and trust issues are being uncovered. Biochips have multiple usage scenarios such as in a biomedical research lab and in a remote location. Depending on the usage scenario, biochemical assay implementation faces different threats. To highlight this issue, we describe three real-life scenarios:

1. A disgruntled employee can tamper with the biochemical experiments to take revenge on colleagues or management (5 cases of AIDS-study sabotage reported, 1986). Recently, a chemist at a water treatment plant was found guilty of tampering with a colleague's water test for months (Jealousy led Montana chemist to taint colleague's water tests, 2019). The usage of biochips in such labs increases the risk of such attacks due to the biochip's easy controllability.
2. An unscrupulous biochip designer, who uses fraudulent or falsified claims, is a threat to users, investors, and regulators. The *Edison* microfluidic blood testing device from *Theranos* faced technical, commercial, and legal challenges over the scientific basis of its technologies (Theranos voids two years of Edison blood test results, 2016). Such incidents gather a lot of negative press and hamper progress in such technologies (Theranos effect, 2019).
3. Studies have flagged security flaws in medical devices, e.g., tampering of controls, denial-of-service, data theft, and ransom attacks (It is insanely easy to hack hospital equipment, 2014). This has led to recall of a large number of the medical devices and a re-evaluation of their regulations (When medical devices get hacked, hospitals often do not know it, 2018). A biochip cyberphysical system (CPS) is similar to current medical devices, which consists of hardware, software, and network connections [14]. As biochips are becoming an integral part of healthcare services, these threats become more pronounced.

The above threats may lead to a loss of revenue and trust or, more importantly, jeopardize the well-being of biochip users [57]. They can cause denial-of-service and wrong bioassay outcomes. Addressing these threats becomes even more important as biochips are being advocated for use in artificial-intelligence-based decision making [42] and with the emergence of miniaturized versions of oneself for medical tests [67].

20.1.2 Biochip Security: Taxonomy

Given the security critical nature of the biochip application, the research community has focused its efforts on discovering its attack space and devising countermeasures. The biochip CPS design flow is shown in Fig. 20.2. Each of the design stage faces threat from malicious actors: (1) a remote user can tamper with the controlling software through a network-based attack; (2) a malicious end user can tamper

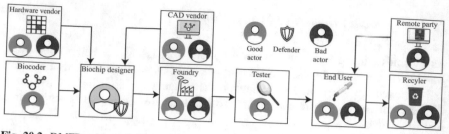

Fig. 20.2 DMFB product design chain with good and bad actors

Fig. 20.3 Taxonomy of the microfluidic biochip security

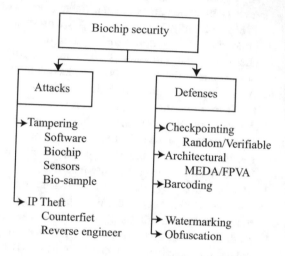

with the biochip platform, the sensors, and the controller; (3) a rogue actor in a fabrication unit can insert stealthy modification in the biochip—called Trojans; and (4) bio-sample vendor can provide tampered samples. These attack can result in the manipulation of the end result, denial-of-service, loss of revenue, and loss of brand value.

Bio-protocol development requires large investments for cross-domain innovations in biochemical analysis, microfluidics, and cyberphysical systems. For example, pharmaceutical companies invest large sums of money and person-hours in a slow and expensive drug development process laced with tough regulations. This process is prone to stealing of sensitive research data (Drug development and intellectual property theft, 2016). For rapid and low-cost drug development, pharmaceutical companies are using various types of microfluidic biochips that minimize the assay time and reagent requirement [26]. This opens a new avenues for IP theft. This not only leads to loss of revenue but also to counterfeit products entering the market. Figure 20.3 summarizes the footprint of the biochip security research so far.

In the rest of the chapter, Sect. 20.2 describes various attacks that threaten the integrity of biochip systems and discusses the corresponding defenses. Section 20.3

explains the security of bio-samples and benchtop experiments. Bio-IP protection is discussed in Sect. 20.4, and Sect. 20.5 concludes this chapter.

20.2 Biochip System Integrity

A fully integrated biochip system consists of a controller, sensor feedback, and network connection [68]. A bioassay description (represented as a sequencing graph [5]) is synthesized to an actuation sequence that realizes the bioassay through various fluidic operations on the bioassay (Fig. 20.4). However, the fluidic operations are susceptible to multiple manufacturing imperfections, which can lead to run-time faults. To detect such defects and for error recovery in the biochip operations [8, 21], run-time monitoring through sensor feedback is required. CCD camera and/or capacitive sensors are used to monitor a droplet location and size on the DMFB [37]. CCD cameras are more popular due to their precision [37]. The image is cropped into sub-images to focus on an area of interest (e.g., an electrode). These sub-images are correlated with a template to monitor the droplet occupancy and size at the desired locations. The DMFB can also be connected to a network for run-time monitoring of assay operations, result analysis, and software update (Laboratory monitoring, 2018).

20.2.1 Threat Model

Biochips have multiple usage scenarios, such as in a biomedical research lab and a remote location. Depending on the usage scenario, biochemical assay implementa-

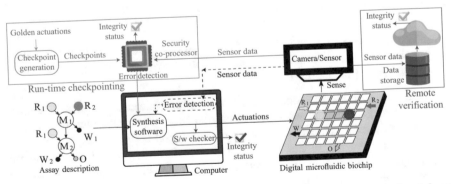

Fig. 20.4 A DMFB cyberphysical system with multiple defenses. The software checker defense is in red, and the remote verification is in blue. The checkpoint-based defense is in green, which replaces the dotted portion of control loop

tion faces different threats. The responses to the ensuing four questions clarify the threat model for biochip security:

Who Presents a Risk and Why? The attacker—who is in a remote location or near the biochip—could be a competitor seeking to bring disrepute to the biochip designer [60]. The proximity attacker can be an insider seeking to harm the end user by manipulating the biochip results [43] or by denying service (DoS attack) [45, 53, 54].

How Does One Launch an Attack? The cyberphysical biochip system comprises controllers, software, network interface, sensors, and pneumatic actuators. The designer can source them from third-party vendors [1, 65]. One can connect the biochip to a network for software updates and process the results online. Informed by this biochip supply chain, the attacker can launch an attack as follows:

1. Exploit the in-built hardware or software Trojan to access the biochip controller [18, 47, 55].
2. Use malware to gain control of the network and to manipulate the control software or stored actuation sequence [15, 27].
3. Compromise the biochip by inducing faults in the controller or actuators using electrical probes or lasers [45, 53, 54].

What Are the Constraints on an Attacker? The attacker manipulates the results of the biochip in a stealthy and untraceable way. To do this, the attacker has to evade detection by the sensors. The defender can monitor the biochip using sensors [35, 37].

Who Are the Trusted Actors? The biochip designer is the defender. The designer trusts the biochip platform and the end user.

20.2.2 Case Study

Based on the above threat model, an attacker can maliciously modify the actuation sequence referred to as actuation tampering. This is elucidated through a case study of glucose assay [2, 33, 50, 51].

Glucose Assay To test the concentration of the glucose level in a patient's sample, serial dilution is used to generate the calibration curve [3]; see the blue solid line in Fig. 20.6. Given any sample, its glucose concentration is measured by comparing the absorbent reaction to the calibration curve. For example, if the absorbent reaction rate of a sample is measured as 0.001, the concentration of this sample is interpolated as 100 mg/dL.

The golden sequencing graph of a glucose assay is shown in Fig. 20.5a, where Chain 1 and Chain 2 are used for generating a series of different concentrations. These concentrations with their absorbent rates are measured to draw the golden

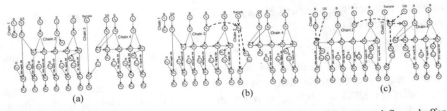

(a)   (b)   (c)

Fig. 20.5 The golden glucose test assay and two malicious attacks. B, GS, and S are buffer, glucose solution, and sample, respectively; the nodes I_i, M_i, D_i, W_i, and S_i correspond to dispensing, merging, detection, the discarding of a waste droplet, and splitting operations, respectively. (a) The graphs for Chain 1 and Chain 2 are used for generating the calibration curve, and the concentration of the sample in Chain 3 is tested and interpolated with the calibration curve. (b) An attack is stealthily introduced in the second splitting operation. The splitting ratio is changed from 1 : 1 to 2 : 1. (c) An attack is inserted during high-level synthesis; the outcome droplet of Chain 3 is discarded and the waste droplet in Chain 2 is transported for detection

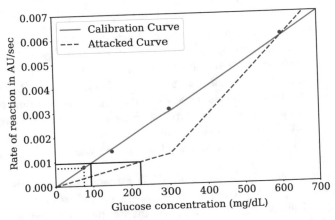

Fig. 20.6 Golden glucose calibration curve and the attacked calibration curve

calibration curve. The sample in Chain 3 is tested to get the absorbent reaction rate, and by interpolating with the golden curve, the concentration of the sample is determined (as 94 mg/dL). An attacker can carry out one of the following attack:

Miscalibration This attack changes the volume ratio of first split in Chain 2, and the concentrations of the child nodes are therefore changed accordingly, see Fig. 20.5b. If the attack succeeds without being detected, the calibration curve is also altered, which is plotted as a red dot line in Fig. 20.6. The glucose concentration of the sample is mistakenly interpreted as 225 mg/dL, and the outcome for the patient may be hypoglycemia because of the insulin overdoes.

Contamination This stealthy attack uses a waste droplet in Chain 2, then discards the valuable sample, and sends to the waste reservoir. The reconstructed graph is shown in Fig. 20.5c, where the two swapped operations are doubly circled in red. In this case, the glucose level is severely underestimated. As a result, the patient remains in hyperglycemia, which may lead to headache, fatigue, blurred vision, or even diabetic coma.

Parameter Manipulation A critical step in bio-protocol development lies in the choice of parameters that optimize the assay. This requires systematic exploration of the design space to understand the interplay between parameters (Understanding variability in DNA-amplification reactions, 2018), which include mixing time, incubation time, mixing ratio, reagent volume, and concentration. The developer, after many trials, determines the parameter values. Any tampering of the parameter values during the execution of the bio-protocol on a biochip can invalidate the outcome.

20.2.3 Checkpointing

To increase trust in such systems, a layer of security needs to be built into the biochip CPS. However, such security measures ought to be applicable in diverse usage scenarios such as in a biomedical lab and in a remote online/offline device. Sensor-based monitoring is a common defense applied in industrial control system security [38]. Biochip researchers have adopted sensor-based run-time monitoring of certain time steps at chosen biochip location—referred as *checkpoints*. Such measures can complement the existing software and network security measures. Table 20.1 presents a brief comparison of defense mechanisms in the context of DMFB security.

The earliest checkpoint defense uses CCD camera to monitor the DMFB at random time steps [57, 58]. The DMFB snapshots are processed to determine the run-time state of the biochip. By comparing the run-time state against the golden state over the entire execution cycles, the bioassay execution is validated (Fig. 20.4).

Table 20.1 Comparison of DMFB defense mechanisms

Defense	Tampering attack detection		Error recovery [37]	Application scenario		
	Software [3]	Hardware[54]		Biomedical lab	Online device	Offline device
Software checker [66]	✓	✗	✗	✓	✓	✓
Run-time checkpointing [55]	✓	✓	✓	✓	✓	✓
Remote verification [4]	✓	✓	✗	✓	✓	✗

Most of the DMFB systems have minimal computing resources to minimize cost. Due to the time required for image capture and processing, continuous run-time monitoring of all DMFB cells is not possible. For example, the work in [55] shows that no more than 20 checkpoints can be examined in an execution cycle. This constraint was derived by considering an image pattern matching algorithm implementation on a mid-range ARM Cortex-M3 microcontroller. To overcome this constraint, heuristic defenses based on *checkpoints* are employed [34, 55]. Here, a spatial and temporal subset of the steps executed on the DMFB is sampled and used to compare against the golden specification. Algorithms based on randomized, weighted, and module-less choices have been proposed to derive the checkpoints [10, 34, 55]. These methods help in boosting the probability of attack detection. Another way of increasing the chances of attack detection is to use platform with constrained capabilities. For this purpose, pin-constrained designs have been explored to maximize the probability of attack detection [56].

20.2.4 Provable Security

Checkpointing provides some assurance of integrity. Due to the monitoring resource limitation, the achievable probability of detecting an attack can be as low as 50% on a realistic embedded DMFB controller. This is not satisfactory and not likely to inspire confidence in users of these DMFB systems. Recently, the Micro-Electrode-Dot-Array (MEDA) DMFB has been developed. From a security perspective, MEDA is promising as it overcomes the resource constraints of a traditional DMFB.

MEDA has a "sea-of-electrodes" (micro-electrodes) that can be dynamically grouped to act as an actuator for droplet movement [62, 63]. Each micro-electrode is integrated with activation circuitry and sensing modules that allow fine-grained control and real-time sensing of a droplet [30]. The sensor data specifies the droplet location, size, and shape—the *droplet map*.

The real-time droplet map can be compared with the golden droplet map for assay validation. A MEDA-based security method can utilize sensor information to automatically recognize fluidic operations, and construct the dependencies between these operations. This reconstructs the implemented bioassay, i.e., assay specification. Such setup enables formal verification by comparing the "implemented" bioassay against the "golden" bioassay [33].

Caveat The fine-grained control and sensing in MEDA is a double edge sword for security. It aids seamless monitoring of the entire biochip. And it also aids an attacker in launching stealthier attacks that were not feasible on a traditional DMFB [51]. MEDA-specific aliquot operation can be used to fine-grained manipulation of glucose assay outcomes [51]. MEDA-specific differential speeds in smaller (faster) and larger (slower) droplets can be used to launch stealthy manipulations, referred to as shadow attack [39, 50].

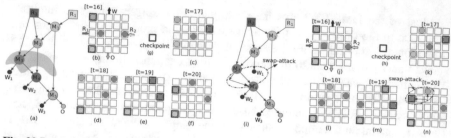

Fig. 20.7 (a) DAG for mixing reagents R_1 and R_2 using *REMIA* [22]. (b–e) DMFB snapshots that implement the highlighted portion of the DAG: (b) Snapshot after the mixing M_3 at time step $t = 16$; (c)–(f) droplet routing operations for subsequent assay operations. The checkpoints for monitoring are shown. (i) Modified DAG due to a swap attack. (j)–(n) Swap attack between the violet and green droplets ($t = 20$) is returned as a counter example to prove that the checkpoints are incorrectly placed

20.2.5 Security Verification

As checkpoints cannot monitor *all* possible behaviors of the DMFB, there is room for undetected attacks. This problem can arguably be solved by increasing the computation capacity and using integrated sensors (MEDA). Nevertheless, a biochip designer should have the tools to weigh the design choices rather than make an informed decision, i.e., either be content with the current low-cost processor or upgrade to a processor with higher computational ability. In other words, we need tools that can confirm if an attack can evade detection for a given resource constraint [46]. This can be achieved by a satisfiability-based exact analysis of a given (checkpoint-based) defense against a given attack. The SAT-based method can:

- verify that the given defense can thwart an attack on the bioassay implemented on the DMFB, or
- find an attack plan to evade the defense (if unsafe), and
- devise a fool-proof checkpoint mechanism using attack plans as counterexamples.

To reason about this formulation, the deductive power of satisfiability solvers can be used [49]. This can be demonstrated through the following example:

Example 20.1 Let us consider the DAG for mixing two input reagents using *REMIA* [22] as shown in Fig. 20.7a. This bioassay implementation on a 5×5 DMFB takes $T = 40$ time steps. Figure 20.7b–f shows the snapshots of the DMFB for time steps $t = 16, 17, \cdots, 20$.

Checkpoints are incorporated by the designer Fig. 20.7(g-h) and are highlighted on the DMFB grid in each time step in Fig. 20.7. For these set of checkpoints, the proposed exact analysis shows that two droplets can be swapped from time step

$t = 16$ to $t = 20$, as shown in Fig. 20.7j–n. The resultant modified bioassay is shown in Fig. 20.7i.

Caveat The checkpoint verification methodology does not consider faults in biochip implementation. The bioassay sequence is not strictly deterministic due to the (faults) error correction routines [31, 37]. Further, bioassay sequence can be modified in run time by conditional loops such as if–then–else statements or loops with non-constant number of iterations. Therefore, there is a need to devise defense and corresponding verification strategy that supports dynamic execution of bioassay.

20.3 Bio-Sample Security

Attacks on the integrity of a microfluidic system can occur even before bioassay execution begins. An attacker can tamper with the sample or the sample ID during sample packaging and shipping. Existing microfluidic system flows assume that the samples under investigation were collected, labeled, and shipped in a trustworthy manner. That is, the samples have not been contaminated or were not exposed to environmental conditions that would lead to fouling of the samples or alteration in its intrinsic biological behavior. In a more adversarial setting, trust in the sample collection requires that the samples were neither deliberately replaced by a malicious actor nor tampered with harmful reagents. Therefore, we must ensure that the samples were not intentionally tampered with (i.e., substituted or contaminated) by a malicious actor.

As an example of an evidence-level threat, we consider the DNA fingerprinting flow: an attacker can falsify blood and saliva samples containing DNA from a person other than the donor of these fluids [7]. Forged evidence can be realized using a strand of hair or drinking cup used by the target person. Forged DNA can then be extracted, amplified, and planted within the collected samples. The attacker may be either an intruder or a malicious technician in the forensics laboratory, and the attack does not require detailed knowledge of the forensic analysis platform.

To prevent this attack, crime scene investigators can use utilities to securely "barcode" the collected evidence cells. In the barcoding routine, evidence cells are supplemented with specially designed DNA molecules. These molecules do not change the properties of the evidence cells, but they allow the evidence cells to have unique DNA signatures at specific regions of the DNA. These barcodes are assumed to be generated from a probabilistic distribution whose parameters are known to only authenticated persons at the front end (molecular barcodes generation) and the back end (deciphering barcoded samples at the forensics laboratory). A cyber-level key management scheme is also required to control the interactions among the participating parties.

Fig. 20.8 Measurement of
DNA amplification using
gel electrophoresis

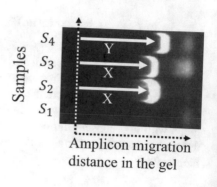

Samples

Amplicon migration
distance in the gel

20.3.1 Case Study: DNA Fingerprinting

We consider DNA fingerprinting as a case study for the proposed molecular barcoding scheme [25].

Deoxyribonucleic acid (DNA) is a molecule that represents the genetic material in all living organisms. It is composed of a chain of structures known as *nucleotides* (also known as *bases*) whose order forms the identity of a DNA sequence. Each nucleotide can be one of four structures: adenine (**A**), cytosine (**C**), guanine (**G**), and thymine (**T**). This sequence forms a single-stranded DNA (ssDNA) that is chemically bonded with another complementary ssDNA to form a double-stranded DNA (dsDNA) that has a double helix structure. Often, dsDNA is just referred to as DNA for short.

Agarose gel electrophoresis is a routinely used method for separating protein or DNA molecules based on their molecular weight, which is proportional to the size of DNA strands [28]. Hence, PCR products (amplicons) are size-separated by the aid of an electric field where negatively charged DNA molecules migrate toward an anode (positive) pole. The shorter the amplicon (i.e., the lower the molecular weight), the further the sample will reach on the gel. Figure 20.8 shows the outcome of gel electrophoresis for four samples used in our study. Samples S_2 and S_3 are samples that exhibit amplicon generation through DNA amplification, and the target amplicons contain a large number of base pairs (bp). Hence, the molecular weight size markers (shown as white bands and named "DNA bands") associated with these samples indicate that a short distance (X) has been migrated. Sample S_4 also exhibits DNA amplification, but the resulting amplicons contain a smaller number of base pairs.[1] Therefore, the DNA band associated with S_4 indicates that S_4 has travelled a longer distance; see Y. Sample S_1 represents an NC pathway, and it does not show a white band since no DNA amplification has occurred.

The process of DNA amplification is precise, and it is sensitive to any modifications applied to the DNA-preparation process [19]. Therefore, the results we obtain

---

[1]A gene mutant with a smaller number of base pairs can be created from another by enabling enzymatic gene deletion.

from agarose gel electrophoresis can be analyzed to capture potential attacks on DNA-fingerprinting flows.

20.3.1.1 Attack Models

DNA-preparation steps for PCR can be regarded as potential attack surfaces that can be exploited by a malicious adversary, who may be interested, for example, in tampering with the final results of a DNA analysis and information forensics. The tampering can result in the complete destruction of evidence (the true DNA-amplification profile), or modification of evidence such that it produces a misleading result. To demonstrate this capability, we carried out a benchtop experiment that executed DNA-amplification analysis on 6 different types of samples. Using these 6 samples, 4 parallel runs of DNA fingerprinting using PCR amplification were performed. The first run represents the golden, i.e., mainstream, implementation of the protocol, where all the protocol settings are adjusted normally. The remaining three runs represent malicious implementation of PCR, where DNA preparation was deliberately altered. We "simulated" three evidence-level attacks to demonstrate the following malicious behavior:

- *Attack 1:* An attack that causes all collected samples to maliciously report equivalent positive signals for DNA amplification, therefore concealing the identity of the collected samples. We refer to this attack as *positive denial-of-service (PDoS) attack.*
- *Attack 2:* This attack causes all samples to maliciously report a negative value of DNA amplification (i.e., no DNA amplification occurs). This damages all the collected samples. We refer to this attack as *negative denial-of-service (NDoS) attack.*
- *Attack 3:* An attack that replaces or switches samples. This attack is *sample forgery.*

Attack 1 and Attack 2 completely destroy the evidence, whereas Attack 3 alters the evidence.

Figure 20.9 shows the results obtained by gel electrophoresis. The DNA-ladder plot has four regions: (1) the top-left region (Region 1) contains the DNA bands associated with the golden PCR reaction; (2) the top-right region (Region 2) represents the DNA bands based on Attack 1 (PDoS); (3) the bottom-left region (Region 3) shows the DNA bands for Attack 2 (NDoS); (4) the bottom-right region (Region 4) contains the DNA bands based on Attack 3 (sample switching). In all these sections, the result of the benchmark (the DNA ladder) is located at the leftmost column (column 1).

By analyzing Region 1 (trusted reaction), we observe that column 2 does not show any DNA bands; this result is expected since this column is associated with the NC tube where DNA amplification does not occur. In contrast, column 3 shows a DNA band, capturing the effect of DNA amplification within the PC tube. The remaining columns show various results of DNA amplification, depending on

Column
number

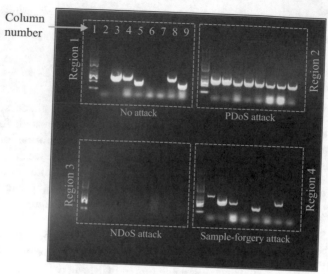

Fig. 20.9 Gel electrophoresis results of the four DNA-fingerprinting runs (Region 1: golden DNA fingerprinting; Regions 2–4: malicious DNA fingerprinting)

whether the target gene expresses at the target amplicon locus and also the length of the amplicon (number of base pairs). The DNA samples represented by columns 4 and 8 exhibit DNA amplification, and they have amplicon mutants that are longer (contain more base pairs) compared to those in the DNA samples represented by columns 5 and 9.

In Region 2, we observe that all the columns exhibit the same high DNA-amplification profile, indicating that all the samples were deliberately manipulated to express the target gene. This behavior shows the impact of maliciously adding PC reagent to all samples, causing a PDoS attack. This result is invalid especially because a white band signal appears at column 1, which is supposed to suppress amplicons generation for the NC. As a result, this result cannot be used by an information forensics lab to distinguish DNA samples.

Similarly, in Region 3, no DNA amplification is reported at all samples since no DNA bands exist. This result indicates that all the samples, including the reference PC sample at column 3, were tampered with to suppress the expression of the target amplicons, causing NDoS attack. This result is also invalid and cannot be used to distinguish DNA samples. Note that both PDoS and NDoS attacks can be easily detected.

Finally, the different DNA bands observed in Region 4 indicate that no PDoS or NDoS attacks were launched. However, the profile of DNA bands shown in Region 4 is different from that of Region 1, indicating that DNA samples were likely switched or replaced with other samples, i.e., subjected to a sample-forgery attack. Note that this attack is hard to detect unless the switching action impacted either the PC or the NC samples, causing abnormal behavior at either column. For instance, in column 1 of Region 4, we observe a DNA band, meaning that DNA amplification has occurred. Since column 1 is associated with the NC sample, which is supposed to suppress DNA amplification, this observation is sufficient to prove that either the NC sample was tampered with/contaminated or the samples (including NC) have been switched.

In practice, Attack 3 (sample forgery) can be stealthy and hard to observe, especially if the attacker is aware of the locations of the NC and PC samples. Therefore, in Sect. 20.3.2, we present a benchtop study that provides an efficient countermeasure technique against Attack 3 based on DNA barcoding.

20.3.2 Molecular Barcoding

At the information forensics laboratory, the embedded DNA barcode has to be amplified first and then compared with a database to ensure that no sample-forgery attacks has occurred.[2] If the sequenced barcode matches the database, then the sample is considered genuine and it can be analyzed. On the other hand, if no barcode or an incorrect barcode is detected, then the associated sample is likely false and it has to be discarded. Figure 20.10 explains the above strategy in steps using a schematic representation. In the first step, a trusted party designs a set of barcodes with varying lengths and develops barcode-specific primers. These barcodes and the associated primers are registered in a secure database along with their secret identification numbers. Only authenticated users and collectors have access to this database.

Each of these classes has only a specific view of the database. In the second step, authenticated collectors obtain the barcodes and their associated secret identification numbers, whereas authenticated analyzers receive primers and their secret identification numbers. The identification of a barcode-specific primer must match the identification of the barcode. Also, such authenticated users are not required to have detailed knowledge of the barcoding sequences. The interactions described thus far, denoted by A and B in Fig. 20.10, ensure that collectors and analyzers are individually trusted. Therefore, the process of molecular barcoding and detection can be trustworthy.

While collectors and analyzers can work independently off of the secret information and material obtained from the trusted party, they still need to interact because barcoded samples are prepared by collectors in field and are delivered to analyzers in information forensics laboratories. A significant advantage of the proposed defense is that both types of users can interact securely and semi-anonymously. Therefore, in the third step, a collector communicates with an analyzer by sending two types of materials: (1) a barcoded DNA sample (biological material), which encapsulates a

---

[2]Sample forgery can be performed by a man-in-the-middle (MITM) attacker who secretly relays and possibly alters transferred samples.

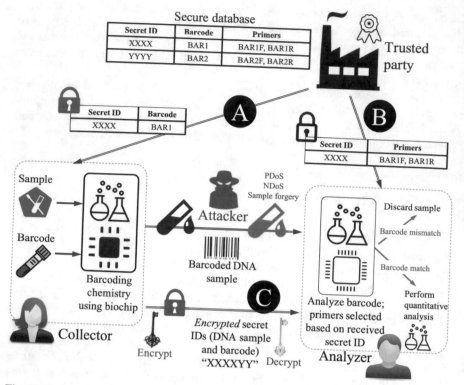

Fig. 20.10 Details of the multi-space security scheme

secret molecular barcode obtained from the trusted party; (2) a public key-encrypted message (information material), denoted by C in Fig. 20.10, which includes secret information about the barcode and the sample identification numbers. The use of public-key cryptography [52] ensures that analyzers can decrypt the message, using a private key, select the right type of primers based on the received identification message, and then perform barcode identification using PCR. After completing the identification routine, the analyzer can verify the genuineness of all collected DNA samples. The work in [25] demonstrated the multi-space security scheme using benchtop experiments.

20.4 Biochip IP Protection

Advancement of microfluidic technologies enables an automated and low-cost platform for implementing benchtop bio-protocols (Automating lab protocols; Fluidigm systems, 2018). A benchtop to microfluidic platform adaptation has three highly interdependent components:

1. setting of the outcome objectives of the bioassay, such as execution time, output quality, and reagent wastage;
2. re-imagining the benchtop operations such as washing, filtering, culturing, and mixing;
3. determining appropriate values for parameters to attain the desired objectives. The developer, through many trials, determines the parameter value's range, which includes mixing time, incubation time, mixing ratio, reagent volume, and concentration.

Thus, bio-protocol development requires a systematic understanding of the interplay between numerous parameters, and it is unraveled through experimental iterations. Further, a bioassay implementation needs to overcome manufacturing defects that contribute to operation-time failures. To expose such failures and to facilitate error recovery, the biochip cyberphysical system incorporates one or more sensors [31, 36, 37]. The sensor feedback control entails sensing the quality of the assay outcome at various stages. Based on the sensor data, control-flow decisions are made [24]. The intermediate bioassay outputs are verified against quality criteria; based on this verification, the relevant bioassay steps are repeated [23].

Companies, such as pharmaceuticals, invest large sums of money and person-hours in a slow and expensive bio-protocol development process laced with tough regulations. For example, a DMFB implementation of a thyroid-stimulating hormone immunoassay requires hundreds of experiments to determine the right bio-protocol parameters [13]. This process is prone to stealing of sensitive research data (Drug development and intellectual property theft, 2016). In 2016, two scientists at a leading pharmaceutical company were indicted for conspiring with a competitor to steal promising drug research secrets (2 GSK scientists indicted in secrets case involving China, 2016; Pharmaceutical giant rocked by ransomware attack, 2017). For rapid and low-cost drug development, pharmaceutical companies are using various types of microfluidic biochips that minimize the assay time and reagent requirement [26]. This opens new avenues for IP theft [11, 17, 48]. Biochip applications inside and outside a lab pose a new challenge to IP protection: *Inside a Lab*—Due to the transparent nature of bio-protocol implementation on the biochip, the bio-protocol sequence can be reverse-engineered. This can be achieved using the biochip snapshots and/or the controlling actuation sequence [11]. This opens the door for bio-IP theft by reverse engineering. *Outside a Lab*— Traditionally, bio-protocols were implemented in controlled laboratory environments. Biochip technology permits the execution of bio-protocols in remote settings on a miniaturized platform. Though this enables new applications such as point-of-care diagnostics (FDA advisors back approval of Baebies' SEEKER analyzer for newborns, 2016), it makes the bio-protocols susceptible to illegal copy and counterfeit production [2, 29].

20.4.1 IP Locking

Locking of bio-IP defends against an overproduction attack, such that an untrusted foundry cannot overproduce biochip hardware and sell it for profit. Since the bio-IP consists of bioassay and biochip hardware, the locking is implemented on a bioassay and biochip hardware.

In [32], daisy chaining of micro-electrodes and the use of one-time programmability in MEDA biochips provide effective bitstream scrambling and IP protection of biochemical protocols. To prevent counterfeit copies, unique microfluidic signatures were explored for a given DMFB platform [17, 29]. In a DMFB, the IP consists of not only hardware layouts but also of the biochemical assays (bioassays) that are intended to be executed on-chip. DMFB designers therefore must defend these protocols against theft. This can be achieved by "locking" the biochemical assays through random insertion of dummy mix-split operations, subject to several design rules [6].

20.4.2 IP Watermarking

Bio-protocol watermarking serves as proof-of-ownership in a court of law. A bio-protocol implementation has inherent variability in domain-specific parameters such as mixing ratio, sensor calibration, and incubation time, or different bio-protocol steps, or different watermarking techniques of varying complexities, are used. These include watermarking for bio-protocol synthesis parameters and the cyberphysical system's control-path parameters. The sample-preparation step of a bio-protocol can be watermarked based on integer linear programming. Here, the bio-protocol IP is watermarked hierarchically by embedding a secret signature [48]. Such a signature can be attributed exclusively to the owner (like a hash), as shown in Fig. 20.11.

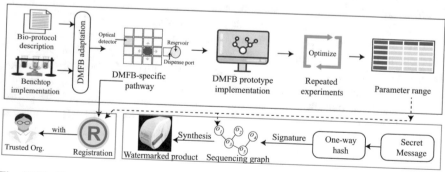

Fig. 20.11 The method for watermarking a bio-protocol implementation on a DMFB

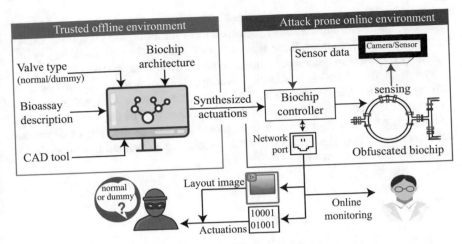

Fig. 20.12 A sieve-valve-based obfuscation technique to prevent reverse engineering of bioassay by a network attack

20.4.3   IP Obfuscation

Bio-protocols implemented on biochips can be easily reverse-engineered. An attacker, through a network attack, can gain access to the biochip snapshots and/or the actuation sequences. Bio-protocol operations (e.g., load, mix) involve a one-to-one mapping between biochip snapshots and actuations. To deter reverse engineering, the actuation sequence is obfuscated by carefully inserting dummy valves in the biochip [44]. The bioassay developer keeps the bioassay description and the dummy-valve locations confidential and uses a CAD tool on a trusted offline computer to synthesize the obfuscated actuation sequence. The obfuscated sequence is loaded on the biochip controller used to conduct the high-value experiments, as shown in Fig. 20.12. In this way, the following crucial IP information can be protected: (1) structural: building blocks that constitute the biochip; (2) behavioral: how each block is operating, e.g., which fluid is selected by a multiplexer; (3) parametric: information about the reagent volume and mix time [44].

20.5   Conclusion

Microfluidic platforms have immense potential in paving the way for rapid and low-cost biochemical analysis. However, the cyberphysical system that enables biochip automation is susceptible to malicious attacks and IP theft. In an era of growing cyber threat, security is a major hurdle in the large-scale adaptation of microfluidic technologies. These emerging applications have different computational and energy resources, and therefore, the safety measures of general-purpose computing cannot be applied as is. Therefore, there is a need to relook at the traditional safety and reliability measures in the microfluidic biochip cyberphysical systems.

The field of microfluidic security overlaps hardware and software control, biology, security, and IP protection. The lack of cross-domain expertise prevents foreseeing the security implications of the design choices. The objective of research so far has been to developing metrics and tools that evaluate the security of various design choices. This helps bridge the cross-domain gap by evaluating various threats and devising countermeasures.

References

1. S.S. Ali, M. Ibrahim, J. Rajendran, O. Sinanoglu, K. Chakrabarty, Supply-chain security of digital microfluidic biochips. IEEE Comput. **49**(8), 36–43 (2016)
2. S.S. Ali, M. Ibrahim, O. Sinanoglu, K. Chakrabarty, R. Karri, Microfluidic encryption of on-chip biochemical assays, in *Proc. BioCAS* (2016), pp. 152–155
3. S.S. Ali, M. Ibrahim, O. Sinanoglu, K. Chakrabarty, R. Karri, Security assessment of cyberphysical digital microfluidic biochips. IEEE/ACM Trans. Comput. Biol. Bioinform. **13**(3), 445–458 (2016)
4. S. Bhattacharjee, A. Banerjee, K. Chakrabarty, B.B. Bhattacharya, Correctness checking of bio-chemical protocol realizations on a digital microfluidic biochip, in *Proc. VLSID* (2014), pp. 504–509
5. S. Bhattacharjee, S. Chatterjee, A. Banerjee, T. Ho, K. Chakrabarty, B.B. Bhattacharya, Adaptation of biochemical protocols to handle technology-change for digital microfluidics. IEEE Trans. CAD Integr. Circuits Syst. **36**(3), 370–383 (2017)
6. S. Bhattacharjee, J. Tang, M. Ibrahim, K. Chakrabarty, R. Karri, Locking of biochemical assays for digital microfluidic biochips, in *Proc. ETS* (2018), pp. 1–6
7. K. Bolden, DNA fabrication, a wakeup call: The need to reevaluate the admissibility and reliability of DNA evidence. Ga. St. UL Rev. **27**, 409–1155 (2011)
8. K. Chakrabarty, Design automation and test solutions for digital microfluidic biochips. IEEE Trans. Circuits Syst. I Regul. Pap. **57**(1), 4–17 (2010)
9. K. Chakrabarty, R.B. Fair, J. Zeng, Design tools for digital microfluidic biochips: Toward functional diversification and more than Moore. IEEE Trans. Comput. Aided Des. Integr. Circuits Syst. **29**(7), 1001–1017 (2010)
10. S. Chakraborty, C. Das, S. Chakraborty, Securing module-less synthesis on cyberphysical digital microfluidic biochips from malicious intrusions, in *Proc. VLSID* (2018), pp. 467–468
11. H. Chen, S. Potluri, F. Koushanfar, BioChipWork: Reverse engineering of microfluidic biochips, in *Proc. ICCD* (2017), pp. 9–16
12. C.D. Chin, V. Linder, S.K. Sia, Commercialization of microfluidic point-of-care diagnostic devices. Lab Chip **12**, 2118–2134 (2012)
13. K. Choi, A. Ng, R. Fobel, D. Chang-Yen, E. Yarnell, E. Pearson, M. Oleksak, A. Fischer, R. Luoma, J. Robinson, J. Audet, A. Wheeler, Automated digital microfluidic platform for magnetic-particle-based immunoassays with optimization by design of experiments. Analytical Chemistry **85**(20), 9638–9646 (2013)
14. N. Dey, A.S. Ashour, F. Shi, S.J. Fong, J.M.R.S. Tavares, Medical cyber-physical systems: A survey. J. Med. Syst. **42**(4), 74 (2018)
15. C. Dong, L. Liu, H. Liu, W. Guo, X. Huang, S. Lian, X. Liu, T. Ho, A survey of DMFBs security: State-of-the-art attack and defense, in *2020 21st International Symposium on Quality Electronic Design (ISQED)* (2020), pp. 14–20

16. R. Fobel, C. Fobel, A.R. Wheeler, DropBot: An open-source digital microfluidic control system with precise control of electrostatic driving force and instantaneous drop velocity measurement. Appl. Phys. Lett. **102**(19), 193513 (2013)
17. C.-W. Hsieh, Z. Li, T.-Y. Ho, Piracy prevention of digital microfluidic biochips, in *Proc. ASPDAC* (2017), pp. 512–517
18. H. He, H. Hu, Field-level digital microfluidic biochips Trojan detection based on hamming distance, in *2020 IEEE 4th Information Technology, Networking, Electronic and Automation Control Conference (ITNEC)*, vol. 1 (2020), pp. 640–643
19. R. Higuchi, C. Fockler, G. Dollinger, R. Watson, Kinetic PCR analysis: real-time monitoring of DNA amplification reactions. Biotechnology **11**(9), 1026 (1993)
20. T.-Y. Ho, J. Zeng, K. Chakrabarty, Digital microfluidic biochips: A vision for functional diversity and more than Moore, in *Proc. of ICCAD* (IEEE Press, 2010), pp. 578–585
21. K. Hu, M. Ibrahim, L. Chen, Z. Li, K. Chakrabarty, R. Fair, Experimental demonstration of error recovery in an integrated cyberphysical digital-microfluidic platform, in *Proc. IEEE BioCAS* (2015), pp. 1–4
22. J.-D. Huang, C. Liu, T. Chiang, Reactant minimization during sample preparation on digital microfluidic biochips using skewed mixing trees, in *Proc. ICCAD* (2012), pp. 377–383
23. M. Ibrahim, K. Chakrabarty, K. Scott, Synthesis of cyberphysical digital-microfluidic biochips for real-time quantitative analysis. IEEE Trans. CAD Integr. Circuits Syst. **36**(5), 733–746 (2017)
24. M. Ibrahim, M. Gorlatova, K. Chakrabarty, The internet of microfluidic things: Perspectives on system architecture and design challenges: Invited paper, in *2019 IEEE/ACM International Conference on Computer-Aided Design (ICCAD)* (2019), pp. 1–8
25. M. Ibrahim, T.-C. Liang, K. Scott, K. Chakrabarty, R. Karri, Molecular barcoding as a defense against benchtop biochemical attacks on DNA fingerprinting and information forensics. IEEE Trans. Inf. Forensics Secur. (2020)
26. N. Khalid, I. Kobayashi, M. Nakajima, Recent lab-on-chip developments for novel drug discovery. Syst. Biol. Med. **9**(4), e1381 (2017)
27. R. Langner, Stuxnet: dissecting a cyberwarfare weapon. IEEE Secur. Priv. **9**(3), 49–51 (2011)
28. P.Y. Lee, J. Costumbrado, C.-Y. Hsu, Y.H. Kim, Agarose gel electrophoresis for the separation of DNA fragments. J. Visualized Exp. JoVE (62) (2012)
29. J. Li, S. Wang, K.S. Li, T. Ho, Digital rights management for paper-based microfluidic biochips, in *2018 IEEE 27th Asian Test Symposium (ATS)* (2018), pp. 179–184
30. Z. Li, T.-Y. Ho, K. Lai, K. Chakrabarty, P. Yu, C. Lee, High-level synthesis for micro-electrode-dot-array digital microfluidic biochips, in *Proc. DAC* (2016), pp. 1–6
31. Z. Li, K.Y. Lai, J. McCrone, P. Yu, K. Chakrabarty, M. Pajic, T. Ho, C. Lee, Efficient and adaptive error recovery in a micro-electrode-dot-array digital microfluidic biochip. IEEE Trans. CAD Integr. Circuits Syst. **37**(3), 601–614 (2018)
32. T. Liang, K. Chakrabarty, R. Karri, Programmable daisychaining of microelectrodes for IP protection in MEDA biochips, in *2019 IEEE International Test Conference (ITC)* (2019), pp. 1–10
33. T.-C. Liang, M. Shayan, K. Chakrabarty, R. Karri, Execution of provably secure assays on MEDA biochips to thwart attacks, in *Proc. ASPDAC* (2019), pp. 51–57
34. C. Lin, J.-D. Huang, H. Yao, T.-Y. Ho, A comprehensive security system for digital microfluidic biochips, in *Proc ITC Asia* (2018), pp. 151–156
35. C. Liu, B. Li, B.B. Bhattacharya, K. Chakrabarty, T.-Y. Ho, U. Schlichtmann, Testing microfluidic fully programmable valve arrays (FPVAs), in *Design, Auto. Test in Europe* (2017), pp. 91–96
36. Y. Luo, K. Chakrabarty, T.-Y. Ho, Dictionary-based error recovery in cyberphysical digital-microfluidic biochips, in *Proc. ICCAD* (2012), pp. 369–376
37. Y. Luo, K. Chakrabarty, T.-Y. Ho, Error recovery in cyberphysical digital microfluidic biochips. IEEE Trans. CAD Integr. Circuits Syst. **32**(1), 59–72 (2013)
38. S. McLaughlin, C. Konstantinou, X. Wang, L. Davi, A. Sadeghi, M. Maniatakos, R. Karri, The cybersecurity landscape in industrial control systems. Proc. IEEE **104**(5), 1039–1057 (2016)

39. S. Mohammed, S. Bhattacharjee, T.-C. Liang, J. Tang, K. Chakrabarty, R. Karri, Shadow attacks on MEDA biochips, in *Proc. ICCAD* (2018), pp. 73:1–73:8

40. A.H.C. Ng, R. Fobel, C. Fobel, J. Lamanna, D.G. Rackus, A. Summers, C. Dixon, M.D.M. Dryden, C. Lam, M. Ho, N.S. Mufti, V. Lee, M.A.M. Asri, E.A. Sykes, M.D. Chamberlain, R. Joseph, M. Ope, H.M. Scobie, A. Knipes, P.A. Rota, N. Marano, P.M. Chege, M. Njuguna, R. Nzunza, N. Kisangau, J. Kiogora, M. Karuingi, J.W. Burton, P. Borus, E. Lam, A.R. Wheeler, A digital microfluidic system for serological immunoassays in remote settings. Sci. Transl. Med. **10**(438) (2018)

41. P. Pop, I.E. Araci, K. Chakrabarty, Continuous-flow biochips: Technology, physical-design methods, and testing. IEEE Des. Test **32**(6), 8–19 (2015)

42. J. Riordon, D. Sovilj, S. Sanner, D. Sinton, E.W. Young, Deep learning with microfluidics for biotechnology. Trends Biotechnol. **37**(3), 310–324 (2019)

43. U. S. Service, CERT, C. Magazine, Deloitte, 2011 cybersecurity watch survey: How bad is the insider threat? (2011). https://apps.dtic.mil/dtic/tr/fulltext/u2/a589979.pdf

44. M. Shayan, S. Bhattacharjee, Y. Song, K. Chakrabarty, R. Karri, Desieve the attacker: Thwarting IP theft in Sieve-Valve-based biochips, in *2019 Design, Automation Test in Europe Conference Exhibition (DATE)* (2019), pp. 210–215

45. M. Shayan, S. Bhattacharjee, Y. Song, K. Chakrabarty, R. Karri, Security assessment of microfluidic fully-programmable-valve-array biochips, in *2019 32nd International Conference on VLSI Design and 2019 18th International Conference on Embedded Systems (VLSID)* (2019), pp. 197–202

46. M. Shayan, S. Bhattacharjee, Y. Song, K. Chakrabarty, R. Karri, Toward secure microfluidic fully programmable valve array biochips. IEEE Trans. Very Large Scale Integr. (VLSI) Syst. **27**(12), 2755–2766 (2019)

47. M. Shayan, S. Bhattacharjee, Y. Song, K. Chakrabarty, R. Karri, Microfluidic Trojan design in flow-based biochips, in *2020 Design, Automation Test in Europe Conference Exhibition (DATE)* (2020), pp. 1037–1042

48. M. Shayan, S. Bhattacharjee, J. Tang, K. Chakrabarty, R. Karri, Bio-protocol watermarking on digital microfluidic biochips. IEEE Trans. Inf. Forensics Secur. (2019)

49. M. Shayan, S. Bhattacharjee, R. Wille, K. Chakrabarty, R. Karri, How secure are checkpoint-based defenses in digital microfluidic biochips? IEEE Trans. Comput. Aided Des. Integr. Circuits Syst. 1–1 (2020)

50. M. Shayan, T. Liang, S. Bhattacharjee, K. Chakrabarty, R. Karri, Towards secure checkpointing for micro-electrode-dot-array biochips. IEEE Trans. Comput. Aided Des. Integr. Circuits Syst. 1–1 (2020)

51. M. Shayan, J. Tang, K. Chakrabarty, R. Karri, Security assessment of micro-electrode-dot-array biochips. IEEE Trans. CAD Integr. Circuits Syst. **38**(10), 1831–1843 (2019)

52. W. Stallings, *Cryptography and Network Security: Principles and Practice* (Pearson Education India, 2003)

53. J. Tang, M. Ibrahim, K. Chakrabarty, R. Karri, Security implications of cyberphysical flow-based microfluidic biochips, in *IEEE ATS.* (2017), pp. 115–120

54. J. Tang, M. Ibrahim, K. Chakrabarty, R. Karri, Security trade-offs in microfluidic routing fabrics, in *Proc. ICCD* (2017), pp. 25–32

55. J. Tang, M. Ibrahim, K. Chakrabarty, R. Karri, Secure randomized checkpointing for digital microfluidic biochips. IEEE Trans. CAD Integr. Circuits Syst. **37**(6), 1119–1132 (2018)

56. J. Tang, M. Ibrahim, K. Chakrabarty, R. Karri, Tamper-resistant pin-constrained digital microfluidic biochips, in *Proc. DAC* (2018), pp. 1–6

57. J. Tang, M. Ibrahim, K. Chakrabarty, R. Karri, Towards secure and trustworthy cyberphysical microfluidic biochips. IEEE Trans. CAD Integr. Circuits Syst. **38**(4), 589–603 (2019)

58. J. Tang, R. Karri, M. Ibrahim, K. Chakrabarty, Securing digital microfluidic biochips by randomizing checkpoints, in *Proc. ITC* (2016), pp. 1–8

59. M. Turetta, F.D. Ben, G. Brisotto, E. Biscontin, M. Bulfoni, D. Cesselli, A. Colombatti, G. Scoles, G. Gigli, L.L. Del Mercato, Emerging technologies for cancer research: Towards personalized medicine with microfluidic platforms and 3d tumor models. Curr. Med. Chem. **25**, 4616–4637 (2018)
60. U.S. Government, Increase in insider threat cases highlight significant risks to business networks and proprietary information (2014). https://www.ic3.gov/media/2014/140923.aspx
61. N. Vergauwe, D. Witters, F. Ceyssens, S. Vermeir, B. Verbruggen, R. Puers, J. Lammertyn, A versatile electrowetting-based digital microfluidic platform for quantitative homogeneous and heterogeneous bio-assays. J. Micromech. Microeng. **21**(5), 054026 (2011)
62. G. Wang, D. Teng, S. Fan, Digital microfluidic operations on micro-electrode array architecture, in *Proc. NEMS* (2011), pp. 1180–1183
63. G. Wang, D. Teng, Y. Lai, Y. Lu, Y. Ho, C. Lee, Field-programmable lab-on-a-chip based on microelectrode dot array architecture. IET Nanobiotechnology **8**(3), 163–171 (2014)
64. G.M. Whitesides, The origins and the future of microfluidics. Nature **442**(7101), 368–373 (2006)
65. T. Xu, K. Chakrabarty, Integrated droplet routing in the synthesis of microfluidic biochips, in *2007 44th ACM/IEEE Design Automation Conference* (2007), pp. 948–953
66. J. Zambreno, A. Choudhary, R. Simha, B. Narahari, N. Memon, N. Memon, Safe-ops: An approach to embedded software security. ACM Trans. Embed. Comput. Syst. **4**(1), 189–210 (2005)
67. Y.S. Zhang, A medical mini-me: one day your doctor could prescribe drugs based on now a biochip version of you reacts. IEEE Spectrum **56**(4), 44–49 (2019)
68. Y. Zhao, T. Xu, K. Chakrabarty, Integrated control-path design and error recovery in the synthesis of digital microfluidic lab-on-chip. ACM J. Emerg. Technol. Comput. Syst. **6**(3), 11:1–11:28 (2010)

Index

Printed in the United States
by Baker & Taylor Publisher Services